AF796153

Batteries for Sustainability

This volume collects selected topical entries from the *Encyclopedia of Sustainability Science and Technology* (ESST). ESST addresses the grand challenges for science and engineering today. It provides unprecedented, peer-reviewed coverage of sustainability science and technology with contributions from nearly 1,000 of the world's leading scientists and engineers, who write on more than 600 separate topics in 38 sections. ESST establishes a foundation for the research, engineering, and economics supporting the many sustainability and policy evaluations being performed in institutions worldwide.

Editor-in-Chief
ROBERT A. MEYERS, RAMTECH LIMITED, Larkspur, CA, USA

Editorial Board
RITA R. COLWELL, Distinguished University Professor, Center for Bioinformatics and Computational Biology, University of Maryland, College Park, MD, USA

ANDREAS FISCHLIN, Terrestrial Systems Ecology, ETH-Zentrum, Zürich, Switzerland

DONALD A. GLASER, Glaser Lab, University of California, Berkeley, Department of Molecular & Cell Biology, Berkeley, CA, USA

TIMOTHY L. KILLEEN, National Science Foundation, Arlington, VA, USA

HAROLD W. KROTO, Francis Eppes Professor of Chemistry, Department of Chemistry and Biochemistry, The Florida State University, Tallahassee, FL, USA

AMORY B. LOVINS, Chairman & Chief Scientist, Rocky Mountain Institute, Snowmass, USA

LORD ROBERT MAY, Department of Zoology, University of Oxford, Oxford, OX1 3PS, UK

DANIEL L. MCFADDEN, Director of Econometrics Laboratory, University of California, Berkeley, CA, USA

THOMAS C. SCHELLING, 3105 Tydings Hall, Department of Economics, University of Maryland, College Park, MD, USA

CHARLES H. TOWNES, 557 Birge, University of California, Berkeley, CA, USA

EMILIO AMBASZ, Emilio Ambasz & Associates, Inc., New York, NY, USA

CLARE BRADSHAW, Department of Systems Ecology, Stockholm University, Stockholm, Sweden

TERRY COFFELT, Research Geneticist, Arid Land Agricultural Research Center, Maricopa, AZ, USA

MEHRDAD EHSANI, Department of Electrical & Computer Engineering, Texas A&M University, College Station, TX, USA

ALI EMADI, Electrical and Computer Engineering Department, Illinois Institute of Technology, Chicago, IL, USA

CHARLES A. S. HALL, College of Environmental Science & Forestry, State University of New York, Syracuse, NY, USA

RIK LEEMANS, Environmental Systems Analysis Group, Wageningen University, Wageningen, The Netherlands

KEITH LOVEGROVE, Department of Engineering (Bldg 32), The Australian National University, Canberra, Australia

TIMOTHY D. SEARCHINGER, Woodrow Wilson School, Princeton University, Princeton, NJ, USA

Ralph J. Brodd
Editor

Batteries for Sustainability

Selected Entries from the Encyclopedia of Sustainability Science and Technology

Springer

Editor
Ralph J. Brodd
Broddarp of Nevada
2161 Fountain Springs Drive
Henderson, NV, USA

This book consists of selections from the Encyclopedia of Sustainability Science and Technology edited by Robert A. Meyers, originally published by Springer Science+Business Media New York in 2012.

ISBN 978-1-4614-5790-9 ISBN 978-1-4614-5791-6 (eBook)
DOI 10.1007/978-1-4614-5791-6
Springer New York Heidelberg Dordrecht London

Library of Congress Control Number: 2012954270

© Springer Science+Business Media New York 2013
This work is subject to copyright. All rights are reserved by the Publisher, whether the whole or part of the material is concerned, specifically the rights of translation, reprinting, reuse of illustrations, recitation, broadcasting, reproduction on microfilms or in any other physical way, and transmission or information storage and retrieval, electronic adaptation, computer software, or by similar or dissimilar methodology now known or hereafter developed. Exempted from this legal reservation are brief excerpts in connection with reviews or scholarly analysis or material supplied specifically for the purpose of being entered and executed on a computer system, for exclusive use by the purchaser of the work. Duplication of this publication or parts thereof is permitted only under the provisions of the Copyright Law of the Publisher's location, in its current version, and permission for use must always be obtained from Springer. Permissions for use may be obtained through RightsLink at the Copyright Clearance Center. Violations are liable to prosecution under the respective Copyright Law.
The use of general descriptive names, registered names, trademarks, service marks, etc. in this publication does not imply, even in the absence of a specific statement, that such names are exempt from the relevant protective laws and regulations and therefore free for general use.
While the advice and information in this book are believed to be true and accurate at the date of publication, neither the authors nor the editors nor the publisher can accept any legal responsibility for any errors or omissions that may be made. The publisher makes no warranty, express or implied, with respect to the material contained herein.

Printed on acid-free paper

Springer is part of Springer Science+Business Media (www.springer.com)

Contents

1 **Batteries, Introduction** 1
Ralph J. Brodd

2 **Battery Cathodes** 5
Marca M. Doeff

3 **Battery Components, Active Materials for** 51
J. B. Goodenough

4 **Electrochemical Supercapacitors and Hybrid Systems** 93
Katsuhiko Naoi

5 **Lead Acid Battery Systems and Technology for Sustainable Energy** 117
Kathryn R. Bullock

6 **Rechargeable Batteries, Separators for** 135
Shriram Santhanagopalan and Zhengming (John) Zhang

7 **Lithium Battery Electrolyte Stability and Performance from Molecular Modeling and Simulations** 195
Grant D. Smith and Oleg Borodin

8 **Lithium Ion Batteries, Electrochemical Reactions in** 239
Paul J. Sideris and Steve G. Greenbaum

9 **Lithium-Ion Batteries, Safety** 285
Brian Barnett, David Ofer, Suresh Sriramulu, and Richard Stringfellow

10 **Lithium-Ion Battery Systems and Technology** 319
Zhengming (John) Zhang and Premanand Ramadass

11 Medical Device Batteries 359
Michael J. Root

12 Nanocarbons for Supercapacitors 393
François Béguin and Encarnación Raymundo-Piñero

13 Nickel-Based Battery Systems 423
Ralph J. Brodd

14 Olivine Phosphate Cathode Materials, Reactivity and Reaction Mechanisms 445
Atsuo Yamada

15 Silicon-Based Anodes for Li-Ion Batteries 471
Ji-Guang Zhang, Wei Wang, Jie Xiao, Wu Xu, Gordon L. Graff, Gary Yang, Daiwon Choi, Deyu Wang, Xiaolin Li, and Jun Liu

Index 505

Chapter 1
Batteries, Introduction

Ralph J. Brodd

Batteries for storage of electricity from solar and wind generation farms are a key element in the success of sustainability. Electric vehicles are the second link in the chain powered by advanced battery systems. This section describes the various devices used in electrochemical energy storage including an overview of electrochemical processes and devices.

Nanocarbons for supercapacitors describes electrochemical capacitors, sometimes called supercapacitors, and are energy storage devices similar to batteries but capable of delivering large amounts of energy in a very short time. These devices rely on the characteristics of the electrical double layer that forms on all conductors when immersed in an electrolyte. The electrical double layer has a high capacitance because the charges are separated on the order of atomic dimensions. Since there is no mass transfer involved in the delivery of current, response time to a change is 10^{-6} s or faster. As a result, electrochemical capacitors can deliver large amounts of energy in short periods of time. The double layer forms in less than 10^{-6} s and responds to changes in a similar time frame. This is about 1,000 times faster than the electrochemical reaction at a battery electrode that has a time constant in the range of 10^{-3} s. The devices find application wherever power is required to service an application.

Electrochemical Supercapacitors and Hybrid Systems describes electrochemical energy storage devices having higher energy density than electrolytic capacitors and higher power density than batteries. These devices rely on the characteristics of the electrical double layer that forms on all polarized conductors when immersed in an electrolyte. The double layer forms in less than 10^{-6} s and responds to changes in a similar time frame. This is about 1,000 times faster than an electrochemical

This chapter was originally published as part of the Encyclopedia of Sustainability Science and Technology edited by Robert A. Meyers. DOI:10.1007/978-1-4419-0851-3

R.J. Brodd (✉)
Broddarp of Nevada, 2161 Fountain Springs Drive, Henderson, NV 89014, USA
e-mail: rbrodd@gmail.com

R.J. Brodd (ed.), *Batteries for Sustainability: Selected Entries from the Encyclopedia of Sustainability Science and Technology*, DOI 10.1007/978-1-4614-5791-6_1,
© Springer Science+Business Media New York 2013

reaction at a battery electrode that has a time constant in the range of 10^{-3} s. As a result, these can deliver large amounts of energy in a very short time. These devices find application where high power delivery is required.

Rechargeable Batteries, Separators for describes the characteristics of porous membranes that hold the electrolyte and physically separate the negative from the positive electrodes in battery systems. The separator is a key element in battery construction as it physically separates and prevents direct contact (shorting) between the positive and negative electrodes. The separator materials are insulators but have the capability to absorb conducting electrolyte solutions to provide electrical continuity between the anode (negative terminal) and cathode (positive terminal) of the battery system. Each new battery system places new requirements placed on the separator material. The recent growth in the battery industry with the introduction of several new battery systems has resulted in the need to create and optimize new separator materials to meet the market demands. Correspondingly, the demand for novelty in separator membranes, to match the newer battery chemistries and geometries, continues to grow.

Lead Acid Battery Systems and Technology for Sustainable Energy describes the basis for the many different commercial lead acid battery designs and electrical requirements ranging from automotive and stationary energy storage to sustainable energy storage and power system regulation. The lead acid batteries constitute the single largest battery market segment. The reliability, availability, and low cost are the key to the success of the lead acid system. It is among the greenest of battery systems because of its established recycling industry. The ability to adapt the system to new applications and to optimize their performance to meet the demands of new applications has been a successful international effort for 150 years and continues to generate improved battery designs and power systems for the future.

While graphite is the material of choice for the anode in present commercial lithium ion batteries, an active search to identify a new high-energy anode material has identified silicon-based anodes for Li-ion batteries as a promising replacement for the graphite anode material. The silicon-lithium alloy has promise to significantly increase the cell capacity up to ten times that of the present graphite lithium anode. The successful transition will require developing method to handle the volume change characteristic of the silicon materials without disintegration of the anode structure. Strategies are based on fundamental mechanistic research to better understand the electrochemical lithiation and delithiation processes of silicon materials during charge-discharge cycling in terms of crystal structure, phase transitions, morphological changes, and reaction kinetics. It is expected that a viable silicon nanostructure anode can be developed with three to five times the capacity of graphite.

Nickel-based battery systems describes the progression of higher-energy nickel cathode battery systems starting with iron for electric vehicle applications; followed by cadmium, metal hydride, hydrogen, and zinc anodes and their high power and good cycle life; and ending with nickel-hydrogen batteries for long life in space applications. At the turn of the twentieth century, simultaneously, Edison in the USA and Junger in Sweden independently developed the nickel-iron rechargeable battery system. The system powered the first electric vehicles in the early 1900s and is still produced today for long-life energy storage applications.

Lithium Ion Batteries, Electrochemical Reactions in details the various advanced experimental techniques that are used to define the characteristics of new high performance materials as well as improve the performance of existing materials for lithium ion batteries. X-ray, neutron and nuclear magnetic resonance techniques, etc. each provide essential complimentary information on solid materials on the crystal structure of new and potential cathode and anode materialsat an atom level leading to improved performance of existing materials as well as identifying potential high performance new materials.

Medical Device Batteries describes wearable and implantable medical devices powered by batteries. Devices include those that are used for cardiac rhythm management (pacemakers, defibrillators, and heart failure devices), hearing loss, bone growth and fusion, drug delivery for therapy or pain relief, nerve stimulation for pain management, urinary incompetence and nervous system disorders, vision, diagnostic measurements and monitoring, and mechanical heart pumps.

Lithium-Ion Battery Systems and Technology batteries have revolutionized battery powered electronic devices with its light weight, high energy storage capability, and long cycle life. Since its introduction in 1991, it has grown to a $4 + billion market in 2010 and dominated powering the modern portable electronic devices. It replaced nickel cadmium and nickel metal hydride batteries and is positioned to power the electric vehicle market in the near future. The prime reasons for its rapid success and proliferation in consumer electronic market are high unit cell voltage, high energy density, and long cycle and shelf life with no memory effect. The significant progress of Li-ion batteries is mainly due to numerous innovations and advancements in materials, designs, and safety. This entry is intended to provide an overview of Li-ion batteries on several aspects. Starting with a brief discussion on its history, commercial success, and working mechanism, all the critical components inside the cell have been discussed with adequate details. The later part of this entry will primarily focus on manufacturing process with detailed discussion on cell-level safety, followed by overview of recent advances and modern trends in Li-ion systems.

Lithium-ion Batteries, safety provides an overview of the safety considerations for Li-ion cells. Presently, Li-ion cells have a record of field failures or safety incident of one incident every ten million cells. The 18650 cell used in portable electronic applications contains sufficient energy to self-heat the cell to over 600°C. This does not include oxidation of the electrolyte solvent by the cathode oxide materials. Adiabatically, including the electrolyte, the temperature is significantly higher. The causes of an incident include overcharge and heating from external sources. All cells have safety devices such as PTC, CID, vents, and safety circuitry. An internal short is the most common trigger for a safety incident.

Lithium Battery Electrolyte Stability and Performance from Molecular Modeling and Simulations provide an example of the power of this experimental technique. Molecular orbital calculations have proven useful and have the capability to identify the details of lithium insertion into battery-related materials and to identify new cathode materials as well as improve the performance of existing materials. Crystallite size has tremendous effect to the thermodynamics and kinetics in intercalation compounds and impacts the diffusion/transport length, effective reaction surface area, surface energy, and interphase energy. Also, the experimental verification of one-dimensional lithium diffusion in Li_xFePO_4 was confirmed.

Battery cathodes provides an overview of the current cathode materials available for use in Li-ion batteries and a discussion of the various battery systems. Li-ion batteries are dual intercalation systems, in which both the cathode and the anode have structures that allow reversible insertion and extraction of lithium cations. In principle, there are numerous materials that undergo reversible intercalation and can serve as electrode materials. Hence, unlike the lead acid battery which describes a specific chemistry, the chemistry of the Li-ion battery is not fixed but determined by the choice of anode and cathode materials.

Olivine Phosphate Cathode Materials, Reactivity and Reaction Mechanisms provide a summary of the characteristics of the phosphate cathode materials. The interest in phosphates was triggered by the work of Goodenough in identifying lithium iron phosphate, $LiFePO_4$, as a promising cathode material for use in Li-ion battery systems. The nanosize crystal structure gives the material unique properties including high surface area, very long cycle life, and, just as important, a low-cost cathode material.

The Battery Components, Active Materials for reviews the role played by the pioneers Volta, Daniell, Davy, Galvani, Faraday, and Davie in providing the basis for batteries as they are known today. The key to identifying new battery materials is found in the fundamental properties of the materials and is the route to improve performance of present materials as well as the identification of new materials with long-life capability.

Chapter 2
Battery Cathodes

Marca M. Doeff

Glossary

Anode – (negative electrode)	This electrode donates electrons during cell discharge.
Battery	A device consisting of one or many electrochemical cells connected together, in which chemical energy is converted into power. These can be further categorized as primary (non-rechargeable) or secondary (rechargeable) systems.
Capacity	The amount of charge that a battery contains, often expressed as mAh or Ah. This depends on the size of the battery and its chemistry. Rated capacity also depends on the current used.
Cathode – (positive electrode)	This electrode accepts electrons during cell discharge.
Cell	One unit of a battery, commonly consisting of an anode, a cathode, an electrolyte, a separator, and two current collectors.
Energy density or specific energy	Energy per unit volume or weight of a material or a device, respectively, often expressed as Wh/L or Wh/kg. Energy is a product of the cell voltage and capacity per unit volume or weight.

This chapter was originally published as part of the Encyclopedia of Sustainability Science and Technology edited by Robert A. Meyers. DOI:10.1007/978-1-4419-0851-3

M.M. Doeff (✉)
Environmental Energy Technologies Division, Lawrence Berkeley National Laboratory, University of California, Berkeley, CA 94720, USA
e-mail: mmdoeff@lbl.gov

R.J. Brodd (ed.), *Batteries for Sustainability: Selected Entries from the Encyclopedia of Sustainability Science and Technology*, DOI 10.1007/978-1-4614-5791-6_2,
© Springer Science+Business Media New York 2013

Intercalation compound (host material, insertion compound)	Originally, this referred specifically to layered structures that can undergo insertion of ions or molecules between the van der Waals gaps, but is now commonly used for any structure that undergoes topotactic insertion reactions. For Li-ion battery materials, it refers specifically to compounds that undergo reductive insertion of lithium ions, such as graphite (used as an anode) or $LiCoO_2$ (used as a cathode).
Jahn–Teller effect	The geometric distortion of nonlinear complexes of certain transition metal ions to remove degeneracy. For example, Mn(III) in octahedral coordination is expected to have an electronic configuration of $t_{2g}^3e_g^1$. Elongation along one axis of the octahedron, for example, decreases the symmetry and removes the degeneracy.
Power density and specific power	Power per unit volume or weight, respectively, often expressed as W/L or W/kg. Power is the product of the current and the operating voltage. This is a function both of the materials used and the cell design.
Practical energy density or specific energy	Based on the entire weight or volume of the device including inert components. It may be only 1/4–1/2 of the theoretical energy density. It may also refer only to the useable portion of the theoretical capacity of the anode or cathode material itself.
Ragone plot	A plot showing the relationship between energy density and power density for any particular battery chemistry. This relationship is a function both of battery design and chemistry for Li-ion batteries.
Solid electrolyte interface (SEI)	A very thin (nanometer scale) layer formed on a lithium or lithiated graphite anode, which develops upon reaction with certain kinds of electrolytic solutions. The SEI is a specific kind of reaction layer that is ionically conductive but electronically insulating. It passivates the electrode, preventing further reaction with the electrolytic solution, and allows reversible operation of the device.
Specific capacity	The amount of charge per unit weight that a battery electrode material contains, often expressed as mAh/g. This is a fundamental characteristic of the material, and depends upon its redox chemistry and structure.

Theoretical energy density, specific energy, capacity	Based on weight or volume of the electrode active materials only.
Topotactic transformation	A transformation in a crystal lattice involving displacement or exchange of atoms, which maintains the basic structure.

Definition of the Subject and Its Importance

In a discharging battery, the cathode is the positive electrode, at which electrochemical reduction takes place. As current flows, electrons from the circuit and cations from the electrolytic solution in the device move toward the cathode. Although these processes are reversed during cell charge in secondary batteries, the positive electrode in these systems is still commonly, if somewhat inaccurately, referred to as the cathode, and the negative as the anode. Because this terminology is widespread throughout the Li-ion battery literature, this usage will be adopted for this article.

Li-ion batteries are dual intercalation systems, in which both the cathode and the anode have structures that allow reversible insertion and extraction of lithium cations. In principle, there are numerous materials that undergo reversible intercalation and can serve as electrode materials. Hence the chemistry of the Li-ion battery is not fixed, unlike the great majority of battery systems. The choice of cathode greatly affects the performance and cost of a Li-ion battery; for example, it is a major determinant of energy density, since it typically has a lower specific capacity than the most common anode material, graphite (372 mAh/g), to which it must be matched.

Introduction

The very high theoretical capacity of lithium (3,829 mAh/g) provided a compelling rationale from the 1970s onward for the development of rechargeable batteries employing the elemental metal as an anode. The realization that some transition metal compounds undergo reductive lithium intercalation reactions reversibly allowed the use of these materials as cathodes in these devices, most notably, TiS_2 [1–3]. Another intercalation compound, $LiCoO_2$, was described shortly thereafter [4, 5] but, because it was produced in the discharged state, was not considered to be of interest by battery companies at the time.

Due to difficulties with the rechargeability of lithium and related safety concerns, however, alternative anodes were sought. The graphite intercalation compound (GIC) LiC_6 was considered an attractive candidate [6] but the high reactivity with commonly used electrolytic solutions containing organic solvents

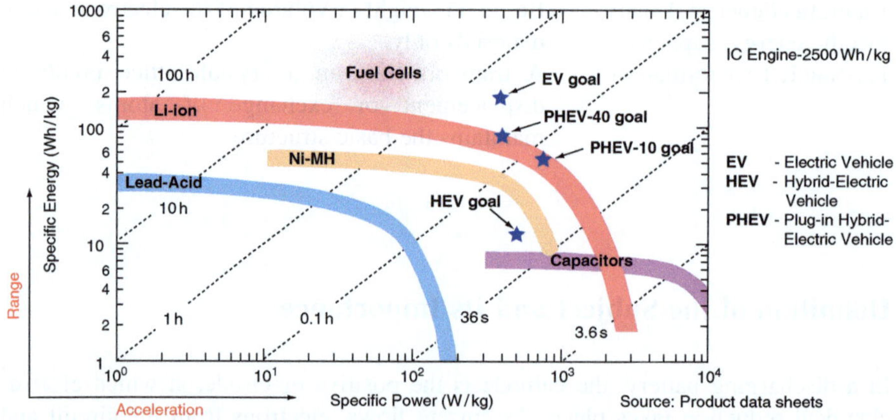

Fig. 2.1 A Ragone plot, showing the relationship between specific power and specific energy for several types of electrochemical devices. Goals set by FreedomCar for HEV, PHEVs, and EVs are also indicated (Used with permission from [10])

was recognized as a significant impediment to its use. The development of electrolytes that allowed the formation of a solid electrolyte interface (SEI) on surfaces of the carbon particles was a breakthrough that enabled commercialization of Li-ion batteries [7]. In 1990, Sony announced the first commercial batteries based on a dual Li-ion intercalation system [8]. These devices are assembled in the discharged state, so that it is convenient to employ a prelithiated cathode such as $LiCoO_2$ with the commonly used graphite anode. After charging, the batteries are ready to power devices.

The practical realization of high energy density Li-ion batteries revolutionized the portable electronics industry, as evidenced by the widespread market penetration of mobile phones, laptop computers, digital music players, and other lightweight devices, since the early 1990s. In 2009, worldwide sales of Li-ion batteries for these applications alone were US$7 billion [9]. Furthermore, their performance characteristics (Fig. 2.1) make them attractive for traction applications such as hybrid electric vehicles (HEVs), plug-in hybrid electric vehicles (PHEVs), and electric vehicles (EVs); a market predicted to be potentially ten times greater than that of consumer electronics. In fact, only Li-ion batteries can meet the requirements for PHEVs as set by the US Advanced Battery Consortium (USABC), although they still fall slightly short of EV goals.

In the case of Li-ion batteries, the trade-off between power and energy shown in Fig. 2.1 is a function both of device design and the electrode materials that are used. Thus, a high-power battery (e.g., one intended for an HEV) will not necessarily contain the same electrode materials as one designed for high energy (i.e., for an EV). As is shown in Fig. 2.1, power translates into acceleration, and energy into range, or miles traveled, for vehicular uses. Furthermore, performance, cost, and abuse-tolerance requirements for traction batteries [11] differ considerably from those for consumer electronics batteries. Vehicular applications are particularly sensitive to cost; currently, Li-ion batteries are priced at about US $1,000/kWh, whereas the USABC goal is US $150/kWh [12]. The three most expensive components of a Li-ion battery, no matter what the configuration, are the cathode, the separator, and the electrolyte [13]. Reduction of cost has been one of the primary driving forces for the investigation of new cathode materials to replace expensive $LiCoO_2$, particularly for vehicular applications. Another extremely important factor is safety under abuse conditions such as overcharge. This is particularly relevant for the large battery packs intended for vehicular uses, which are designed with multiple cells wired in series arrays. Premature failure of one cell in a string may cause others to go into overcharge during the passage of current. These considerations have led to the development of several different types of cathode materials, as will be covered in the next section. Because there is not yet one ideal material that can meet requirements for all applications, research into cathodes for Li-ion batteries is, as of this writing, a very active field.

Characteristics of Battery Cathode Materials

Modern cathode materials for Li-ion batteries are generally prepared in the lithiated (discharged) state, so that they can be paired with delithiated anodes such as graphite. For ease of handling, it is desirable that the material be reasonably air-stable at room temperature. Furthermore, graphite anodes impose a penalty of approximately 0.1 V in average cell potential compared to Li metal. To compensate for this, and also to maximize energy density, cathodes intended for use in Li-ion batteries have higher average potentials versus Li/Li^+ than earlier materials such as TiS_2 and vanadates [14] developed for Li metal batteries. Furthermore, the requirement for high specific capacity generally restricts choices to compounds containing first-row transition metals (usually Mn, Fe, Co, and Ni). Environmental and toxicity concerns have precluded the development of most V or Cr-containing materials, although electroactive compounds containing these metals exist. Today's technologically important cathodes fall into two broad categories: metal oxides and polyanionic compounds.

Figure 2.2 shows representations of the crystal structures of the most commonly used cathode materials for Li-ion batteries, and Table 2.1 summarizes their general properties. Figure 2.3 shows typical discharge profiles of selected materials in Li half-cell configurations.

Layered Transition Metal Oxides

$LiCoO_2$ was the first of the layered transition metal oxides to be commercialized, and is still used today in batteries for consumer devices. It has the structure shown in Fig. 2.2a, where Co and Li, located in octahedral sites, occupy alternating layers

Fig. 2.2 Structures of common cathode materials: (**a**) The layered structure of $LiCoO_2$ with *c*-axis oriented vertically. The octahedrally coordinated Li ions in *3a* sites are represented as spheres and CoO_6 (Co in *3b* sites) as octahedra; (**b**) the cubic structure of $LiMn_2O_4$ spinel, with tetrahedrally coordinated Li ions (in *8a* sites) represented as spheres, and MnO_6 (Mn in *16d* sites) as octahedra; and (**c**) the olivine structure of $LiFePO_4$, looking down the *b*-axis. Octahedrally coordinated Li ions are represented as spheres, and FeO_6 and PO_4 as octahedra and tetrahedra, respectively

Table 2.1 Characteristics of commercial Li-ion battery cathode materials

Material	Structure	Potential versus Li/Li^+, average V	Specific capacity, mAh/g	Specific energy, Wh/kg
$LiCoO_2$	Layered	3.9	140	546
$LiNi_{0.8}Co_{0.15}Al_{0.05}O_2$ (NCA)	Layered	3.8	180–200	680–760
$LiNi_{1/3}Co_{1/3}Mn_{1/3}O_2$ (NMC)	Layered	3.8	160–170	610–650
$LiMn_2O_4$ and variants (LMO)	Spinel	4.1	100–120	410–492
$LiFePO_4$ (LFP)	Olivine	3.45	150–170	518–587

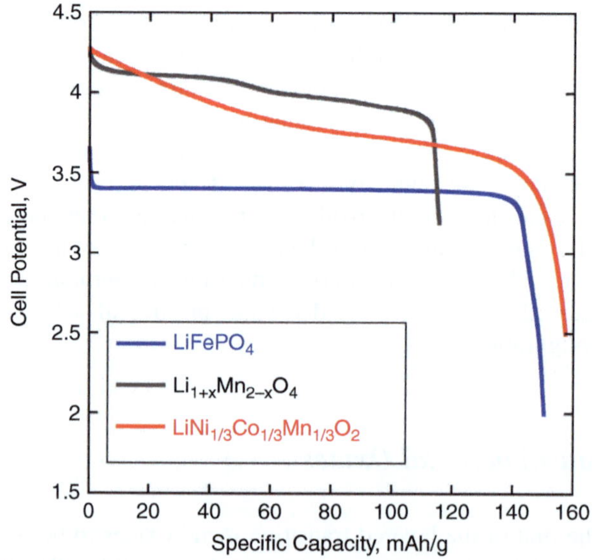

Fig. 2.3 Discharge profiles of lithium cells containing $LiFePO_4$, $Li_{1+x}Mn_{2-x}O_4$, or $LiNi_{1/3}Co_{1/3}Mn_{1/3}O_2$ electrodes

along the 111 planes in a cubic close-packed (ccp) oxygen array, to form a structure with overall hexagonal symmetry (space group R-3m). This type of stacking arrangement is called O3 in layer notation, indicating that there are three transition metal layers per unit cell and the Li ions are octahedrally coordinated.

Delithiation proceeds topotactically [15, 16] and is reversible over the composition range $1 \geq x \geq \sim 0.5$ for x in Li_xCoO_2, giving a practical specific capacity of about 140 mAh/g below 4.2 V versus Li/Li^+. Extraction of lithium to values of $x < \sim 0.5$ results in higher practical capacities initially, but often increases cycling losses [17]. This has been attributed to side reactions involving particle surfaces, which increase cell impedance, and to structural instability associated with phase changes at very low values of x in Li_xCoO_2 [18–21]. Coating $LiCoO_2$ particles, or rigorous heat treatment to remove surface species [22–24], results in improved cycling below 4.5 V versus Li/Li^+, but full delithiation is still not possible without cycling losses.

A layered compound with the nominal composition of $LiNiO_2$ has also been extensively studied for battery applications [25–28]. Its lower cost compared to $LiCoO_2$, and the potential for higher energy density were driving forces for its development. Although the structure is similar to $LiCoO_2$, $LiNiO_2$ readily exhibits non-stoichiometry. During synthesis, there is a tendency toward loss of lithium and reduction of some Ni to the +2 oxidation state. The Ni^{2+} migrates to Li^+ *3a* sites, due to the similarities in size between the two types of ions. The actual composition can be written as $Li_{1-z}Ni_{1+z}O_2$ with $0 < z < 0.2$, or as $(Li_{1-z}Ni_z^{2+})_{3a}(Ni_z^{2+}Ni_{1-z}^{3+})_{3b}O_2$ [29]. The degree of disorder is influenced by the synthetic conditions, and the electrochemical properties (e.g., the first cycle reversibility) are affected by the degree of non-stoichiometry [30, 31]. The difficulty in synthesizing high-quality $LiNiO_2$ was one impediment to its widespread adoption as a cathode material in Li-ion batteries, although nearly ideal structures can be obtained if sufficient care is taken [29]. (However, perfectly stoichiometric $LiNiO_2$ materials probably do not exist). Partial substitution of Ni with Co [32] was later found to be effective at reducing the cationic disorder on *3a* sites, leading to compositions such as $LiNi_{0.8}Co_{0.2}O_2$. Cobalt also helps to reduce oxygen loss at high states-of-charge, improving safety.

The thermal instability of Li_xNiO_2 at high states-of-charge [33–36] also raised a great deal of concern about the safety of this material when used as a cathode. The properties of $LiNiO_2$ have been improved via coating and doping with Mg, leading to a very high-capacity electrode material [37] with better thermal properties. However, the most commonly used electrode material related to $LiNiO_2$ is $LiNi_{0.8}Co_{0.15}Al_{0.05}O_2$ or NCA, which is now commercially produced (see Table 2.1 for general properties). The presence of Al in NCA improves both the thermal [33–36] and electrochemical properties [38, 39]. The high specific capacity and good power capability of this material make it attractive for vehicular applications although it is still not considered as inherently safe as other candidates such as LFP (see Table 2.2 and the discussion of olivines below).

Several ternary Li-Fe-O phases exist with differing arrangements of cations in cubic close-packed oxygen arrays [40]. However, $LiFeO_2$ with the R-3m structure

Table 2.2 Relative merits of selected commercial Li-ion battery cathodes

Advantages	Disadvantages
LMO ($LiMn_2O_4$ and variants)	
Low cost	Mn solubility issue, affecting cycle life
Excellent high rate performance	Low capacity
High operating voltage	
No resource limitations	
Moderate safety (oxygen release)	
LFP ($LiFePO_4$ and variants)	
Moderately low cost	Low operating voltage
Excellent high rate performance	Low capacity, especially for substituted variants
No resource limitations	Controlling patents
Very slow reaction with electrolyte	
Excellent safety (no oxygen release)	
NMC ($LiNi_{1/3}Co_{1/3}Mn_{1/3}O_2$) and variants	
High capacity	High cost of Ni and Co
High operating voltage	Potential resource limitations
Slow reaction with electrolytes	Relatively new in performance
Moderate safety (oxygen release)	Controlling patents
NCA ($LiNi_{0.8}Co_{0.15}Al_{0.05}O_2$)	
Performance is well established	High cost of Ni and Co
Slow reaction with electrolytes	Potential resource limitations
High capacity	Controlling saft patents
High voltage	
Excellent high rate performance	

is metastable and generally must be prepared via indirect methods such as ion exchange of $NaFeO_2$. The electrochemical properties of this material [41] and most other polymorphs [42, 43] are not, however, suitable for lithium-ion batteries due to low and highly sloping voltage profiles or poor cycling properties. Li_5FeO_4 with the anti-fluorite structure has a potential profile that may be compatible in a Li-ion battery configuration, but delithiation does not appear to proceed via simple oxidative deintercalation [44]. This material has been proposed for use as a lithium-ion source for lithium-ion batteries (i.e., to lithiate graphite anodes so that cathodes in the charged state may be used).

Ternary Li-Mn-O phases with Mn in the +3 or +4 oxidation state crystallize as spinels, the rock salt structure Li_2MnO_3, or as orthorhombic $LiMnO_2$ with a corrugated structure (o-$LiMnO_2$) but not as O3 layered structures. Lithium manganese oxide spinels (LMO) are technologically important cathodes and will be considered in a later section (*vide infra*). The Na-Mn-O system contains numerous polymorphs, including layered structures. $NaMnO_2$ has the same stacking arrangement as $LiCoO_2$, but is monoclinically distorted (space group C2/m), due to the abundance of Jahn–Teller distorted Mn^{3+} ions. Ion exchange of this compound yields a layered $LiMnO_2$ (designated O'3, with the prime indicating the monoclinic distortion) [45], but it rapidly converts to spinel upon electrochemical cycling, as does orthorhombic $LiMnO_2$ [46]. O'3-$LiMnO_2$, o-$LiMnO_2$, and spinel manganese oxide all have cubic close-packed oxygen arrays and differ only in the cation arrangement. Upon electrochemical delithiation of the first two materials, disproportion of Mn^{3+} into Mn^{2+} and Mn^{4+} ions occurs and Mn^{2+} ions subsequently migrate into vacant sites in the lithium layers via low-energy pathways, facilitating structural rearrangement to spinel [47].

Lithium-deficient layered Li_xMnO_{2+y} ($x \approx 0.7$, $y \geq 0.05$), which has an O2 rather than O3 stacking arrangement, does not convert to spinel upon cycling [48, 49], because the oxygen array is not ccp. In addition, non-stoichiometric lithium manganese oxides that are intergrowths of O2 and O3 phases show better resistance to conversion than the pure O3 structure [50, 51]. The presence of transition metal vacancies in these compounds (which reduce the concentration of the Jahn–Teller Mn^{3+} ions) results in low rate capability, because the vacancies tend to trap nearby lithium ions, impeding their mobility [52].

Other lithium manganese oxides with tunnel structures [53–55] exhibit interesting electrochemical properties, particularly those based on the $Na_{0.44}MnO_2$ structure. This material has excellent cycling characteristics [56] and rate capability [57], and does not convert to spinel, although the practical capacity is limited by voltage considerations (the average potential at which lithium is extracted is the highest of any known manganese oxide [58]). All of these tunnel and O2 or O2/O3 layered materials, however, must be prepared via ion exchange of their sodium manganese oxide structural analogs, complicating their preparation. Furthermore, the lithium deficiency of the ion-exchanged materials limits the capacity in Li-ion battery configurations, as all of the cycleable lithium must originate from the cathode.

Li_2MnO_3 can be considered a layered structure similar to $LiCoO_2$, but with 111 planes alternately occupied by Li and $Li_{1/3}Mn_{2/3}$. Ordering of Li and Mn in the transition metal layer imposes a monoclinic superstructure, but the stacking arrangement is essentially identical to that of the R-3m structures discussed above. Mn in this compound is tetravalent and all lithium sites are occupied, precluding either oxidative deintercalation or reductive intercalation. Nevertheless, several researchers have observed electrochemical activity of Li_2MnO_3 when it is charged in lithium half-cells to high potentials [59]. Recent evidence [60] suggests that lithium deintercalation occurs with simultaneous oxygen loss during the initial charge, as well as H^+/Li^+ exchange. The converted material becomes electroactive and can subsequently be lithiated during cell discharge. Acid leaching of Li_2MnO_3 also yields an electroactive manganese oxide phase [61]. The reaction involves loss of Li_2O from the structure and ion exchange, to produce layered $H_{1-x}Li_x[Li_{0.33}Mn_{0.67}]O_2$ [62].

The compound $Li_{1.2}Mn_{0.4}Cr_{0.4}O_2$ [63] is essentially a solid solution of layered Li_2MnO_3 and $LiCrO_2$ with an O3 stacking arrangement [64]. Very high discharge capacities (~200 mAh/g) can be obtained based on $Cr^{3+} \leftrightarrow Cr^{6+}$ redox processes. Conversion to spinel phases does not occur because Mn remains in the +4 oxidation state throughout charge and discharge, nor does the electrode need chemical or electrochemical activation prior to use. Nevertheless, concerns over the environmental impact of hexavalent chromium have prevented further development of this material.

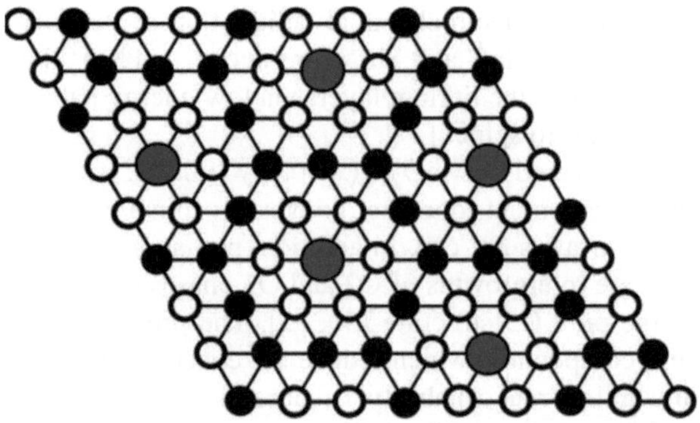

Fig. 2.4 A transition metal layer in $LiNi_{0.5}Mn_{0.5}O_2$ showing the flower pattern. Li is represented by the large gray circles, Mn by small white circles, and Ni by black circles (Used with permission from Ref. [71])

The search for improved layered oxide materials containing low-cost transition metals ultimately led to the discovery of $LiNi_{0.5}Mn_{0.5}O_2$ [65–67] as well as compounds with the general composition $Li[Ni_xLi_{(1/3-2x/3)}Mn_{(2/3-x/3)}]O_2$ [68]. Computational modeling [69] and spectroscopic investigations [70] show that the Ni and Mn in these materials are in the +2 and +4 oxidation states, respectively. During normal charge processes in an electrochemical cell, nickel is ultimately oxidized to the +4 oxidation state, and Mn remains tetravalent. $LiNi_{0.5}Mn_{0.5}O_2$ does not convert to spinel upon cycling, unlike the metastable O3-$LiMnO_2$, because no trivalent Mn is ever produced in the structure.

The defect chemistry of $LiNi_{0.5}Mn_{0.5}O_2$ also distinguishes it from $LiNiO_2$ (where nickel is primarily in the +3 oxidation state initially) discussed above. A feature of $LiNi_{0.5}Mn_{0.5}O_2$ is anti-site mixing [65–67, 71, 72] in which some Ni^{2+} ions are located on the Li *3a* sites, and some Li^+ ions on transition metal *3b* sites, that is, $(Li_{1-x}Ni_x)_{3a}(Li_xNi_{0.5-x}Mn_{0.5})_{3b}O_2$, $x{\approx}0.08-0.12$. Mn ions in the transition metal layers preferentially surround the Li ions located in 3*a* sites to form Li_2MnO_3-like clusters, and Ni ions occupy sites adjacent to Mn, to form a "flower pattern" (Fig. 2.4). The anti-site mixing is thought to be a thermodynamically favored feature of the ion ordering and thus intrinsic to the structure. Unlike with $LiNiO_2$, simple manipulation of conditions during direct synthesis is therefore not likely to result in a near-ideal layered structure.

$Li[Ni_xLi_{(1/3-2x/3)}Mn_{(2/3-x/3)}]O_2$ compounds can be considered solid solutions of Li_2MnO_3 and $LiNi_{0.5}Mn_{0.5}O_2$. Although the discharge capacity between 4.4 and 3.0 V versus Li/Li^+ drops as the Ni content decreases, the opposite trend is observed after cell charge to 4.8 V and subsequent discharge to 2.0 V [73]. In addition, the Li_2MnO_3-rich materials exhibit a plateau at about 4.5 V during charge, the length of which is inversely proportional to x. This is thought to be deintercalation of Li ions with concomitant loss of oxygen (for a net loss of Li_2O), similar to what happens when Li_2MnO_3 is charged in an electrochemical cell. This process can result in very high discharge capacities upon subsequent cycles (e.g., 230 mAh/g for $x = 1/3$).

An advantage to these materials and related Li-rich materials (where excess lithium replaces some of the transition metal ions on *3b* sites, but the Mn content still equals that of the Ni content) is decreased cation mixing. This improves rate capability [74] as compared to $LiNi_{0.5}Mn_{0.5}O_2$.

The presence of a relatively large number of Ni ions in the Li *3a* sites of the $LiNi_{0.5}Mn_{0.5}O_2$ structure due to the anti-site mixing has a negative impact on the Li diffusivity, resulting in a low-rate cathode material. $LiNi_{0.5}Mn_{0.5}O_2$, prepared by low-temperature ion exchange of the nearly perfectly layered $NaNi_{0.5}Mn_{0.5}O_2$ analog, is capable of sustaining higher discharge currents than samples prepared by conventional methods [75]. However, the most successful method used to address this issue has been to incorporate some cobalt into the structure [76]. The best known of these compounds is $Li[Ni_{1/3}Co_{1/3}Mn_{1/3}]O_2$, sometimes referred to as NMC (Tables 2.1 and 2.2). This compound is gradually replacing $LiCoO_2$ in consumer batteries (in some cases, a mixture of the two are used as the cathode) and is under consideration for some vehicular applications. The voltage profile of NMC versus Li/Li^+ is gradually sloping (Fig. 2.3), as is typical of many layered compounds, and provides a somewhat higher capacity than $LiCoO_2$ below 4.3 V versus Li/Li^+ (typically about 160 mAh/g compared to 140). Capacities in excess of 200 mAh/g can be achieved when cells are charged to higher voltage limits, although this usually results in diminished cycle life.

The oxidation states of Ni, Co, and Mn in as-made $Li[Ni_{1/3}Co_{1/3}Mn_{1/3}]O_2$ are +2, +3, and +4, respectively [77]. First principle calculations suggest that Ni^{2+} is ultimately oxidized to Ni^{4+} during extraction of the first two-thirds of the lithium during electrochemical charging in lithium cells, and oxidation of Co^{3+} to Co^{4+} occurs only during removal of the last one-third [78] (i.e., at high cell potentials). As in $LiNi_{0.5}Mn_{0.5}O_2$, Mn remains inactive throughout normal operating cell voltages. The redox activity of Ni during the extraction of lithium from $Li[Ni_{1/3}Co_{1/3}Mn_{1/3}]O_2$ has been verified by in situ X-ray absorption spectroscopic (XAS) studies, although the details of the participation of Co in the electrochemistry are less clear [79–83]. Charge compensation associated with Co is currently thought to occur, at least partially, at oxygen sites, due to the covalency of the Co-O bonds.

The improved electrochemical properties and better structural, chemical, and thermal stability [84, 85] of $Li[Ni_{1/3}Co_{1/3}Mn_{1/3}]O_2$ compared to $LiCoO_2$ or NCA render it an attractive replacement cathode material for consumer batteries. Nevertheless, the presence of a significant amount of costly cobalt, albeit in reduced quantities compared to $LiCoO_2$, may still make it too expensive for most vehicular applications. Compounds having the general formula $Li[Ni_xCo_{1-2x}Mn_x]O_2$ with $x = 0.4$ or 0.45 have recently been developed by Whittingham and coworkers specifically to address the cost issue [86–89]. Although the amount of anti-site mixing increases with larger *x*, these materials exhibit very good electrochemical performances. The effect of increasing the Ni content of these materials on the thermal stability, especially at high states of charge, remains to be determined, however.

Another approach has been to partially substitute another metal such as aluminum for cobalt [89, 90]. Although the amount of substitution must be kept low to maintain the high specific capacities, both the thermal stability of the delithiated materials [91] and the electrochemical properties [92, 93] appear to be enhanced.

All of the NMCs discussed above contain equal amounts of Ni and Mn, which ensure that the oxidation states of the two metals are primarily +2 and +4, respectively. Recent attempts to increase Mn content in stoichiometric materials having the formula $Li[Ni_{0.45-x}Co_{0.1}Mn_{0.45+x}]O_2$ yielded phase mixtures for values of $x > 0.05$ [94] and a solid solution phase for $x = 0.05$. In this compound, $Li[Ni_{0.4}Co_{0.1}Mn_{0.5}]O_2$, the extra manganese is in the +3 oxidation state. Unfortunately, this material exhibits poorer capacity retention upon electrochemical cycling and lower rate capability than $Li[Ni_{0.45}Co_{0.1}Mn_{0.45}]O_2$. It is, however, possible to form solid solutions between $Li[Ni_{1/3}Co_{1/3}Mn_{1/3}]O_2$ and Li_2MnO_3 to make Li- and Mn-rich materials where the excess Mn is in the +4 oxidation state. As with the related $Li[Ni_xLi_{(1/3-2x/3)}\,Mn_{(2/3-x/3)}]O_2$ compounds, these materials can be electrochemically activated during charge to high potentials in lithium cells. The activation process (deintercalation of Li, concomitant loss of oxygen, and H^+/Li^+ ion exchange) yields electrodes with discharge capacities as high as 290 mAh/g [95, 96]. The very high capacities make these materials interesting for both consumer and automotive applications, and they are now being actively developed (e.g., by Envia Systems, Newark, CA).

Manganese Oxide Spinels

The marked tendency for many manganese oxide phases to convert to spinel structures during lithiation in electrochemical cells, or upon heating with a lithium source, attests to the stability of this structure in the Li-Mn-O system. The stoichiometric spinel, $LiMn_2O_4$, is easily synthesized in air from a variety of Li and Mn-containing precursors. Figure 2.2b shows its cubic structure (space group Fd-3m), where Li ions occupy tetrahedral *8a* sites and Mn is located in octahedral *16d* sites in a ccp array of oxygen anions. In addition, there are vacant tetrahedral and octahedral interstitial sites in the three-dimensional structure that provide pathways for lithium diffusion.

Early work [97] showed that it is possible to remove lithium from $LiMn_2O_4$ using a mild acid treatment. During this process, Mn is oxidized from an average +3.5 oxidation state to +4, and λ-MnO_2, which retains the spinel framework, is formed (Eq. 2.1). In addition, some dissolution of Mn occurs:

$$2LiMn_2O_4 + 4H^+(aq) \rightarrow 2Li^+(aq) + Mn^{2+}(aq) + 2H_2O + 3\,\lambda\text{-}MnO_2 \qquad (2.1)$$

Oxidative extraction of lithium from the tetrahedral *8a* sites of $LiMn_2O_4$ to form λ-MnO_2 in a lithium cell [98, 99] was subsequently demonstrated to occur slightly above 4 V. It is also possible to insert lithium into the vacant octahedral sites of

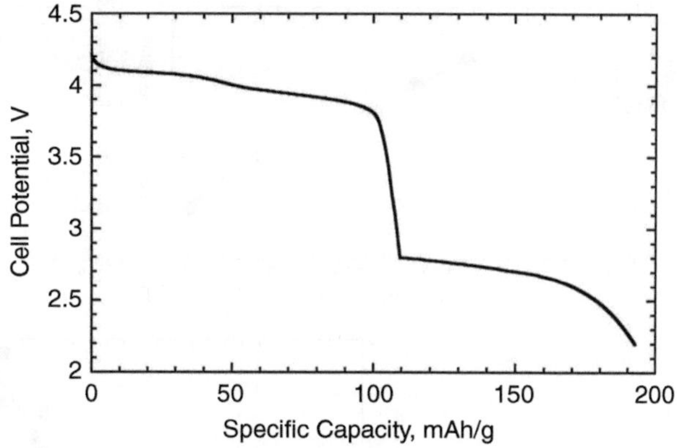

Fig. 2.5 Discharge profile of a Li/modified-$LiMn_2O_4$ cell at 0.1 mA/cm^2. Cell was initially discharged to 2 V and then charged to 4.3 V. This was the third discharge between 4.3 and 2.0 V

$LiMn_2O_4$ either chemically or electrochemically, with concomitant reduction of Mn. When carried out in a cell with a lithium anode, this process occurs below 3 V, and the discharge profile is flat, indicative of a two-phase reaction. The tetragonally distorted phase $Li_2Mn_2O_4$ (space group $I4_1/amd$) is formed due to the cooperative Jahn–Teller effect, which occurs when more than half the manganese is in the form of d^4 Jahn–Teller Mn^{3+} ions. The phase transition results in a 16% increase in the c/a ratio [100] ($a = c = 8.248$ Å in $LiMn_2O_4$; $a = 8.007$ Å, $c = 9.274$ Å in $Li_2Mn_2O_4$). In theory, lithium cells containing $LiMn_2O_4$ can either be charged or discharged initially, and then cycled over a composition range of $0 \leq x \leq 2$ in $Li_xMn_2O_4$ to give a total specific capacity of 285 mAh/g. Figure 2.5 shows the voltage profile of a Li/modified $LiMn_2O_4$ cell assembled in the author's laboratory, after discharging into the 3 V region and full recharge. Both 4 and 3 V plateaus are evident; although the overall capacity is somewhat lower than the theoretical value.

The large anisotropic volume change associated with formation of tetragonal $Li_2Mn_2O_4$ results in particle disintegration and loss of connection within the composite electrode. This causes such a rapid loss of capacity that, in practice, the cycle lives of cells containing $LiMn_2O_4$ discharged even a few times into the 3 V region are greatly compromised. In contrast, the extraction of lithium from $LiMn_2O_4$ above 4 V versus Li/Li^+ maintains the cubic structure and results in smaller volume changes. For this reason, discharge is typically limited to the 4 V plateau (Fig. 2.3), which exhibits much better reversibility. The theoretical capacity associated with this plateau is 148 mAh/g, although it is difficult to fully extract all the lithium, resulting in somewhat lower practical values.

From the early 1990s onward, intensive efforts were devoted to the development of $LiMn_2O_4$ for lithium-ion batteries, particularly at Bellcore Laboratories [101–103]. Although limiting discharges to the 4 V plateau resulted in much better cycling than when both the 3 and 4 V plateaus were utilized, gradual capacity fading was still observed [104]. This behavior was attributed to several factors, including irreversible side reactions with the electrolyte due to the high potential, loss of oxygen from the delithiated spinel, dissolution of Mn, and production

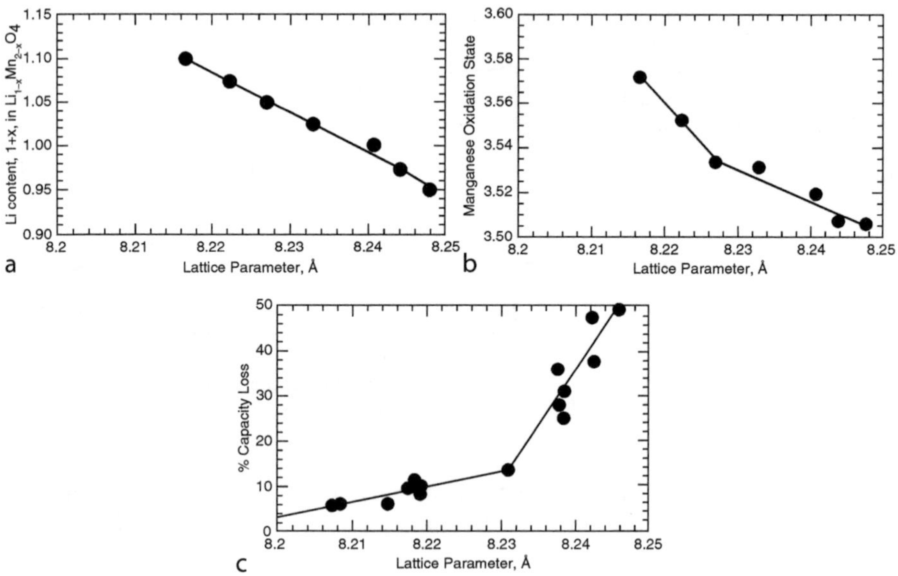

Fig. 2.6 (**a**) The variation of the lattice parameter, *a*, with the lithium content of $Li_{1+x}Mn_{2-x}O_4$; (**b**) the variation of lattice parameter with the average Mn oxidation state in $Li_{1+x}Mn_{2-x}O_4$; and (**c**) relationship between capacity loss after 120 cycles and the lattice parameter in $Li_{1+x}Mn_{2-x}O_4$ (Used with permission from Ref. [113])

of tetragonal $Li_2Mn_2O_4$ at particle surfaces, particularly at high discharge rates [100, 105].

Partial substitution of Mn to form $LiM_xMn_{2-x}O_4$ improves capacity retention [106–112]. Of particular interest are the lithium-substituted materials having the general formula $Li_{1+x}Mn_{2-x}O_4$. As lithium is substituted for manganese on the *16d* sites (x is increased), the average oxidation state of the latter rises. This effectively decreases the 4 V capacity in proportion to the amount of substitution but results in a higher average oxidation state of the Mn at the end-of-discharge. This ameliorates the tendency to form the tetragonally distorted phase under nonequilibrium conditions such as rapid discharging, because this occurs only when the Mn oxidation state falls below an average of 3.5. Because Mn is fully oxidized before all the lithium can be extracted from *8a* sites, side reactions are suppressed. Finally, dissolution of Mn decreases because this phenomenon is dependent on the concentration of trivalent ions.

Li substitution and the attendant increase in Mn oxidation state in the as-made $Li_{1+x}Mn_{2-x}O_4$ compounds decrease the lattice parameter, *a* (Fig. 2.6a and b). The magnitude of this value correlates closely with the amount of capacity loss upon cycling (Fig. 2.6c), so that it can be used to predict the cycling behavior of the spinel materials [113]. In particular, improvements are most marked when $a \leq 8.23$ Å. Because the cycling behavior is so much better than $LiMn_2O_4$, nearly all commercial manganese oxide spinel materials (LMOs) made today are lithium substituted. These have theoretical specific capacities of 100–120 mAh/g (Table 2.1), somewhat lower than that for $LiMn_2O_4$.

The maximum possible substitution in $Li_{1+x}Mn_{2-x}O_4$ corresponds to $x = 1/3$ (the compound $Li_{4/3}Mn_{5/3}O_4$ or $Li_4Mn_5O_{12}$), at which point all of the Mn is in the +4 oxidation state and there is no 4 V capacity, because oxidative extraction of lithium cannot occur. Nevertheless, it is possible to insert lithium reversibly into octahedral *16c* sites at about 2.9 V versus Li/Li^+ until a composition of $Li_7Mn_5O_{12}$ is reached, for a total of 156 mAh/g [114, 115]. The cubic symmetry is maintained until about $Li_{6.5}Mn_5O_{12}$, at which point Mn is reduced to an average oxidation state of 3.5 and the cooperative Jahn–Teller effect induces a tetragonal distortion. Because this process occurs only at the end-of-discharge, $Li_4Mn_5O_{12}$ exhibits much better cycling behavior on the 3 V plateau than does $LiMn_2O_4$. However, the low operating voltage and capacity, and the inability to extract lithium makes $Li_4Mn_5O_{12}$ impractical for Li-ion batteries.

It is also possible to prepare cation-deficient or defect spinels, $Li_{1-x}Mn_{2-2x}O_4$ [114]. The cation vacancies increase the average oxidation state of Mn and decrease the capacity above 4 V in favor of that below 3 V, proportionally with x. In the end-member compound $Li_2Mn_4O_9$ (corresponding to $x = 0.11$ in $Li_{1-x}Mn_{2-2x}O_4$), all Mn is tetravalent, and lithium cannot be extracted, so that there is no capacity at 4 V. Lithium can be inserted below 3 V versus Li/Li^+, to a maximum composition of $Li_5Mn_4O_9$. Although the theoretical capacity (213 mAh/g) is higher than that of $Li_4Mn_5O_{12}$, approximately 1/3 of it is associated with the formation of the Jahn–Teller distorted tetragonal phase.

In principle, defect spinels with values of $x < 0.11$ should cycle on the 4 V plateau better than $LiMn_2O_4$ for the same reasons as the lithium-rich stoichiometric spinels. In practice, it is difficult to control the degree of non-stoichiometry during synthesis, so these materials are not of interest commercially.

In spite of the improved performance exhibited by the lithium-substituted spinels, capacity fading upon cycling is still observed in electrochemical cells, particularly at elevated temperatures (55°C) [116–121]. This has negative implications particularly for vehicular applications, where batteries may be subjected to a wide variety of operating and storage conditions, including very warm environments. The dissolution is associated with Mn^{3+} disproportionation (Eq. 2) in the presence of acidic components of the $LiPF_6$/organic carbonate electrolyte solutions used in Li-ion batteries.

$$2Mn^{3+}(\text{solid}) \rightarrow Mn^{4+}(\text{solid}) + Mn^{2+}(\text{solution}) \qquad (2.2)$$

Loss of manganese leads to the formation of a defect spinel structure with reduced or no 4 V capacity [104, 116–121]. Additionally, proton exchange [116–121], phase separation, film formation, and precipitation of MnO and MnF_2 may occur, increasing cell impedance and exacerbating the capacity fading. More significantly, dissolved Mn^{2+} can cross over to the anode, become reduced, and precipitate as the metal, increasing the charge-transfer resistance and disrupting the critical SEI layer [122, 123]. The graphite anode is markedly more sensitive to this effect than lithium is. Thus, assessment of spinel cathodes to determine the effectiveness of an approach to improve resistance to dissolution should ultimately be carried out in a full cell configuration rather than in lithium half-cells.

Some methods designed to address the problem of spinel dissolution include coating of particles [124, 125] and the use of new non-fluorinated salts such as lithium bis(oxalato)borate (LiBOB) which do not generate HF in situ [126–128]. Interestingly, dissolution of manganese appears to be suppressed in electrodes where LMO is mixed with an NMC [129]. Although the cycling behavior of Li-ion cells with spinel electrodes is still inferior to that of devices containing alternative commercial cathode materials (Table 2.2), other aspects of performance such as rate capability and safety [33–36, 130], the wide availability of manganese precursors, and the potential for low cost make LMO attractive especially for vehicular applications. Thus, at least one battery manufacturer focused on vehicle technologies (e.g., Enerdel, Indianapolis, IN) is pursuing the development of Li-ion batteries with spinel cathodes and hard carbon anodes (which are less sensitive to the effects of manganese dissolution than graphite).

Several lithium-containing oxide spinels having transition metals other than Mn can also be synthesized. One example is $LiCo_2O_4$, which can be prepared from a low-temperature form of $LiCoO_2$ [131]. (The latter, known as LT-$LiCoO_2$, has a structure intermediate between that of a spinel and layered compound). Another is $LiNi_2O_4$ [25]. Unlike $LiMn_2O_4$, the electrochemical characteristics of most of these spinels are inferior to those of the layered analogs, so that they are not of technological interest as cathode materials. However, $Li_4Ti_5O_{12}$ (LTO), which, because of its voltage properties, functions as an anode material, demonstrates excellent cycling performance. Although the energy density is lower than that of graphite, it does not require the formation of an SEI. Thus, it is also an attractive match for LMO cathode materials. Batteries with LTO anodes and LMO or other high-voltage cathodes may find utility in HEVs, which do not require as high an energy density as pure EVs.

The intensive search for substituted lithium manganese spinels with better cycling performance led to the discovery of several materials having capacity near 5 V versus Li/Li^+. These include spinels substituted with Cr, Fe, Cu, Co, or Ni [132–136]. In most cases, these materials exhibit a plateau near 4 V as well as a second plateau near 5 V, with the magnitude of the latter increasing with larger x in $LiM_xMn_{2-x}O_4$ (where M = Cr, Fe, Cu, Co, or Ni). The 4 V capacity is associated with the usual Mn^{3+}/Mn^{4+} spinel redox couple, whereas the higher voltage plateau is associated with redox processes of the substituting metal. Interestingly, the compound $LiNi_{1/2}Mn_{3/2}O_4$ has very little 4 V capacity, unlike other $LiM_{1/2}Mn_{3/2}O_4$ spinels [137] (Fig. 2.7).

In the case of $LiNi_{1/2}Mn_{3/2}O_4$, nearly all of the Mn is tetravalent [136], so that further oxidation is not possible. The specific capacity of 147 mAh/g at 4.7 V is associated with Ni^{2+}/Ni^{4+} redox processes. The high energy density and high voltage makes this material attractive for vehicular applications, because fewer cells need to

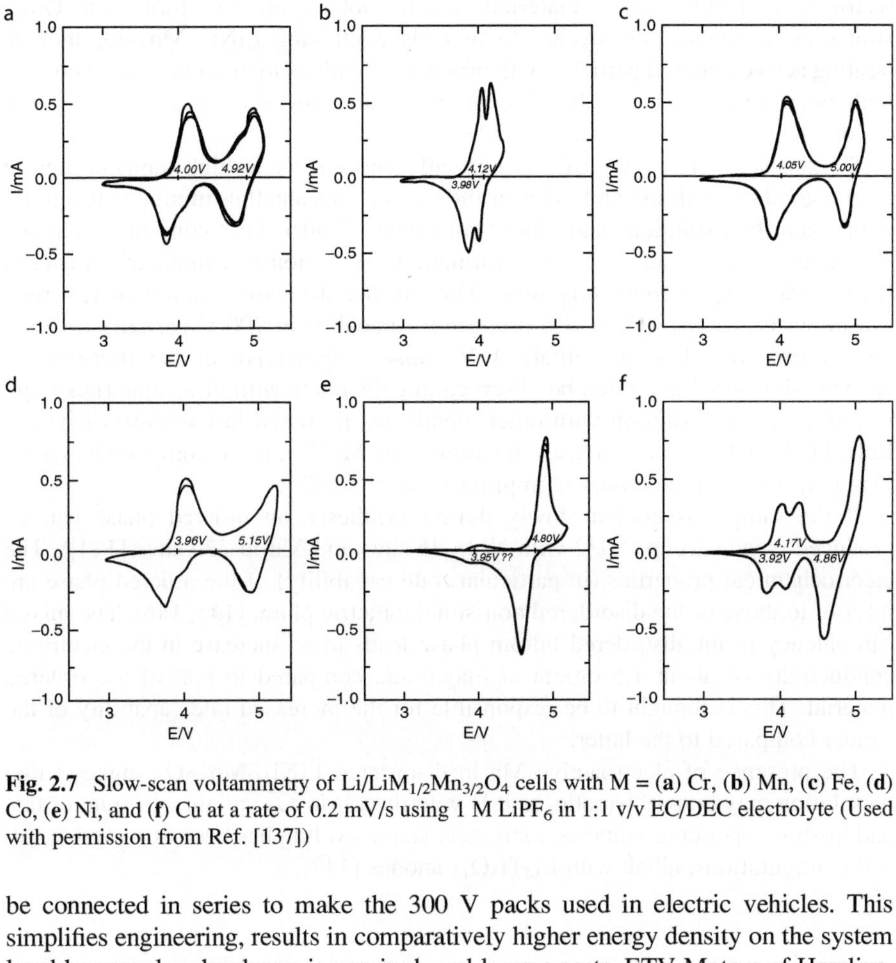

Fig. 2.7 Slow-scan voltammetry of $Li/LiM_{1/2}Mn_{3/2}O_4$ cells with M = (**a**) Cr, (**b**) Mn, (**c**) Fe, (**d**) Co, (**e**) Ni, and (**f**) Cu at a rate of 0.2 mV/s using 1 M $LiPF_6$ in 1:1 v/v EC/DEC electrolyte (Used with permission from Ref. [137])

be connected in series to make the 300 V packs used in electric vehicles. This simplifies engineering, results in comparatively higher energy density on the system level because less hardware is required, and lowers costs. ETV Motors of Herzliya, Israel, is one company actively developing $LiNi_{1/2}Mn_{3/2}O_4$ batteries for this purpose.

The very high potential at which this cathode operates, however, presents considerable challenges for cycle life. Although conventional carbonate-based electrolytes used in Li-ion batteries have, in principle, excellent oxidative stability, the presence of impurities such as water may contribute to side reactions that decrease resistance to irreversible oxidation, resulting in shortened battery lifetimes. In addition, cell components such as carbon additives used to improve conductivity in composite cathodes, or surfactants that aid in wetting may oxidize irreversibly at high potentials, and current collectors may corrode. Thus, cycling is usually restricted to a voltage range below about 4.3 V versus Li/Li^+ (4.2 V in cells with graphite anodes). Stringent purification of electrolytes and the use of special protective additives are allowing gradual extension of this range. In addition, the development of new electrolytes containing ionic liquids or sulfolanes is expected to enable high-voltage cell operation, although the choice of anode may then be restricted to LTO or other materials that do not require SEI formation. Other strategies to prolong the cycle life of cells containing $LiNi_{1/2}Mn_{3/2}O_4$ include treating active material particles with protective coatings to minimize direct contact with electrolyte solutions [138–140], or partial substitution of Ni or Mn with other cations such as Ti, Mg, or Zn [141–143].

Synthesis of $LiNi_{1/2}Mn_{3/2}O_4$ is normally carried out at high temperatures to ensure good crystallinity and to lower the surface area and thus minimize reactivity with electrolyte solutions and other components. Under these conditions, oxygen loss may occur, leading to the formation of a non-stoichiometric material ($LiNi_{1/2}Mn_{3/2}O_{4-x}$) containing some Mn^{3+} in the structure. Additionally, a rock salt impurity, $Li_xNi_{1-x}O$, is also produced. Annealing at 700°C partially reverses the oxygen loss, but the telltale 4 V capacity indicative of the presence of electroactive Mn^{3+} can often be observed in cells made with these materials (e.g., in Fig. 2.7e). Substitution with other metals, as discussed in the papers listed in Refs. [141–143], often decreases the amount of Mn^{3+} in these compounds, which may play a role in the observed improvements in cycling.

If the sample is cooled slowly during synthesis, an ordered phase can be produced (space group $P4_332$ with Ni in *4b* sites and Mn in *12d* sites [144]). The electrochemical properties (in particular, rate capability) of the ordered phase are inferior to those of the disordered non-stoichiometric phase [145, 146]. The mixed Mn valency in the disordered Fd-3m phase leads to an increase in the electronic conductivity of about 1.5 orders of magnitude compared to that of the ordered material. This is thought to be responsible for the increased rate capability of the former compared to the latter.

The presence of electroactive Mn in disordered $LiNi_{1/2}Mn_{3/2}O_4$ suggests that dissolution may be problematic, as it is with the $LiMn_2O_4$ variants. In spite of this and the high operating voltages, extremely stable cycling has been observed in full cell configurations, albeit with $Li_4Ti_5O_{12}$ anodes [147].

LiFePO₄ and Other Phospho-olivines

In 1997, Goodenough and coworkers reported on the electrochemical properties of a new class of cathode materials known as the phospho-olivines [148], which adopt the orthorhombic structure (space group Pnma) shown in Fig. 2.2c. In the case of $LiFePO_4$ (triphylite), Li^+ and Fe^{2+} occupy octahedral sites, and P is located in tetrahedral sites in a somewhat distorted hexagonal close-packed (hcp) oxygen array. The FeO_6 octahedra share corners and LiO_6 octahedra share edges along tunnels down the *b*-axis, through which the Li ions can diffuse. Extraction and reinsertion of Li from $LiFePO_4$ proceeds at about 3.45 V in a lithium cell with a theoretical specific capacity of 170 mAh/g. The potential is independent of the composition x in Li_xFePO_4 (i.e., the voltage profile is flat, Fig. 2.3) indicating that a two-phase reaction occurs. In situ X-ray diffraction and Mossbauer experiments

Fig. 2.8 Photographs of $LiFePO_4$ samples containing varying amounts of carbon (Used with permission from Ref. [153])

on Li/$LiFePO_4$ cells [149] show that the two relevant phases are triphylite and $FePO_4$, also known as heterosite (Eq. 2.3).

$$LiFePO_4\,(\text{triphylite}) \leftrightarrow FePO_4\,(\text{heterosite}) + e^- + Li^+ \qquad (2.3)$$

Initial reports on the electrochemical characteristics of $LiFePO_4$ emphasized its poor rate capability and the lower than expected utilization in lithium cells even when discharged at low current densities (e.g., only about 60% in Ref. [149]). This was attributed to the low electronic conductivities of both the triphylite and heterosite phases [150]. Coating particles with carbon ameliorates the low conductivity and result in better rate performance [151]. This is most conveniently achieved by including a carbon source during initial synthesis of the $LiFePO_4$ sample. The presence of carbon or carbon-containing precursors during calcination in an inert atmosphere has several additional beneficial effects; it retards grain growth, resulting in small particles that allow rapid extraction of Li ions, and prevents formation of Fe^{3+}-containing impurities. It also allows the use of easy-to-handle, low-cost, Fe^{3+}-containing starting materials such as iron nitrate, which are reduced during the heating process [152].

In practice, many lab-synthesized samples of $LiFePO_4$ contain carbon, due to the use of precursors such as iron oxalate [153]. Even small amounts of carbon can result in drastic sample color changes from the off-white of native $LiFePO_4$ (Fig. 2.8). The electrochemical performance is also greatly affected by the carbon content, although this also depends on the distribution of carbon over particle surfaces and details of its structure [154–157]. Carbons produced in situ during synthesis of $LiFePO_4$ are disordered due to the relatively low calcination temperatures (generally, 600–700°C). Disordered carbons are less conductive than graphite, and conductivity depends on the size and number of graphene domains. The use of graphitization catalysts (often containing iron) improves the conductivity of coatings by several orders of magnitude, and can also result in co-production of carbon nanotubes or fibers, which wire particles together [158, 159]. The result is that very small amounts of carbon, often less than 2 wt.%, are sufficient to produce high-rate $LiFePO_4$ samples. This is important because too much light-weight carbon adversely affects the tap densities of $LiFePO_4$ composites, further decreasing the already somewhat low energy density [160].

The changes in sample color, increases in bulk conductivities, and improved electrochemical performance induced by the presence of very small amounts of carbon (or other highly colored impurities) complicate the interpretation of results of doping experiments intended to increase the intrinsic conductivity of $LiFePO_4$. A color change from off-white to black or gray should not be regarded as proof-positive of successful doping and increased intrinsic electronic conductivity, for the reasons described above. Even when carbonaceous precursors are not used, the use of organic solvents or plastic containers during grinding or other processing steps may introduce carbon. Early claims of successful low-level aliovalent substitution [161] on the Li sites of $LiFePO_4$, resulting in greatly enhanced conductivities and electrochemical performance, may have been a fortuitous consequence of well-distributed electronically conductive impurities such as carbon and metal-rich phosphides [162–166]. Recent studies of the defect chemistry of $LiFePO_4$ show that substitution of multivalent ions on the Li site is very limited, and always accompanied by lithium vacancies, so that no mixed valency for iron occurs [167]. Earlier computational modeling of the defect, dopant, and Li transport properties of $LiFePO_4$ [168] indicates that substitution of multivalent ions on Li or Fe sites (other than divalent ions on the latter) is not energetically favored. It was predicted that the most likely type of defect is the Li-Fe anti-site pair, in which Li^+ ions are located on Fe^{2+} sites and are balanced by an equal number of Fe^{2+} ions on Li sites. Recent work, however, shows that there is a marked asymmetry between the number of Fe ions on Li sites and vice versa [169–171]. Very few lithium atoms can be accommodated on iron sites. More commonly, Fe ions on Li sites are accompanied by a small number of vacancies, to form $[Li_{1-2x}Fe_x]FePO_4$. The presence of relatively immobile multivalent ions on Li sites interferes with fast Li diffusion, since $LiFePO_4$ is essentially a one-dimensional ion conductor [172–174]. Likewise, a large number of anti-site defects are likely to impact electrochemical performance adversely [175, 176] as is seen in Refs. [169–171]. At any rate, it is apparent that near room temperature, only a relatively small concentration of vacancies or interstitial defects is possible.

Carbon coatings, although convenient for the reasons stated above, are not absolutely necessary to assure good electrochemical performance [177], as long as particle sizes are kept small and particle size distributions are narrow. Nanostructuring reduces diffusion distances and ameliorates the effects of low electronic conductivity, allowing full discharge at even moderate to high rates. Subtle changes in the discharge characteristics of $LiFePO_4$ samples occur as particle size is decreased, most notably, a stronger dependence of voltage upon composition (x in Li_xFePO_4) close to the end-of-charge and end-of-discharge, implying solid-solution behavior [178, 179]. Other researchers have observed XRD, neutron diffraction, and electrochemical evidence of a small degree of lithium non-stoichiometry in larger Li_xFePO_4 particles at both composition extremes [180, 181], as illustrated in Fig. 2.9.

The existence of a small degree of non-stoichiometry (and thus mixed valence states for iron) may explain why this system is electroactive, given that both end members have such poor electronic conductivity. It should be noted, however, that kinetic effects, surface impurities, and the presence of defects [179, 182–184] also influence the shape of the $Li/LiFePO_4$ discharge profile. This makes it hard to

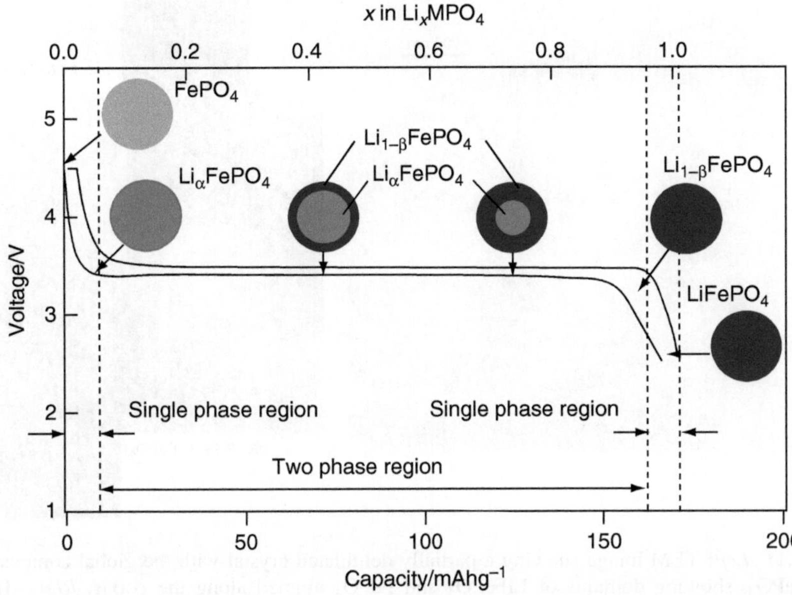

Fig. 2.9 Scheme of the discharge processes of Li_xFePO_4, showing solid-solution behavior close to $x = 0$ and $x = 1$ (Used with permission from Ref. [180])

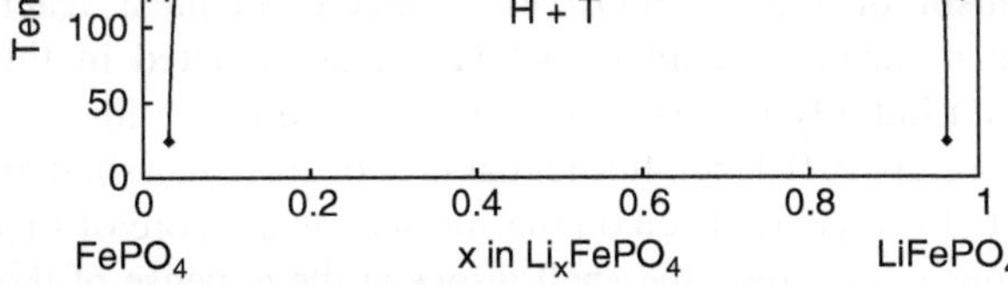

Fig. 2.10 Phase diagram of the Li_xPO_4 system, where T, H, and D denote triphylite, heterosite, and disordered (but not amorphous) phases, respectively (Used with permission from Ref. [186])

determine what the exact compositional ranges of solid-solution behavior are at room temperature from the electrochemical data alone. Complications from the effects of strain also make the interpretation of diffraction data difficult.

Experiments conducted on heated $LiFePO_4/FePO_4$ mixtures of varying concentrations [185, 186] show that solid solutions form from $0 \leq x \leq 1$ in Li_xFePO_4 at elevated temperatures, however. A representative phase diagram is shown in Fig. 2.10. The solubility limits of lithium in heterosite and vacancies in

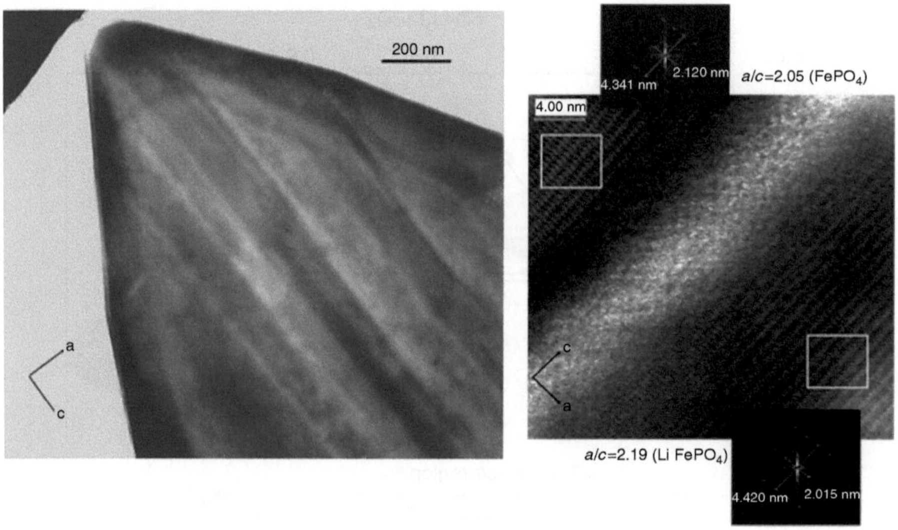

Fig. 2.11 *Left*: TEM image showing a partially delithiated crystal with the global composition $Li_{0.5}FePO_4$, showing domains of $LiFePO_4$ and $FePO_4$ aligned along the *c*-axis. *Right*: High-resolution TEM image with Fourier transforms of the boxed areas, identifying the two phases (Used with permission from Ref. [189])

triphylite are approximated in the figure, because of the uncertainties outlined above. Additionally, due to slow kinetics, the exact temperature boundaries are not known, explaining the differences observed by various researchers. Depending upon the initial concentration, and details of the crystal sizes and morphologies, metastable lithium-deficient phases can persist for remarkably long times after cooling [187].

The mechanism of the $LiFePO_4/FePO_4$ electrochemical reaction has been described using a shrinking core model [188], as depicted in Fig. 2.9. During discharge, lithium initially inserts into $FePO_4$ to form the solid-solution $Li_{\alpha}FePO_4$ phase, where α is close to 0. Upon further passage of current, a shell of a lithium-rich material ($Li_{1-\beta}FePO_4$) is produced on the surface, over a core of $Li_{\alpha}FePO_4$. As the reaction continues to progress, the shell grows at the expense of the core, and the interface between the two phases shrinks, until complete conversion to $Li_{1-\beta}FePO_4$ occurs. Further lithiation produces $LiFePO_4$ at the end-of-discharge. The entire process is reversed upon charge. The anisotropy of lithium diffusion in $LiFePO_4$ [172–174], however, complicates the interpretation of the shrinking core model. A transmission electron microscopy (TEM) study of partially delithiated large plate-like particles [189] show that ordered domains of $FePO_4$ and $LiFePO_4$ alternate in the *ac* plane, separated by narrow disordered (but not amorphous) regions, where lithium mobility is enhanced. The phase transformation proceeds in the direction of the *a*-axis at dislocation lines that run parallel to the *c*-axis, consistent with one-dimensional Li^+ diffusion in the *b*-direction (Fig. 2.11). This somewhat resembles an early "mosaic" model proposed by Andersson et al. as an alternative to the shrinking core model [190]. These observations also suggest that the best rate performance can be achieved with particles that are thin in the direction of Li diffusion (*b*-axis), but that nanostructuring in other directions may not be required.

In another recent study [191] using high-resolution electron energy loss spectroscopy (EELS) on a chemically prepared sample, a different distribution of the two phases in a partially delithiated particle was observed. In this case, $LiFePO_4$ was present only at the edges with $FePO_4$ located in the interior (but not on surfaces of the *ac* planes) of plate-like particles significantly smaller than those studied in Ref. [189]. This is consistent with a progressive emptying or filling of Li channels oriented along the *b*-axis, with the front between the two phases moving perpendicular to the *ac* plane. No disordered regions at the phase boundaries were detected, however. This lack of evidence for solid-solution behavior argues against a shrinking core mechanism, which requires that at least a small degree of non-stoichiometry exist in the system.

Variations in particle sizes and morphologies, as well as in the concentration of defects, may well influence the progression of the delithiation reaction during chemical oxidation and explain these differing results. In electrochemical cells with composite electrodes, the reaction may nucleate preferentially at sites where contact with conductive carbon additives or coatings and the electrolyte solution is the closest, leading to a different pattern of phase distributions than is seen in the chemically delithiated samples, where nucleation can occur at numerous sites simultaneously. In partially delithiated Li_xFePO_4 samples recovered from electrochemical cells, no mixed-phase nanoparticles were observed, leading the researchers to surmise that the reaction front propagates much more rapidly than nucleation [192]. A recent kinetic study also indicates that the $LiFePO_4/FePO_4$ transformation is controlled by a phase-boundary reaction [193]. However, newly discovered evidence of intercrystallite ionic transport, leading to a redox reaction between delithiated and lithiated particles and subsequent equilibration [194] complicates the interpretation of these results. Moreover, areas of composite electrodes with poor electronic connectivity may be electrochemically inactive, leading to islands of unreacted $LiFePO_4$. In that case, all of the lithium will be extracted from $LiFePO_4$ in better connected areas, and mixtures of $FePO_4$ and $LiFePO_4$ will be observed, as in Ref. [192].

While the actual mechanism of the $LiFePO_4/FePO_4$ transformation remains a subject of debate, it seems clear that the shrinking core model does not apply to the primary particles themselves. The model may, however, still be applicable on a larger scale, for example, to the secondary particles comprised of agglomerates.

Over the past decade, better control of synthesis parameters, the use of conductive coatings, and nanostructuring have transformed $LiFePO_4$ from a poorly performing material to one of the most attractive cathodes for Li-ion batteries. $LiFePO_4$ appears to be particularly well suited to high-power applications including power tools and vehicle propulsion. A recent study shows that extraordinarily high charge and discharge currents can be sustained in cells designed so that a modified $LiFePO_4$ limits rate [195]. (While these devices are not practical for real-world use, they serve to demonstrate the intrinsic rate capability of $LiFePO_4$

without interference from mass transfer limitations in other cell components). There have been reports [196, 197] of iron dissolution in $LiPF_6$-containing electrolytes leading to capacity fading at elevated temperatures (analogous to the problems observed with manganese oxide spinels), although this may be due to the presence of iron-containing surface impurities such as Fe_2P rather than to the $LiFePO_4$ itself. The thermal properties of the $LiFePO_4/FePO_4$ system are a particularly attractive feature, leading to enhanced safety (LFP, Table 2.2). Upon heating, $FePO_4$ produced by charging $LiFePO_4$ transforms to the thermodynamically favored trigonal form (α-quartz structure). While this reaction is irreversible and leads to loss of capacity, it progresses without evolution of oxygen [198]. This is in contrast to transition metal oxides, all of which lose oxygen to some degree at high states-of-charge. This makes this cathode especially appealing for traction applications, where safety is paramount.

Because $LiFePO_4$ is less dense than the layered oxides or spinels (the crystallographic density is 3.6 g/cm^3 compared to about 5 g/cm^2 for the oxides), and the potential versus lithium is lower, the energy density is less than that of the oxides (Table 2.2). The specific energy, in contrast, is competitive with LMO (Table 2.1), but for many applications, the energy per unit volume matters more than per weight. The discharge potential of the isostructural $LiMnPO_4$ is about 0.5 V versus Li/Li^+ higher than that of $LiFePO_4$ [199], which increases the specific energy by about 15%. Unfortunately, this material is even less dense than $LiFePO_4$ (3.4 g/cm^3), and the conductivity appears to be several orders of magnitude lower [200, 201]. Synthetic procedures used to produce nanoparticulate $LiMnPO_4$ and variants include sol-gel, precipitation, combustion techniques, spray pyrolysis, and hydrothermal or solvothermal (polyol process) methods [176, 202–208]. For samples calcined in air for the sake of convenience, a fair amount of carbon is generally added after synthesis to offset the low conductivity. Coating can also be achieved via incorporation of carbonaceous precursors or elemental carbon during heat treatment under inert atmosphere, as with $LiFePO_4$. Electrochemical performance improves as the average primary particle size decreases [209]. Partial Mg substitution on the Mn site is also beneficial [208, 210–212]. This effect has been attributed to stabilization against strain associated with the Jahn–Teller distortion of Mn^{3+} ions in the delithiated phase, as well as decreased volume changes upon cycling. The best results reported so far have been for solvothermally prepared powders (polyol process) embedded in a carbon matrix [213]. Still, the rate capability of $LiMnPO_4$ is not as impressive as that of $LiFePO_4$, and many of the strategies employed to improve the electrochemistry (Mg substitution, addition of large amounts of carbon, nanostructuring) reduce the already fairly small energy density advantage. The disappointing performance has been variously attributed to poor transport properties in either the bulk $LiMnPO_4$ or $MnPO_4$ phases, the instability of the $MnPO_4$ phase itself, or surface effects [201, 214, 215]. While the voltage profile of $LiMnPO_4$ is flat during charge and discharge, indicating a two-phase process, evidence of solid-solution behavior near the $MnPO_4$ limit has been observed [216]. As with the $LiFePO_4/FePO_4$ system, the non stoichiometry may enable the electrochemical activity. However, no solid-solution behavior near the $LiMnPO_4$ end of the phase diagram has yet been detected.

Although the performance characteristics of the $LiMnPO_4/MnPO_4$ system may improve with further work, one troubling aspect of this electrode is its thermal instability in the charged state. Unlike $LiFePO_4$, $MnPO_4$ evolves oxygen gas upon heating to 200°C and decomposes to form $Mn_2P_2O_7$ [217, 218]. The instability of $MnPO_4$ is probably related to the strong Jahn–Teller distortion of the Mn^{3+} ion. $LiMnPO_4$ is therefore less attractive than $LiFePO_4$ in terms of safety, and does not have better performance characteristics than the metal oxides. Charging is also somewhat problematic, because high over-potentials are required due to the kinetic limitations. Thus, it seems unlikely that this cathode material will prove useful for traction batteries in the near term, although other applications may be feasible.

Phospho-olivine compounds containing Co or Ni also exist. $LiCoPO_4$ exhibits electrochemical activity at about 4.8 V versus Li/Li^+ [219, 220]. $LiNiPO_4$ is predicted to discharge at about 5.1 V in a lithium cell, but appears to be nearly completely inactive [221]. Neither of these is likely to be used as cathodes for lithium-ion batteries in the near term in part because of the very high voltages needed to charge the materials, which present a number of practical problems.

Solid solutions containing a mixture of transition metals can be prepared over a wide composition range [176]. $Li[Mn_yFe_{1-y}]PO_4$ compounds exhibit two plateaus in the discharge profile at about 4.1 and 3.45 V versus Li/Li^+ with their relative magnitudes proportional to the amounts of Mn and Fe, respectively [148]. In general, for $y < 0.8$, the mixed metal phospho-olivines appear to undergo more facile electrochemical reactions than the pure end-member phase, $LiMnPO_4$ [215, 222]. A study of the delithiation reactions of these phases [223] suggested that the system is single phase when $x < 1-y$, but biphasic over the 4.1 V plateau (corresponding to the Mn^{2+}/Mn^{3+} redox couple). A later synchrotron diffraction study [224] showed two two-phase regions roughly coinciding with the two plateaus and a narrow solid-solution region between $x = 0.55$ and 0.67 for $Li_xMn_{0.6}Fe_{0.4}PO_4$, corresponding to the steeply sloping portion of the discharge profile between the plateaus.

Other mixed transition metal phospho-olivines have been prepared containing Ni and Co [220] and Fe, Mn, and Co [225–227]. In the case of the $Li[Ni_yCo_{1-y}]PO_4$, only one plateau was observed at 4.8 V, corresponding to the Co^{2+}/Co^{3+} redox couple. The only effect of the Ni substitution is to reduce the overall capacity, demonstrating the poor electro-activity of Ni in the phospho-olivine structure. In lithium cells containing $LiMn_{1/3}Fe_{1/3}Co_{1/3}PO_4$, on the other hand, three distinct electrochemical features are observed, corresponding to redox processes of the three types of transition metals (Fig. 2.12). Interestingly, the reaction appears to be single phase over the entire composition range, $0 \leq x \leq 1$ in $LiMn_{1/3}Fe_{1/3}Co_{1/3}PO_4$, and the redox potentials of the Fe^{2+}/Fe^{3+} and Co^{2+}/Co^{3+} couples are slightly shifted compared to the corresponding $LiMPO_4$ electrodes containing only one metal. The manganese redox processes appear to be more sluggish than those of iron and cobalt, and at higher discharge rates, the electrochemical feature at 4.1 V disappears. Tailoring of the voltage profile by incorporating multiple metals in the

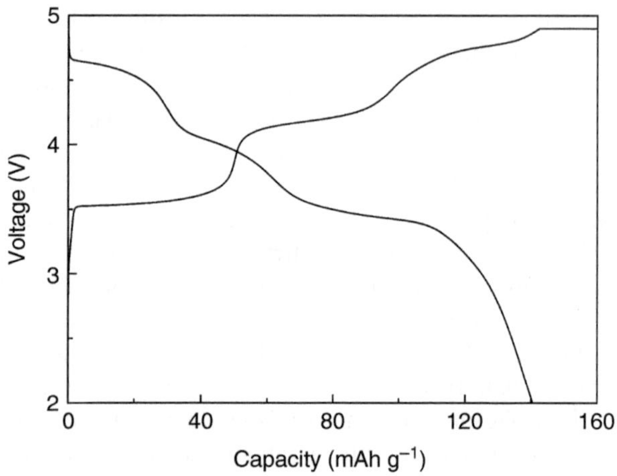

Fig. 2.12 Charge and discharge of a Li/$LiMn_{1/3}Fe_{1/3}Co_{1/3}PO_4$ cell, showing three distinct plateaus corresponding to the redox of Co (4.8 V), Mn (4.1 V), and Fe (3.45 V) (Used with permission from Refs. [225–227])

olivine structure may have practical use as a state-of-charge indicator for vehicular batteries (acting much like the gas gauge in conventional combustion engine automobiles).

Other Phosphates and Poly-anionic Compounds

There are many other transition metal-containing poly-anionic compounds with structures amenable to lithium insertion processes. Among these are numerous materials with the Nasicon structure (Nasicon stands for sodium superionic conductor and originally referred to $Na_{1+x}Zr_2P_{3-y}Si_yO_{12}$ compounds [228]). These have the general formula $A_xM_2(XO_4)_3$, where A is an alkali metal or alkaline earth, M is a transition metal or combination of transition metals, and X is P, As, Si, or S or a combination of these. M is octahedrally coordinated by oxygen, and X is tetrahedrally coordinated. All corners of the XO_4 tetrahedra are shared with MO_6 octahedra and vice versa; these link to form open channels in which the alkali or alkaline earth ions are located. There are also closely related three-dimensional framework structures with differing arrangements of corner-sharing polyhedra and distributions of alkali metal cations, such as the monoclinic $Li_3Fe_2(PO_4)_3$ shown in Fig. 2.13.

Transport of alkali metal ions through the tunnels in Nasicons can be extremely rapid, particularly at elevated temperatures, although the electronic conductivities are low. For these reasons, these materials were originally proposed for use as solid ionic conductors (e.g., to replace β″ alumina in high temperature Na/S batteries). In spite of their low electronic conductivities, researchers recognized that Nasicon structures with redox-active transition metals and related three-dimensional framework compounds could function as electrode materials as early as the late 1980s [229–231] and numerous materials were investigated [232, 233]. In many cases, the electrochemical

Fig. 2.13 The structure of monoclinic $Li_3Fe_2(PO_4)_3$ looking down the unique axis. P is located in the blue tetrahedra, and Fe in the *green* octahedra. Li ions are represented as spheres

properties are unsuitable for Li-ion batteries. For example, Li cannot be extracted from monoclinic $Li_3Fe_2(PO_4)_3$, only inserted, and the discharge occurs at a rather low voltage (2.8 V versus Li/Li^+). In contrast, up to two lithiums per formula unit can be extracted reversibly from monoclinic $Li_3V_2(PO_4)_3$, corresponding to a capacity of about 130 mAh/g at an average potential of 3.8 V [234–236]. Complex multiphasic behavior is exhibited upon redox, resulting in multiple plateaus in the discharge profile. In spite of significant volume changes upon lithium insertion and extraction, the material shows very good cycling behavior.

Fluorophosphates are relatively rare, but have recently been studied for Li-ion applications. The lithium in $LiVPO_4F$ [237–239] can be extracted reversibly at a potential of about 4.2 V versus Li/Li^+. Interestingly, lithium can also be inserted at about 1.8 V. Thus, a graphite-free lithium-ion battery can be built, using $LiVPO_4F$ as both the anode and the cathode! $Na_3V_2(PO_4)_2F_3$ [240], Na_2FePO_4F [241], and $LiFePO_4F$ [242, 243] have also been investigated. The first two compounds were used as-is in electrochemical cells, and underwent ion exchange in situ. Although these cathodes eventually stabilize when used in lithium metal cell configurations and the cycling behavior is good, the presence of sodium in full cells with graphite anodes may disrupt the SEI layer and is therefore undesirable. Thus, these compounds will have to be ion-exchanged prior to use. In the case of $LiFePO_4F$, the investigators saw some evidence that a small amount of lithium could be extracted (with concomitant oxidation of Fe^{3+} to Fe^{4+}), but the main electrochemical reaction was insertion of lithium and reduction of Fe^{3+} to Fe^{2+}.

Other electroactive phosphorus-containing compounds include iron(III) hydroxyl phosphates with lipscombite or tavorite structures [244, 245], β- and ε-$VOPO_4$ [246–248], and $LiVP_2O_7$ [249]. Although the Nasicons, three-dimensional framework structures, and the aforementioned phosphates and diphosphates all have interesting electrochemical properties, most are not competitive with $LiFePO_4$ or the commercialized metal oxides for a variety of reasons. The iron-containing materials generally discharge at too low a voltage for use in Li-ion battery

Fig. 2.14 Views of a β-polymorph of Li_2MSiO_4 looking down the *a*-axis (*left*), the *b*-axis (*middle*), and the *c*-axis (*right*). MO_4 (*yellow*) and SiO_4 (*pink*) tetrahedra alternate and connect to form layers through which lithium ions (*green spheres*) can diffuse

configurations and many have fairly low capacities. Some would need to be lithiated prior to use. The vanadium-containing materials cycle well, and have sufficiently high voltages and capacities to ensure high energy density, but raise concerns about toxicity and environmental impact.

Interest has also recently been directed toward a family of silicates with the general composition Li_2MSiO_4, where M is a divalent cation such as Fe^{2+}, Mn^{2+}, and Co^{2+} or a combination of these [250–257]. The chemical formula implies that it may be possible to extract two Li ions per transition metal unit, giving a very high theoretical specific capacity of about 330 mAh/g. These materials adopt structures in which Li, Si, and the transition metal ions all occupy tetrahedral sites in a distorted close-packed oxygen array. The particular polymorph obtained depends upon the nature of the transition metal as well as the temperature and method of preparation. In the β-polymorphs (low-temperature forms), the tetrahedra share only corners, and all point in the same direction. In γ-polymorphs (high temperature forms), triads of tetrahedra are arranged so that the central one points in the opposite direction and shares edges with the two outer tetrahedra. Numerous tetrahedral structures exist, with different ordering of the ions and varying degrees of distortion. Several different views of an electroactive β-polymorph (space group $Pmn2_1$) are shown in Fig. 2.14, which illustrates its quasi-layered nature. Accurate structure determination of these phases is complicated by the presence of impurities in many preparations and variations in synthetic procedures, which may lead to subtle structural differences [256, 258].

About one Li^+ per formula unit can be extracted from Li_2FeSiO_4 in electrochemical cells at low current densities and at 60°C [250, 251]. The initial extraction takes place at about 3.1 V versus Li/Li^+, but subsequent discharges and charges occur at 2.8 V. This indicates that a structural change takes place, the exact nature of which is not entirely clear. After this initial change, however, relatively good cycling is observed, with an overall capacity of about 130 mAh/g (corresponding to somewhat less than 1 Li^+ per Li_2FeSiO_4). There is no experimental evidence that more than one Li^+ can be extracted, and first principle calculations indicate that it is not possible to do so at potentials compatible with conventional organic liquid electrolytes [259].

The differences between the potentials at which the first and second lithium ions can be extracted are much smaller when M=Mn, Co, or Ni in Li_2MSiO_4 than when it is Fe, however. As with the olivine system, the redox potentials for Li_2MnSiO_4 electrochemical processes (at least for extraction of the first lithium) are expected to be within the oxidative stability limits of organic electrolytes, while those of Li_2CoSiO_4 and Li_2NiSiO_4 are probably not. However the electrochemical behavior of Li_2MnSiO_4 appears to be much worse than that of Li_2FeSiO_4 [251, 254]. Belharaouk et al. [253] were able to achieve initial discharge capacities of about 135 mAh/g at room temperature, over a wide voltage range, by using carbon-coated submicron-sized particles, but this is still far less than predicted for complete extraction of all the lithium. Discharge capacities as high as about 250 mAh/g, however, can be obtained when mixed metal compounds $Li_2Mn_xFe_{1-x}SiO_4$ [252, 256] are used as cathodes in lithium cells. Unfortunately, the quality of the energy obtained is poor because the voltage profiles are steeply sloping over a range of about 4.5–1.0 V versus Li/Li^+. At such low potentials, relatively little power can be obtained from the electrode. The extreme slope may be partly due to the severe kinetic and electronic limitations of the materials, leading to large over-potentials in lithium cells. If so, a combination of the carbon-coating and nanostructuring strategies used to improve the olivines may ameliorate the poor voltage characteristics and lead to new cathodes with nearly double the energy currently available.

Conversion Electrodes

A different class of electrode materials is based not upon insertion processes but on the displacement reactions of binary metal compounds [260, 261] (Eq. 2.4):

$$nLi^+ + ne^- + M^{n+}X_m \leftrightarrow M + nLiX_{m/n}(X = O, F, N, S) \qquad (2.4)$$

Conversion electrodes have long been used in primary systems (e.g., Li/CF_x batteries), but have only been proposed for use in rechargeable devices recently, when it was recognized that the reactions could be made reversible for nanoparticulate materials. While the specific capacities can be enormous, the voltage characteristics make most of these electrodes more suitable for use as anodes rather than as cathodes. A notable exception seems to be that of metal fluorides, which are electroactive anywhere from 2 to 4 V versus Li/Li^+, depending on the identity of the metal, with specific capacities as high as 600 mAh/g. It is typical of conversion electrodes to exhibit significant hysteresis upon recharge, which may be intrinsic to the materials and processes. The mechanisms of conversion reactions are complex and involve bond breaking and considerable rearrangement, unlike that of insertion electrodes. The reassembly of the original phase upon recharge does not necessarily involve the same kinetic pathways and intermediates as the production of the elemental metal and LiF during discharge. Thus, it may not be possible to remove the hysteresis entirely either by engineering of devices or by improving the material properties to ensure faster rate capability. This property has an adverse effect on the round-trip efficiencies, and may make conversion electrodes unsuitable for Li-ion batteries unless a solution can be found to this problem.

Future Directions

Because none of the cathode materials discussed above are ideal for every application, the search for new insertion electrodes continues. Batteries are relatively simple devices, but the simultaneous requirements for high energy density, long cycle life, safety, and low cost impose stringent restrictions upon the choices available for cathode materials. The highly successful example of $LiFePO_4$ has directed researchers toward the exploration of other poly-anionic compounds that may share its excellent thermal stability at high states of charge. It is now evident that nanostructuring and/or conductive coatings can overcome kinetic and electronic limitations, allowing materials that would once have been thought of as unsuitable for electrodes to be considered. Changes in the phase behavior of nanoparticulate materials (in comparison to the conventionally sized powders) can render compounds electroactive that, in bulk form, appear to be unpromising due to their insulating nature. The shortened diffusion distances for lithium in nanoparticles may enable high rate capability in powders that appear to have severe rate limitations when micron-sized or larger. While this allows the menu of potential cathodes to be greatly expanded, the concept of "going nano" should be approached with caution. It is most successful when it is applied to materials that undergo redox well within the electrolyte stability range, such as with $LiFePO_4$. The increased surface area of nanoparticulate powders compared to conventionally sized materials exacerbates any tendency toward irreversible reaction with electrolytes. In addition to the very serious safety concerns this raises, the paradoxical result may be to lower rate capability due to the increased presence of resistive reaction layers on cathode particle surfaces [262]! Nano-sizing may also adversely impact practical specific energies because processing into electrodes becomes more difficult. It is often necessary, for example, to add more carbon to the composite cathode to ensure good electrical connectivity, but this contributes dead weight. High surface area particles also do not pack well, leading to decreases in the tap densities and lower energy densities. Ensuring that nanoparticles are regularly shaped rather than irregularly can ameliorate these tendencies, however [263]. Another potentially useful approach is to synthesize mesoporous materials [264] in which the particles are micron-sized but contain pores of controlled sizes and distributions so as to maximize contact between the electrolyte and active material. This effectively shortens diffusion distances, while minimizing the tap density effects and maintaining the ease of electrode processing.

The use of nano metric or very thin coatings on cathode particle surfaces to improve the electronic conductivity of a material, to decrease dissolution, suppress oxygen loss, or to extend the operating voltage range is somewhat less problematic than simply nano-sizing cathode powders. In principle, this may allow using a wider variety of materials, including high-voltage/high-energy electrodes safely, without compromising cycle life or requiring the use of exotic electrolytes. Here the challenge is to identify and prepare coatings that prevent contact with electrolyte, yet allow the transport of lithium ions. They must also remain essentially defect-free over many cycles in which the cathode particles may undergo significant volume changes. A related concept is the preparation of "core-shell" or compositionally graded materials in which two or more compounds are combined in the same particle [265]. The core of the material can be a high-capacity material and the outer shell one that has lower energy but better thermal stability, for example. A difficulty is the tendency for the shell to become detached from the core if the volume changes upon cycling are not carefully matched. A compositionally graded material is less subject to these stresses, but more complex to prepare.

Researchers continue to look for materials that can deliver higher capacities and energies than those currently available. The layered-layered composite materials (i.e., the solid solutions between Li_2MnO_3 and $LiNi_{0.5}Mn_{0.5}O_2$ or $LiNi_xCo_{1-2x}Mn_xO_2$), which are very high capacity once they are activated in electrochemical cells, are a recent promising development. Cathodes that can undergo more than one oxidation state change per metal center reversibly could substantially increase the capacity of lithium-ion batteries. These include the Li_2MSiO_4 compounds and conversion electrodes discussed above, but these require much more work before they can be envisioned for use in lithium-ion batteries.

The challenges presented by these new materials illustrate the delicate balance of performance characteristics, cost, and safety that the battery electrode designer must be able to meet. There is now renewed interest among researchers in lithium metal batteries, because of the very high energy densities that might be achieved. However, now instead of intercalation electrodes, developers are reconsidering the use of sulfur and air (oxygen) cathodes for lithium metal batteries intended for vehicles. The extremely high theoretical capacities of these materials (e.g., over 1,600 mAh/g for sulfur) in lithium batteries are certainly tantalizing, but the difficulties associated with their use have proven daunting. In the case of sulfur, LiS_x intermediates dissolve in the electrolyte solutions, eventually migrating to the lithium anode and precipitating as Li_2S. The development of the protected lithium electrode (PLE) by PolyPlus Battery Company (Berkeley, CA) has extended the cycle life significantly, and Li/S batteries are now used for some military applications. In PLEs, the lithium anode is covered by a thin dense protective film (e.g., a lithium-ion conducting ceramic such as Lisicon, a ceramic with the Nasicon structure), which not only prevents contact with soluble intermediates but also with the electrolyte. An interlayer between the lithium and the protective layer is also needed to prevent reduction of the latter. Thus, in principle, it is possible to operate lithium cells with a wide variety of electrolytes, including aqueous ones. A lithium/water battery has even been developed for niche applications, although it is intended for use only as a primary, at present.

While in principle it should be possible to operate a lithium/air battery using a PLE to prevent contact of lithium with adventitious water and CO_2 in the atmosphere, rechargeable systems have so far eluded success. The reduction of oxygen itself is somewhat sluggish, necessitating the use of a catalyst. The Li_2O_2, which is the reaction product when nonaqueous electrolytes are used, precipitates out during discharge and eventually clogs the pores of the air electrode structure. While some researchers have claimed to have successfully recharged a lithium/air battery, the large hysteresis observed suggests that a side-reaction involving the electrolyte is responsible for the charging behavior instead of reoxidation of Li_2O_2 to oxygen.

Ultimately, the success or failure of lithium/sulfur or air systems, however, hinges upon convincing demonstration of safe and reliable operation of the lithium anode over the many cycles required for applications requiring rechargeable devices.

Acknowledgments This work was supported by the Assistant Secretary for Energy Efficiency and Renewable Energy, Office of Vehicle Technologies of the US Department of Energy under Contract No. DE-AC02-05CH11231. The author would like to thank Dr. Kinson Kam, Dr. Jordi Cabana, and Mr. Thomas Conry for helpful discussion.

Bibliography

Primary Literature

1. Whittingham MS, Gamble FR (1975) The lithium intercalates of the transition metal dichalcogenides. Mat Res Bull 10:363–372
2. Whittingham MS (1976) The role of ternary phases in cathode reactions. J Electrochem Soc 123:315–320
3. Whittingham MS (1978) Chemistry of intercalation compounds: metal guests in chalcogenide hosts. Prog Solid Stat Chem 12:41–99
4. Mizushima K, Jones PC, Wiseman PJ, Goodenough JB (1980) Li_xCoO_2 ($0 \leq x \leq 1$): a new cathode material for batteries of high energy density. Mat Res Bull 15:783–799
5. Goodenough JB (2007) Cathode materials: a personal perspective. J Power Sources 174:996–1000
6. Dresselhaus MS, Dresselhaus G (1981) Intercalation compounds of graphite. Adv Phys 30:139–326
7. Fong R, von Sacken U, Dahn JR (1990) Studies of lithium intercalation into carbons using nonaqueous electrochemical cells. J Electrochem Soc 137:2009–2013
8. Nagaura T, Tozawa K (1990) Lithium ion rechargeable battery. Prog Batteries Sol Cells 9:209–219
9. Akridge J, Brodd R (2010) Li-ion markets. In: Pacific power symposium, Waikoloa, HI
10. Srinivasan V (2008) Batteries for vehicular applications. In: AIP conference proceedings physics of sustainable energy, Berkeley, CA, vol 1044, pp 283–296
11. Karden E, Ploumen S, Fricke B, Miller T, Snyder K (2007) Energy storage devices for future hybrid electric vehicles. J Power Sources 168:2–11
12. Chalk SG, Miller JF (2006) Key challenges and recent progress in batteries, fuel cells, and hydrogen storage for clean energy systems. J Power Sources 159:73–80
13. Gaines L, Cuenca R (2000) Costs of lithium-ion batteries for vehicles. Report, Center for Transportation Research, Argonne National Laboratory, Argonne

14. Murphy DW, Christian PA (1979) Solid state electrodes for high energy batteries. Science 205:651–656
15. Ohzuku T, Ueda A (1994) Solid-state redox reactions of $LiCoO_2$ (R-3m) for 4 volt secondary lithium cells. J Electrochem Soc 141:2972–2977
16. Reimers JN, Dahn JR (1992) Electrochemical and in situ diffraction studies of lithium intercalation in Li_xCoO_2. J Electrochem Soc 139:2091–2097
17. Chen Z, Lu Z, Dahn JR (2002) Staging phase transitions in Li_xCoO_2. J Electrochem Soc 149: A1604–A1609
18. Aurbach D, Markovsky B, Rodkin A, Levi E, Cohen YS, Kim H-J, Schmidt M (2002) On the capacity fading of $LiCoO_2$ intercalation electrodes: the effect of cycling, storage, temperature, and surface film forming additives. Electrochim Acta 47:4291–4306
19. Amatucci GG, Tarascon JM, Klein LC (1996) CoO_2, The end member of the Li_xCoO_2 solid solution. J Electrochem Soc 143:1114–1123
20. Van der Ven A, Aydinol MK, Ceder G (1998) First-principles evidence for stage ordering in Li_xCoO_2. J Electrochem Soc 145:2149–2155
21. Xia H, Lu L, Meng YS, Ceder G (2007) Phase transitions and high voltage electrochemical behavior of $LiCoO_2$ thin films grown by pulsed laser deposition. J Electrochem Soc 154: A337–A342
22. Cho J, Kim G (1999) Enhancement of thermal stability of $LiCoO_2$ by $LiMn_2O_4$ coating. Electrochem Solid-State Lett 2:253–255
23. Kim B, Kim C, Kim T-G, Ahn D, Park B (2006) The effect of $AlPO_4$-coating layer on the electrochemical properties in $LiCoO_2$ thin films. J Electrochem Soc 153:A1773–A1777
24. Chen Z, Dahn JR (2004) Methods to obtain excellent capacity retention in $LiCoO_2$ cycled to 4.5V. Electrochim Acta 49:1079–1090
25. Thomas MGSR, David WIF, Goodenough JB (1985) Synthesis and structural characterization of the normal spinel $Li[Ni_2O_4]$. Mat Res Bull 20:1137–1146
26. Dahn JR, von Sacken U, Michal CA (1990) Structure and electrochemistry of $Li_{1\pm y}NiO_2$ and a new Li_2NiO_2 phase with the $Ni(OH)_2$ structure. Sol State Ionics 44:87–97
27. Broussely M, Perton F, Biensan P, Bodet JM, Labat J, Lecerf A, Delmas C, Rougier A, Pérès JP (1995) Li_xNiO_2, a promising cathode for rechargeable lithium batteries. J Power Sources 54:109–114
28. Dahn JR, von Sacken U, Juzkow MW, Al-Janaby H (1991) Rechargeable $LiNiO_2$/carbon cells. J Electrochem Soc 138:2207–2211
29. Rougier A, Gravereau P, Delmas C (1996) Optimization of the composition of the $Li_{1-z}Ni_{1+z}O_2$ electrode materials: structural, magnetic, and electrochemical studies. J Electrochem Soc 143:1168–1175
30. Pérès JP, Delmas C, Rougier A, Broussely M, Perton F, Biensan P, Willman P (1996) The relationship between the composition of lithium nickel oxide and the loss of reversibility during the first cycle. J Phys Chem Solids 57:1057–1060
31. Delmas C, Pérès JP, Rougier A, Demourgues A, Weill F, Chadwick A, Broussely M, Perton F, Biensan P, Willman P (1997) On the behavior of the Li_xNiO_2 system: an electrochemical and structural overview. J Power Sources 68:120–125
32. Ohzuku T, Ueda A, Nagayama M, Iwakoshi Y, Komori H (1993) Comparative study of $LiCoO_2$, $LiNi_{1/2}Co_{1/2}O_2$ and $LiNiO_2$ for 4 volt secondary lithium cells. Electrochim Acta 38:1159–1167
33. Dahn JR, Fuller EW, Obrovac M, von Sacken U (1994) Thermal stability of Li_xCoO_2, Li_xNiO_2 and λ-MnO_2 and consequences for the safety of Li-ion cells. Sol State Ionics 69:265–270
34. Ohzuku T, Ueda A, Kouguchi M (1995) Synthesis and characterization of $LiAl_{1/4}Ni_{3/4}O_2$ (R-3m) for lithium-ion (Shuttlecock) batteries. J Electrochem Soc 142:4033–4439
35. Arai H, Sakurai Y (1999) Characteristics of Li_xNiO_2 obtained by chemical delithiation. J Power Sources 81–82:401–405

36. Albrecht S, Kümpers J, Kruft M, Malcus S, Vogler C, Wahl M, Wohlfahrt-Mehrens M (2003) Electrochemical and thermal behavior of aluminum- and magnesium-doped spherical lithium nickel cobalt mixed oxides $Li_{1-x}(Ni_{1-y-z}Co_yM_z)O_2$ (M=Al, Mg). J Power Sources 119–121:178–183
37. Onnerud PT, Shi JJ, Dalton SL, Lampe-Onnerud C (2008) Lithium metal oxide materials and methods of synthesis and use. US Patent 2008/0286460 A1
38. Chen CH, Liu J, Stoll ME, Henriksen G, Vissers DR, Amine K (2004) Aluminum-doped lithium nickel cobalt oxide electrodes for high-power lithium-ion batteries. J Power Sources 128:278–285
39. Kostecki R, Lei J, McLarnon F, Shim J, Striebel K (2006) Diagnostic evaluation of detrimental phenomena in high-power lithium-ion batteries. J Electrochem Soc 153:A669–A672
40. Tabuchi M, Tsutsui S, Masquelier C, Kanno R, Ado K, Matsubara I, Nasu S, Kageyama H (1998) Effect of cation arrangement on the magnetic properties of lithium ferrites ($LiFeO_2$ prepared by hydrothermal reaction and post-annealing method). J Solid State Chem 140:159–167
41. Ado K, Tabuchi M, Kobayashi H, Kageyama H, Nakamura O, Inaba Y, Kanno R (1997) Preparation of $LiFeO_2$ with alpha-$NaFeO_2$-type structure using a mixed-alkaline hydrothermal method. J Electrochem Soc 144:L177–L180
42. Kanno R, Shirane T, Kawamoto Y, Takeda Y, Takano M, Ohashi M, Yamaguchi Y (1996) Synthesis, structure, and electrochemical properties of a new lithium iron oxide, $LiFeO_2$, with a corrugated layer structure. J Electrochem Soc 143:2435–2442
43. Morales J, Santos-Peña J (2007) Highly electroactive nanosized α-$LiFeO_2$. Electrochem Commun 9:2116–2120
44. Johnson CS, Kang S-H, Vaughey JT, Pol SV, Balasubramanian M, Thackeray MM (2010) Li_2O Removal from Li_5FeO_4: a cathode precursor for lithium-ion batteries. Chem Mater 22:1263–1270
45. Armstrong AR, Bruce PG (1996) Synthesis of layered $LiMnO_2$ as an electrode for rechargeable lithium batteries. Nature 381:499–500
46. Thackeray MM (1997) Manganese oxides for lithium batteries. Prog Solid State Chem 25:1–71
47. Reed J, Ceder G (2004) Role of electronic structure in the susceptibility of metastable transition-metal oxide structures to transformation. Chem Rev 104:4513–4534
48. Paulsen JM, Thomas CL, Dahn JR (1999) Layered Li-Mn-oxide with the O2 structure: a cathode material for Li-Ion cells which does not convert to spinel. J Electrochem Soc 146:3560–3565
49. Eriksson TA, Lee YJ, Hollingsworth J, Reimer JA, Cairns EJ, Zhang X-F, Doeff MM (2003) Influence of substitution on the structure and electrochemistry of layered manganese oxides. Chem Mater 15:4456–4463
50. Dollé M, Patoux S, Doeff MM (2005) Layered manganese oxide intergrowth electrodes for rechargeable lithium batteries. 1. substitution with Co or Ni. Chem Mater 17:1036–1043
51. Patoux S, Dollé M, Doeff MM (2005) Layered manganese oxide intergrowth electrodes for rechargeable lithium batteries. 2. substitution with Al. Chem Mater 17:1044–1054
52. Robertson AD, Armstrong AR, Bruce PG (2000) Influence of ion exchange conditions on the defect chemistry and performance of cobalt doped layered lithium manganese oxide based intercalation compounds. Chem Commun 20:1997–1998
53. Doeff MM, Peng MY, Ma Y, De Jonghe LC (1994) Orthorhombic Na_xMnO_2 as a cathode material for secondary sodium and lithium polymer batteries. J Electrochem Soc 141:L145–L147
54. Doeff MM, Richardson TJ, Kepley L (1996) Lithium insertion processes of orthorhombic Na_xMnO_2-base electrode materials. J Electrochem Soc 143:2507–2516
55. Hu F, Doeff MM (2004) Electrochemical characterization of manganese oxide cathode materials based on $Na_{0.4}MnO_2$. J Power Sources 129:296–302
56. Armstrong AR, Huang H, Jennings RA, Bruce PG (1998) $Li_{0.44}MnO_2$: an intercalation electrode with a tunnel structure and excellent cyclability. J Mater Chem 8:255–259

57. Doeff MM, Anapolsky A, Edman L, Richardson TJ, De Jonghe LC (2001) A high-rate manganese oxide for rechargeable lithium battery applications. J Electrochem Soc 148: A230–A236
58. Akimoto J, Awaka J, Takahashi Y, Kijima N, Tabuchi M, Nakashima A, Sakaebe H, Tatsumi K (2005) Synthesis and electrochemical properties of $Li_{0.44}MnO_2$ as a novel 4 V cathode material. Electrochem Solid-State Lett 8:A554–A557
59. Kalyani P, Chitra S, Mohan T, Gopukumar S (1999) Lithium metal rechargeable cells using Li_2MnO_3 as the positive electrode. J Power Sources 80:103–106
60. Yu DYW, Yanagida K, Kato Y, Nakamura H (2009) Electrochemical activities in Li_2MnO_3. J Electrochem Soc 156:A417–A424
61. Rossouw MH, Thackeray MM (1991) Lithium manganese oxides from Li_2MnO_3 for rechargeable lithium battery applications. Mat Res Bull 26:463–473
62. Paik Y, Grey CP, Johnson CS, Kim J-S, Thackeray MM (2002) Lithium and deuterium NMR studies of acid-leached layered lithium manganese oxides. Chem Mater 14:5109–5115
63. Storey C, Kargina I, Grincourt Y, Davidson IJ, Yoo YC, Seung DY (2001) Electrochemical characterization of a new high capacity cathode. J Power Sources 97–98:541–544
64. Ammundsen B, Paulsen J, Davidson I, Liu R-S, Shen C-H, Chen J-M, Jang L-Y, Lee J-F (2002) Local structure and first cycle redox mechanism of layered $Li_{1.2}Cr_{0.4}Mn_{0.4}O_2$ cathode material. J Electrochem Soc 149:A431–A436
65. Ohzuku T, Makimura Y (2001) Layered lithium insertion material of $LiNi_{1/2}Mn_{1/2}O_2$: a possible alternative to $LiCoO_2$ for advanced lithium-ion batteries. Chem Lett 8:744–745
66. Rossen E, Jones CDW, Dahn JR (1992) Structure and electrochemistry of $Li_xMn_yNi_{1-y}O_2$. Solid State Ionics 57:311–318
67. Makimura Y, Ohzuku T (2003) Lithium insertion material of $LiNi_{1/2}Mn_{1/2}O_2$ for advanced lithium-ion batteries. J Power Sources 119–121:156–160
68. Lu Z, MacNeil DD, Dahn JR (2001) Layered cathode materials $Li[Ni_xLi_{(1/3-2x/3)}Mn_{(2/3-x/3)}]O_2$ for lithium-ion batteries. Electrochem Solid State Lett 4:A191–A194
69. Reed J, Ceder G (2002) Charge, potential, and phase stability of layered $Li(Ni_{0.5}Mn_{0.5})O_2$. Electrochem Solid-State Lett 5:A145–A148
70. Yoon W-S, Grey CP, Balasubramanian M, Yang X-Q, McBreen J (2003) In situ x-ray absorption spectroscopic study on $LiNi_{0.5}Mn_{0.5}O_2$ cathode material during electrochemical cycling. Chem Mater 15:3161–3169
71. Van der Ven A, Ceder G (2004) Ordering in $Li_x(Ni_{0.5}Mn_{0.5})O_2$ and its relation to charge capacity and electrochemical behavior in rechargeable lithium batteries. Electrochem Commun 6:1045–1050
72. Yoon W-S, Paik Y, Yang X-Q, Balasubramanian M, McBreen J, Grey CP (2002) Investigation of the local structure of the $LiNi_{0.5}Mn_{0.5}O_2$ cathode material during electrochemical cycling by x-ray absorption and NMR spectroscopy. Electrochem Solid State Lett 5: A263–A266
73. Lu Z, Beaulieu LY, Donaberger RA, Thomas CL, Dahn JR (2002) Synthesis, structure, and electrochemical behavior of $Li[Ni_xLi_{1/3-2x/3}Mn_{2/3-x/3}]O_2$. J Electrochem Soc 149: A778–A791
74. Kang S-H, Park S-H, Johnson CS, Amine K (2007) Effects of Li content on structure and electrochemical properties of $Li_{1+x}(Ni_{0.5}Mn_{0.5})_{1-x}O_2$ ($0 \leq x \leq 0.15$) electrodes in lithium cells (1.0-4.8 V). J Electrochem Soc 154:A268–A274
75. Kang K, Meng YS, Bréger J, Grey CP, Ceder G (2006) Electrodes with high power and high capacity for rechargeable lithium batteries. Science 311:977–980
76. Yabuuchi N, Ohzuku T (2003) Novel lithium insertion material of $LiCo_{1/3}Ni_{1/3}Mn_{1/3}O_2$ for advanced lithium-ion batteries. J Power Sources 119–121:171–174
77. Koyama Y, Tanaka I, Adachi H, Makimura Y, Ohzuku T (2003) Crystal and electronic structures of superstructural $Li_{1-x}[Co_{1/3}Ni_{1/3}Mn_{1/3}]O_2$ ($0 \leq x \leq 1$). J Power Sources 119–121:644–648

78. Hwang BJ, Tsai YW, Carlier D, Ceder G (2003) A combined computational/experimental study on $LiNi_{1/3}Co_{1/3}Mn_{1/3}O_2$. Chem Mater 15:3676–3682
79. Kim J-M, Chung H-T (2003) The first cycle characteristics of $Li[Ni_{1/3}Co_{1/3}Mn_{1/3}]O_2$. Electrochim Acta 49:937–944
80. Kobayashi H, Arachi Y, Emura S, Kageyama H, Tatsumi K, Kamiyama T (2005) Investigation on lithium de-intercalation mechanism for $Li_{1-y}Ni_{1/3}Mn_{1/3}Co_{1/3}O_2$. J Power Sources 146:640–644
81. Tsai YW, Hwang BJ, Ceder G, Sheu HS, Liu DG, Lee JF (2005) In-situ x-ray absorption spectroscopic study on variation of electronic transitions and local structure of $LiNi_{1/3}Co_{1/3}Mn_{1/3}O_2$ cathode material during electrochemical cycling. Chem Mater 17:3193–3199
82. Kim MG, Shin HJ, Kim J-H, Park S-H, Sun Y-K (2005) XAS investigation of inhomogeneous metal-oxygen bond covalency in bulk and surface for charge compensation in Li-ion battery cathode $Li[Ni_{1/3}Co_{1/3}Mn_{1/3}]O_2$ material. J Electrochem Soc 152:A1320–A1328
83. Yoon W-S, Balasubramanian M, Chung KY, Yang X-Q, McBreen J, Grey CP, Fischer DA (2005) Electrochemical Li-ion deintercalated $Li_{1-x}Co_{1/3}Ni_{1/3}Mn_{1/3}O_2$ electrode system by combination of soft and hard x-ray absorption spectroscopy. J Am Chem Soc 127:17479–17487
84. Choi J, Manthiram A (2005) Role of chemical and structural stabilities on the electrochemical properties of layered $LiNi_{1/3}Mn_{1/3}Co_{1/3}O_2$ cathodes. J Electrochem Soc 152:A1714–A1718
85. Belharouak I, Lu W, Vissers D, Amine K (2006) Safety characteristics of $Li(Ni_{0.8}Co_{0.15}Al_{0.05})O_2$ and $Li(Ni_{1/3}Co_{1/3}Mn_{1/3})O_2$. Electrochem Commun 8:329–335
86. Ngala JK, Chernova NA, Ma M, Mamak M, Zavalij PY, Whittingham MS (2004) The synthesis, characterization and electrochemical behavior of the layered $LiNi_{0.4}Mn_{0.4}Co_{0.2}O_2$ compound. J Mater Chem 14:214–220
87. Xiao J, Chernova NA, Whittingham MS (2008) Layered mixed transition metal oxide cathodes with reduced cobalt content for lithium ion batteries. Chem Mater 20:7454–7464
88. Chernova NA, Ma M, Xiao J, Whittingham MS, Breger J, Grey CP (2007) Layered $Li_xNi_yMn_yCo_{1-2y}O_2$ cathodes for lithium ion batteries: understanding local structure via magnetic properties. Chem Mater 19:4682–4693
89. Liu D, Wang Z, Chen L (2006) Comparison of structure and electrochemistry of Al- and Fe-doped $LiNi_{1/3}Co_{1/3}Mn_{1/3}O_2$. Electrochim Acta 51:4199–4203
90. Hu S-K, Chou T-C, Hwang B-J, Ceder G (2006) Effect of Co content on performance of $LiAl_{1/3-x}Co_xNi_{1/3}Mn_{1/3}O_2$. J Power Sources 160:1287–1293
91. Zhou F, Zhao X, Lu Z, Jiang J, Dahn JR (2008) The effect of Al substitution on the reactivity of delithiated $LiNi_{1/3}Mn_{1/3}Co_{(1/3-z)}Al_zO_2$ with non-aqueous electrolyte. Electrochem Commun 10:1054–1057
92. Wilcox J, Patoux S, Doeff M (2009) Structure and electrochemistry of $LiNi_{1/3}Co_{1/3-y}M_yMn_{1/3}O_2$ (M = Ti, Al, Fe) positive electrode materials. J Electrochem Soc 156:A192–A198
93. Wilcox JD, Rodriguez EE, Doeff MM (2009) The impact of aluminum and iron substitution on the structure and electrochemistry of $Li(Ni_{0.4}Co_{0.2-y}M_yMn_{0.4})O_2$. J Electrochem Soc 156: A1011–A1018
94. Xiao J, Chernova NA, Whittingham MS (2010) Influence of manganese content on the performance of $LiNi_{0.9-y}Mn_yCo_{0.1}O_2$ ($0.45 \leq y \leq 0.60$) as a cathode material for li-ion batteries. Chem Mater 22:1180–1185
95. Thackeray MM, Kang S-H, Johnson CS, Vaughey JT, Benedek R, Hackney SA (2007) Li_2MnO_3-stabilized $LiMO_2$ (M=Mn, Ni, Co) electrodes for lithium-ion batteries. J Mater Chem 17:3112–3135
96. Johnson CS, Li N, Lifief C, Thackeray MM (2007) Anomalous capacity and cycling stability of $xLi_2MnO_3 \bullet (1-x)LiMO_2$ electrodes (M=Mn, Ni, Co) in lithium batteries at 50°C. Electrochem Commun 9:787–795
97. Hunter JC (1981) Preparation of a new crystal form of manganese dioxide: λ-MnO_2. J Sol State Chem 32:142–147
98. Thackeray MM, David WIF, Bruce PG, Goodenough JB (1983) Lithium insertion into manganese spinels. Mat Res Bull 18:461–472

99. Goodenough JB, Thackeray MM, David WIF, Bruce PG (1984) Lithium insertion/extraction reactions with manganese oxides. Rev Chim Miner 21:435–455
100. Thackeray MM (1999) Spinel electrodes for lithium batteries. J Am Ceram Soc 82:337–3354
101. Tarascon JM, Wang E, Shokoohi FK, McKinnon WR, Colson S (1991) The spinel phase of $LiMn_2O_4$ as a cathode in secondary lithium cells. J Electrochem Soc 138:2859–2864
102. Tarascon JM, McKinnon WR, Coowar F, Bowmer TN, Amatucci G, Guyomard D (1994) Synthesis conditions and oxygen stoichiometry effects on Li insertion into the spinel $LiMn_2O_4$. J Electrochem Soc 141:1421–1431
103. Guyomard D, Tarascon JM (1994) The carbon/$Li_{1+x}Mn_2O_4$ system. Sol State Ionics 69:222–237
104. Xia Y, Zhou Y, Yoshio M (1997) Capacity fading on cycling of 4 V Li/$LiMn_2O_4$ cells. J Electrochem Soc 144:2593–2600
105. Amatucci G, Tarascon J-M (2002) Optimization of insertion compounds such as $LiMn_2O_4$ for li-ion batteries. J Electrochem Soc 149:K31–K46
106. Gummow RJ, de Kock A, Thackeray MM (1994) Improved capacity retention in rechargeable 4 V lithium/lithium manganese oxide (spinel) cells. Sol State Ionics 69:59–67
107. Shaju KM, Subba Rao GV, Chowdari BVR (2002) Spinel phases, $LiM_{1/6}Mn_{11/6}O_4$ (M=Co, CoAl, CoCr, CrAl), as cathodes for lithium-ion batteries. Sol State Ionics 148:343–350
108. Tucker MC, Reimer JA, Cairns EJ (2002) A 7Li NMR study of capacity fade in metal-substituted lithium manganese oxide spinels. J Electrochem Soc 149:A574–A585
109. Grush MM, Horne CR, Perera RCC, Ederer DL, Cramer SP, Cairns EJ, Callcott TA (2000) Correlating electronic structure with cycling performance of substituted $LiMn_2O_4$ electrode materials: a study using the techniques of soft x-ray absorption and emission. Chem Mater 12:659–664
110. Myung S-T, Komaba S, Kumagai N (2001) Enhanced structural stability and cyclability of Al-doped $LiMn_2O_4$ spinel synthesized by the emulsion drying method. J Electrochem Soc 148:A482–A489
111. Ariyoshi K, Iwata E, Kuniyoshi M, Wakabayashi H, Ohzuku T (2006) Lithium aluminum manganese oxide having spinel framework structure for long-life lithium-ion batteries. Electrochem Sol State Lett 9:A557–A560
112. Amatucci GG, Pereira N, Zheng T, Tarascon J-M (2001) Failure mechanism and improvement of the elevated temperature cycling of $LiMn_2O_4$ compounds through the use of the $LiAl_xMn_{2-x}O_{4-z}F_z$ solid solution. J Electrochem Soc 148:A171–A182
113. Whittingham MS (2004) Lithium batteries and cathode materials. Chem Rev 104:4271–4301
114. Thackeray MM, de Kock A, Rossouw MH, Liles D, Bittihn R, Hoge D (1992) Spinel electrodes from the Li-Mn-O system for rechargeable lithium battery applications. J Electrochem Soc 139:363–366
115. Kim J, Manthiram A (1998) Low temperature synthesis and electrode properties of $Li_4Mn_5O_{12}$. J Electrochem Soc 145:L53–L55
116. Pasquier Du, Blyr A, Courjal P, Larcher D, Amatucci G, Gérand B, Tarascon J-M (1999) Mechanism for limited 55°C storage performance of $Li_{1.05}Mn_{1.95}O_4$ electrodes. J Electrochem Soc 146:428–436
117. Wen SJ, Richardson TJ, Ma L, Striebel KA, Ross PN, Cairns EJ (1996) FTIR spectroscopy of metal oxide insertion electrodes. J Electrochem Soc 143:L136–L138
118. Eriksson T, Gustafsson T, Thomas JO (2002) Surface structure of $LiMn_2O_4$ electrodes. Electrochem Sol State Lett 5:A35–A38
119. Akimoto J, Takahashi Y, Kijima N (2005) Direct observation of the bulk degradation of $Li_{1.1}Mn_{1.9}O_4$ single crystals after high-temperature storage. Electrochem Sol State Lett 8:A361–A364
120. Kobayashi H, Sakaebe H, Komoto K, Kageyama H, Tabuchi M, Tatsumi K, Kohigashi T, Yonemura M, Kanno R, Kamiyama T (2003) Structure and physical property changes of de-lithiated spinels for $Li_{1.02-x}Mn_{1.98}O_4$ after high-temperature storage. Sol State Ionics 156:309–318

121. Quinlan FT, Sano K, Willey T, Vidu R, Tasaki K, Stroeve P (2001) Surface characterization of the spinel $Li_xMn_2O_4$ cathode before and after storage at elevated temperatures. Chem Mater 13:4207–4212
122. Tsunekawa H, Tanimoto S, Marubayashi R, Fujita M, Kifune K, Sano M (2002) Capacity fading of graphite electrodes due to the deposition of manganese ions on them in Li-ion batteries. J Electrochem Soc 149:A1326–A1331
123. Amine K, Liu J, Kang S, Belharouak I, Hyung Y, Vissers D, Henriksen G (2004) Improved lithium manganese oxide spinel/graphite Li-ion cells for high-power applications. J Power Sources 129:14–19
124. Cho J, Kim GB, Lim HS, Kim C-S, Yoo S-I (1999) Improvement of structural stability of $LiMn_2O_4$ cathode material on 55°C cycling by sol-gel coating of $LiCoO_2$. Electrochem Sol State Lett 2:607–609
125. Wang EI (1998) Method of treating lithium manganese oxide spinel. US Patent 5,783,328
126. Xu W, Angell CA (2001) LiBOB and its derivatives weakly coordinating anions, and the exceptional conductivity of their nonaqueous solutions. Electrochem Sol State Lett 4:E1–E4
127. Xu K, Zhang S, Jow TR, Xu W, Angell CA (2002) LiBOB as salt for lithium-ion batteries a possible solution for high temperature operation. Electrochem Solid State Lett 5(1): A26–A29
128. Chen Z, Amine K (2006) Capacity fade of $Li_{1+x}Mn_{2-x}O_4$-based lithium-ion cells. J Electrochem Soc 153:A316–A320
129. Kitao H, Fujihara T, Takeda K, Nakanishi N, Nohma T (2005) High-temperature storage performance of Li-ion batteries using a mixture of Li-Mn spinel and Li-Ni-Co-Mn oxide as a positive electrode material. Electrochem Sol State Lett 8:A87–A90
130. MacNeil DD, Dahn JR (2001) The reaction of charged cathodes with nonaqueous solvents and electrolytes II. $LiMn_2O_4$ charged to 4.2 V. J Electrochem Soc 148:A1211–A1215
131. Thackeray MM (1995) Structural considerations of layered and spinel lithiated oxides for lithium ion batteries. J Electrochem Soc 142:2558–2563
132. Sigala C, Guyomard D, Verbaere A, Piffard Y, Tournoux M (1995) Positive electrode materials with high operating voltage for lithium batteries: $LiCr_yMn_{2-y}O_4$ ($0 \leq y \leq 1$). Sol State Ionics 81:167–170
133. Kawai H, Nagata M, Tabuchi M, Tukamoto H, West AR (1998) Novel 5V spinel cathode $Li_2FeMn_3O_8$ for lithium ion batteries. Chem Mater 10:3266–3268
134. Ein-Eli Y, Howard WF (1997) $LiCu_x{}^{II}Cu_y{}^{III}Mn_{[2-(x+y)]}{}^{III,\ IV}O_4$: 5V cathode materials. J Electrochem Soc 144:L205–L207
135. Kawai H, Nagata M, Tukamoto H, West AR (1998) A novel cathode $Li_2CoMn_3O_8$ for lithium ion batteries operating over 5V. J Mater Chem 8:837–839
136. Zhong Q, Bonakdarpour A, Zhang M, Gao Y, Dahn JR (1997) Synthesis and electrochemistry of $LiNi_xMn_{2-x}O_4$. J Electrochem Soc 144:205–213
137. Ohzuku T, Takeda S, Iwanaga M (1999) Solid-state redox potentials for $Li[Me_{1/2}Mn_{3/2}]O_4$ (Me: 3d-transition metal) having spinel-framework structures: a series of 5 volt materials for advanced lithium-ion batteries. J Power Sources 81–82:90–94
138. Sun Y-K, Lee Y-S, Yoshio M, Amine K (2002) Synthesis and electrochemical properties of ZnO-coated $LiNi_{0.5}Mn_{1.5}O_4$ spinel as 5V cathode material for lithium secondary batteries. Electrochem Sol State Lett 5:A99–A102
139. Liu J, Manthiram A (2009) Kinetics study of the 5V spinel cathode $LiMn_{1.5}Ni_{0.5}O_4$ before and after surface modifications. J Electrochem Soc 156:A833–A838
140. Patoux S, Daniel L, Bourbon C, Lignier H, Pagano C, Le Cras F, Jouanneau S, Martinet S (2009) High voltage spinel oxides for Li-ion batteries: from the materials research to the application. J Power Sources 189:344–352
141. Kim J-H, Myung S-T, Yoon CS, Oh I-H, Sun Y-K (2004) Effect of Ti substitution of $LiNi_{0.5}Mn_{1.5-x}Ti_xO_4$ and their electrochemical properties as lithium insertion material. J Electrochem Soc 151:A1911–A1918

142. Arunkumar TA, Manthiram A (2005) Influence of lattice parameter differences on the electrochemical performance of the 5V spinel $LiMn_{1.5-y}Ni_{0.5-z}M_{y+z}O_4$ (M=Li, Mg, Fe, Co, and Zn). Electrochem Sol State Lett 8:A403–A405
143. Ooms FGB, Kelder EM, Schoonman J, Wagemaker M, Mulder FM (2002) High-voltage $LiMg_{\delta}Ni_{0.5-\delta}Mn_{1.5}O_4$ spinels for Li-ion batteries. Sol State Ionics 152–153:143–153
144. Gryffoy D, Vandenberghe RE, Legrand E (1991) A neutron diffraction study of some spinel compounds containing octahedral Ni and Mn at a 1:3 ratio. Mater Sci Forum 79–82:785–790
145. Takahashi K, Saitoh M, Sano M, Fujita M, Kifune K (2004) Electrochemical and structural properties of a 4.7V-class $LiNi_{0.5}Mn_{1.5}O_4$ positive electrode material prepared with a self-reaction method. J Electrochem Soc 151:A173–A177
146. Kunduraci M, Amatucci GG (2006) Synthesis and characterization of nanostructured 4.7V $Li_xMn_{1.5}Ni_{0.5}O_4$ spinels for high-power lithium-ion batteries. J Electrochem Soc 153: A1345–A1352
147. Wu HM, Belharouak I, Deng H, Abouimrane A, Sun Y-K, Amine K (2009) Development of $LiNi_{0.5}Mn_{1.5}O_4/Li_4Ti_5O_{12}$ system with long cycle life. J Electrochem Soc 156:A1047–A1050
148. Padhi AK, Nanjundaswamy KS, Goodenough JB (1997) Phospho-olivines as positive-electrode materials for rechargeable lithium batteries. J Electrochem Soc 144:1188–1194
149. Andersson AS, Kalska B, Häggstrom L, Thomas JO (2000) Lithium extraction/insertion in $LiFePO_4$: an x-ray diffraction and Mössbauer spectroscopy study. Sol State Ionics 130:41–52
150. Tarascon J-M, Armand M (2001) Issues and challenges facing rechargeable lithium batteries. Nature 414:359–376
151. Ravet N, Chouinard Y, Magnan JF, Besner S, Gauthier M, Armand M (2001) Electroactivity of natural and synthetic triphylite. J Power Sources 97–98:503–507
152. Barker J, Saidi MY, Swoyer JL (2003) A carbothermal reduction method for the preparation of electroactive materials for lithium ion applications. J Electrochem Soc 150:A684–A688
153. Doeff MM, Wilcox JD, Kostecki R, Lau G (2006) Optimization of carbon coatings on $LiFePO_4$. J Power Sources 163:180–184
154. Dominko R, Bele M, Gaberscek M, Remskar M, Hanzel D, Pejovnik S, Jamnik J (2005) Impact of the carbon coating thickness on the electrochemical performance of $LiFePO_4$/C composites. J Electrochem Soc 152:A607–A610
155. Huang H, Yin S-C, Nazar LF (2001) Approaching theoretical capacity of $LiFePO_4$ at room temperature at high rates. Electrochem Sol State Lett 4:A170–A172
156. Doeff MM, Hu Y, McLarnon F, Kostecki R (2003) Effect of surface carbon structure on the electrochemical performance of $LiFePO_4$. Electrochem Sol State Lett 6:A207–A209
157. Zaghib K, Shim J, Guerfi A, Charest P, Striebel KA (2005) Effect of carbon source as additives in $LiFePO_4$ as positive electrode for li-ion batteries. Electrochem Sol State Lett 8: A207–A210
158. Wilcox JD, Doeff MM, Marcinek M, Kostecki R (2007) Factors influencing the quality of carbon coatings on $LiFePO_4$. J Electrochem Soc 154:A389–A395
159. Doeff MM, Wilcox JD, Yu R, Aumentado A, Marcinek M, Kostecki R (2008) Impact of carbon structure and morphology on the electrochemical performance of $LiFePO_4$/C composites. J Sol State Electrochem 12:995–1001
160. Chen Z, Dahn JR (2002) Reducing carbon in $LiFePO_4$/C composite electrodes to maximize specific energy, volumetric energy, and tap density. J Electrochem Soc 149:A1184–A1189
161. Chung S-Y, Bloking JT, Chiang Y-M (2002) Electronically conductive phospho-olivines as lithium storage electrodes. Nat Mater 1:123–128
162. Thackeray MM (2002) An unexpected conductor. Nat Mater 1:81–82
163. Ravet N, Abouimrane A, Armand M (2003) From our readers. Nat Mater 2:702–703
164. Herle PS, Ellis B, Coombs N, Nazar LF (2004) Nano-network electronic conduction in iron and nickel olivine phosphates. Nat Mater 3:147–152
165. Delacourt C, Wurm C, Laffont L, Leriche J-B, Masquelier C (2006) Electrochemical and electrical properties of Nb- and/or C-containing $LiFePO_4$ composites. Sol State Ionics 177:333–341

166. Rho Y-H, Nazar LF, Perry L, Ryan D (2007) Surface chemistry of $LiFePO_4$ studied by Mössbauer and x-ray photoelectron spectroscopy and its effect on electrochemical properties. J Electrochem Soc 154:A283–A289
167. Wagemaker M, Ellis BL, Lützenkirchen-Hecht D, Mulder FM, Nazar LF (2008) Proof of supervalent doping in olivine $LiFePO_4$. Chem Mater 20:6313–6315
168. Islam MS, Driscoll DJ, Fisher CAJ, Slater PR (2005) Atomic-scale investigation of defects, dopants, and lithium transport in the $LiFePO_4$ olivine-type battery material. Chem Mater 17:5085–5092
169. Maier J, Amin R (2008) Defect chemistry of $LiFePO_4$. J Electrochem Soc 155:A339–A344
170. Axmann P, Stinner C, Wohlfahrt-Mehrens M, Mauger A, Gendron G, Julien CM (2009) Nonstoichiometric $LiFePO_4$: defects and related properties. Chem Mater 21:1636–1644
171. Amin R, Maier J (2008) Effect of annealing on transport properties of $LiFePO_4$: towards a defect chemical model. Sol State Ionics 178:1831–1836
172. Morgan D, Van der Ven A, Ceder G (2004) Li conductivity in Li_xMPO_4 (M=Mn, Fe, Co, Ni) olivine materials. Electrochem and Sol State Lett 7:A30–A32
173. Li J, Yao W, Martin S, Vaknin D (2008) Lithium ion conductivity in single crystal $LiFePO_4$. Sol State Ionics 179:2016–2019
174. Amin R, Maier J, Balaya P, Chen DP, Lin CT (2008) Ionic and electronic transport in single crystalline $LiFePO_4$ grown by optical floating zone technique. Sol State Ionics 179:1683–1687
175. Gardiner GR, Islam MS (2010) Anti-site defects and ion migration in the $LiFe_{0.5}Mn_{0.5}PO_4$ mixed-metal cathode material. Chem Mater 22:1242–1248
176. Chen J, Vacchio MJ, Wang S, Chernova N, Zavalij PY, Whittingham MS (2008) The hydrothermal synthesis and characterization of olivines and related compounds for electrochemical applications. Sol State Ionics 178:1676–1693
177. Delacourt C, Poizot P, Levasseur S, Masquelier C (2006) Size effects on carbon-free $LiFePO_4$ powders the key to superior energy density. Electrochem Sol State Lett 9:A352–A355
178. Meethong N, Huang H-YS, Carter WC, Chiang Y-M (2007) Size-dependent lithium miscibility gap in nanoscale $Li_{1-x}FePO_4$. Electrochem Sol State Lett 10:A134–A138
179. Gibot P, Casas-Cabanas M, Laffont L, Levasseur S, Carlach P, Hamelet S, Tarascon J-M, Masquelier C (2008) Room-temperature single-phase Li insertion/extraction in nanoscale Li_xFePO_4. Nat Mater 7:741–747
180. Yamada A, Koizumi H, Sonoyama N, Kanno R (2005) Phase change in Li_xFePO_4. Electrochem Sol State Lett 8:A409–A413
181. Yamada A, Koizumi H, Nishimura S-I, Sonoyama N, Kanno R, Yonemura M, Nakamura T, Kobayashi Y (2006) Room-temperature miscibility gap in Li_xFePO_4. Nat Mater 5:357–360
182. Hamelet S, Gibot P, Casas-Cabanas M, Bonnin D, Grey CP, Cabana J, Leriche J-B, Rodriguez-Carvajal CM, Levasseur S, Carlach P, Van Thournout M, Tarascon J-M, Masquelier C (2009) The effects of moderate thermal treatments under air on $LiFePO_4$-based nano powders. J Mater Chem 19:3979–3991
183. Yu DYW, Donoue K, Kadohata T, Murata T, Matsuta S, Fujitani S (2008) Impurities in $LiFePO_4$ and their influence on material characteristics. J Electrochem Soc 155:A526–A530
184. Dokko K, Shiraishi K, Kanamura K (2005) Identification of surface impurities on $LiFePO_4$ particles prepared by a hydrothermal process. J Electrochem Soc 152:A2199–A2202
185. Delacourt C, Poizot P, Tarascon J-M, Masquelier C (2005) The existence of a temperature-driven solid solution in Li_xFePO_4 for $0 \leq x \leq 1$. Nat Mater 4:254–260
186. Dodd JL, Yazami R, Fultz B (2006) Phase diagram of Li_xFePO_4. Electrochem Sol State Lett 9:A151–A155
187. Chen G, Song X, Richardson TJ (2007) Metastable solid-solution phases in the $LiFePO_4$/$FePO_4$ system. J Electrochem Soc 154:A627–A632
188. Srinivasan V, Newman J (2004) Discharge model for the lithium iron-phosphate electrode. J Electrochem Soc 151:A1517–A1529

189. Chen G, Song X, Richardson TJ (2006) Electron microscopy study of the $LiFePO_4$ to $FePO_4$ phase transition. Electrochem Sol State Lett 9:A295–A298
190. Andersson AS, Thomas JO (2001) The source of first cycle capacity loss in $LiFePO_4$. J Power Sources 97–98:498–502
191. Laffont L, Delacourt C, Gibot P, Wu MY, Kooyman P, Masquelier C, Tarascon JM (2006) Study of the $LiFePO_4$/$FePO_4$ two-phase system by high-resolution electron energy loss spectroscopy. Chem Mater 18:5520–5529
192. Delmas C, Maccario M, Croguennec L, Le Cras F, Weill F (2008) Lithium deintercalation in $LiFePO_4$ nanoparticles via a domino-cascade model. Nat Mater 7:665–671
193. Allen JL, Jow TR, Wolfenstine J (2007) Kinetic study of the electrochemical $FePO_4$ to $LiFePO_4$ phase transition. Chem Mater 19:2108–2111
194. Lee KT, Kan WH, Nazar LF (2009) Proof of intercrystallite ionic transport in $LiMPO_4$ electrodes (M=Fe, Mn). J Am Chem Soc 131:6044–6045
195. Kang B, Ceder G (2009) Battery materials for ultrafast charging and discharging. Nature 458:190–193
196. Amine K, Liu J, Belharouak I (2005) High-temperature storage and cycling of C-$LiFePO_4$/graphite Li-ion cells. Electrochem Commun 7:669–673
197. Koltypin M, Aurbach D, Nazar L, Ellis B (2007) On the stability of $LiFePO_4$ olivine cathodes under various conditions (electrolyte solutions, temperatures). Electrochem Solid-State Lett 10:A40–A44
198. Yang S, Song Y, Zavalij PY, Whittingham MS (2002) Reactivity, stability and electrochemical behavior of lithium iron phosphates. Electrochem Commun 4:239–244
199. Li G, Azuma H, Tohda M (2002) $LiMnPO_4$ as the cathode for lithium batteries. Electrochem Sol State Lett 5:A135–A137
200. Rissouli K, Benkhouja K, Ramos-Barrado JR, Julien C (2003) Electrochemical conductivity in lithium orthophosphates. Mater Sci Eng B 98:185–189
201. Delacourt C, Laffont L, Bouchet R, Wurm C, Leriche J-B, Morcrette M, Tarascon J-M, Masquelier C (2005) Toward understanding of electrical limitations (electronic, ionic) in $LiMPO_4$ (M=Fe, Mn) electrode materials. J Electrochem Soc 152:A913–A921
202. Delacourt C, Poizot P, Morcrette M, Tarascon J-M, Masquelier C (2004) One step low-temperature route for the preparation of electrochemically active $LiMnPO_4$ powders. Chem Mater 16:93–99
203. Yang J, Xu JJ (2006) Synthesis and characterization of carbon-coated lithium transition metal phosphates $LiMPO_4$ (M=Fe, Mn, Co, Ni) prepared via a nonaqueous sol-gel route. J Electrochem Soc 153:A716–A723
204. Kwon N-H, Drezen T, Exnar I, Teerlinck I, Isono M, Graetzel M (2006) Enhanced electrochemical performance of mesoparticulate $LiMnPO_4$ for lithium ion batteries. Electrochem and Solid State Lett 9:A277–A280
205. Kim TR, Kim DH, Ryu HW, Moon JH, Lee JH, Boo S, Kim J (2007) Synthesis of lithium manganese phosphate nanoparticle and its properties. J Phys Chem Sol 68: 1203–1206
206. Bakenov Z, Taniguchi I (2010) Electrochemical performance of nanocomposite $LiMnPO_4$/C cathode materials for lithium batteries. Electrochemistry Commun 12:75–78
207. Xiao J, Xu W, Choi D, Zhang J-G (2010) Synthesis and characterization of lithium manganese phosphate by a precipitation method. J Electrochem Soc 157:A142–A147
208. Doeff MM, Chen J, Conry TE, Wang R, Wilcox J, Aumentado A (2010) Combustion synthesis of nanoparticulate $LiMg_xMn_{1-x}PO_4$ ($x = 0, 0.1, 0.2$) carbon composites. J Mater Res 25:1460–1468
209. Drezen T, Kwon N-H, Bowen P, Terrlinck I, Isono M, Exnar I (2007) Effect of particle size on $LiMnPO_4$ cathodes. J Power Sources 174:949–953
210. Chen G, Wilcox JD, Richardson TJ (2008) Improving the performance of lithium manganese phosphate through divalent cation substitution. Electrochem Sol State Lett 11:A190–A194

211. Shiratsuchi T, Okada S, Doi T, Yamaki J-I (2009) Cathodic performance of $LiMn_{1-x}M_xPO_4$ (M = Ti, Mg, and Zr) annealed in an inert atmosphere. Electrochim Acta 54:3145–3151
212. Bakenov Z, Taniguchi I (2010) $LiMg_xMn_{1-x}PO_4$/C cathodes for lithium batteries prepared by a combination of spray pyrolysis with wet ballmilling. J Electrochem Soc 157:A430–A436
213. Martha SK, Markovsky B, Grinblat J, Gofer Y, Haik O, Zinigrad E, Aurbach D, Drezen T, Wang D, Deghenghi G, Exnar I (2009) $LiMnPO_4$ as an advanced cathode material for rechargeable lithium batteries. J Electrochem Soc 156:A541–A552
214. Yonemura M, Yamada A, Takei Y, Sonoyama N, Kanno R (2004) Comparative kinetic study of olivine Li_xMPO_4 (M=Fe, Mn). J Electrochem Soc 151:A1352–A1356
215. Yamada A, Chung S-C (2001) Crystal chemistry of the olivine-type $Li(Mn_yFe_{1-y})PO_4$ and $(Mn_yFe_{1-y})PO_4$ as possible 4 V cathode materials for lithium batteries. J Electrochem Soc 148:A960–A967
216. Chen G, Richardson TJ (2009) Solid solution phases in the olivine-type $LiMnPO_4/MnPO_4$ system. J Electrochem Soc 156:A756–A762
217. Kim S-W, Kim J, Gwon H, Kang K (2009) Phase stability study of $Li_{1-x}MnPO_4$ ($0 \leq x \leq 1$) cathode for Li rechargeable battery. J Electrochem Soc 156:A635–A638
218. Chen G, Richardson TJ (2010) Thermal instability of olivine-type $LiMnPO_4$ cathodes. J Power Sources 195:1221–1224
219. Okada S, Sawa S, Egashira M, Yamaki J-I, Tabuchi M, Kageyama H, Konishi T, Yoshino A (2001) Cathode properties of phospho-olivine $LiMPO_4$ for lithium secondary batteries. J Power Sources 97–98:430–432
220. Wolfenstine J, Allen J (2004) $LiNiPO_4$–$LiCoPO_4$ solid solutions as cathodes. J Power Sources 136:150–153
221. Deniard P, Dulac AM, Rocquefelte X, Grigorova V, Lebacq O, Pasturel A, Jobic S (2004) High potential positive materials for lithium-ion batteries: transition metal phosphates. J Phys Chem Sol 65:229–233
222. Chang X-Y, Wang Z-X, Li X-H, Zhang L, Guo H-J, Peng W-J (2005) Synthesis and performance of $LiMn_{0.7}Fe_{0.3}PO_4$ cathode material for lithium ion batteries. Mater Res Bull 40:1513–1520
223. Yamada A, Kudo Y, Liu K-Y (2001) Phase diagram of $Li_x(Mn_yFe_{1-y})PO_4$ ($0 \leq x, y \leq 1$). J Electrochem Soc 148:A1153–A1158
224. Bramnik NN, Bramnik KG, Nikolowski K, Hinterstein M, Baehtz C, Ehrenbert H (2005) Synchrotron diffraction study of lithium extraction from $LiMn_{0.6}Fe_{0.4}PO_4$. Electrochem Sol State Lett 8:A379–A381
225. Park Y-U, Kim J, Gwon H, Seo D-H, Kim S-W, Kang K (2010) Synthesis of multicomponent olivine by a novel mixed transition metal oxalate coprecipitation method and electrochemical characterization. Chem Mater 22:2573–2581
226. Gwon H, Seo D-H, Kim S-W, Kim J, Kang K (2009) Combined first-principle calculations and experimental study on multi-component olivine cathode for lithium rechargeable batteries. Adv Funct Mater 19:1–8
227. Seo D-H, Gwon H, Kim S-W, Kim J, Kang K (2010) Multicomponent olivine cathode for lithium rechargeable batteries: a first-principles study. Chem Mater 22:518–523
228. Goodenough JB, Hong Y-P, Kafalas JA (1976) Fast Na^+-ion transport in skeleton structures. Mater Res Bull 11:203–220
229. Delmas C, Nadiri A (1988) The Nasicon-type titanium phosphates $ATi_2(PO_4)_3$ (A=Li, Na) as electrode materials. Sol State Ionics 28–30:419–423
230. Delmas C, Cherkaoui F, Nadiri A, Hagenmuller P (1987) A Nasicon-type phase as intercalation electrode: $NaTi_2(PO_4)_3$. Mat Res Bull 22:631–639
231. Manthiram A, Goodenough JB (1989) Lithium insertion into $Fe_2(SO_4)_3$ frameworks. J Power Sources 26:403–408
232. Nanjundaswamy KS, Padhi AK, Goodenough JB, Okada S, Ohtsuka H, Arai H, Yamaki J (1996) Synthesis, redox potential evaluation and electrochemical characteristics of NASICON-related-3D framework compounds. Sol State Ionics 92:1–10

233. Masquelier C, Padhi AK, Nanjundaswamy KS, Goodenough JB (1998) New cathode materials for rechargeable lithium batteries: the 3-D framework structures $Li_3Fe_2(XO_4)_3$ (X=P, As). J Sol State Chem 135:228–234
234. Saïdi MY, Barker J, Huang H, Swoyer JL, Adamson G (2002) Electrochemical properties of lithium vanadium phosphate as a cathode material for lithium-ion batteries. Electrochem Sol State Lett 5:A149–A151
235. Morcrette M, Leriche J-B, Patoux S, Wurm C, Masquelier C (2003) In situ x-ray diffraction during lithium extraction from rhombohedral and monoclinic $Li_3V_2(PO_4)_3$. Electrochem Sol State Lett 6:A80–A84
236. Yin S-C, Strobel PS, Grondey H, Nazar LF (2004) $Li_{2.5}V_2(PO_4)_3$: a room-temperature analogue to the fast-ion conducting high-temperature γ-phase of $Li_3V_2(PO_4)_3$. Chem Mater 16:1456–1465
237. Gover RKB, Burns P, Bryan A, Saidi MY, Swoyer JL, Barker J (2006) $LiVPO_4F$: A new active material for safe lithium-ion batteries. Sol State Ionics 177:2635–2638
238. Barker J, Gover RKB, Burns P, Bryan A, Saidi MY, Swoyer JL (2005) Structural and electrochemical properties of lithium vanadium fluorophosphate, $LiVPO_4F$. J Power Sources 146:516–520
239. Barker J, Gover RKB, Burns P, Bryan A (2005) A symmetrical lithium-ion cell based on lithium vanadium fluorophosphate, $LiVPO_4F$. Electrochem Sol State Lett 8:A285–A287
240. Gover RKB, Bryan A, Burns P, Barker J (2006) The electrochemical insertion properties of sodium vanadium fluorophosphate, $Na_3V_2(PO_4)_2F_3$. Sol State Ionics 177:1495–1500
241. Ellis BL, Makahnouk WRM, Makimura Y, Toghill K, Nazar LF (2007) A multifunctional 3.5 V iron-based phosphate cathode for rechargeable batteries. Nat Mater 6:749–753
242. Ramesh TN, Lee KT, Ellis BL, Nazar LF (2010) Tavorite lithium iron fluorophosphate cathode materials: phase transition and electrochemistry of $LiFePO_4F$–Li_2FePO_4F. Electrochem Sol State Lett 13:A43–A47
243. Recham N, Chotard J-N, Jumas J-C, Laffont L, Armand M, Tarascon J-M (2010) Ionothermal synthesis of Li-based fluorophosphates electrodes. Chem Mater 22:1142–1148
244. Song Y, Zavalij PY, Chernova NA, Whittingham MS (2005) Synthesis, crystal structure, and electrochemical and magnetic study of new iron (III) hydroxyl-phosphates, isostructural with lipscombite. Chem Mater 17:1139–1147
245. Marx N, Croguennec L, Carlier D, Bourgeois L, Kubiak P, Le Cras F, Delmas C (2010) Structural and electrochemical study of a new crystalline hydrated iron(III) phosphate $FePO_4 \cdot H_2O$ obtained from $LiFePO_4(OH)$ by ion exchange. Chem Mater 22:1854–1861
246. Gaubicher J, Le Mercier T, Chabre Y, Angenault J, Quarton M (1999) Li/β-$VOPO_4$: a new 4V system for lithium batteries. J Electrochem Soc 146:4375–4379
247. Kerr TA, Gaubicher J, Nazar LF (2000) Highly reversible Li insertion at 4V in ε-$VOPO_4$/α-$LiVPO_4$ cathodes. Electrochem Sol State Lett 3:460–462
248. Song Y, Zavalij PY, Whittingham MS (2005) ε-$VOPO_4$: electrochemical synthesis and enhanced cathode behavior. J Electrochem Soc 152:A721–A727
249. Barker J, Gover RKB, Burns P, Bryan A (2005) $LiVP_2O_7$: a viable lithium-ion cathode material? Electrochem Sol State Lett 8:A446–A448
250. Nytén A, Abouimrane A, Armand M, Gustafsson T, Thomas JO (2005) Electrochemical performance of Li_2FeSiO_4 as a new Li-battery cathode material. Electrochem Commun 7:156–160
251. Dominko R, Bele M, Gaberscek M, Meden A, Remskar M, Jamnik J (2006) Structure and electrochemical performance of Li_2MnSiO_4 and Li_2FeSiO_4 as potential Li-battery cathode materials. Electrochem Commun 8:217–222
252. Gong ZL, Li YX, Yang Y (2006) Synthesis and characterization of $Li_2Mn_xFe_{1-x}SiO_4$ as a cathode material for lithium ion batteries. Electrochem Sol State Lett 9:A542–A544
253. Belharouak I, Abouimrane A, Amine K (2009) Structural and electrochemical characterization of Li_2MnSiO_4 cathode material. J Phys Chem C 113:20733–20737

254. Lyness C, Delobel B, Armstrong AR, Bruce PG (2007) The lithium intercalation compound Li_2CoSiO_4 and its behaviour as a positive electrode for lithium batteries. J Chem Soc Chem Commun 46:4890–4892
255. Armstrong AR, Lyness C, Ménétrier M, Bruce PG (2010) Structural polymorphism in Li_2CoSiO_4 intercalation electrodes: a combined diffraction and NMR study. Chem Mater 22:1892–1900
256. Dominko R (2010) Silicates and titanates as high-energy cathode materials for Li-ion batteries. In: Proceedings of SPIE: energy harvesting and storage: materials, devices and applications II, Orlando
257. Zaghib K, Salah AA, Ravet N, Mauger A, Gendron F, Julien CM (2006) Structural, magnetic and electrochemical properties of lithium iron orthosilicate. J Power Sources 160:1381–1386
258. Nishimura S-I, Hayase S, Kanno R, Yashima M, Nakayama N, Yamada A (2008) Structure of Li_2FeSiO_4. J Am Chem Soc 130:13212–13213
259. Arroyo-de Dompablo ME, Armand M, Tarascon JM, Amador U (2006) On-demand design of polyoxianionic cathode materials based of electronegativity correlations: an exploration of the Li_2MSiO_4 system (M=Fe, Mn, Co, Ni). Electrochem Commun 8:1292–1298
260. Li H, Balaya P, Maier J (2004) Li-storage via heterogeneous reaction in selected binary metal fluorides and oxides. J Electrochem Soc 151:A1878–A1885
261. Bervas M, Mansour AN, Yoon W-S, Al-Sharab JF, Badway F, Cosandey F, Klein LC, Amatucci GG (2006) Investigation of the lithiation and delithiation conversion mechanisms of bismuth fluoride nanocomposites. J Electrochem Soc 153:A799–A808
262. Aurbach D (2000) Review of selected electrode-solution interactions which determine the performance of Li and Li ion batteries. J Power Sources 89:206–218
263. Jaiswal A, Horne CR, Chang O, Zhang W, Kong W, Wang E, Chern T, Doeff MM (2009) Nanoscale $LiFePO_4$ and $Li_4Ti_5O_{12}$ for high rate Li-ion batteries. J Electrochem Soc 156:A1041–A1046
264. Ren Y, Armstrong AR, Jiao F, Bruce PG (2010) Influence of size on the rate of mesoporous electrodes for lithium batteries. J Am Chem Soc 132:996–1004
265. Sun Y-K, Myung S-T, Park B-C, Prakash J, Belharouak I, Amine K (2009) High energy cathode material for long-life and safe lithium batteries. Nat Mater 8:320–324

Books and Reviews

Ammundsen B, Paulsen J (2001) Novel lithium-ion cathode materials based on layered manganese oxides. Adv Mater 13:943–956

Brodd RJ, Bullock KR, Leising RA, Middaugh RL, Miller JR, Takeuchi E (2004) Batteries, 1977 to 2002. J Electrochem Soc 151:K1–K11

Bruce PG (2008) Energy storage beyond the horizon: rechargeable lithium batteries. Solid State Ionics 179:752–760

Ellis BL, Lee KT, Nazar LF (2010) Positive electrode materials for Li-ion and Li batteries. Chem Mater 22:691–714

Fergus JW (2010) Recent developments in cathode materials for lithium ion batteries. J Power Sources 195:939–954

Goodenough JB, Kim Y (2010) Challenges for rechargeable Li batteries. Chem Mater 22:587–603

Huggins RA (2009) Advanced batteries, materials sciences aspects. Springer, New York

Nazri GA, Pistoia O (eds) (2003) Lithium batteries: science and technology. Kluwer, Norwell

Ohzuku T, Brodd RJ (2007) An overview of positive-electrode materials for advanced lithium-ion batteries. J Power Sources 174:449–456

Palacin MR (2009) Recent advances in rechargeable battery materials: a chemist's perspective. Chem Soc Rev 38:2565–2575

Tarascon J-M (2010) Key challenges in future Li-battery research. Phil Trans R Soc Lond A 368:3227–3241

Yamada A, Hosoya M, Chung S-C, Kudo Y, Hinokuma K, Liu K-Y, Nishi Y (2003) Olivine-type cathodes achievements and problems. J Power Sources 119–121:232–238

Chapter 3
Battery Components, Active Materials for

J.B. Goodenough

Glossary

Battery efficiency	Output energy/input energy for storage of electrically energy as chemical energy.
Electrolyte window	Separation of electrolyte LUMO and HOMO energies of a liquid electrolyte and of conduction and valence bands of a solid electrolyte.
Fermi energy	Electrochemical potential of a solid.
HOMO	Highest occupied molecular orbital.
Insertion compound	A host solid into which a working ion (guest) can be inserted/extracted reversibly over a solid-solution range.
LUMO	Lowest unoccupied molecular orbital.
Polarization, $\eta = V_{OC} - V(q, I)$	Loss of battery voltage at a state of charge q due to resistance to ion transfer inside battery cell where an electronic current I is flowing outside of battery.
Rate of charge/discharge, nC	Time, (60/n) min, for complete discharge or charge of a battery or cell; it is also a measure of the current.
Redox couple	Cation $M^{(m+1)}/M^{m+}$ mixed-valent energy applicable to localized-electron configurations.
SEI layer	Solid/electrolyte interface (passivation) layer at an electrode having its Fermi energy outside the electrolyte window.
Separator	

This chapter was originally published as part of the Encyclopedia of Sustainability Science and Technology edited by Robert A. Meyers. DOI:10.1007/978-1-4419-0851-3

J.B. Goodenough (✉)
Texas Materials Institute, Mechanical Engineering, University of Texas,
1 University Station, Austin, TX 78712, USA
e-mail: jgoodenough@mail.utexas.edu

R.J. Brodd (ed.), *Batteries for Sustainability: Selected Entries from the Encyclopedia of Sustainability Science and Technology*, DOI 10.1007/978-1-4614-5791-6_3,
© Springer Science+Business Media New York 2013

	A solid layer permeable to the working ion that separates anode from cathode to prevent electron transfer between them inside a battery cell.
Spin state	Refers to spin of a redox couple, which may be reduced (low-spin state) from its free-ion value (high-spin state) by a ligand-field splitting of energies of atomic orbitals that is larger than the Hund intraatomic-exchange splitting of electron spins.
State of charge	Amount of chemical energy in a battery or cell relative to the total energy available.
Tap density	A measure of the volume density, which increases with compaction of the active electrode particles.
Working ion	Ion carrying ionic current inside a battery cell.

Definition

The active materials of a battery are the chemically active components of the two electrodes of a cell and the electrolyte between them.

A battery consists of one or more electrochemical cells that convert into electrically energy the chemical energy stored in two separated electrodes, the anode and the cathode. Inside a cell, the two electrodes are kept apart by an inert separator that is permeable to a liquid electrolyte or by a solid electrolyte. The electrolyte allows transfer from the anode to the cathode of only the working cation, which carries the positive charge associated with the chemical reaction between the electrodes; the electrolyte and separator force the electrons of the reaction to flow outside the battery where they can be stopped (open circuit) or discharged from the anode to the cathode through a load where they do work. In a rechargeable battery, the chemical energy can be restored by the application of a reverse charging current. The active materials determine such parameters as the electric-power capability of a battery, its energy density, its calendar and cycle life, its cost, and its safety. Each battery application has a different set of requirements. Tailoring of the active materials to the demands of a particular application is an ongoing process. These applications range from the powering of heart pacemakers to portable electronic devices to electric vehicles to electrically-energy storage for the grid. The active materials used in batteries for some of these different applications are discussed.

Introduction

The conversion of chemical energy to electricity was first demonstrated in 1800 by Volta, who constructed a battery – the *voltaic pile* – from alternating plates of silver and zinc separated by a cloth soaked in a salt solution. In 1806, Davy used the electricity from a voltaic pile to convert electrical energy into chemical energy by separating electrochemically alkali metals from their hydroxides. The electrochemical cell was invented in 1836 by Daniell, who immersed Zn in an aqueous $ZnSO_4$ solution and copper in a $CuSO_4$ solution in separate tubes; the two tubes were joined by a salt bridge containing a porous glass frit. This cell was called a *galvanic cell* in memory of Galvani, who had discovered in the 1790s that two different metals and a salt solution could cause muscle contractions in a frog's leg. Galvanic cells were developed to supply electricity before electric generators were available. The gifted experimentalist, Michael Faraday, began his career as an apprentice in Davy's laboratory; he later went on to lay the foundations for the science of electrochemistry and the development of the modern battery. It was Faraday who was to refer to the two different metals of a voltaic pile and of a galvanic cell as electrodes; the reductant was called the anode and the oxidant the cathode. The salt solution connecting the two electrodes by an ionic current was the electrolyte. Negatively charged ions are referred to as anions and positively charged ions are referred to as cations because they are attracted, respectively, to the anode and the cathode when electrons flow from the anode to the cathode.

Batteries

A *battery* consists of one or more electrically connected *electrochemical cells* that store chemical energy in their two *electrodes*, the *anode* and the *cathode*; the battery converts the chemical energy into electrical energy on *discharge*. The electric output of a battery is a *discharge current* I at a voltage V to give an electric-power output $P = IV$. The power capability is expressed as *specific* or *gravimetric power* in watts per kilogram (W/kg) or *volumetric power density* in watts per liter (W/L). The *energy density* of a battery is the amount of *electrical energy* QV that is available from a fully charged battery per unit weight or volume; it is expressed as *specific* or *gravimetric* energy in watt-hours per kilogram (Wh/kg) or as volumetric energy density in watt-hours per liter (Wh/L). Q is the total charge transported by I over the time Δt of a complete discharge

$$Q = \int_0^{\Delta t} I dt = \int_0^{Q} (dq/dt)\, dt \tag{3.1}$$

and Q/weight (Ah/kg or mAh/g) is the *capacity* of the battery. The voltage of a battery can be increased by connecting individual electrochemical cells in series; the current per cell for a given battery current can be decreased and therefore the time for a full discharge can be increased by connecting cells in parallel. The larger the voltage and capacity of an individual cell, the fewer the cells in the battery.

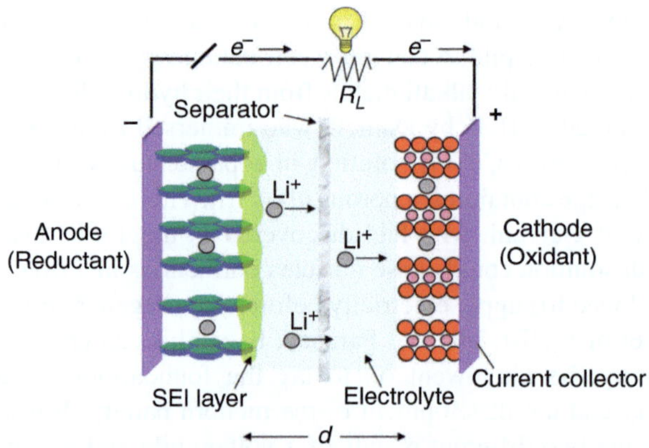

Fig. 3.1 Schematic showing components of $LiC_6/Li_{1-x}CoO_2$ cell. Internal Li^+-ion and external electronic (e^-) currents I_i and I are shown for discharge with switch closed. The SEI passivation layer is formed because the energy E_{FA} of the LiC_6 anode is higher than the LUMO of the carbonate electrolyte (see Fig. 3.3)

A *primary battery* is discarded after providing electrical power until its chemical energy is exhausted. A *secondary battery*, also called a *rechargeable battery* or an *accumulator*, can have its spent chemical energy restored by the application of a *charging current* in a direction opposite to that of the discharge current. The *cycle life* of a secondary battery is the number of discharge/charge cycles it can support before losing 20% of its capacity. The energy loss per cycle is the difference between the electric energy used to charge the battery and that gained on discharge; the electrical energy gained on discharge divided by the electrical energy spent on charge is the *efficiency* of a secondary battery for storage of electrical energy.

Figure 3.1 illustrates schematically the components of an individual electrochemical cell. The *anode* is a reductant and the cathode is an *oxidant*. Inside the cell, the two electrodes are kept apart by a physical barrier, a *separator*, that prevents any electron transfer between them inside the cell. However, the two electrodes are connected electrically inside the cell by an *electrolyte* that conducts a *working ion* and is an electronic insulator. A liquid electrolyte permeates a chemically inactive, porous separator; a solid electrolyte also acts as a separator. The working ion is normally a cation: H^+, Li^+, or Na^+. Chemical reaction between the two electrodes of a cell involves both electron and working-ion transfer between the electrodes. The electrolyte forces the electrons to flow outside the cell while allowing the working ion to be transported inside the cell. Therefore, a battery has a *negative terminal* that collects electrons via a *current collector* from all the cell anodes for delivery to an external circuit and a *positive terminal* that receives electrons from the external circuit and delivers them via a current collector to all the cell cathodes during a battery discharge.

External connection of the battery terminals allows an electronic discharge current I to pass through an external load of resistance R_L, where it does work, to the cathode where it neutralizes the ionic current I_i inside the battery. Thus, on discharge the

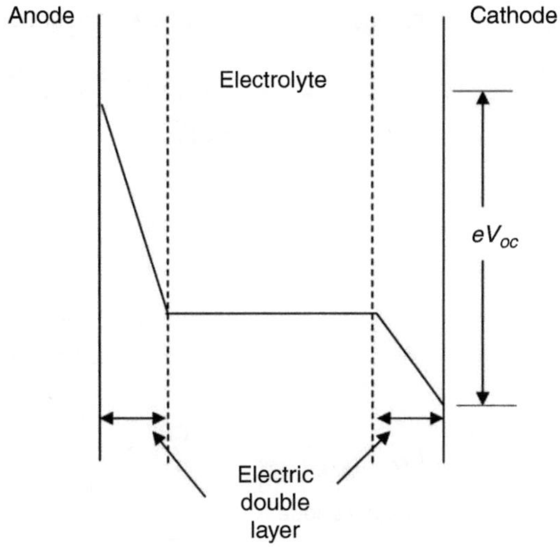

Fig. 3.2 Voltage levels in the various sections of a cell at open circuit. The double layers at the anode and the cathode have different structures and compositions

battery transforms its chemical energy into electrical energy at the load. The much higher electronic conductivity σ_e compared to the working-ion conductivity σ_i means that a cell is fabricated with a small thickness/area ratio. Disconnection of the two battery terminals stops not only the external discharge current, but also the internal ionic current since, without charge neutralization, an internal electric field between the cell electrodes is built up to stop further flow of the working ion. This electric field is expressed as a voltage between the terminals of a cell at open circuit, the *open-circuit voltage* V_{oc}, with a positive charge at the cathodic positive terminal and a negative charge at the anodic negative terminal. Mobile ions in the electrolyte form *electrical double layers* at the surfaces of the electrodes to neutralize the electrode charge, and the voltage drop inside the cell is across the double layers, Fig. 3.2.

This entry describes strategies used by the chemists to identify appropriate active materials, electrodes and electrolyte, for the individual electrochemical cells of a battery.

Fundamental Constraints

Maintaining Electrode–Electrolyte Contact

Ion transfer across the electrode/electrolyte interfaces requires retention of a good electrode–electrolyte contact during discharge and charge. Since ion and electron transfers between the electrodes change the volumes of the electrodes, it is advantageous to have a liquid or polymer electrolyte contacting solid crystalline electrodes; a solid electrolyte is needed to keep the electrodes apart if one or both of the electrodes is a liquid or a gas.

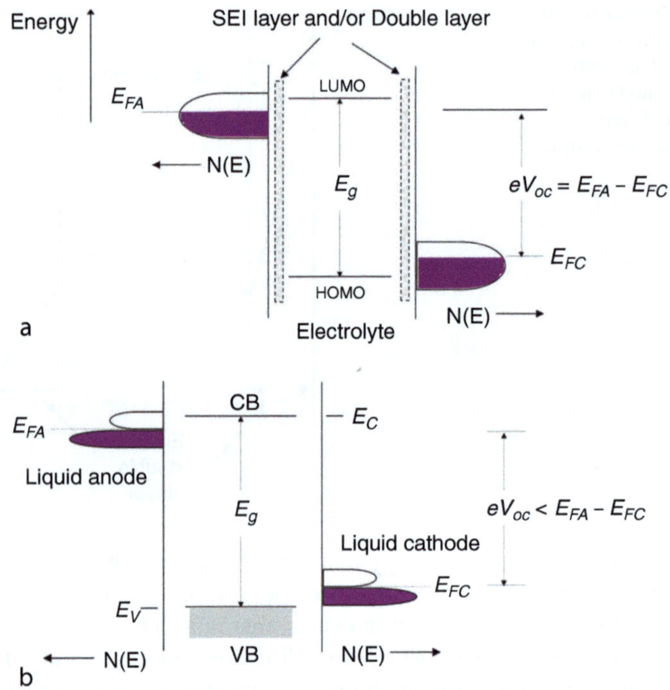

Fig. 3.3 Definition of the electrolyte window E_g for (**a**) liquid and (**b**) solid electrolytes. The E_{FA} and/or E_{FC} of solid electrolytes may lie in either a band of one-electron states as in (**a**) or a multielectron redox couple as in (**b**); for liquid electrodes, they lie in a mixed-valence redox couple

Electrolyte Window

The *window* of a liquid electrolyte is the energy gap E_g between its lowest unoccupied molecular orbital (LUMO) and its highest occupied molecular orbital (HOMO). The window of a solid electrolyte is the energy gap E_g between the bottom of the empty electronic conduction band, E_c, and the top of the filled electronic valence band, E_v. Figure 3.3 illustrates how the window of an electrolyte constrains the V_{oc} of a cell.

Figure 3.3a shows the case of solid electrodes having anode and cathode Fermi energies E_{FA} and E_{FC} in their electronically conductive state. An E_{FA} or E_{FC} may lie in a partially filled conduction band or in a mixed-valent redox couple. An E_{FA} above the electrolyte LUMO will reduce the electrolyte and an E_{FC} below the electrolyte HOMO will oxidize the electrolyte. Therefore, thermodynamic stability restricts the open-circuit voltage to

$$eV_{oc} = E_{FA} - E_{FC} \leq E_g \quad (3.2)$$

where e is the magnitude of the charge of an electron. However, charge transfer at an electrode/electrolyte interface may create a double layer and/or a passivation layer that blocks a subsequent ionic transfer at open circuit to stop any further electrolyte reduction or oxidation and therefore to give a *kinetic stability* for a limited $(eV_{oc} - E_g) > 0$. A passivation layer is referred to as a *solid-electrolyte interphase* (SEI); it should not block ion or electron transfer under closed-circuit conditions.

Solid electrolytes, Fig. 3.3b, are chosen to have a window $E_g > eV_{oc}$.

Ionic Conductivity

Kinetic limitations on the rate of working-ion transfer between electrodes cause the cell voltage V to drop relative to V_{oc}. The voltage drop

$$\eta(q, I) = V_{OC} - V \tag{3.3}$$

is called the *polarization* or the *overvoltage*; q is the state of charge. The polarization is due to the resistance to ionic motion inside the cell.

Ions move diffusively. Therefore, the ionic conductivity is given by

$$\sigma_i = (nq^2/kT)D_i \sim Nc(1 - c)\exp(-\Delta H_m/kT) \tag{3.4}$$

where $n = cN$ is the density of carriers of charge q; $D_i = D_O \exp(-\Delta G_m/kT)$ is the ionic diffusion coefficient containing a Gibbs free energy $\Delta G_m = \Delta H_m - T\Delta S_m$ for an ion to move from one position to another. The motional enthalpy ΔH_m is an activation energy for ion motion, and $(1-c)$ is the fraction of sites that are empty into which ions can move.

In a strong acidic or alkaline aqueous electrolyte, $n = cN$ is large and ΔH_m is small so long as the water molecules are free to rotate as in a liquid; the H^+ ions hop within hydrogen bonds, but molecular rotation is needed to subsequently reorient the hydrogen to a new bond direction (Grotthus motion). In a liquid aqueous electrolyte, the H^+-ion conductivities are generally $\sigma_H \approx 1$ S/cm at 20°C. However, the window of an aqueous electrolyte is only 1.23 eV, which has led to consideration of nonaqueous electrolytes. Since H^+ ions are not mobile in nonaqueous liquid electrolytes, the Li^+ ion is normally used in these electrolytes. Few organic solvents dissolve a sufficient concentration of lithium salts with good separation of the Li^+ ions from the anion of the salt. The Li^+-ion conductivities in the best nonaqueous liquid electrolytes are $10^{-3} < \sigma_{Li} < 10^{-2}$ S/cm at 20°C.

Solid cathodes of secondary-battery cells are *insertion compounds*. An insertion compound has a *host* crystalline structure into which a *guest* working ion can be inserted/extracted reversibly into/from its interstitial space while the ionic charge is being neutralized by reduction/oxidation of the host structure with electrons

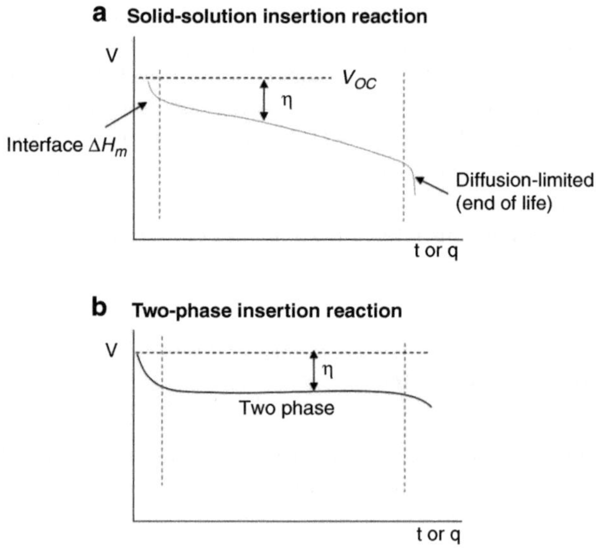

Fig. 3.4 Schematic profiles for (**a**) a solid-solution insertion reaction and (**b**) a two-phase insertion reaction showing the polarization $\eta = V_{OC} - V$. *Note*: Two-phase reactions retain the higher voltage found at the onset of the two-phase region

coming from/sent to the external circuit. These electrodes are mixed ionic/electronic conductors. During an insertion or extraction reaction, the essential architecture of the host structure remains intact throughout the process, but two types of insertion reaction need to be distinguished: (1) a single-phase solid-solution reaction between the guest and the host; (2) a two-phase reaction. A first-order transition between two phases results in the coexistence of the two phases; a first-order transition may not be reversible, but where the host undergoes an ion-displacement transition that separates a phase rich in the working ion from one that is poor in the working ion, the two-phase reaction can be reversible. In a single-phase solid-solution reaction, the output voltage V of the cell decreases monotonically with its state of charge as is illustrated schematically in Fig. 3.4a. The *state of charge* of a cell refers to the fraction of the full-charge chemical energy existing in a partially charged cell. Where the host structure undergoes a first-order structural change during discharge, the output voltage V of the cell is independent of the state of charge over the two-phase region as is illustrated in Fig. 3.4b. This behavior follows from the Gibbs phase rule, which states that the number of thermodynamic parameters that are free to vary is

$$F = C - P + 2 \tag{3.5}$$

where P is the number of coexisting phases and C is the number of components, guest and host, taking part in the reaction. Where two phases coexist, $C = P$ and $F = 2$. Since temperature and pressure are specified, there is no degree of freedom left, so other parameters such as voltage have to be constants. In an insertion reaction, the host interstitial space into which the working ion can be inserted may be 1D, 2D, or 3D (D = dimensional) depending on the host structure.

In Fig. 3.4a, three ion-transfer processes are distinguishable: (1) an *activation polarization* due to the energy ΔH_m required to transfer ions across the electrode/electrolyte interfaces, (2) a change in $E_{FA}-E_{FC}$ and (3) an end-of-life *concentration polarization*. The slope in region (2) of a similar $V(q)$ vs. I curve gives the *ohmic polarization*. The activation polarization is manifest on the initiation of discharge; the ohmic polarization is manifest by the slope of the $V(I)$ profile in mid-discharge. At the end of life, either the concentration of working ions reaching the cathode is depleted and/or the concentration $(1-c)$ N of the cathode acceptor sites at the electrolyte/electrode interface is depleted to where the ionic diffusion current is no longer able to keep up with a fixed discharge current I. The increase in η causes a decrease in V at the end of life, and this decrease occurs at a larger state of charge the higher is I and, from Eq. 3.4, the lower is the temperature T. The end-of-life drop-off in V is also referred to as the *diffusion-limited* regime of the $V(t)$ profile. The loss of capacity at higher I and lower T is *reversible*; capacity is restored at lower discharge currents I and higher temperatures T where the working-ion profile across the electrolyte/electrode interface approaches more closely the equilibrium profile. Thus, the reversible Q of Eq. (3.1) decreases with increasing I; $Q=Q(I)$. Clearly, a fast diffusion of the working ion not only in the electrolyte, but also in any insertion-compound electrode is needed for a secondary power battery.

Energy Density

The specific energy QV/weight of a cell depends not only on the voltage V of the cell, but also on the total charge Q per kilogram that is transported between the electrodes. In order to achieve a high cell voltage, the E_{FA} of the anode and the E_{FC} of the cathode need to be matched to the LUMO and the HOMO, respectively, of an electrolyte with a large window E_g between its LUMO and HOMO. In a secondary battery, Q is the total charge that is transported reversibly between the electrodes. In a secondary battery in which the cells are assembled in the discharged state, Q is limited to the number and charge of working ions that can be extracted reversibly from the cathode host. However, where the anode develops a non-blocking SEI layer, Li^+ ions are trapped irreversibly in the SEI layer during the initial charge, which reduces the capacity available from the cathode on subsequent charge/discharge cycles. To eliminate this loss of cathode capacity, it would be necessary to form the SEI layer on the anode before assembling the cells of a battery.

In addition to formation of an SEI layer on the anode, chemical reactions between the cathode and the electrolyte or within the cathode can deplete the amount of active cathode material. These reactions introduce an *irreversible* capacity loss that is expressed as a *capacity fading* with the number of charge/discharge cycles. The *cycle life* of a cell is the number of discharge/charge cycles a cell can undergo without too great an irreversible capacity loss to be useful; the *service life* of a cell is the length of time over which the cell retains more than 80% of its capacity. The service life depends on the cycle life, that is, the cell use as well as on the age of the cell.

Morphology

In order to minimize the reversible capacity loss at higher discharge/charge currents, solid electrodes are made porous so as to maximize the electrode/electrolyte interface area per unit planar area of the electrode. Solid porous electrodes are commonly a complex composite of the active material in the form of small particles coexisting with and contacting a current-passing additive such as carbon, metal particles, or a conductive polymer; the conductive additive percolates through the composite mass to connect electronically the active particles to one another and to the electrode current collector. This composite mass is commonly held together by a chemically inactive binder such as Teflon. Typically, such a composite electrode has about 30% porosity percolating through it so as to allow impregnation of the pores by a liquid electrolyte in order to have the electrolyte contact every active particle. The resulting electrode/ electrolyte interface area is complex, and the ionic currents within the composite electrode are tortuous. Modeling the current flows within an electrode as well as the composite morphology has become an important tool in the design of a composite electrode and of the distribution of primary currents in the metallic current collectors.

Heat

The net useful energy released for a given chemical reaction between the electrodes is the enthalpy ΔH of the Gibbs free energy

$$\Delta G = \Delta H - T\Delta S \tag{3.6}$$

where $T\Delta S$ is the heat associated with the entropy change ΔS of the reaction at an absolute temperature T. For a fixed T, this heat is reversible, that is, it is exothermic on discharge and endothermic on charge or *vice versa*. In electrical terms,

$$\Delta S = nFdV/dT \tag{3.7}$$

where n is the number of electrons per mole of reactants participating in the discharge reaction and F is the Faraday constant, that is, the amount of charge transferred when one equivalent weight of mass reacts. From Eqs. 3.6 and 3.7, the reversible heat on discharge is $nFT(dV/dT)$, which may be positive or negative, exothermic or endothermic. In addition, the internal resistance R_b of a cell creates heat I^2R_b that must be dissipated. *Thermal runaway* occurs if the internal heat causes an electrode reaction with the electrolyte to become self-sustaining, the reaction entering an autocatalytic mode. This situation is responsible for fires associated with battery operations, and an internal electronic short circuit between the electrodes occasioned by dendrite growth from an anode across the electrolyte to the cathode with repeated cycling can trigger such an event where the electrolyte is flammable.

Cost

Cost as well as safety, capacity, and power capability are constraints on any commercial device. Lower cost requires not only low-cost, nontoxic materials, and simple battery-stack management, but also a long service life, ease of manufacture, and a minimization of the number of cells to achieve a specified voltage and discharge time. Safety of a secondary battery at an acceptable cost favors either a nonflammable electrolyte or a solid-electrolyte separator that blocks short-circuiting from any dendrite formation on a solid anode where dendrite formation threatens short-circuiting of cells on repeated cycling. A long service life requires elimination of irreversible chemical reactions within the individual cells.

The cost and service life of secondary batteries for electric vehicles must compete against the cost and service life of the internal combustion engine. In their favor is elimination of the cost to the environment of distributed emissions of the greenhouse gas CO_2; and if charging of the battery of the electric vehicle can be done with wind or radiant solar energy, all CO_2 emissions associated with the vehicle can be eliminated.

Electrode Reactions

Primary batteries [1] having liquid electrolytes generally use a solution or a solution-precipitation reaction of an element at the anode and a displacement or insertion reaction at the cathode. Most secondary batteries use insertion reactions at a solid cathode and may also use insertion reactions at the anode.

Some common elemental solution or solution-precipitation reactions are:

$$\begin{aligned}
&\mathrm{Li}^0 - e^- = \mathrm{Li}^+ \\
&\mathrm{Na}^0 - e^- = Na^+ \\
&\mathrm{Zn}^0 - 2e^- + \mathrm{HCl} = \mathrm{ZnCl}_2 + 2\mathrm{H}^+ \\
&\mathrm{Zn}^0 - 2e^- + 2\mathrm{H}_2\mathrm{O} = \mathrm{Zn(OH)}_2 + 2\mathrm{H}^+ \\
&\mathrm{Cd}^0 - 2e^- + 2\mathrm{H}_2\mathrm{O} = \mathrm{Cd(OH)}_2 + 2\mathrm{H}^+ \\
&\mathrm{Pb}^0 - 2e^- + \mathrm{H}_2 SO = \mathrm{PbSO}_4 + 2\mathrm{H}^+
\end{aligned} \tag{3.8}$$

The E_{FA} of Li^0 lies above the LUMO of a nonaqueous liquid-carbonate electrolyte containing a lithium salt. Although a non-blocking SEI passivation layer is created at the surface, lithium is not plated back uniformly during a charge; it forms a rough, mossy surface that, on repeated discharge/charge cycles, develops dendrites that can grow across the electrolyte to contact the cathode. Therefore, with a liquid electrolyte Li^0 is only used as an anode in a primary battery or in half-cells used to test the performance of candidate cathode materials for secondary batteries. However, it may be used in a secondary cell having a solid Li^+-ion electrolyte that blocks dendrite growth across it and is not reduced on contact with a dendrite.

The E_{FA} of Zn^0 lies 0.76 eV above the $2H^+/H_2$ LUMO of an aqueous electrolyte, which has previously prohibited its use in a secondary battery. However, it is used successfully as the anode of a primary cell having either an HCl acidic or KOH alkaline electrolyte. The Cd^{2+}/Cd^0 couple at 0.40 eV above the $2H^+/H_2$ couple is better matched to the aqueous LUMO, so Cd has been the preferred anode for a secondary alkaline battery with a KOH electrolyte. A lower electrochemical potential of an alkaline versus an acidic electrolyte allows the double layer in an alkaline electrolyte to lower the E_{FA} of Cd to below the $2H^+/H_2O$ LUMO. The Pb^{2+}/Pb^0 couple is only 0.13 eV above $2H^+/H_2$, so it provides a reversible anode reaction in a sulfuric-acid electrolyte. In all cases with aqueous electrolytes, the anode–electrolyte reaction releases mobile working H^+ ions to react with the cathode.

In a displacement reaction, the working ion displaces a cation of the electrode as is illustrated by the cathodic discharge reaction

$$2Li^+ + 2e^- + CuS = Li_2S + Cu^0 \tag{3.9}$$

Normally, such a first-order reaction is not easily reversed, and to date cathodes undergoing such a reaction are only found in primary batteries. However, they represent a Q corresponding to two electrons per electrode cation to give a large capacity. Although displacement reactions that leave a host framework intact may be reversible, for example,

$$2Li^+ + 2e^- + Fe_3O_4 = Li_2Fe_2O_4 + Fe^0 \tag{3.10}$$

the associated volume changes have made these reactions impractical unless they are buffered by an amorphous carbon matrix.

In a reversible insertion reaction, the working ion is inserted as a guest into a host structure forming either a solid solution in a single-phase reaction or two phases in a first-order reaction in which ionic displacements of the host segregate the electrode into guest-rich and guest-poor phases. For example, two anode reactions are the single-phase reaction of a metal hydride MH

$$MH - xe^- = MH_{1-x} + xH^+ \tag{3.11}$$

and the lithiated graphite two-phase reaction

$$LiC_6 - e^- = 6C + Li^+ \tag{3.12}$$

in which graphene sheets are shifted in stages relative to one another on Li insertion (Fig. 3.9). In anodes, the guest is inserted on charge and extracted on discharge. As a cathode, the guest is inserted on discharge and extracted on charge.

Primary Batteries

Zinc–Manganese Batteries

Zinc–manganese batteries [1] dominate the primary-battery market. These batteries are of two types depending on the pH of their aqueous electrolyte. Each uses a carbon-MnO_2 cathode; the carbon provides electronic conductivity. The MnO_2 is best prepared as αMnO_2 or γMnO_2, which have structures, Fig. 3.5, with 1D channels into which H^+ ions can be inserted. The Leclanché cell, Zn/HCl/MnO_2, is acidic, and the Zn anode undergoes a Zn-solution reaction; Zn^{2+} ions enter a $ZnCl_2$ + NH_4Cl salt solution to displace $2H^+$ ions to the cathode. In the alkaline cell, Zn/

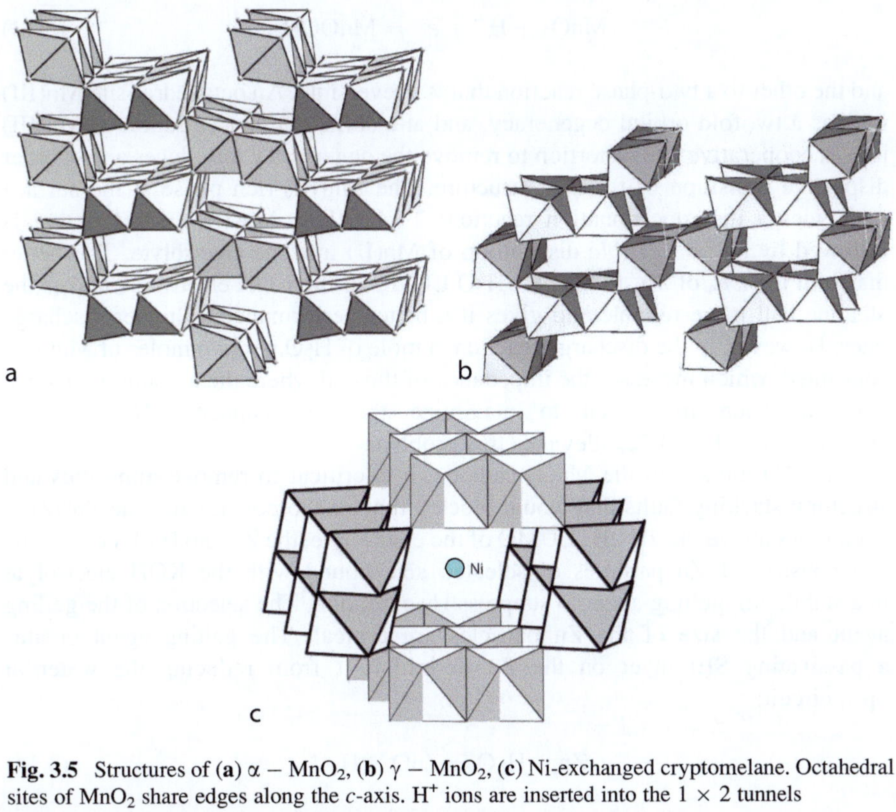

Fig. 3.5 Structures of **(a)** $\alpha - MnO_2$, **(b)** $\gamma - MnO_2$, **(c)** Ni-exchanged cryptomelane. Octahedral sites of MnO_2 share edges along the *c*-axis. H^+ ions are inserted into the 1×2 tunnels

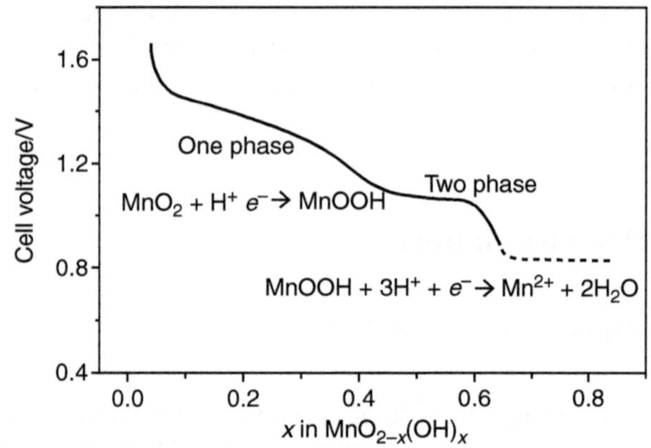

Fig. 3.6 Discharge voltage profile of the $MnO_{2-x}(OH)_x$ cathode of a Zn/MnO_2 primary battery showing solid-solution (one-phase) and two-phase compositional ranges

KOH/MnO_2, the Zn anode reacts with the aqueous electrolyte to form $Zn(OH)_2$ + $2H^+$; the $Zn(OH)_2$ subsequently transforms to $ZnO\downarrow + H_2O$. As shown in Fig. 3.6, the discharge voltage of a zinc–manganese battery exhibits two ranges, one corresponding to a single-phase solid-solution reaction

$$MnO_2 + H^+ + e^- = MnOOH \tag{3.13}$$

and the other to a two-phase reaction that is irreversible. An octahedral-site Mn(III) ion has a twofold orbital degeneracy, and at a sufficient concentration of Mn(III) ions, a cooperative site distortion to remove the degeneracy introduces a first-order displacive transition of the host structure. The Mn(III)-rich phase at the surface undergoes a disproportionation reaction: 2 Mn(III) = Mn(II) + Mn(IV) that is followed by an *irreversible* dissolution of Mn(II) into the electrolyte. The better match of the E_{FA} of Zn^0 to the $2H^+/H_2O$ LUMO of the KOH electrolyte makes the alkaline cell more reliable and gives it a better performance at higher discharge rates. However, in the discharge reaction, a mole of H_2O per two moles of MnO_2 is consumed, which increases the impedance of the cell where the amount of electrolyte has been minimized to maximize the cell capacity. Ni-exchanged cryptolmelane, Fig. 3.5c, alleviates this problem.

With 1D tunnels in the MnO_2 cathode, it is critical to remove impurities and structural stacking faults that would block a tunnel. Moreover, since the Zn^0/Zn^{2+} couple lies above the $2H^+/H_2$ LUMO of the electrolyte, the Zn anode of the alkaline cell consists of Zn particles of selected size bound with the KOH electrolyte in a stabilizing gelling agent to suppress H_2 evolution. The selection of the gelling agent and the size of the Zn particles are critical. The gelling agent creates a passivating SEI layer on the Zn to inhibit it from reducing the water at open circuit:

$$Zn + H_2O = ZnO + H_2 \uparrow \tag{3.14}$$

Zinc–Air and Silver–Zinc Batteries

The alkaline zinc–air battery has the highest energy density of all aqueous batteries since only the Zn-powder anode is contained in the cell; the other reactant, oxygen, is available from the surrounding air. The capacity of a cell is dictated by the amount of fine Zn powder that is available for discharge. Nevertheless, a light cathode structure is present; it consists of a porous carbon or nickel mesh containing both the liquid electrolyte to convey H^+ ions and MnO_2 or other particles as catalyst in the pores. Gaseous O_2 is given access to the porous carbon where it penetrates the pores to be reduced on the MnO_2 at a triple-phase boundary (TPB) by the reaction

$$O_2 + 4H^+ + 4e^- = 2H_2O \tag{3.15}$$

However, the oxygen–reduction reaction Eq. 15 forms an intermediate peroxide:

$$\begin{aligned} &O_2 + H_2O + 2e^- = O_2H^- + OH^- \\ &O_2H^- + H_2O + 2e^- = 3OH^- \\ &4H^+ + 4OH^- = 4H_2O \end{aligned} \tag{3.15a}$$

and breakdown of the peroxide H_2O_2 and/or its radical O_2H^- is rate-limiting. These cells are available in sizes small enough to fit into the ear to power a hearing aid.

The silver–zinc battery also uses a KOH alkaline electrolyte. In this battery, the cathode undergoes the displacement reaction

$$2H^+ + 2e^- + Ag_2O = 2Ag^0 + H_2O \tag{3.16}$$

This battery is primarily used in button cells for small portable devices such as watches.

Lithium Primary Batteries

Primary lithium cells use a lithium-metal anode and a nonaqueous electrolyte having a larger window than the 1.23 eV of an aqueous electrolyte. Both liquid and solid Li^+-ion electrolytes can be used.

Liquid Electrolytes. The carbonates dissolve a fair concentration of lithium salts, normally $LiClO_4$ or $LiSO_3CF_3$ in propylene carbonate-dimethyl ether in primary cells. Since the Li^0/Li^+ couple lies about 1 eV above the LUMO of the carbonate, dimethyl ether is added to react with the Li^0 to form a passivating SEI layer. With a passivating layer on the Lithium anode and an electrolyte window $E_g \approx 3.5$ eV, cell voltages of 4.0 V or higher are possible. With a higher voltage V, the lithium primary batteries provide a larger energy density, but Li^+-ion transport across the SEI layer and a σ_{Li} in the nonaqueous electrolyte significantly lower than the σ_H in a strong acidic or alkaline aqueous electrolyte reduces the rate capability of a Lithium primary battery at 20°C. The solid-cathode insertion compounds CF_x and MnO_2 are thermodynamically stable and provide a long shelf life; Li/CF_x and Li/MnO_2 batteries have a V = 3.0 V and are designed for relatively low-rate applications.

The FeS_2 cathode of a Li/FeS_2 battery undergoes a two-step displacement reaction

$$4Li^+ + 4e^- + FeS_2 = 2Li_2S + Fe^0 \tag{3.17}$$

in which an amorphous Li_2FeS_2 is first formed by the breaking of the disulfide $(S_2)^{2-}$ ions of FeS_2. This battery only provides V = 1.6 V in the first step, but at higher temperatures it gives a high-rate performance superior to that of the alkaline Zn/MnO_2 cells.

The Li/I_2 cell is used preferably for implanted heart pacemakers; the di-iodine bond is broken at V = 2.7 V in the cathode reaction

$$2Li^+ + 2e^- + I_2 = 2LiI \tag{3.18}$$

Solid Electrolytes. A *protected Lithium anode* is under development for both primary and secondary batteries that promise much larger capacities. This strategy is illustrated by the Li/seawater primary battery in which a Lithium anode is immersed in a nonaqueous electrolyte, the *anolyte*, that is separated from seawater contacting a cathode current collector by a Li^+-ion solid-electrolyte separator. The seawater acts as a liquid cathode. Except for contact with a negative post, the Lithium anode and its anolyte are sealed in a compartment containing a Li^+-ion solid-electrolyte wall that interfaces the seawater. The anolyte is chemically stable to both the Lithium and the solid electrolyte; the solid electrolyte must not be reduced on contact with the Li^0 anode. Moreover, either the seal or the compartment must be compliant to allow for the change in volume of the Lithium on discharge. The seawater is not contained in an *open cell*; it is contained within a battery in a *closed cell*. The Li^+ ions from the anode react with water at the cathode current collector:

$$Li^+ + e^- + H_2O = LiOH + \tfrac{1}{2}H_2 \uparrow \tag{3.19}$$

to give a constant voltage V = 2.7 V. The efficiency of a Li/seawater secondary cell is not good, but other liquid cathodes, for example, $Fe(CN)_6^{3-}/Fe(CN)_6^{4-}$ in water, offer a more efficient electrical-energy storage than the seawater cathode and, with a solid electrolyte, permit a flow through of only the liquid cathode (see last section for flow-through rechargeable batteries).

Secondary Batteries with Aqueous Electrolytes

The *lead–acid battery* $Pb/H_2SO_4/PbO_2$ [1] provides a fast cell discharge at 2.0 V, it is relatively low-cost, and 98% of the batteries used in the USA are recycled. This battery has dominated the market for rechargeable batteries, but Pb is heavy and the practical energy density of the battery is only about 25% of its limited theoretical energy density (Wh/kg). Therefore, other secondary batteries are used for handheld devices and contend for the electric-vehicle market.

The Pb anode is matched to the $2H^+/H_2O$ LUMO of the aqueous H_2SO_4 electrolyte; during discharge, the Pb^{2+} ion forms $PbSO_4$ on entering the electrolyte to displace $2H^+$ to the PbO_2 cathode. On reduction of the PbO_2 by electrons from the anode, H^+ ions are inserted into the PbO_2 to form $PbO_{2-x}(OH)_x$. A kinetic stability impedes reaction of the reduced PbO_2 with the electrolyte, but electrons introduced into the Pb-6*s* band disproportionate into electron-rich $Pb(OH)_2$ regions within the PbO_2 matrix

$$PbO_{2-x}(OH)_x = \tfrac{1}{2}xPb(OH)_2 + (1 - \tfrac{1}{2}x)PbO_2 \tag{3.20}$$

followed by a slow reaction

$$Pb(OH)_2 + H_2SO_4 = PbSO_4 + 2H_2O \tag{3.21}$$

At open circuit, electrode reactions that charge the electrodes lead to a slow oxidation of the electrolyte with H_2 evolution at the anode and O_2 evolution at the cathode. These reactions represent an irreversible *self-discharge*. Once the electrolyte is introduced, the battery has a poor shelf life. Under development are acidic aqueous electrolytes in which Pb(II) is soluble rather than condensing into the solid $PbSO_4$. This development of the lead–acid cell promises a flow battery not requiring a separation membrane. The separation membrane of redox-flow batteries (see last section) remains a challenging problem for the aqueous redox-flow technology.

The *cadmium–nickel* cell, Cd/KOH/NiOOH, has its Cd anode and $NiO_{1-x}(OH)_{1+x}$, $0 \leq x \leq 1$, cathode well-matched to the LUMO and the HOMO of the alkaline KOH electrolyte; the small, sealed cell gives a fast discharge and charge at a discharge voltage of 1.2 V. Like the Pb anode of the lead–acid battery, the Cd anode undergoes a solution-precipitation reaction in which Cd^{2+} reacts with the water of the electrolyte, Eq. 3.4, to form reversibly solid $Cd(OH)_2$ and displace $2H^+$ ions to the cathode. The cathode is the layered, monoclinic β-NiOOH of Fig. 3.7; it converts to hexagonal β-$Ni(OH)_2$ in a two-phase reaction on insertion of H^+ to give a plateau in the $V(x)$ discharge curve at 1.2 V versus Cd/$Cd(OH)_2$. In β-NiOOH, the oxygen stacking along the *c*-axis within a unit cell is ABCA, Fig. 3.7; the H^+ ions are located preferentially in the trigonal-prismatic sites between A-A stackings. Overcharging introduces water into the H-depleted

Fig. 3.7 Layered structure of monoclinic β – NiOOH

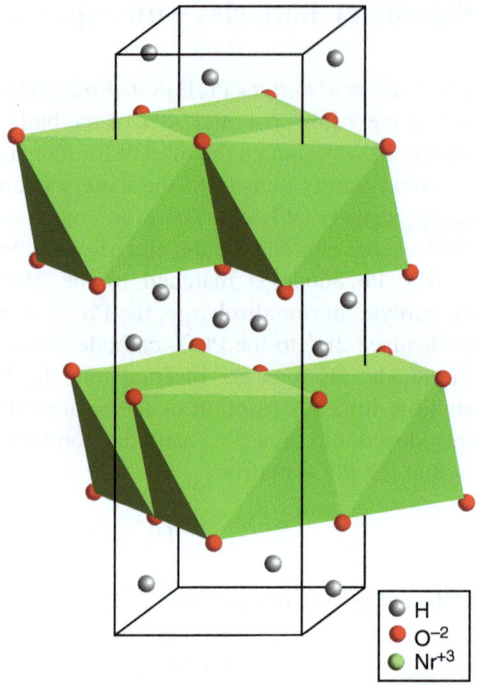

galleries to give γ-NiOOH·y$H_2O_?$. Best performance of the β-NiOOH-β-$Ni(OH)_2$ cathode is found with small particles having a large defect content. The gelatinous $Cd(OH)_2$ remains close to the anode; and during charge, the $Cd(OH)_2$ at the anode reverts to Cd + H_2O as the cathode releases $2H^+$ per Cd^{2+} ion to the electrolyte. At the end of charge, all the Cd^{2+} ions are reduced back to Cd and all the nickel is present as Ni(III). The Ni(IV)/Ni(III) couple cannot be accessed with a further increase in the charging voltage; instead, the electrolyte is oxidized:

$$2H_2O = 2H_2 + O_2 \tag{3.22}$$

This cell was the first rechargeable battery to be used in handheld devices.

With a suitable gelling agent to passivate carefully sized Zn particles, a Zn/KOH/NiOOH secondary battery may be marketed.

The metal hydride–nickel cell, MH_x/KOH/NiOOH, also makes use of the NiOOH cathode; it replaces the Cd anode with a metal hydride, MH_x, insertion alloy having its Fermi energy E_{FA}, like that of Cd, well-matched to the LUMO of the aqueous KOH electrolyte. However, the H^+-ion diffusivity in the hydride is slower than that in the gelatinous $Cd(OH)_2$, so it has a poorer rate capability at lower temperatures. On the other hand, it eliminates the heavy, toxic Cd to give a higher energy density (Wh/kg) and provides a safe power source for electric vehicles. However, a Zn/KOH/NiOOH cell of good cycle life may prove competitive.

Nonaqueous Electrolytes for Li-Ion Secondary Batteries

The demand for a rechargeable battery with a greater specific energy requires a higher output voltage and therefore an electrolyte with a larger window E_g between its LUMO and HOMO than the 1.23 eV of an aqueous electrolyte. Since H^+ ions are not mobile over an appreciable temperature range above and below 20°C in the absence of water, the Li^+ and Na^+ ions are the next logical candidates for the working ion.

Early work with Na^+ ions concentrated on solid electrolytes. Although solid electrolytes have a large window, they do not retain a large-area interface with a solid electrode that changes its volume significantly with the state of charge of the battery. Although all solid-state microbatteries are made, a solid electrolyte preferably contacts a liquid, either a liquid electrode or a liquid electrolyte. The sodium–sulfur battery uses liquid electrodes and operates above 300°C [2]. On the other hand, a composite electrolyte having a solid-electrolyte separator and different liquid electrolytes either side, as in the Li/seawater primary battery, can provide a feasible room-temperature secondary battery [3]. For example, the Li^+-ion solid-electrolyte $Li_{1.3}Ti_{1.7}Al_{0.3}(PO_4)_3$ with the framework structure of hexagonal $Fe_2(SO_4)_3$, Fig. 3.8, can have interstitial Li giving $\sigma_{Li} > 10^{-4}$ S/cm at 20°C [4]. Rechargeable batteries based on this strategy are now being investigated. However, this commercially available inorganic solid electrolyte contains Ti(IV), which is too easily reduced. Another solid electrolyte with a $\sigma_{Li} > 10^{-4}$ S/cm needs to be identified.

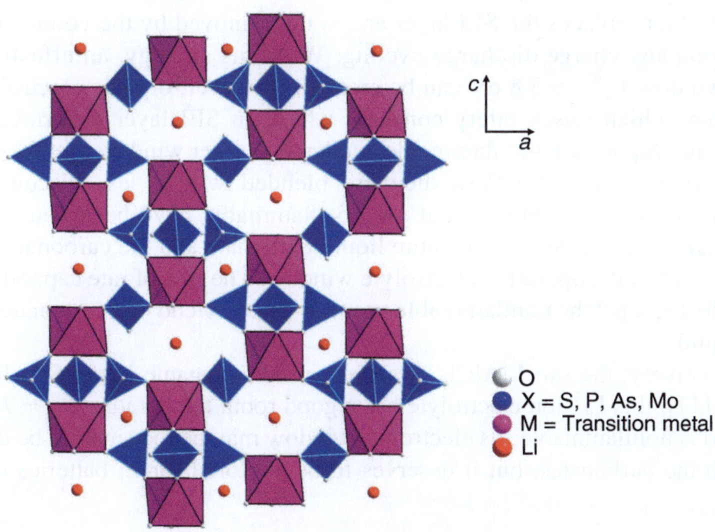

Fig. 3.8 NASICON (NA SuperIonic CONductor) framework structure, which is the same as that of hexagonal $Fe_2(SO_4)_3$

A solid polymer electrolyte can retain contact over a large surface area of a solid electrode provided the changes in electrode volume with state of charge of a cell are modest. Polyethylene oxides (PEOs) containing a lithium salt are low-cost, non-toxic, Li^+-ion electrolytes with good chemical stability [5], but a $\sigma_{Li} < 10^{-5}$ S/cm at 20°C is too low for a power battery and the HOMO versus Lithium is below 4.0 V. The introduction of the oxide particles Al_2O_3, TiO_2, SiO_2, or ZrO_2 creates a more amorphous polymer matrix by inhibiting chain crystallization and attracting Li^+ from its salt; the result is an enhanced σ_{Li} and Li^+-ion transference number $t = \sigma_{Li}/\sigma$, where σ is the total anion plus Li^+-ion conductivity; but σ_{Li} is still not comparable to that of the best liquid electrolytes, [6]. Therefore, most attention has been given to Li^+-ion batteries containing liquid electrolytes.

The most widely used liquid electrolytes consist of Lithium salts dissolved in organic carbonates: propylene carbonate (PC), dimethyl or diethyl carbonate (DMC or DEC), ethyl methyl carbonate (EMC), and mixtures of DMC or DEC with ethylene carbonate (EC). The EC component provides a passivating SEI layer on a carbon anode where the E_{FA} of the anode lies above the LUMO of the carbonates, which is about 1.1 eV below the E_{FA} of Lithium. The carbonates are reasonably good solvents for Li salts and they have a relatively low viscosity, which results in a low activation energy for Li^+-ion diffusion to give a $\sigma_{Li} > 10^{-3}$ S/cm at 20°C and a ready penetration into a porous electrode. The HOMO is about 4.3 eV below the E_{FA} of Lithium, which gives an electrolyte window $E_g \approx 3.7$ eV. A carbonate electrolyte decomposes at a voltage $V > 4.9$ V versus Lithium; an SEI layer can stabilize the Ni, Mn oxides with cathode voltages in the interval $4.3 < V < 4.9$ V, but at the expense of a larger activation energy for Li^+-ion transfer across it. However, this ΔH_m can be reduced either by doping to remove Ni from the surface or by coating the particles with a thin layer of a mixed Li^+-ion/electronic conductor that replaces the SEI layer and is not removed by the volume changes that accompany charge/discharge cycling. With this strategy, an effective electrolyte window $E_g^{eff} \approx 3.8$ eV can be created. However, organic electrolytes are flammable, which raises safety concerns where an SEI layer is formed on the anode. Ionic liquids are not flammable and have a larger window, but they are too viscous to be practical unless they are blended with a less viscous liquid. Carbonate-ionic liquid blends that are not flammable have been demonstrated, but they are more viscous than a pure liquid carbonate and the carbonate component determines the operative electrolyte window. The loss of rate capability must be weighed against the nonflammable property in any blend of a carbonate with an ionic liquid.

Alternatively, the salt $LiAlCl_4$ dissolved in the inorganic liquid SO_2 has been proposed [7]; this Li^+-ion electrolyte has a good room temperature $\sigma_{Li} = 7 \times 10^{-2}$ S/cm and is nonflammable; its electrolyte window may be too small to be competitive with the carbonates, but it deserves to be explored for Li batteries of lower voltage.

Insertion Compounds for Li-Ion Secondary Batteries

Anodes with Li^+-Ion Liquid Electrolytes

In a liquid-carbonate electrolyte, dendrites form on an elemental Lithium anode that can grow across the electrolyte to short-circuit a cell on repeated cycling. Therefore, carbon or an alloy buffered by carbon is used with a liquid electrolyte [8].

Graphite has the layered structure of Fig. 3.9a and Li can be inserted rapidly and reversibly between the layered sheets to create sites capped by six C atoms and form LiC_6, Fig. 3.9b, c. Similar reactions with graphitic sheets occur for the several different forms of carbon; but with a 2D insertion that shifts neighboring graphene sheets relative to one another, a more amorphous type of carbon is preferred over large crystallites of graphite. A carbon anode into which Li^+ ions from the cathode are inserted on an intial charge, Eq. 3.12, is commonly used in Li^+-ion rechargeable batteries. However, the E_{FA} of the reduced carbon in LiC_6 lies above the LUMO of

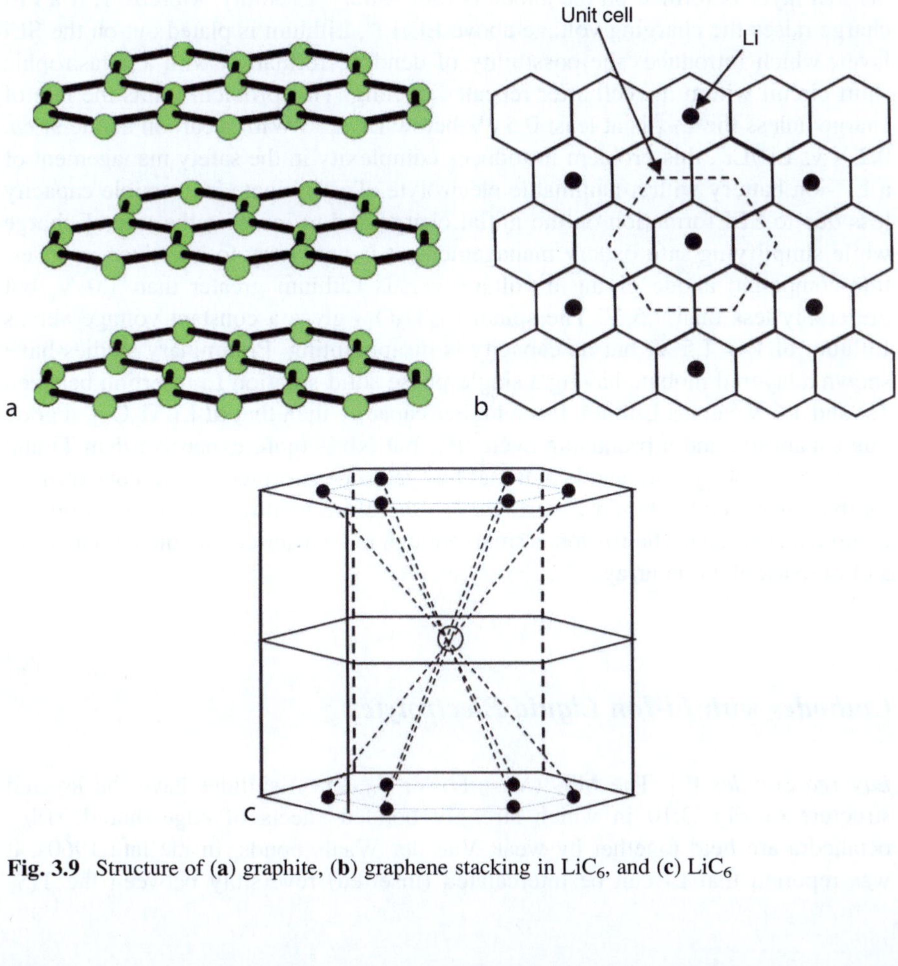

Fig. 3.9 Structure of (**a**) graphite, (**b**) graphene stacking in LiC_6, and (**c**) LiC_6

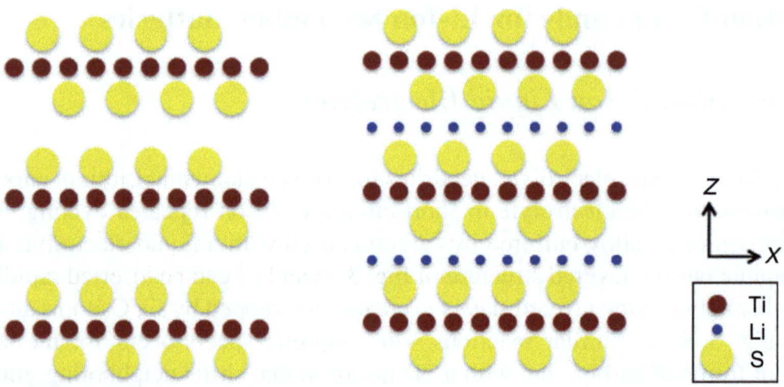

Fig. 3.10 Layered TiS_2 and $LiTiS_2$ structures

a carbonate electrolyte, so on the initial charge it consumes irreversibly Li^+ ions from the cathode in the formation of a non-blocking SEI layer on the carbon unless the SEI layer is formed on the anode before battery assembly. Moreover, if a fast charge raises the charging voltage above Li^+/Li^0, Lithium is plated out on the SEI layer, which introduces the possibility of dendrite formation with a catastrophic short circuit within the cell after repeated cycling. This problem limits the rate of charge unless the E_{FA} is at least 0.5 eV below Li^+/Li^0. With a carbon anode at *ca.* 0.2 V vs Li^+/Li^0, this problem introduces complexity in the safety management of a Li^+-ion battery with a flammable electrolyte. To eliminate irreversible capacity loss due to SEI formation on the initial charge and to increase the rate of charge while simplifying safe battery management, it is necessary to introduce an insertion-compound anode giving a voltage versus Lithium greater than 1.0 V, but preferably less than 1.5 V. The spinel $Li_4Ti_5O_{12}$ gives a constant voltage versus Lithium of $V = 1.5$ V, but its capacity is disappointing. Preliminary studies have shown a layered niobate having a single-phase solid-solution Li insertion between 1.2 and 1.6 V versus Lithium has a higher capacity than that of $Li_4Ti_5O_{12}$, a good rate capability, and a promising cycle life; but Nb is more expensive than Ti and a fully satisfactory, safe anode material has yet to be accepted. The secondary Li^+-ion batteries using a liquid electrolyte are therefore identified by their insertion-compound cathode. The Li^+ ion is small enough to be inserted into the interstices of a close-packed anion array.

Cathodes with Li-Ion Liquid Electrolytes

Layered Sulfides [9]. The MS_2 (M = Ti, V, $V_{0.5}Cr_{0.5}$) sulfides have the layered structure of Fig. 3.10 in which strongly bonded sheets of edge-shared $MS_{6/3}$ octahedra are held together by weak Van der Waals bonds. In the late 1960s, it was reported that Li can be intercalated (inserted) reversibly between the TiS_2

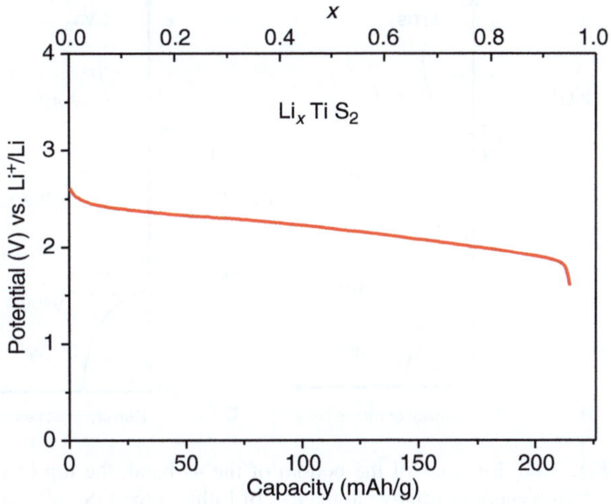

Fig. 3.11 V(*x*) versus Lithium profile for Li_xTiS_2 and $Li_{2x}[Ti_2]S_4$, $0 \leq x \leq 1$

layers to give the solid-solution reaction Li_xTiS_2 ($0 \leq x \leq 1$) with the voltage profile of Fig. 3.11 for a Li/$LiClO_4$ in PC/TiS_2 cell. However, catastrophic failures due to dendrite formation on the Lithium anode forced consideration of an anode that would lower the cell voltage to where the cell would not be competitive with secondary cells having an aqueous electrolyte. Nevertheless, this early report raised awareness that the designer of a cathode material for a Li^+-ion rechargeable battery should look to insertion compounds capable of providing a larger voltage versus Lithium than 2.5 V. However, the designer must not only identify a host structure into/from which Li can be inserted/extracted reversibly, but also consider the limiting E_{FC} of the host structure and whether its E_{FC} can be matched to the HOMO of the electrolyte. For this latter consideration, it is instructive to consider the limiting E_{FC} of a layered sulfide.

Figure 3.12a illustrates schematically the energy density of one-electron states of TiS_2 versus their energy ε relative to the E_{FA} of a Lithium anode. The outer *s* and *p* electrons of the Ti and S atoms are primarily involved in the strong Ti–S bonding. The ionic and covalent components of this bonding create a filled broad band of one-electron states that have a primarily S-3*p* character separated by an energy gap from an empty broad band of states that have a primarily Ti-4*s* character. These bands have similar energies in all the MS_2 layered compounds, but the bottom of the M-4*s* band falls progressively with increasing atomic number of the M atom. Empty Ti-3*d* states lie in the energy gap between these two bands. The cubic component of the site symmetry of a Ti(IV) cation splits the σ-bonding and π-bonding 3*d* orbitals by an energy Δ_c; the σ-bonding states, not shown, lie above the bottom of the Ti-4*s* band and the π-bonding states lie in the energy gap between the broad Ti-4*s* and S-3*p* bands. The trigonal component of the site symmetry splits the π-bonding 3*d* states into a lower, orbitally nondegenerate a_1 band and a higher, orbitally twofold-degenerate e_π band. The Fermi energy of stoichimetric TiS_2 lies in a small energy gap between the bottom of the Ti-a_1 band and the top of the S-3*p*

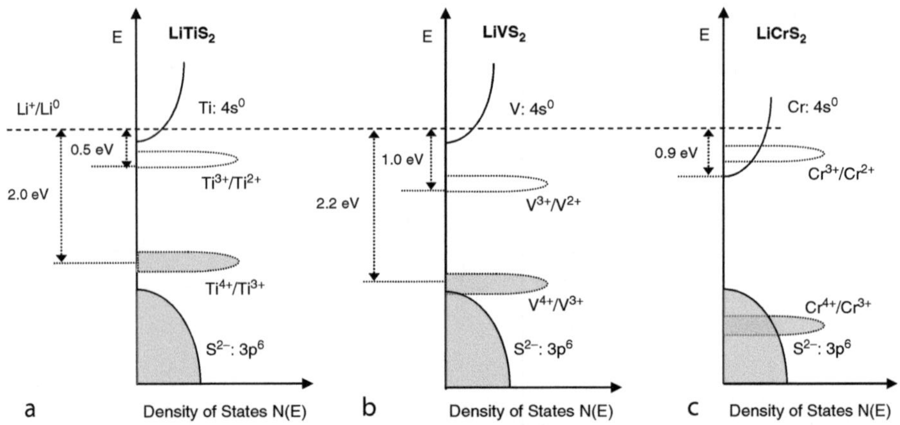

Fig. 3.12 Energies of the bottom of the 4*s* band, the top of the S-3*p* bands, and π-bonding d-electron energies relative to the E_{FA} of Lithium for TiS_2, VS_2, and $LiCrS_2$. In $Li_{1+x}MS_2$, all the Li occupy tetrahedral sites of the Li layer and the M(III)/M(II) levels are raised to give $V = 0.5$ V and 1.0 V, respectively, for $Li_{1+x}TiS_2$ and $Li_{1+x}VS_2$; the Cr(III)/Cr(II) level is raised above the bottom of the 4*s* band so that Li insertion into CrS_2 displaces Cr^0

bands. Intercalation of Li^+ ions is neutralized by electrons entering the Ti-a_1 band; as E_{FC} rises with x in Li_xTiS_2, the voltage versus Lithium drops smoothly in this single-phase solid-solution reaction, Fig. 3.11. The Ti^{4+}/Ti^{3+} couple of Fig. 3.12a actually lies in an itinerant-electron a_1 band that is half-filled in $LiTiS_2$; however, displacement of Li to the tetrahedral sites of the Li layer in $Li_{1+x}TiS_2$ raises the Ti^{3+}/Ti^{2+} couple to just below the Ti-4*s* band.

As the atomic nuclear charge increases from Ti to V to Cr, the 3*d* electrons become more tightly bound to the nucleus, which lowers their energies. The E_{FC} of VS_2 is *pinned* at the top of the S-3*p* bands, Fig. 3.12b. Pinning of E_{FC} occurs because, as the cation-*d*-like states move from above to below the top of the bonding anion-*p* bands, the empty antibonding states remain above the anion-*p* bonding states; but the empty antibonding states change their character from primarily cation-*d* to primarily anion-*p* while retaining their *d*-orbital symmetry. On the other hand, if the cation-*d* states fall too far below the top of the anion-*p* bands, the holes in the antibonding states become trapped in purely anion-*p* antibonding states of a S–S bond to create the $(S_2)^{2-}$ ion. The Cr(IV) formal valence state cannot be accessed because a localized-electron Cr(IV)/Cr(III) redox couple lies too far below the top of the S-3*p* bands. However, $V_{0.5}Cr_{0.5}S_2$ can be stabilized in the layered MS_2 structure, and reduction of this host by the intercalation of Li into $Li_xV_{0.5}Cr_{0.5}S_2$ accesses the V(V) to V(III) formal valences [10]. A flat $V \approx 2.7$ V for $x < 0.5$ in $Li_xV_{0.25}Cr_{0.75}S_2$ signals that the top of the S-3*p* bands is about 2.7 eV below the E_{FA} of Lithium. This limitation on the voltage in these sulfides motivated study of the extraction of Li from similarly layered oxides since the top of the O-2p bands would be below the top of the S-3*p* bands.

Layered Oxides. Most solid cathode hosts are transition-metal oxides in which E_{FC} lies in either a mixed-valent redox couple or a narrow band of *d*-electron states. The PbO_2 cathode of the lead–acid battery is an exception; its E_{FC} is located in a narrow Pb-6*s* band well-separated from the broad, empty Pb-6*p* band because Pb is heavy.

A layered transition-metal MO_2 oxide analogous to TiS_2 does not exist; the electrostatic repulsive forces between the oxygen layers are stronger than any Van der Waals bonding, which is much weaker in an oxide than in the chalcogenides. However, where the transition-metal M atom has no *d* electrons, or at most one *d* electron, and is too small to be stable in an octahedral site, it may undergo a ferroic M-atom displacement to form a short M = O bond with one oxygen of the octahedral site and a long $O \cdots M$ bond on the opposite side. These displacements create site dipoles that can stabilize a layered oxide as is illustrated by the structure of V_2O_5, Fig. 3.13a. On the other hand, several $LiMO_2$ oxides form rock-salt structures in which the Li^+ and M(III) ions occupy alternate layers of M(III) octahedral sites, Fig. 3.13c. The sheets of edge-shared $MO_{6/3}$ octahedra are analogous to those of TiS_2, but the stacking between sheets gives hexagonal-close-packed sulfide ions in TiS_2, cubic-close-packed oxide ions in $LiMO_2$. The layers can also be stacked to give trigonal-prismatic sites between the layers. Moreover, Li can be either extracted, $Li_{1-x}MO_2$, or inserted, $Li_{1+x}MO_2$. Insertion of more than one Li per formula unit displaces Li^+ ions from octahedral to tetrahedral sites in the Li layer to form Li_2MO_2 in a two-phase reaction. The two-phase insertion reaction $LiVS_2$-Li_2VS_2 gives a voltage of 1.0 V versus Lithium, and an SEI layer is formed on the electrode. The $Li_{1-x}MO_2$ oxides are of interest as cathodes.

Extraction of Li from $LiMO_2$ oxides creates M(IV)/M(III) mixed valence on the M atoms. Where the M atoms have a localized spin *S* associated with a cation configuration $3d^n$, the mobile electron of the mixed-valent system moves diffusively in a hopping transfer $d^{n+1} + d^n = d^n + d^{n+1}$ between like atoms on equivalent lattice sites. In this case, the d^n/d^{n+1} configuration of an M(IV)/M(III) mixed-valent system may be considered a redox couple because the time for a hop is long compared to the period of a lattice vibration that traps the mobile electron or hole at a single site. Where the empty *d*-electron states lie close to the top of an anion-*p* band, the electrons may occupy a narrow band of itinerant-electron states as in TiS_2 and VS_2. Electrons in a partially filled band of itinerant-electron states give a metallic conductivity, those in redox couples require an activation energy for a hop to a like neighbor.

The energies of successive redox couples, for example, M(IV)/M(III) to M(III)/M(II), are separated by an on-site electron–electron electrostatic repulsive energy. In the case of Cr(III) in Fig. 3.12c, the cubic crystal-field splitting is added to this electrostatic on-site energy to raise the Cr(III)/Cr(II) energy to above the bottom of the Cr-4*s* band while the Cr(IV)/Cr(III) couple lies below the top of the S-3*p* bands. Insertion of Li into $LiCrS_2$ results in $Li_2S + Cr^0$; extraction of Li results in $(S_2)^{2-}$ formation. On the other hand, the electrons in the narrow V-a_1 band with E_{FC} pinned at the top of the S-3*p* bands allows access to both the V(V)/V(IV) and V(IV)/V(III) mixed-valence states without any energy gap between them [10].

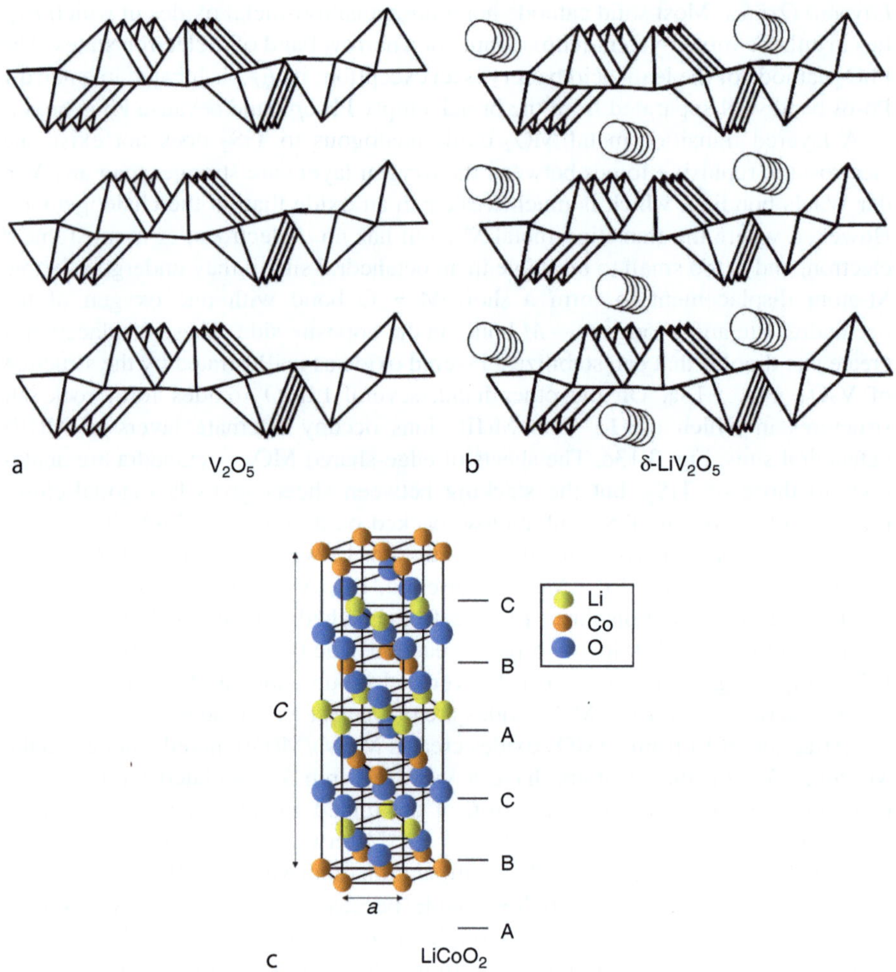

Fig. 3.13 The structures of (**a**) V_2O_5, (**b**) δ-LiV_2O_5, and (**c**) $LiMO_2$

The Co(III) ions in $LiCoO_2$ are in their low-spin $S = 0$ state, but the initial Co(IV) ions appearing on removal of Li in $Li_{1-x}CoO_2$ are in their high-spin $S = 5/2$ state [11]; the holes are trapped at a single Co(IV) site. However, in the range $0.05 \leq x \leq 0.45$, the voltage is flat at $V \cong 4.0$ V versus Lithium, which signals a two-phase reaction between a Li-rich phase containing high-spin Co(IV) and a Li-poor phase containing low-spin Co(IV/III) ions with holes in an itinerant-electron π^* band [12]. For $x \geq 0.55$, a new phase with different stacking of the CoO_2 sheets begins to appear as oxygen is evolved and/or H^+ ions from the organic electrolyte are inserted so as to retain the Co(IV)/Co ratio at about 0.45 [13]. The outgassing of oxygen shows that the holes in the π^* band have sufficient O-2p character, at least at a grain surface, to become trapped in peroxide ions $(O_2)^{2-}$ followed by $2(O_2)^{2-} = 2O^{2-} + O_2\uparrow$.

Clearly, the $Li_{1-x}CoO_2$ electrode reaches its *intrinsic limit* at a ratio Co(IV)/Co ≈ 0.45; this limit is above the HOMO of the electrolyte, so the electrolyte is not being decomposed aside from loss of inserted H^+ ions. The intrinsic limit restricts the practical capacity of $Li_{1-x}CoO_2$ to only 50% of the theoretical capacity, that is, to a $Q = 140$ mAh/g.

A cell with a carbon anode forms an LiC_6 phase during an initial charge from the discharged cathode $LiCoO_2$, and formation of an SEI layer on the reduced carbon during the initial charge reduces further the available capacity of the cell unless the SEI layer on the anode is formed before the cell assembly. Moreover, by fabricating the cell with a discharged cathode, it is the capacity of the cathode that determines the capacity of the cell. Nevertheless, the E_{FA} of the carbon anode is only 0.2 eV below the E_{FA} of Lithium, so the cell has a high specific energy QV because of its $V \approx 3.8$ V. The cell also has a high volumetric energy density, which is why it captured the market for laptop computers, cell telephones, and other handheld devices. The *tap density* of an electrode refers to the density of active particles that can be packed into a given volume, and a high tap density of $LiCoO_2$ is responsible for the high volumetric energy density.

Cobalt is expensive and toxic, so it was natural to try substitution of Ni for Co. However, it proves difficult to prepare well-ordered $LiNiO_2$. The maximum Ni substitution for Co in a layered phase with well-ordered Li^+ ions is $LiCo_{0.15}Ni_{0.85}O_2$ [14]. Since the holes in $Li_{0.5}CoO_2$ occupy a π^* band whereas those on low-spin Ni (III) occupy a σ^* band of higher energy, the $LiCo_{0.15}Ni_{0.85}O_2$ system gives a loss of O_2 and/or H^+ insertion only for $(1-x) < 0.35$, which gives a $Q = 200$ mAh/g. Addition of a small amount of Al in $LiAl_{0.05}Co_{0.15}Ni_{0.8}O_2$ improves the ordering of the Li and optimizes the cyclability of this cathode.

The transfer of M atoms into vacant sites of the Li layer on Li extraction requires their transfer via a tetrahedral site of the interstitial space. The low-spin Co(IV/III) and Ni(IV/III) cations have a strong octahedral-site preference, so transfer to the vacancies in the Li layer is kinetically inhibited below 300°C. However, ions like Fe(III), Mn(II), and Cr(VI) have a strong tetrahedral-site preference and are easily displaced to the interstitial tetrahedral sites on Li removal. Therefore, $LiFeO_2$ does not retain its structure on Li extraction. The disproportionation reactions 2Mn(III) = Mn(II) + Mn(IV) and 3Cr(IV) = 2Cr(III) + Cr(VI) introduce Mn(II) and Cr(VI) ions into the interstitial space to block insertion of Li back into the structure. Therefore, in order to eliminate Co and to take advantage of the pinning of the Ni(IV)–Ni(II) valence states at the top of the O-2p bands, the $LiNi_{0.5}^{(II)}Mn_{0.5}^{(IV)}O_2$ structure was investigated [15]; the Mn(IV) ion has a strong octahedral-site preference and the Mn(V)/Mn(IV) redox couple lies well below the top of the O-2p bands where the disproportionation reaction 3Mn(IV) = 2Mn(III) + Mn(VI) is inhibited. However, the mean size of the Ni(II) and Mn(IV) ions is too large for good ordering of the Li^+ ions in $Li(Ni_{0.5}Mn_{0.5})O_2$ to be easily obtained. Therefore, considerable attention has been given to $Li(Ni_{0.5-x}Co_{2x}Mn_{0.5-x})O_2$ with $x \leq 1/6$ where the Ni(III)/Ni(II) and Ni(IV)/Ni(III) mixed valences are accessed without a step in the voltage profiles on passing from one to the other because of E_{FC} pinning. Moreover, solid solutions of these $LiMO_2$ oxides with $Li(Li_{1/3}Mn_{2/3})O_2$, commonly designated Li_2MnO_3, as a randomly intercalated second phase exhibit, after an initial charge, a high reversible capacity of 250 mAh/g even though Li_2MnO_3 itself is not electrochemically active [16]. The voltage profiles of these solid solutions exhibit two distinct regions during the first charge, an initial sloping range corresponding to the oxidation of the Ni(II) to Ni(IV) and the Co(III) to Co(IV) followed by a plateau at $V \approx 5.5$ V. The plateau at 4.5 V appears to signal oxidation of the electrolyte. The excess Li_2MnO_3 increases the Mn(IV) concentration, which lowers further the top of the O-2- bands and thus increases the intrinsic voltage limit of the layered oxide. The passivating SEI layer formed from the electrolyte at voltages $V > 4.3$ V inhibits Li^+-ion transfer across the electrode/electrolyte interface, but coating with an oxide layer that transports Li^+ ions suppresses formation of the electrolyte SEI layer to allow fast cycling. Nevertheless, these layered oxides must compete against spinel cathodes containing Ni(II) and Mn(IV).

Vanadium Oxides. Several intermediate phases appear between V_2O_3 and V_2O_5; Li insertions into V_2O_5 and V_6O_{13} were investigated in the 1970s, and an amorphous or low-crystallinity VO_x with a polyethylene oxide (PEO) electrolyte is used in a Li/PEO/VO_x secondary battery for stationary storage of electrical energy [17].

Figure 3.13a shows the structure of orthorhombic V_2O_5; corner-shared paired chains of distorted VO_6 octahedra parallel to the *c*-axis share edges to create *c*-axis zigzag chains. Ferroic *c*-axis displacements of V atoms in the paired chains create asymmetric *c*-axis bonds 0 ... V = O, which places the V atoms in square-pyramidal sites. These chains share corners in the *a*–*b* planes with antiferroelectric coupling across shared edges to create the *a*–*b* layers of Fig. 3.13a. Li^+ ions can be inserted reversibly into the *b*-axis tunnels to form δ-LiV_2O_5 via an intermediate ε-$Li_xV_2O_5$ phase.

In the range $0 \leq x \leq 1$ of $Li_xV_2O_5$, reduction of the vanadium on Li^+-ion insertion only tilts the square-pyramidal sites of the $a - b$ planes, but three phases can be distinguished: α $(0 \leq x \leq 0.1)$, ε $(0.35 \leq x \leq 0.7)$, and δ $(0.9 \leq x \leq 1.0)$. The reversible insertion of Li into V_2O_5 gives a voltage profile versus Lithium with two plateaus, one at 3.4 V for the $\alpha + \varepsilon$ range $0.1 \leq x \leq 0.35$ and the other at 3.2 V for the $\varepsilon + \delta$ range $0.7 \leq x \leq 0.9$. On further Li insertion, the voltage drops abruptly to 2.4 V followed by a monotonic decrease with x to about 1.5 V over the range $1 < x < 3$. Cycling through the first-order transition at $x = 1$ damages the crystallites and lowers the cyclability. However, insertion into amorphous or low-crystallinity V_2O_5 allows faster and better cyclability. Insertion of 1.5 Li per V atom gives a relatively large capacity.

The structure of V_6O_{13} shows how the particles of low-crystallinity or amorphous VO_x can accommodate Li^+-ion insertion into 1D tunnels. Figure 3.14 shows the structures of V_2O_5 and V_6O_{13} projected, respectively, onto its (001) and (010) plane, which illustrates how the V_6O_{13} structure can be derived from that of V_2O_5. Removal of oxygen from every third *a*–*c* oxygen plane of V_2O_5 is followed by a shear, which leaves tunnels for Li^+-ion insertion as in V_2O_5. The shear planes need not be ordered as in V_6O_{13}.

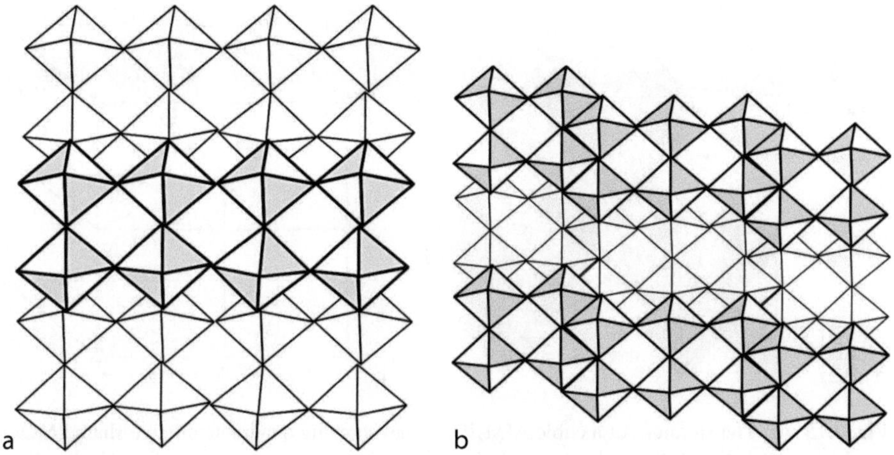

Fig. 3.14 Comparison of structures of V_2O_5 projected onto its (001) plane and V_6O_{13} onto its (010) plane

These observations have led to the synthesis of a two-dimensional (2D) pseudocrystalline phase containing segments of three planes similar to those of V_2O_5 that are displaced from one another by a slip plane. This structure provides an insertion of up to two Li atoms per formula unit with a monotonic voltage decrease with x, except for a small step at $x = 1.5$, over the voltage range $2.5 < V < 3.4$ V versus Lithium. Nevertheless, there is a gradual capacity fade with cycle number.

The voltage profile of $Li_xV_6O_{13}$ shows small steps at $x = 1$, 2, and 3 as the voltage decreases from 2.8 to 2.4 V versus Lithium and a drop from 2.4 to 2.1 V in the interval $3 < x < 4$ that is followed by a gradual drop to 2.0 V as x approaches 6. The steps appear to correspond to Li^+-ion ordering with electrons entering $V - 3d$ bands rather than localized redox couples; the electronic states change character from $V - 3d$ with a large $O - 2p$ component to a dominant $V - 3d$ character with increasing x. Capacity fading on cycling makes this material not competitive as a cathode for a secondary battery.

Spinels. The $A[M_2]O_4$ spinels have a cubic-close-packed array of oxygen atoms with M-site cations in half of the octahedra, the 16*d* sites of space group $Fd\overline{3}m$ of Fig. 3.15. The $[M_2]O_4$ array represents a strongly bonded framework with A atoms in the interstitial tetrahedral sites 8*a*. The empty 16*c* octahedral sites share faces with the 8*a* sites to form, with the 8*a* sites, a 3D-interconnected interstitial space. On Li^+-ion insertion into an empty 16*c* site, the coulombic repulsive force between the Li^+-ion and the near-neighbor A-site cations pushes all the A-site cations in a cascade into the 16*c* sites [18].

In the case of magnetite, $Fe[Fe_2]O_4$, insertion of one Li per formula unit produces an ordered rock-salt array with Fe remaining on all the 16*d* sites and Li and Fe randomly distributed over the 16*c* sites. Further insertion of Li extrudes Fe reversibly

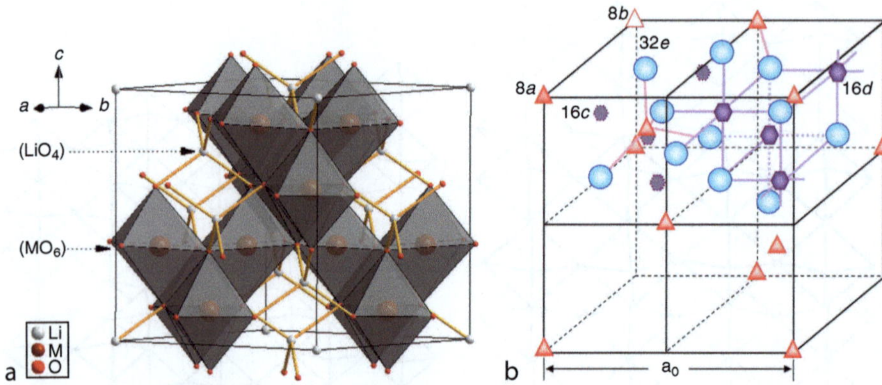

Fig. 3.15 (**a**) The structure of a cubic $A[M_2]O_4$ spinel showing quadrants of edge-shared $MO_{6/3}$ octahedra alternating with quadrants of tetrahedral-site A cations on traversing the <100 > axes. (**b**) Two quadrants of the cubic spinel structure showing the diamond array of the A-site sublattice *8a*; M cations are in *16d* and oxygen in *32e*

from the 16*c* sites to form the ordered rock-salt structure $Li_2[Fe_2]O_4$ and amorphous Fe^0. From this observation, it was clear that Li insertion into the system $Li_{1+y}[Mn_2]O_4$ $(0 < y \leq 1)$ would place Li^+ ions in the octahedral 16*c* sites and Li extraction in the system $Li_{1-y}[Mn_2]O_4$ $(0 < y \leq 1)$ would leave the Li^+ ions in tetrahedral *8a* sites. Moreover, with only Li^+ ions in the 3D interstitial space, a fast Li^+-ion transport could be anticipated. The voltage profile of Fig. 3.16 for the system $Li_x[Mn_2]O_4$, $0 \leq x \leq 2$, shows a 1-V step at $x = 1$ where the Li^+ ions shift from tetrahedral to octahedral sites. This remarkable 1-eV shift in the energy of the Mn(IV)/Mn(III) couple shows that the energy of an active redox couple is sensitive to the Madelung ionic electrostatic energy of the oxide.

In contrast, the voltage profile of Li insertion into the thiospinel framework $[Ti_2]S_4$ shows no step at $x = 1$ [19]; the profile of $Li_x[Ti_2]S_4$, $0 \leq x \leq 2$, is nearly identical to that for the layered system Li_xTiS_2, which shows that the Li^+ ions occupy only octahedral 16*c* sites for all *x* in $Li_x[Ti_2]S_4$. Preparation of the thiospinel $Li[Ti_2]S_4$ is not straightforward; the thiospinel framework $[Ti_2]S_4$ is prepared by chemical extraction of tetrahedral-site Cu^+ from the thiospinel $Cu[Ti_2]S_4$. In both the layered TiS_2 and the spinel framework $[Ti_2]S_4$, the octahedral $TiS_{6/3}$ sites share only edges, not corners, with neighboring $TiS_{6/3}$ sites; only the shape of the interstitial space has been changed from 2D to 3D. The 2D interstitial space of $LiTiS_2$ allows intercalation of other species than the Li^+ ions whereas the 3D interstitial space of the strongly bonded spinel framework does not.

The 1-V step in the voltage profile of $Li_x[Mn_2]O_4$ limits the capacity of this spinel to one Li^+ per two Mn atoms at either 3.0 V or 4.0 V versus Lithium. The flat 3.0 V plateau reflects a cubic to tetragonal transition of the $[Mn_2]O_4$ framework due to a cooperative Jahn–Teller site distortion where more than half of the 16*d* sites are occupied by Mn(III) ions. Octahedral-site, high-spin Mn(III) ions have a localized spin $S = 2$ with one electron occupying an orbitally twofold-degenerate pair of

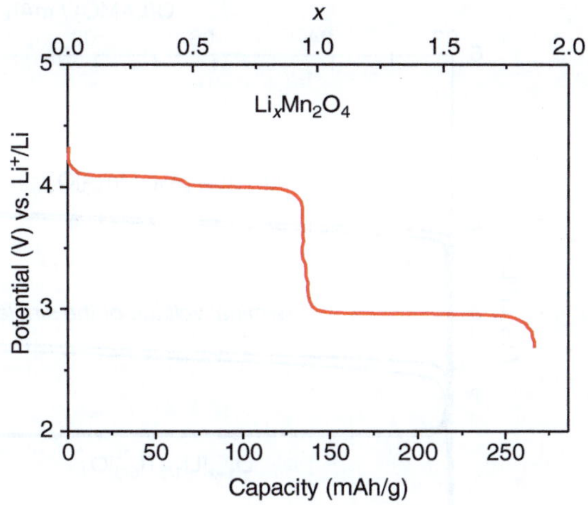

Fig. 3.16 The V(x) versus Lithium profile for $Li_x[Mn_2]O_4$ showing a 1 eV step in the M(IV)/M(III) redox couple at $x = 1$

σ-bonding e_σ orbitals; a site distortion to tetragonal symmetry removes the degeneracy, stabilizing an occupied $(3z^2-r^2)$ orbital at the expense of an empty (x^2-y^2) orbital of the e_σ pair. A cooperative site distortion minimizes the elastic energy per site that it costs to distort a site from cubic symmetry.

The 8a sites consist of two interpenetrating face-centered-cubic arrays to form the diamond lattice. The small step in the 4.0-V plateau of Fig. 3.16 reflects an ordered $Li_{0.5}[Mn_2]O_4$ phase in which the Li^+ ions occupy only one of the two interpenetrating face-centered-cubic 8a arrays; two-phase separations occur either side of $x = 0.5$. The two-phase separation due to Li^+-ion order enhances a surface disproportionation reaction 2Mn(III) = Mn(II) + Mn(IV) that is followed by Mn(II) dissolution into the electrolyte. The Mn(II) crosses over to the anode to poison Li^+-ion mobility across the anode SEI layer. As a result, operation on the 4.0-V plateau is plagued by a reduction in capacity that increases with cycling. The Mn(II) dissolution represents a chemical reaction of the electrode with the electrolyte that is to be distinguished from oxidation of the electrolyte by an E_{FC} below the electrolyte HOMO. The progressive increase with cycling in the irreversible loss of capacity is referred to as a capacity fade; it limits the cycle life of the cell. Substitution of Ni or Al and some Li for Mn suppresses any Jahn–Teller framework distortion near $x = 1$; it also disorders the Li^+ ions at $x = 0.5$ to reduce further the Mn(III)-ion concentration at the surface and hence the Mn(II) dissolution, but at the expense of the capacity of the cathode. Figure 3.17a shows charge/discharge curves for a $Li_{1-x}[Al_{0.1}Li_{0.1}Mn_{1.8}]O_4/Li_{1+x}[Li_{1/3}Ti_{5/3}]O_4$ cell that can undergo safely a fast recharge. At 55°C, this cell retains 80% of its rechargeable capacity after 3,600 cycles. Batteries of five cells connected in series deliver a nominal 12 V.

In the layered $Li_{1-x}Co_{0.15}Ni_{0.85}O_2$ system, pinning of E_{FC} at the top of the O-2p bands gave an intrinsic limit of 3.8 V versus Lithium for the Ni(IV)/Ni(III) couple. Removal of nearly all the Li from the layered oxide $LiNi_{0.5}Mn_{0.5}O_2$ containing Mn

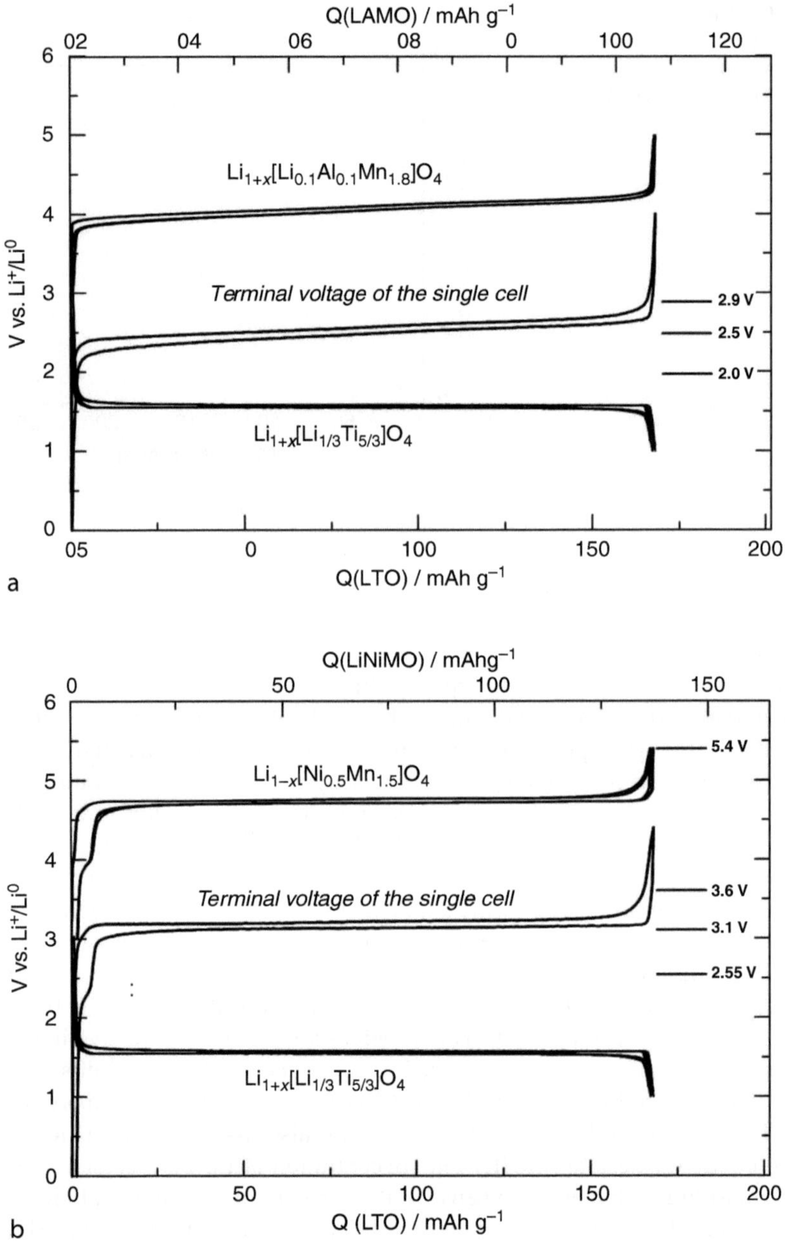

Fig. 3.17 Charge/discharge V(x) versus Li^+/Li^0 for (**a**) $Li_{1-x}[Al_{0.1}Li_{0.1}Mn_{1.8}]O_4$ and (**b**) $Li_{1-x}[Ni_{0.5}Mn_{1.5}]O_4$ with $Li_{1+x}[Li_{1/3}Ti_{5/3}]O_4$ to give terminal single-cell voltages (By courtesy of Prof. T. Ohzuku)

(IV) ions shows not only that Mn(IV) increases the voltage limit of the Ni(IV)/Ni (III) couple by lowering the top of the O-2p bands, but also that where E_{FC} is pinned, oxidation of Ni(II) to Ni(IV) occurs without a step in the voltage profile. Moreover, pinning of E_{FC} gives the flat voltage profile of Fig. 3.17b; at room temperature, Li is removed over the range $0 < x < 1$ at $V \approx 4.7$ V, which is just below the carbonate-electrolyte decomposition voltage above 4.9 V versus Lithium. Batteries with only four $Li_{1-x}[Ni_{0.5}Mn_{1.5}]O_4/Li_{1+x}[Li_{1/3}Ti_{5/3}]O_4$ cells connected in series can deliver a nominal 12 V. Lowering the top of the O-2*p* bands in the spinel structure increases the intrinsic voltage limit of the oxide. In contrast, the Ti(IV) ion of $Li_{1-x}[Ni_{0.5}Ti_{1.5}]O_4$ prevents access to even the Ni(III) oxidation state. Unfortunately, although the SEI layer on $Li[Ni_{0.5}Mn_{1.5}]O_4$ appears to stabilize the cathode at room temperature, raising the temperature to 60°C can introduce a capacity fade on cycling. Coating of the surface of the active oxide particles with another oxide permeable to Li^+ ions has been shown to reduce the activation energy for Li^+-ion transfer across the SEI layer as well as the capacity fade at 60°C. The spinel $Li_{0.5-x}[Ni_{0.5-x}Cr_{2x}Mn_{1.5-x}]O_4$ ($x \cong 0.05$) appears to stabilize the cathode at 60°C without sacrifice of the capacity since the Cr(IV)/Cr(III) couple is accessible and small Cr concentration inhibit the disproportionation of Cr(IV) into Cr(VI) and Cr (III) and are sufficient to remove Ni from the particle surface. The development of 12 V batteries based on spinel insertion compounds is targeted for plug-in hybrid vehicles and electrical-energy storage in solar-powered homes. The goal is a life in excess of 10 years and a driving range per charge in excess of 200 miles at an acceptable dollar and environmental cost.

NASICON Framework [20]. Use of a solid Na^+-ion electrolyte in the Na–S battery stimulated a search in the early 1970s for an improved Na^+-ion conductor. This search concentrated on open framework structures that were electronic insulators and hosts to mobile Na^+ ions. One of these framework structures was that of the system $Na_{1+3x}Zr_2(P_{1-x}Si_xO_4)_3$ with $0 \leq x \leq 1$ having the structure of hexagonal $Fe_2(SO_4)_3$ pictured in Fig. 3.8. This system with $x \approx 2/3$ was named NASICON to signify it is a NA SuperIonic CONductor, and the host framework is now referred to in the literature as the NASICON framework. Substitution of Na^+ ions by Li^+ ions and substitution of a smaller Ti(IV) for Zr(IV) gives LISICON having the same framework; up to 5 Li^+ ions per formula unit can be accommodated in this framework. Substitution of tetrahedral polyatom anions for oxygen opens up the interstitial space for fast Li^+-ion motion, and substitution for Zr of a smaller transition-metal atom having a lower-energy redox couple makes this framework a candidate host for the cathode of a Li^+-ion secondary battery. In this host, the octahedral-site cations share corners with the polyatom tetrahedral anions and *vice versa*, which isolates the transition-metal cations from one another. Therefore, the 3*d* configurations at the transition-metal ions are localized, so the successive redox couples are separated from one another by a finite energy gap. This gap may be large, as in the separation of the Cr(IV)/Cr(III) and Cr(III)/Cr(II) couples pictured in Fig. 3.12b or it may be small as between V(V)/V(IV) and V(IV)/V(III) couples in $Li_{3-x}V_2(PO_4)_3$, vide infra. Since this framework can accommodate up to 5 Li atoms without influencing significantly the redox energies on the octahedral-site cations, the framework lends itself to an examination not only of the influence on the redox energies of the counter cation within the polyatom anion, but also the relative energies of different redox couples for a fixed tetrahedral anion [21].

The Fe(III)/Fe(II) redox energy of $Fe_2(SO_4)_3$ is at 3.6 eV, those of isostructural $Fe_2(MoO_4)_3$ and $Fe_2(WO_4)_3$ are at 3.0 eV below the E_{FA} of Lithium. This finding demonstrates that the stronger the covalent bonding in the polyatom anion, the weaker is the bonding of the oxygen to the octahedral-site cation and therefore the more stable is the transition-metal redox energy. This electrochemical result provides a direct measure of the magnitude of this *inductive effect* [22]. It shows that a redox energy can be tuned not only by the position of the guest ion in the interstitial space of the spinel $Li_x[Mn_2]O_4$ or the redox energy of a near-neighbor cation coexisting in an equivalent lattice site as in $Li(Ni_{0.5}Mn_{0.5})O_2$ versus Li $(Ni_{0.8}Co_{0.2})O_2$, but also by the counter cation in a polyatom anion. Accordingly, the Fe(III)/Fe(II) redox couple in $Li_{3+x}Fe_2(PO_4)_3$ is 2.8 eV below the E_{FA} of Lithium. The 0.8 eV shift of the Fe(III)/Fe(II) redox couple on changing from $(SO_4)^2$ to $(PO_4)^{3-}$ is the same for all the redox energies of the NASICON framework.

As an illustration of how the structure lends itself to locating the relative positions of the available redox energies, Fig. 3.18a shows the voltage profile for $Li_{1+x}FeNb(PO_4)_3$ showing discrete steps on passing from the Fe(III)/Fe(II) to the Nb (V)/Nb(IV) couple to the Nb(IV)/Nb(III) couple at 2.8 eV, 2.3 eV, and 1.8 eV, respectively, below the E_{FA} of Lithium. The Ti(IV)/Ti(III) couple at 2.5 eV in $Li_xTiNb(PO_4)_3$ overlaps the Nb(V)/Nb(IV) couple at 2.5 eV to give the smooth voltage profile from 2.5 to 2.2 V shown in Fig. 3.18b. The relative energies of the redox couples with $(SO_4)^{2-}$ anions remain the same, but all are shifted by about 0.8 eV to lower energy. The voltage profile of a carbon-coated $Li_{3-x}V_2(PO_4)_3$ cathode versus Lithium shows a good cycle life over the interval $0 \leq x \leq 3$ with two-phase reactions at 3.59 V in the interval $0.5 \leq x \leq 1.0$ and 3.67 in the interval $1.0 \leq x \leq 2.0$ corresponding to the V(IV)/V(III) couple and 4.06 V over $2.0 \leq x \leq 3.0$ of the V(V)/V(IV) couple. The 0.08 eV step for the V(IV)/V(III) couple is due to ordering of the Li^+ ions into the unique interstitial site per formula unit between M-atom sites of the framework; the larger 0.4 eV step at $x = 2.0$ is the separation of the V(IV)/V(III) and V(V)/V(IV) couples. Carbon coating of the particles is needed to facilitate these two-phase reactions; the large separation of the transition-metal ions makes any electron transport diffusive, which limits the power capability with this framework unless small, carbon-coated particles are used.

Ordered Olivines $LiMPO_4$. As in a spinel [23], in the olivine structure of Mg_2SiO_4, the cations occupy half of the octahedral and one-eighth of the tetrahedral sites of a slightly distorted, close-packed anion array, but the olivine anion array is close-packed-hexagonal rather than close-packed-cubic. Moreover, two types of occupied octahedral sites are distinguishable, which leads to an ordering of Li^+ and M(II) ions in the $LiMPO_4$ olivines whereas they are disordered in the spinel V $[LiM]O_4$. The ordered MPO_4 array forms a 3D strongly bonded framework in which octahedral-site M atoms share corners within $a-c$ planes of an orthorhombic

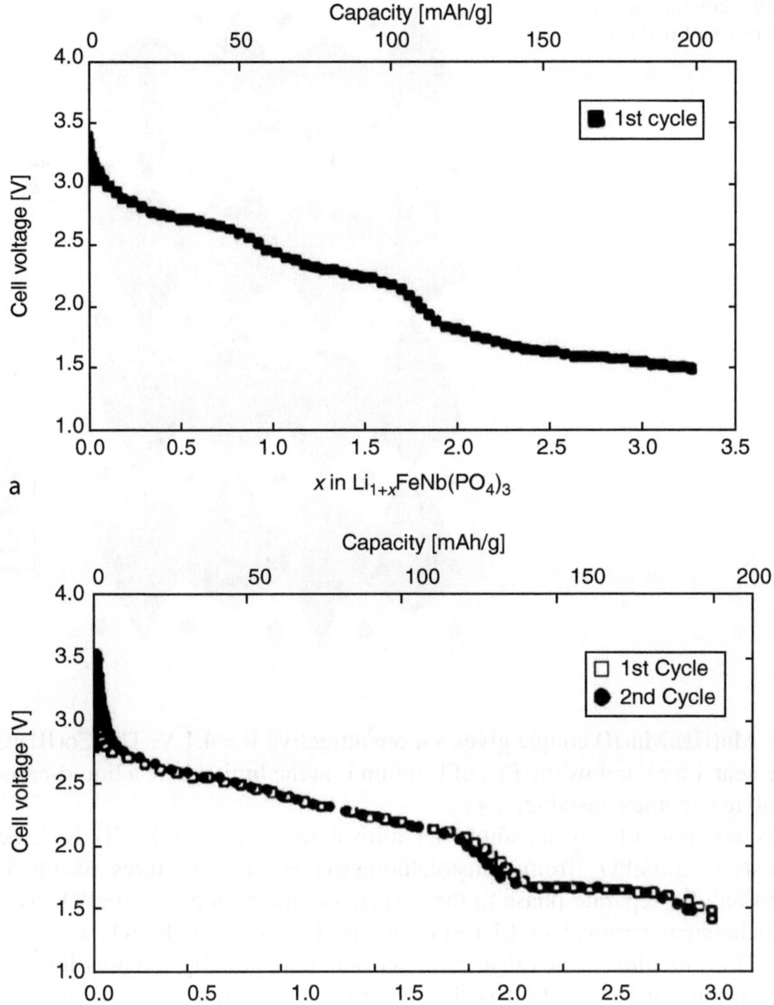

Fig. 3.18 Discharge V(x) versus Li^+/Li^0 for (**a**) $Li_{1+x}FeNb(PO_4)_3$ and (**b**) $Li_xTiNb(PO_4)_3$

unit cell, Fig. 3.19; these $a-c$ planes are bridged by the PO_4 tetrahedra and the guest Li^+ ions occupy 1D a-axis tunnels of edge-shared octahedra. Motion of the Li^+ ions to neighboring empty octahedral sites outside of the tunnels requires a larger activation energy than motion along the tunnels.

The M atoms of $LiMPO_4$ may be Mg, Mn, Fe, Co, Ni, or mixtures thereof. Of these possibilities, $LiFePO_4$ is the most promising cathode material; $Li_{1-x}FePO_4$ provides a flat $V = 3.45$ V versus Lithium. Removal of Li from $LiMnPO_4$ is difficult,

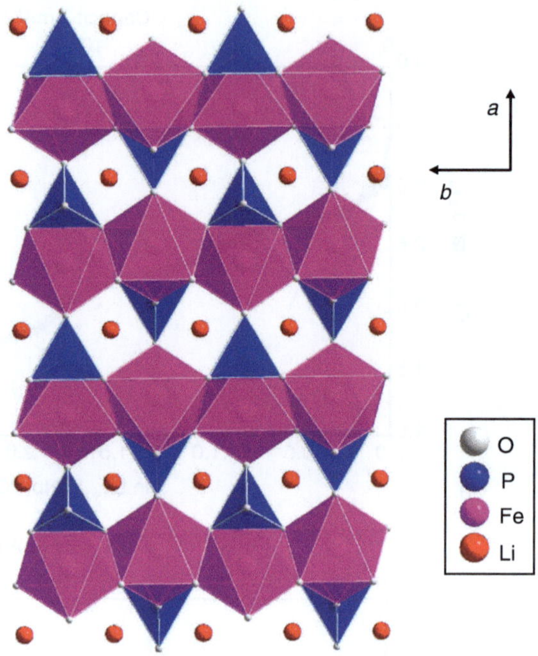

Fig. 3.19 Structure of the ordered olivine $LiFePO_4$

but the Mn(III)/Mn(II) couple gives a more attractive $V = 4.1$ V. The Co(III)/Co(II) couple near 4.8 eV below the E_{FA} of Lithium is at the limit where a liquid-carbonate electrolyte becomes unstable.

It is not possible to substitute an aliovalent cation for Fe(II) in $LiFePO_4$. Synthesis of $LiFePO_4$ from nonstoichiometric starting mixtures results in the dissolution of a separate phase to the surface of stoichiometric $LiFePO_4$ particles. Electrochemical removal of Li results in the formation of $FePO_4$ as a separate phase. This situation means that the electronic conductivity remains low and the voltage profile of $Li_{1-x}FePO_4$ is flat. In order for Li^+ ions to move in or out of a particle rapidly with the electron of an Fe(III)/Fe(II) couple, it is necessary to have an electronically conducting coat on a small nanosize particle that also not only makes electronic contact with the current collector, but also is permeable to Li^+ ions. This coat may be a phase rich in Li that is dissolved from the $LiFePO_4$ particle or carbon; the coat makes electronic contact with the current collector either by added inert carbon or by contact to an electronically conducting polymer that also contacts the current collector and has an active redox energy that overlaps the energy of the Fe(III)/Fe(II) couple of $LiFePO_4$. Polypyrrole (PPy) and polyannaline (PANI) are two such polymers [24]. Figure 3.20 illustrates the voltage profile versus Lithium of carbon-coated $LiFePO_4$ with PPy for different charge/discharge rates. With a $V = 3.45$ V well below the 4.3 V limit imposed by the HOMO of the electrolyte, this cathode is safe; it also has an excellent cycle life. This cathode is used in secondary Li^+-ion power batteries and is an attractive contender for the

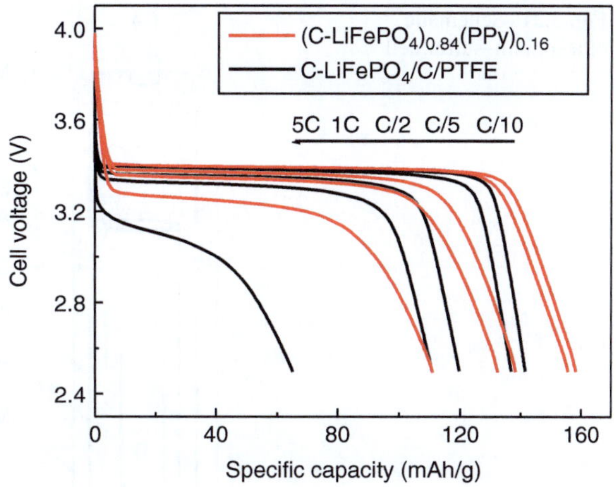

Fig. 3.20 Discharge V(x) versus Li^+/Li^0 for C-coated Li_xFePO_4 with/without PPy for different rates for full discharge from C/10 = 10 h to 5 C =12 min

battery of a plug-in hybrid vehicle despite a redox energy that is 1 eV higher than would appear to be optimal for a carbonate electrolyte.

It should be noted that $LiFePO_4$ nanoparticles embedded in a carbon matrix make a composite that shows promise for the cathode of an electrochemical capacitor of higher energy density than a purely carbon cathode.

Secondary Batteries with Cathodes That Are Not Insertion Compounds

The cell of a *sodium–sulfur* battery, illustrated schematically in Fig. 3.21, consists of two liquid electrodes, molten sodium and sulfur, separated by a ceramic Na^+-ion electrolyte; it operates in the 300–375°C range to maintain molten not only the sodium anode, but also both the sulfur cathode and the immiscible-in-sulfur sodium polysulfide (Na_2S_5 to Na_2S_3) products of the reaction of Na^+ ions from the electrolyte with the sulfur [2]. Because sulfur is an electronic insulator, the reaction of the Na^+ ions with the sulfur occurs at the surface of a carbon felt introduced into the sulfur to bring electrons to a large carbon–sulfur interface area. The negative post of the battery is attached directly to the molten sodium contained in a ceramic-electrolyte tube that is closed at one end and open to a molten-sodium reservoir at the other end; the reservoir keeps the Na^+-ion electrolyte covered with Na throughout a discharge. The sulfur with its carbon felt bathes the outside of the ceramic-electrolyte tube, the carbon felt contacting a metal container to which the positive post is attached. The critical component of the cell is the Na^+-ion ceramic electrolyte.

The Na^+-ion electrolyte for a Na–S battery needs to consist of a framework host with mobile Na^+ ions in the interstitial space, but it must also have the bottom of the conduction band of the host above the E_{FA} of elemental Sodium. These criteria are

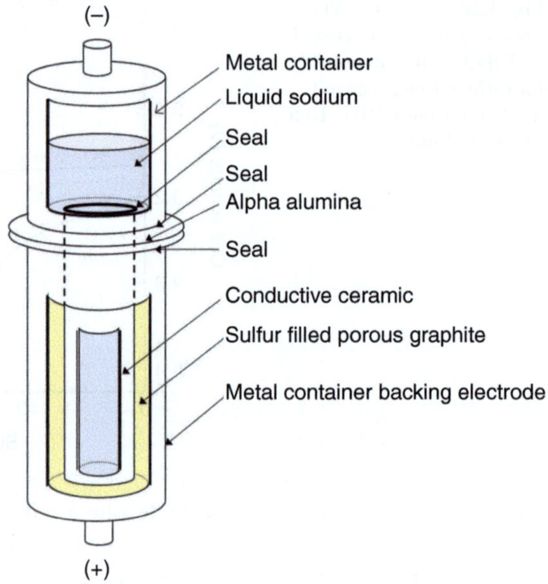

Fig. 3.21 Schematic of a sodium–sulfur cell

met in the Na-β and Na-β″ aluminas; but their frameworks have a 2D interstitial space and, therefore, an anisotropic thermal expansion. This anisotropic property is undesirable for a polycrystalline, thin ceramic membrane that is subject to repeated thermal cycling between room temperature and 350°C. Nevertheless, ceramic engineering of mixed β and β″ aluminas has resulted in a viable Na–S battery for stationary electrical-energy storage [25].

Stoichiometric Na-β alumina, $Na_2O \bullet 11Al_2O_3$, contains $Al_{11}O_{16}$ spinel slabs perpendicular to the [111] axis; the slabs are connected by oxygen bridges within intervening NaO planes containing intersecting 1D tunnels occupied by the Na^+ ions. In stoichiometric Na-β alumina, there are two inequivalent Na^+-ion sites in the NaO planes, one of which is occupied and the other empty. The inequivalence of the Na^+-ion sites raises the activation energy ΔH_m for Na^+-ion motion, but the introduction of excess Na^+ either in $(1 + x)Na_2O \bullet 11Al_2O_3$ or by substituting some Li^+ or Mg^{2+} for Al^{3+} in the spinel blocks reduces ΔH_m in the β phase; in the β″ phase, the Li^+ or Mg^{2+} substitutions shift the spinel slabs relative to one another to make equivalent the Na^+-ion sites in the NaO layers, thereby reducing ΔH_m. A two-phase mixture of β and β″ phases with excess Na^+ ions gives a viable ceramic membrane.

The *ZEBRA cell*, which is under development by the General Electric Co., uses a molten-sodium anode and a solid β,β″-alumina solid electrolyte as in the sodium-sulfur cell, but the positive electrode is large-surface-area nickel rather than molten sulfur with a large-surface-area current collector. The electrolyte on the cathode side of the ZEBRA solid electrolyte is an aqueous $NaAlCl_4$ containing NaCl and NaI as well as a little FeS. The FeS and NaI are added to limit growth of the Ni particles and to aid the overall cathode reaction, which is

$$Ni^0 + 2Cl^- - 2e^- = NiCl_2$$

on charge and on discharge

$$NiCl_2 + 2e^- = Ni^0 + 2Cl^-$$

at a V_{OC} = 2.58 V. A *Lithium–Sulfur* battery also uses a sulfur cathode; its redox reaction is

$$S_8 + 16Li^+ + 16e^- = 8Li_2S \tag{3.23}$$

in which the electrons from the anode on discharge are brought to the reaction by carbon [26]. The two-phase reaction proceeds in two steps, first to the polysulfides Li_2S_8, Li_2S_4, and Li_2S_2 at 2.4 V and then to Li_2S at 2.0 V. Best results to date have been obtained with a CMK-3 carbon impregnated by sulfur and electrolyte. The CMK-3 carbon consists of carbon nanorods kept apart by carbon bridges to give channels into which molten sulfur is introduced. On solidification, channels are opened for the impregnation of electrolyte to give Li^+ ions access to a large surface area of amorphous sulfur. The cathode reaction is relatively slow, and the efficiency of electrical-energy storage is not optimal. Nevertheless, the lower power density and the low voltage are compensated by a capacity of over 1,000 mAh/g. The low voltage requires use of a Li^0 anode. A protected Lithium anode (see Lithium/seawater battery) is a logical candidate for this battery, but a soluble Li_2S_n shuttle molecule may suppress build-up of the dendrites on the Li^0 anode. The jury is still out on whether this battery will be competitive for a low-power application.

Lithium–Metal Salt secondary batteries are analogous to the Lithium–seawater primary battery [3]. A Li^+-ion solid electrolyte separates a nonaqueous anolyte and an aqueous cathode. For example, a Lithium anode with a carbonate anolyte and an aqueous $Fe(CN)_6^{3-}/Fe(CN)_6^{4-}$ cathode has been shown to give a flat voltage $V \approx 3.4$ V with an efficiency that increases with the molar ratio of iron cyanide in the cathode solution [27]. This promising approach requires development of a Li-ion solid electrolyte having a $\sigma_{Li} > 10^{-4}$ S/cm at room temperature that is stable to an acidic cathode solution and is not reduced by contact with a Li^0 dendrite on the anode side.

Lithium-air batteries [28] may also use a solid separator that will block dendrite growth from the anode to the cathode but allows permeation of the Li^+ ion between an anolyte and a catholyte. The simplest such separator would be a solid Li^+-ion solid electrolyte, but a porous glass containing the liquid electrolyte has been used where the anolyte and the catholyte are identical. As in the Zn-air primary battery, a porous carbon containing an oxygen–reduction catalyst on the pore walls and the liquid electrolyte in the pores provides the structure needed to facilitate the catalytic reaction of Li^+ ions with the gaseous O_2 cathode. The cathodic reaction

$$2Li^{+} + 2e^{-} + O_2 = Li_2O_2 \tag{3.24}$$

is reversible, but the voltage difference between charge and discharge currents is large even with the best catalyst, Pt–Au nanoparticles [29], which makes the battery inefficient for electrical-energy storage.

The *Redox-Flow Battery* [30] is a candidate for medium- and large-scale electrical-energy stationary storage. The cells of these batteries contain different redox energies in a liquid anode and a liquid cathode separated by either a solid electrolyte or a microporous polymer or glass separator. Reversible electrode processes take place at the surface of graphite-felt current collectors as in the sulfur cathode of a Na–S cell. The reactant redox couples flow across the carbon-felt conductors, entering the electrochemical cell from external reservoirs, Fig. 3.22. In theory, the capacity of a cell is determined by the volumes of the two electrolyte tanks. Under development is a cell containing vanadium ions in sulfuric-acid solution. The anode compartment contains V^{2+} ions, the cathode compartment $VO_2{}^{+}$; the cell reactions are

$$V^{2+} - e^{-} = V^{3+} \qquad \text{Anode}$$

$$VO_2{}^{+} + 2H^{+} + e^{-} = VO^{2+} + H_2O \quad \text{Cathode}$$

to take advantage of the 1.65 eV difference between the V(III)/V(II) and V(V)/V (IV) redox couples. Alternative couples include, for example, Fe^{3+}/Fe^{2+} and Cr^{3+}/Cr^{2+}. These cells can achieve 80–90% energy-storage efficiencies in large installations at a low cost per kWh. They promise a long cycle life, are relatively easy to maintain, and can be fully cycled without harm to the battery. However, in order to realize the commercial potential of these batteries, several challenges must be met, including development of electrodes resistant to oxidation in its electrolyte, membrane stability and prevention of redox crossover through the separator membrane while maintaining a fast H^{+}-ion transport across the membrane, easier flow across the current collectors, and scale-up optimization.

Future Directions

The Li-ion battery has already introduced the wireless revolution by powering the cell telephone and the laptop computer as well as their derivatives. Modest increases in anode capacity will improve the volume energy density for handheld electronic devices.

The Li-ion battery is also powering portable electric-power tools and small electric vehicles. However, its widespread application to the plug-in hybrid electric vehicle and the all-electric vehicle will depend on lowering the cost while increasing the driving range between recharges. To reach this target, it will be necessary to

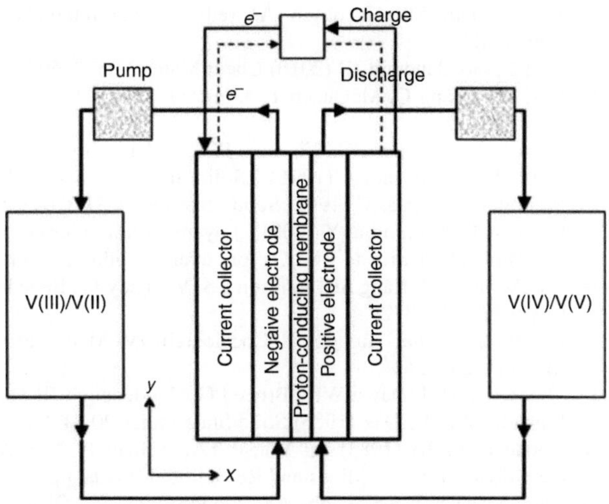

Fig. 3.22 Schematic of the all-vanadium flow-through battery

develop new cathode strategies. These strategies appear to depend on the development of a suitable solid electrolyte and sealants that can prevent crossover of different liquid electrolytes either side of the solid electrolyte.

Electrical-energy storage in batteries for the grid for increasing the efficiency of nuclear energy and for enabling the large-scale introduction of wind and solar energy will probably need the much larger capacities offered by flow-through batteries. In the interim, rebirth of the Na/S battery may provide some large-scale electrical-energy storage.

Realization of the great societal benefits to be derived from the new battery technologies has galvanized an extensive research activity that promises continued development.

Bibliography

Primary Literature

1. Winter M, Brodd RJ (2004) Chem Rev 104:4245–4269
2. Mennicke S (1992) In: Balkanski M, Takahashi T, Tuller HL (eds) Solid state ionics. North-Holland, Amsterdam, pp 3–15
3. Visco SJ, Nimon E, DeJonghe L, Katz B, Petrov A (2007) US Patent 2007/0037058
4. Aono H, Sugimoto E, Sadaoka Y, Imanaka N, Adachi G (1989) J Electrochem Soc 136:590–591
5. Armand MB (1986) Ann Rev Mater Sci 16:245–261
6. Croce F, Appetecci GB, Persi L, Scrosati B (1998) Nature 394:456–458
7. Zinck L, Borck M, Ripp C, Hambitzer G (2006) J Appl Electrochem 36:1291–1295
8. Dahn JR, VonSacken U, Juzkow MW, Al-Janaby H (1991) J Electrochem Soc 138:2207–2211

9. Whittingham MS, Jacobson AJ (eds) (1982) Intercalation chemistry. Academic Press, New York
10. Kim Y, Goodenough JB (2010) Chem Mater 22:587–603
11. Hert JT, Huang Q, McQueen T, Klimczuk T, Bos JWG, Viciu L, Cava RJ (2008) Phys Rev B 77:075119
12. Mizushima K, Jones PC, Wiseman PJ, Goodenough JB (1980) Mater Res Bull 15:783–799
13. Gupta R, Manthiram A (1996) J Solid State Chem 121:483–491
14. Saadoune I, Delmas C (1992) Solid State Ionics 53:370–375
15. Ohzuka T, Makimura Y (2001) Layered lithium insertion material of LiNi1/2Mn1/2O_2: a possible alternative to $LiCoO_2$ for advanced lithium-ion batteries. Chem Lett 8:744–745
16. Thackeray MM, Kang SH, Johnson CS, Vaughey JT, Benedek R, Hackney SA (2007) J Mater Chem 17:3112–3125
17. Delmas C, Cognac-Auradon H, Cocciontelli JM, Ménétrier M, Doumerc JP (1994) Solid State Ionics 69:257–264
18. Thackeray MM, David WIF, Bruce PG, Goodenough JB (1983) Mater Res Bull 18:461–472
19. Sinha S, Murphy DW (1986) Solid State Ionics 20:81–84
20. Goodenough JB (1983) In: Wheat TA, Ahmad A, Kuriakose AK (eds) Progress in solid electrolytes. Energy, Mines and Resources, Canada, pp 53–76
21. Nanjundaswamy KS, Padhi AK, Goodenough JB, Okada S, Ohtsuka H, Arai H, Yamaki J (1996) Solid State Ionics 92:1–10
22. Manthiram A, Goodenough JB (1989) J Power Sources 26:403–408
23. Padhi AK, Nanjundaswamy KS, Goodenough JB (1997) J Electrochem Soc 144:1188–1194
24. Huang Y, Goodenough JB (2008) Chem Mater 20:7237–7241
25. van Gool W (ed) (1973) Fast ion transport in solids. North-Holland/American Elsevier, Amsterdam/New York, pp 559–590
26. Ji X, Lee KT, Nazar LF (2009) Nature Mater 8:500–506
27. Lu Y, Goodenough JB (2011) J Mater Chem. doi: 10.1039/c0jm04222f
28. Debart A, Paterson AJ, Bao J, Bruce PG (2008) Angew Chem Int Ed 47:4521–4524
29. Lu Y-C, Xu Z, Gasteiger HA, Chen S, Hamad-Schifferli K, Shao-Horn Y (2010) Platinum-gold nanoparticles: a highly active bifunctional electrocatalyst for rechargeable lithium-air batteries. J Am Chem Soc 132:12170–12171. doi:10.102/ja1036572
30. Rychick M, Skyllas-Kazacos M (1988) Characteristics of new all-vanadium redox flow battery. J Power Sources 22:59–67

Books and Reviews

Balbuena PB, Wang Y (eds) (2010) Lithium ion batteries: solid-electrolyte interphase. Imperial College Press/World Scientific, London/Singapore
Bockris JO'M, Reddy AKN (1970) Modern electrochemistry. Plenum, New York
Bruce PG (ed) (1995) Solid state electrochemistry. Cambridge University Press, Cambridge
Goodenough JB, Abruña HD, Buchanan MV (eds) (2007) Basic research needs for electrical energy storage. Office of BES, DOE, Washington http://www.sc.doe.gov/bes/reports/files/EES_rpt.pdf
Hagenmuller P, Van Gool W (eds) (1978) Solid electrolytes. Academic, New York
Huggins RA (2009) Advanced batteries. Springer, New York
Maier J (2004) Physical chemistry of ionic materials. Wiley, Chichester
Rieger PH (1994) Electrochemistry, vol 2. Chapman and Hall, New York
Wakihara M, Yamamoto O (eds) (1998) Lithium ion batteries. VCH, Weinheim
Zhang SS (ed) (2007) Advanced materials and methods for lithium-ion batteries. Transworld Research Network, Trivandrum

Chapter 4
Electrochemical Supercapacitors and Hybrid Systems

Katsuhiko Naoi

Glossary

Activated carbon	Also called activated charcoal or activated coal is a form of carbon that has been processed to make it extremely porous and thus to have a very large surface area available for EDLC, adsorption, or chemical reactions.
Carbon nanotube	Carbon nanotubes (CNTs), not to be confused with carbon fiber, are allotropes of carbon with a cylindrical nanostructure. Nanotubes have been constructed with length-to-diameter ratio of up to 132,000,000:1, significantly larger than any other material. These cylindrical carbon molecules have novel properties, making EDLC performances excellent in power capability.
Electrochemical capacitor (EDLC)	An electric double-layer capacitor, also known as supercapacitor, pseudocapacitor, electric double layer capacitor (EDLC), supercapacitor or ultracapacitor is an electrochemical capacitor with relatively high energy density. Compared to conventional capacitors the energy density is typically on the order of thousands of times greater than an electrolytic capacitor. In comparison with conventional batteries or fuel cells, EDLCs have lower energy density but a much higher power density.
Energy density	

This chapter was originally published as part of the Encyclopedia of Sustainability Science and Technology edited by Robert A. Meyers. DOI:10.1007/978-1-4419-0851-3

K. Naoi (✉)
Institute of Symbiotic Science and Technology, Tokyo University of Agriculture & Technology, 2-24-16 Naka-cho, 184–8558 Koganei, Tokyo, Japan
e-mail: k-naoi@cc.tuat.ac.jp

R.J. Brodd (ed.), *Batteries for Sustainability: Selected Entries from the Encyclopedia of Sustainability Science and Technology*, DOI 10.1007/978-1-4614-5791-6_4,
© Springer Science+Business Media New York 2013

	or Specific energy is defined as the energy per unit mass or volume.
Hybrid (asymmetric) capacitor	A hybrid capacitor consists of a battery-like (faradic) electrode and a capacitor-like (non-faradic) electrode, producing higher working voltage and capacitance. With these systems, one can certainly achieve twice or triple enhancements in energy density compared to the conventional EDLCs.
Lithium-ion capacitor	A Lithium-Ion Capacitor (LIC) is a hybrid type of capacitor. Activated carbon is used as cathode. The anode of the LIC consists of carbon material which is pre-doped with lithium ion. This pre-doping process lowers the potential of the anode and allows a high output voltage and higher energy density.
Nanohybrid capacitor	A new lithium-ion–based hybrid capacitor using the lithium titanate ($Li_4Ti_5O_{12}$) negative intercalation electrode that can operate at unusually high current densities. The high-rate $Li_4Ti_5O_{12}$ negative electrode has a unique nano-structure consisting of extremely small nano-crystalline $Li_4Ti_5O_{12}$ nucleated and grafted onto carbon nano-fiber anchors (nc-$Li_4Ti_5O_{12}$/CNF).
Power density	Power density (or *volume power density* or *volume specific power*) is the amount of power (time rate of energy transfer) per unit volume.

Definition of the Subject

An electrochemical capacitor or electric double-layer capacitor (EDLC), also known as supercapacitor, pseudocapacitor, electrochemical double layer capacitor, or ultracapacitor is an energy storage device with high power and relatively high energy density. Compared to conventional electrolytic capacitors the energy density is typically 3 orders of magnitude greater. In comparison with conventional batteries or fuel cells, EDLCs show lower energy density but have a much higher power density. EDLCs have a variety of commercial applications, notably in energy-smoothing and momentary-load devices. They have applications as energy-storage devices used in vehicles, and applications like wind power solar energy systems where extremely fast charging required.

Historical Background

GE, General Electric engineers experimenting with devices using porous carbon electrodes first observed the EDLC effect in 1957. In 1966, SOHIO Standard Oil of

Ohio developed the modern version of the devices, utilizing two layers of activated charcoal separated by a thin porous insulator. This basic cell design remains the basis of EDLCs. SOHIO failed to commercialize their device, licensing the technology to NEC, who finally marketed "supercapacitors" in 1978, as back-up of a computer memory. Around the mid-1990s various advances in materials science and refinement of the existing systems led to rapidly improving performance and an equally rapid reduction in cost. The first industrial applications were carried out for supporting the energy supply to robots.

In 2005, aerospace company, Diehl Luftfahrt Elektronik GmbH adopted EDLCs to power emergency actuation systems for doors and evacuation slides in airliners, including the new Airbus 380 jumbo jet. As of 2007 all solid state micrometer-scale electric double-layer capacitors based on advanced superionic conductors had been developed for low-voltage electronics.

Introduction

Environmental destruction are getting more and more serious in the global scale. It is urgent for human beings to cope with environmental and energy problems. The national development of each country greatly depends on economic activities oriented to environmental improvement business. Inevitably, in the future, it will be necessary to build an economic society taking into consideration the global ecology, that is, a society in which economic development is closely and directly related with eco-friendly activities such as carbon-reducing and energy-saving movements. Energy storage devices such as batteries and electrochemical capacitors hold an important key to solve the problems. The energy storage devices are some of the best fruits of our environmental technologies, and they have large potential for promoting our international contribution. Specifically, electrochemical capacitors are green devices having such excellent characteristics that they have long lifetimes and can be rapidly charged and discharged [7, 28, 31]. In particular, when combined with renewable energy plants (wind power, wave power and solar cell), they remarkably show their advantage and effectiveness [6]. In view of that, the electrochemical capacitors have been vigorously researched and developed in these days, and are expected to improve their energy densities. Thus, the technology is regarded as a promising art for storing electricity, and hence is expanding its possibilities.

Electrochemical Capacitor and Battery

Batteries store and deliver energy on the basis of chemical reactions, and hence their capacities depend on the redox reactions. In contrast, the energy storage or the charging–discharging mechanism by electrochemical capacitors are based on

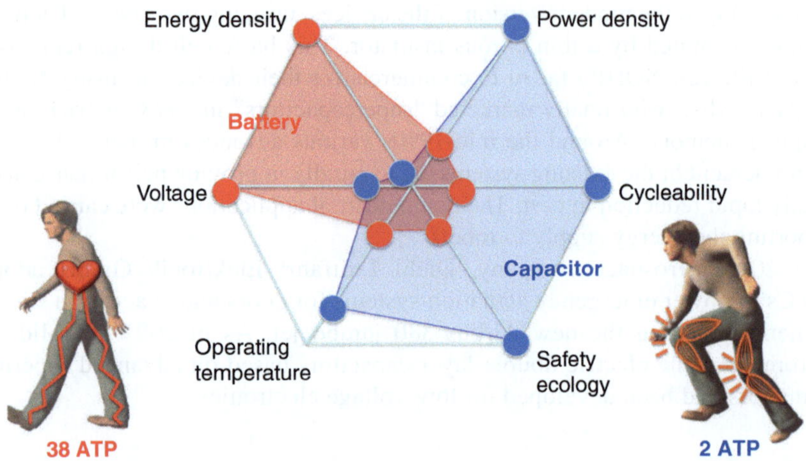

Fig. 4.1 Beautiful contrast of battery and capacitor

physical adsorption/desorption behaviors of ions at the interface of electrode/electrolyte, namely, electric double-layer [3, 6, 7, 28, 31]. Accordingly, the electrochemical capacitors are often called "electric double-layer capacitor (EDLC)," and more recently they are called "Supercapacitor" or "Ultracapacitor" as they increase their performances like six characteristics described below:

1. High power densities capable of instantly charging and discharging.
2. Long lifetimes capable of more than 100,000 charge–discharge cycles. Free from maintenance capable of using for a long time until abandoned.
3. Excellent safety and reliability issues. Low heat generation during operation. Recoverable even after short-circuiting.
4. Wide operational temperature ranges. More than 10°C lower and 20°C higher temperature limits as compared with any battery systems.
5. No fear of material shortage. Activated carbon materials can be obtained from any organic substance.
6. Free of heavy metals and halogens and Low environmental load.

As described above, the electrochemical capacitors have various advantages such as high power, high charge–discharge cycleabilities, high safety, and favorable working temperature ranges. The above characteristics are remarkably contrastive with those of lithium ion batteries as shown in Fig. 4.1. The other cannot replace one of these devices. The contrasting two types of energy devices as represented by lithium ion batteries and electrochemical capacitors can be compared to two different energy-providing systems in a human body, i.e., aerobic reaction system and anaerobic reaction system. It is very interesting to see a human body from the viewpoint of its energy system. In a human body, ATP (adenosine triphosphate) efficiently takes charge of energy production and supply. The ATP is mainly produced by way of two routes (i.e., aerobic reactions and anaerobic reactions). The aerobic reactions, which proceed with oxygen obtained by breathing, slowly produce no less than 38 ATP molecules (from one glucose molecule), to provide large energy. On the other hand, the anaerobic reactions can rapidly produce two ATP molecules, to provide small but instant energy according to necessity. The combination of these systems realizes such smooth energy supply as can correspond to any movement of human body. In the same way, the combinations of lithium ion batteries and capacitors give very high synergy effects. Consequently, energy management can be conducted under any situation by use of the two contrasting energy devices in exquisite combination.

In many uses, lithium ion batteries and capacitors play complementary roles. Since the synergy effects obtained by their combinations are very high, practical examples thereof are increasing in various fields. Further, the capacitors may be combined with devices other than the lithium ion batteries, for example, with fuel batteries or motor/inverters. At present, researchers pay their attention to smart systems employing the combinations of the capacitors and those devices. In consideration of necessity for effective and eco-friendly energy measures, it will be one of the large strategic aims to develop the composite energy storage systems.

Operation Voltage of Electrochemical Capacitors

However, since the EDLCs generally have low energy densities, their uses are limited and they cannot fully meet various performance demands required by the recent markets as shown in Fig. 4.2. Particularly in the field of automobiles, new energy devices are strongly desired to have hybrid characteristics between lithium ion batteries and EDLCs and thereby which can be suitably employed in idle reduction systems. Accordingly, it is expected for them to form a large market [28]. In order to satisfy the performance demands, it is often said to be necessary for the EDLCs to enhance their energy density to 20–30 Wh L^{-1}, which is approximately twice or more than present EDLCs, namely, 5–10 Wh L^{-1}. For realizing the above high energy density, hybrid capacitor systems comprising non-aqueous redox materials are being dynamically researched and developed in recent years [5, 6, 10, 16, 23, 31]. The present chapter deals with the recent contributions to get this high energy density by focusing on two major hybridized cell configurations in organic media.

As is described above, increasing energy density is one of the most crucial matters. For conventional EDLC systems, designed with two symmetrical activated carbon electrodes, increasing voltage is more effective because the energy density increases in proportion to the squared voltage. Thus, it is essential to develop higher electrochemical durability at the electrode/electrolyte interface. Currently, the maximum voltage of an activated carbon-based EDLC (AC/AC) is limited to 2.5–2.7 V in Propylene carbonate (PC) and Acetonitrile (AN)

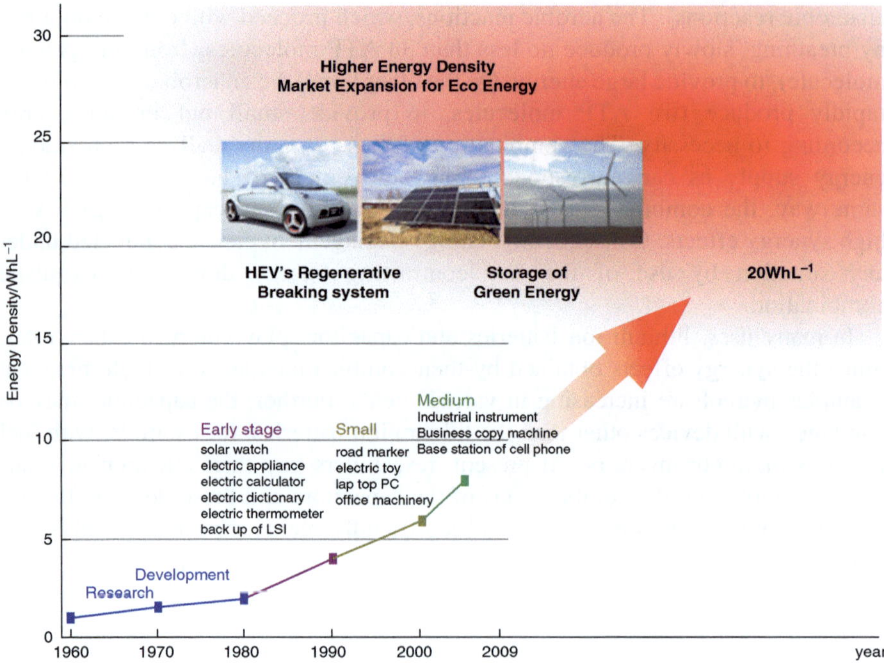

Fig. 4.2 Target value of the next-generation electrochemical capacitor

electrolytes. At applied voltage over 2.7 V, there occurs a significant decrease in capacitance and an increase in internal resistance. In fact the float test revealed that the AC/AC capacitor lose its capacitance ($\Delta C < -13\%$ after 30 days) at 2.5 V, whereas at 2.9 V the capacitance loss became more significant ($\Delta C = -28\%$) for the same duration. The threshold cell voltages between 2.5 and 2.9 V certainly triggers a consecutive and fatal degradation. The undesired faradic process (failure modes) that leads to a capacitance fade is the most critical factor that determines the life of conventional EDLCs and needs to be further investigated as in the recent review article by Simon et al. [28]. Anyway this withstanding-voltage limitation (2.5–2.7 V) certainly exists because of a gas evolution at higher voltages and is a barrier for further enhancement of the energy density of the conventional activated-carbon–based EDLCs.

Attempts to Increase Energy Density of Capacitors

There is presently a major effort to increase the energy density of EDLCs up to a target value in the vicinity of 20–30 Wh kg^{-1} [7]. Many studies have been undertaken in order to enhance the energy density (E) of EDLC. There are mainly

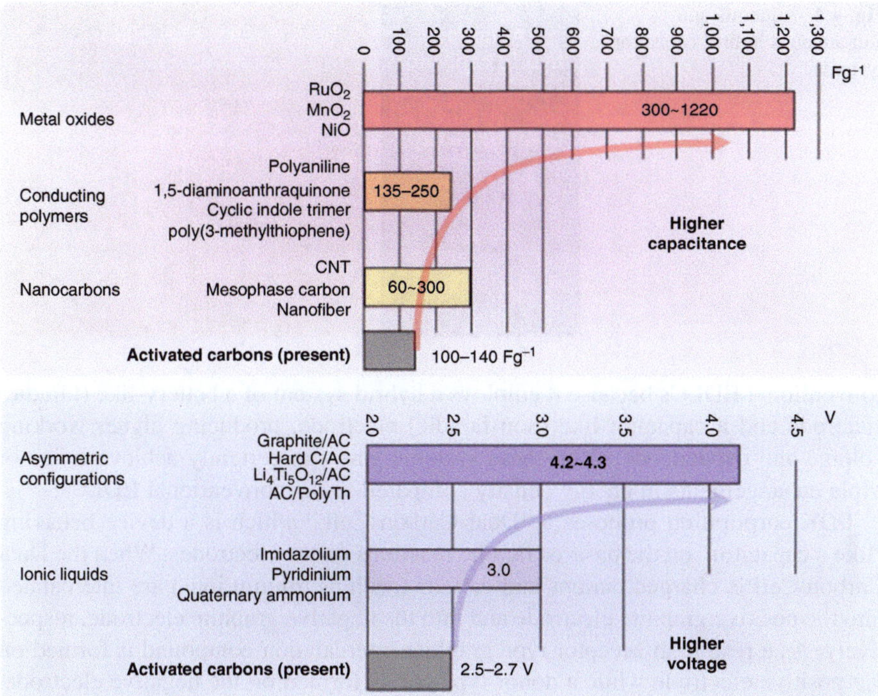

Fig. 4.3 Energy density enhancement of electrochemical capacitors

three approaches (Fig. 4.3): the first is to change the electrode (by higher-capacitance carbons or redox) and the second is to change electrolyte (by durable new electrolyte or ionic liquids).

① High working voltage(V)
② High double layer capacitance(C)

$$E = 1/2\,(CV^2)$$ (c : Capacitance (F); v : Working voltage(V))

Hybrid or Asymmetric Capacitor Systems

The third and the most important approach to meet the goal that is under serious investigation is to develop asymmetric (hybrid) capacitors. As shown in Fig. 4.3 various hybrid capacitor systems are possible by coupling redox-active materials (e.g., graphite [10, 31], metal oxides [5, 14], conducting polymers [17, 18]), and an activated carbon (AC). There are suggested two different systems, one is aqueous and the other is non-aqueous. Some of them are listed below in Fig. 4.4. These approaches can overcome the energy density limitation of the

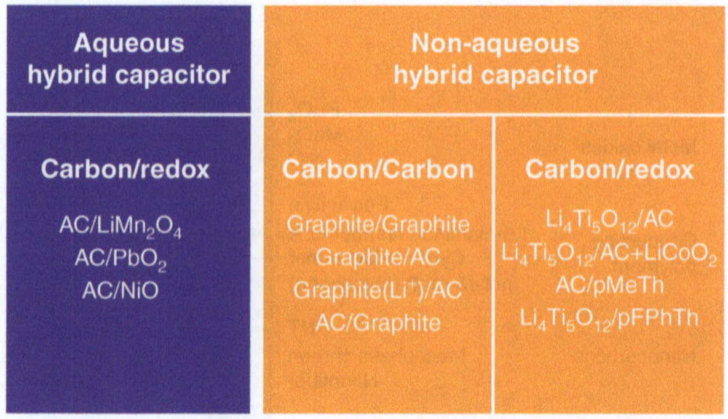

Fig. 4.4 Aqueous and non-aqueous hybrid capacitor systems

conventional EDLCs because it employs a hybrid system of a battery-like (faradic) electrode and a capacitor-like (non-faradic) electrode, producing higher working voltage and capacitance. With these systems, one can certainly achieve twice or triple enhancements in energy density compared to the conventional EDLCs.

FDK corporation proposes a "Dual Carbon Cell," which is a device behaving "like a capacitor" on the basis of faradic reactions on the electrodes. When the Dual Carbon Cell is charged, anions and cations (such as lithium ions) are intercalated into the positive graphite electrode and into the negative graphite electrode, respectively. As a result, an acceptor type graphite intercalation compound is formed on the positive electrode while a donor type one is formed on the negative electrode. Thus, the Dual Carbon Cell is charged or discharged like a capacitor is. It is reported that, if the positive electrode is made of graphite doped with boron, the Dual Carbon Cell can have improved load characteristics.

Meanwhile, many other hybrid capacitors are also proposed each of which utilizes a charge transfer reaction (faradic reaction) on either positive or negative electrode whereby one could have a good scenario attaining higher voltage and higher energy density. Figure 4.5 represents the voltage profiles of the conventional EDLC system (upper), Nanogate (middle), and LIC (lower). Examples of them include a "nanogate" capacitor or a "nanostorage" capacitor (Fig. 4.5, middle), in either of which the negative electrode is an activated carbon but the positive electrode is an amorphous carbon capable of undergoing an anion-intercalation reaction at a high potential (not lower than 4.5 V vs. Li/Li^+), while a normal capacitor comprises both positive and negative electrodes made of activated carbon (Fig. 4.5, upper).

Lithium-Ion Capacitor (LIC)

Among high-energy hybrid capacitors comprising non-aqueous redox materials, a hybrid system called "lithium-ion capacitor (abbreviated as LIC)" has particularly attracted the attention in these days [10, 31]. The LIC is a hybrid

Fig. 4.5 Voltage profiles of a conventional EDLC, Nanogate, and LIC systems

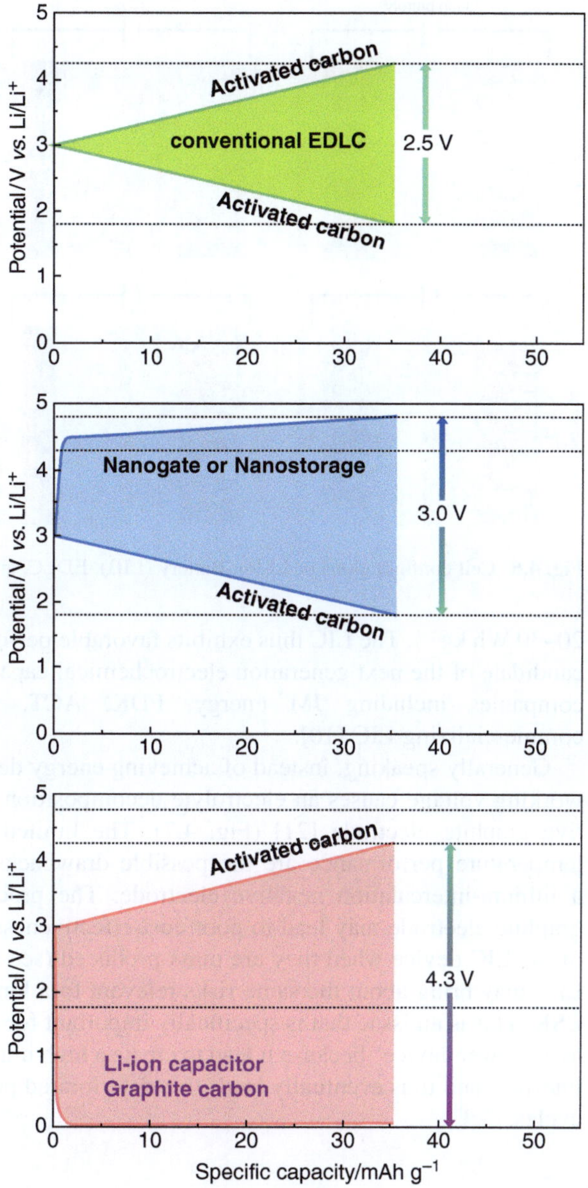

capacitor in which the positive and negative electrodes are made of activated carbon and of graphite pre-doped with lithium ions, respectively (Fig. 4.6).

Since the negative electrode of graphite undergoes the reaction at a potential a little over 0 V vs. Li/Li^+, the LIC has a high working voltage of 3.8 V to 4.0 V. This high working voltage enables the lithium-ion capacitor to realize both a high power density of approximately 5 kW kg^{-1} and an energy density of approximately

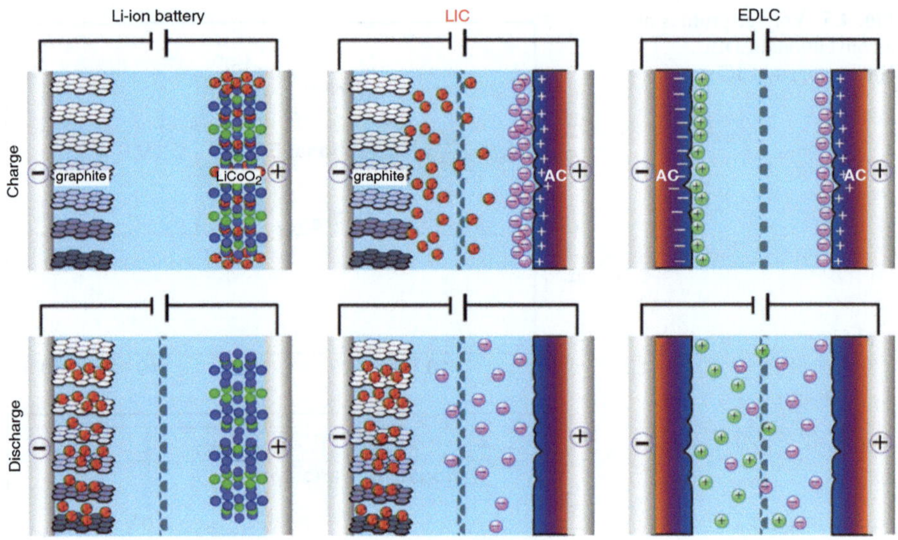

Fig. 4.6 Cell configurations of Li-ion Battery (LIB), EDLC, and LIC

20–30 Wh kg^{-1}. The LIC thus exhibits favorable performance, and is regarded as a candidate of the next generation electrochemical capacitor. Hence, some Japanese companies including JM energy, FDK, ACT, etc. have already started commercializing LIC [10].

Generally speaking, instead of achieving energy density improvement, the high working voltage causes an electrolyte decomposition problem especially at negative graphite electrode [21] (Fig. 4.7). The limited charging rate and the low temperature performance are the possible drawbacks of the LIC because it has a lithium-intercalation negative electrode. The process of pre-lithiation of the graphite electrode may lead to poor cost-effectiveness or instability in the quality of the LIC device when they are mass-produced (see Fig. 4.8). This "pre-doing of Li^{+}" may bring about the same risks relevant to a long-term stability to keep low ESR. This is an issue that is specifically important for an electrochemical capacitor as a "power device" because it lead to create a high impedance electrode/electrolyte interface and thus eventually leads to a deteriorated power performance for longer cycles.

Nanohybrid Capacitor (NHC)

Very recently, Naoi's group and Nippon Chemi-con developed a hybrid capacitor system that certainly achieves a high energy density, high stability, and high safety. This is called "Nanohybrid Capacitor" (abbreviated as NHC) using a super-high-rate nano-structured lithium titanate ($Li_4Ti_5O_{12}$)/carbon composite

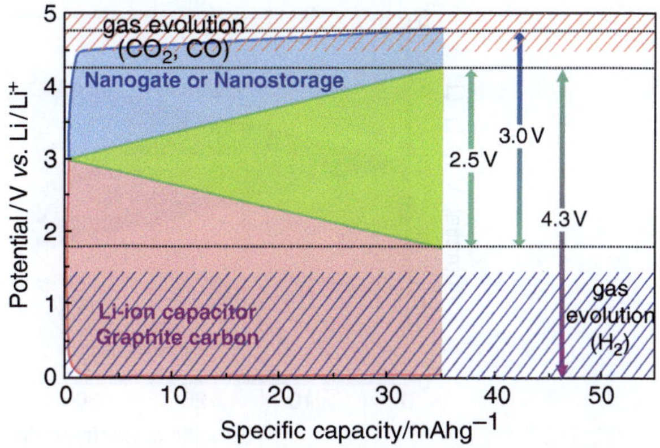

Fig. 4.7 Reliability and electrolyte decomposition at positive and negative electrodes of LIC systems

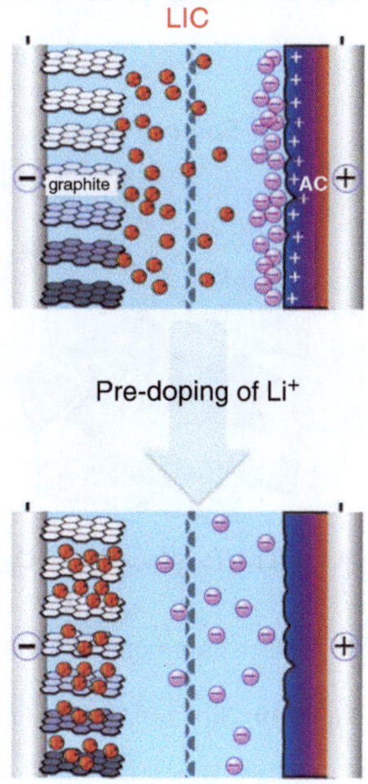

Fig. 4.8 Necessary pre-doing of Li$^+$ for LIC systems

negative electrode. They have kept their eyes on lithium titanate ($Li_4Ti_5O_{12}$) as a stable and safe redox material capable of increasing the energy densities of hybrid capacitors without sacrificing interfacial characteristics. The $Li_4Ti_5O_{12}$ operates at a potential (1.55 V *vs* Li/Li$^+$) out of the range where the electrolyte solution may be decomposed, and hence play key roles for providing capacitor systems as stable and safe as EDLCs (see Fig. 4.9). The $Li_4Ti_5O_{12}$ as a redox material for hybrid capacitors has the following essential advantages in energy density, in stability, and in safety:

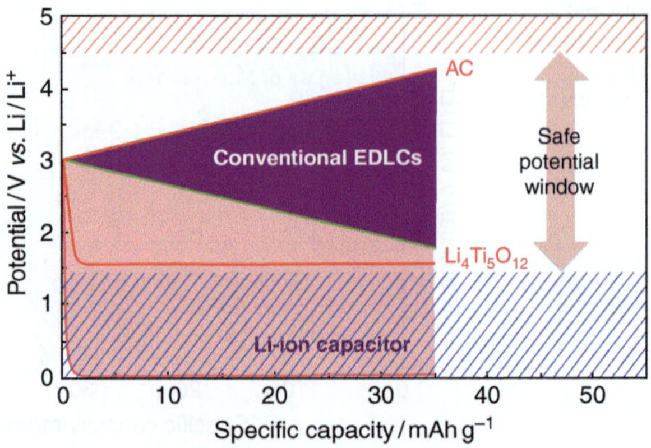

Fig. 4.9 Voltage profile of conventional EDLC, LIC, and NHC

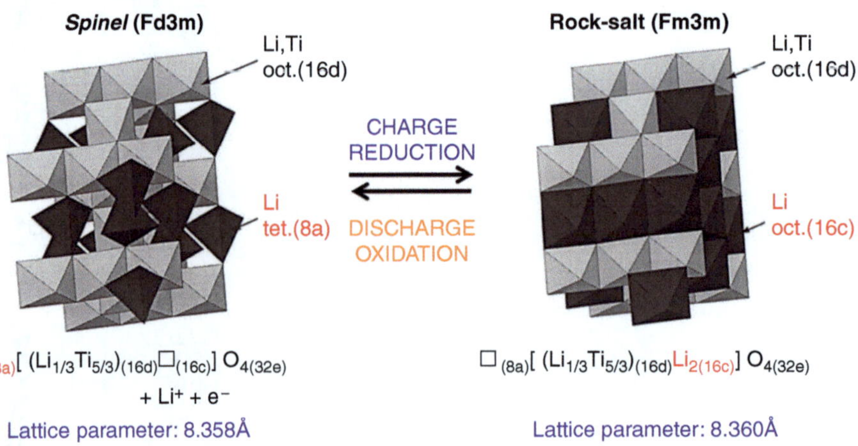

Fig. 4.10 Structural and volumetric changes at the charged and discharged states of $Li_4Ti_5O_{12}$

1. It exhibits a high coulombic efficiency close to 100% during charge–discharge cycle
2. It has a theoretical capacity (175 mAh g^{-1}) four times higher than as activated carbon
3. It undergoes charging and discharging at a constant potential of 1.55 V vs. Li/Li^+, where the electrolyte solution is free from the fear of decomposition (no SEI formation and no gas evolution) [20, 27]
4. As shown in Fig. 4.10, it changes the volume to a very small degree (0.2%) in charging or discharging in intercalating or deintercalating Li^+ ions (zero-strain insertion)
5. Inexpensive raw material

Table 4.1 Comparison of LIC and NHC in their cell configurations

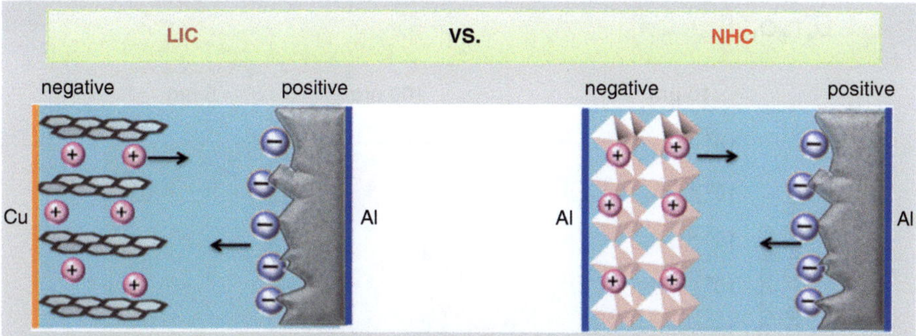

LIC		NHC
AC	Positive electrode	AC, CNT
Graphite/HC/PAC	Negative eletrode	nc-$Li_4Ti_5O_{12}$/CB, CNF, CNT
Li pre-doping necessary	Manufac. procedure	No Li pre-doping
Li-salt / EC, PC	Electrolyte	Li-salt / PC, DMC, AN
Higher than EDLC	Internal resistance	EDLC comparable
Al (pos), Cu hole (neg)	Current collector	Al (pos & neg)
–10°C	Low-temp. perform.	–40°C

As summarized in Table 4.1, a capacitor cell comprising $Li_4Ti_5O_{12}$ is, therefore, expected to be free from necessity for pre-doping of lithium ions (because of the above advantage (1)), to be capable of having a high energy density (because of the advantage (2)), and to be excellent in stability and in safety (because of the advantages (3) and (4)).

Amatucci et al. firstly introduced the $Li_4Ti_5O_{12}$/AC system as safer hybrid systems [1, 2]. However, the conventional $Li_4Ti_5O_{12}$ has the greatest problem of low power characteristics that stem from inherent poor Li^+ diffusion coefficient ($<10^{-6}\,cm^2\,s^{-1}$) [29] and poor electronic conductivity ($<10^{-13}\,\Omega^{-1}\,cm^{-1}$) [8]. Accordingly $Li_4Ti_5O_{12}$ had a big problem of slow output characteristics and was not fully developed for the application in electrochemical capacitors at that time. For the purpose of solving the problem of poor output performance, some measures can be thought of. For example, the $Li_4Ti_5O_{12}$ may be ground down to nano-sized particles and may be combined with an electro-conductive material to prepare a composite.

In the nano-sized $Li_4Ti_5O_{12}$ particles, it is expected that Li^+ ions diffuse and electrons migrate in distances reduced by 1/1,000 (see Figs. 4.11 and 4.12) or less as compared with the distances in normal micron-order $Li_4Ti_5O_{12}$ particles. Further, in the composite, electron paths are effectively formed in the electrode of nano-sized $Li_4Ti_5O_{12}$ particles (Fig. 4.13). On the basis of these expectations, a composite (nc-$Li_4Ti_5O_{12}$/CNF) has been synthesized by means of a new method referred to as "Ultracentrifugal force (UC) nanohybridization method" [19, 24]. Specifically, the composite (nc-$Li_4Ti_5O_{12}$/CNF) comprises nano-crystalline $Li_4Ti_5O_{12}$ particles hyper-dispersed on carbon nano-fibers (CNFs) [13] having high electro-conductivity ($25\,\Omega^{-1}\cdot cm^{-1}$). They utilized the composite as the negative electrode active material,

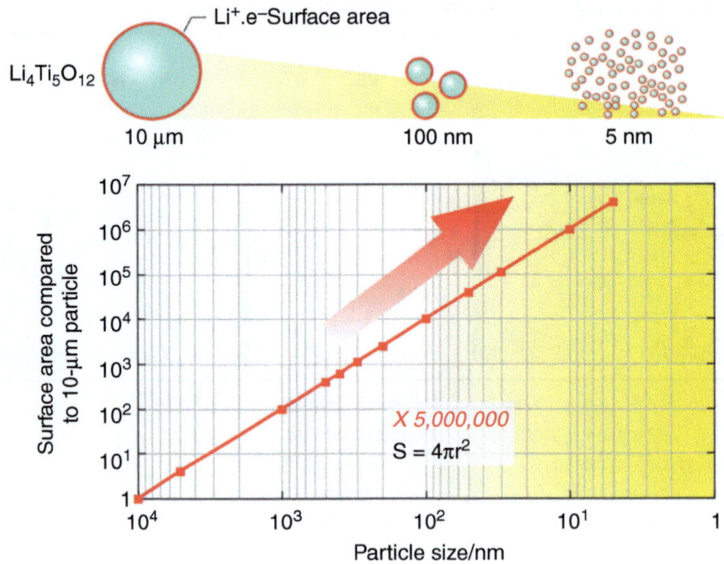

Fig. 4.11 Increased accessibility and surface area as a function of the particle size of $Li_4Ti_5O_{12}$

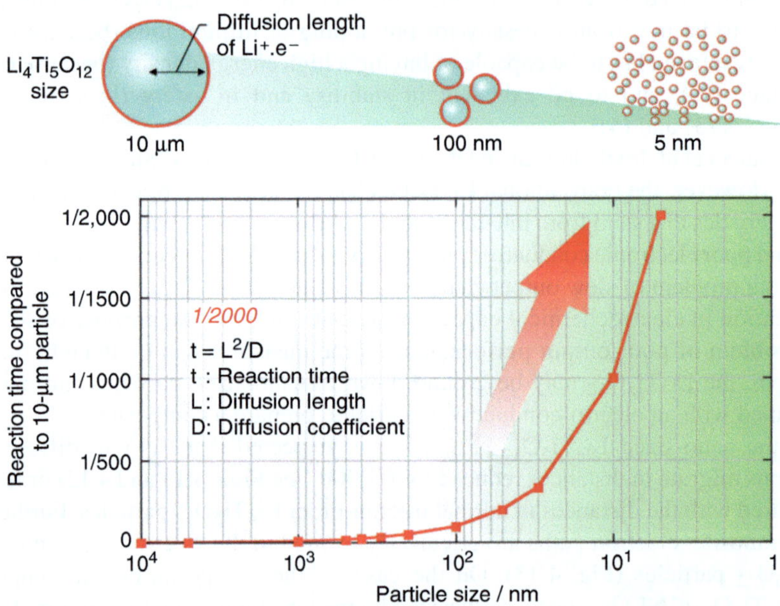

Fig. 4.12 Decreased diffusion length as a function of the particle size of $Li_4Ti_5O_{12}$

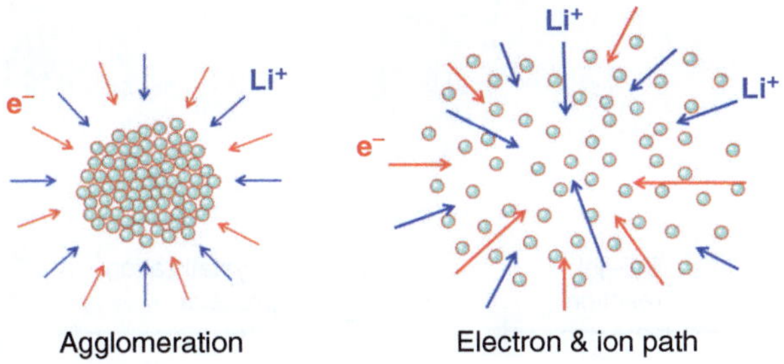

Fig. 4.13 Hyper-dispersion of the nano-size particles of $Li_4Ti_5O_{12}$

and thereby succeeded in producing a novel hybrid capacitor (nanohybrid capacitor) realizing both high power and high energy density.

The nc-$Li_4Ti_5O_{12}$ negative electrode was developed to have a unique nano-structure that can operate at unusually high current densities. Nano-crystalline $Li_4Ti_5O_{12}$ attached onto carbon nano-fibers were prepared by a unique technique (UC method) of a mechano-chemical sol–gel reaction under ultracentrifugal force field [13, 19, 24], followed by an instantaneous heat treatment for very short duration (see Fig. 4.14a). The UC method is induced to form, anchor, and graft the nano-Li-Ti-O precursors on the CNF matrices. The subsequent instantaneous heat treatment is of prime importance to precisely achieve all of the following: high crystallization of $Li_4Ti_5O_{12}$, [13] inhibition of the CNF oxidation decomposition during the annealing at high temperature, [31] and suppression of the agglomeration of the $Li_4Ti_5O_{12}$ particles. These processes are quite simple and require only a few minutes. Actually, the power characteristic of the prepared composite (nc-$Li_4Ti_5O_{12}$/CNF) made a new bench mark (charge-discharge time = 12 s) [4, 11, 12, 15, 24, 32] which exceeds greatly the shortest charge–discharge time that has ever been attained (see Fig. 4.14b).

XRD analysis was performed to confirm the formation of the nc-$Li_4Ti_5O_{12}$ and the presence of the CNF in the nc-$Li_4Ti_5O_{12}$/CNF composites. Figure 4.15a shows the XRD patterns of the prepared nc-$Li_4Ti_5O_{12}$/CNF composite and pristine CNF. The composite has several sharp diffraction peaks at $2\theta = 18, 35, 42, 57$, and $63°$. These peaks correspond to (111), (311), (400), (511), and (440) planes of a face-centered cubic spinel structure with $Fd\bar{3}m$ space group [22, 30], respectively, indicative of the formation of crystalline $Li_4Ti_5O_{12}$. Annealing at 900°C under vacuum well crystallizes the nano-particles of $Li_4Ti_5O_{12}$. A broad peak at around $2\theta = 24.5^\circ$ is observed which corresponds to the (002) plane of the pristine CNF [33]. This means that the CNF exists in the annealed composite and preserves its graphene layer structure. The fact that there are no other peaks observed corresponding to some possible impurities such as TiO_2, Li_2CO_3, and Li_2TiO_3 [9, 26] suggests that there are only two species, crystalline $Li_4Ti_5O_{12}$ and CNF.

Thermogravimetric (TG) measurement of the nc-$Li_4Ti_5O_{12}$/CNF composite was performed under air to estimate the residual weight ratio of the CNF. The obtained TG curve is shown in Fig. 4.15b. The weight loss at 400–600°C resulted

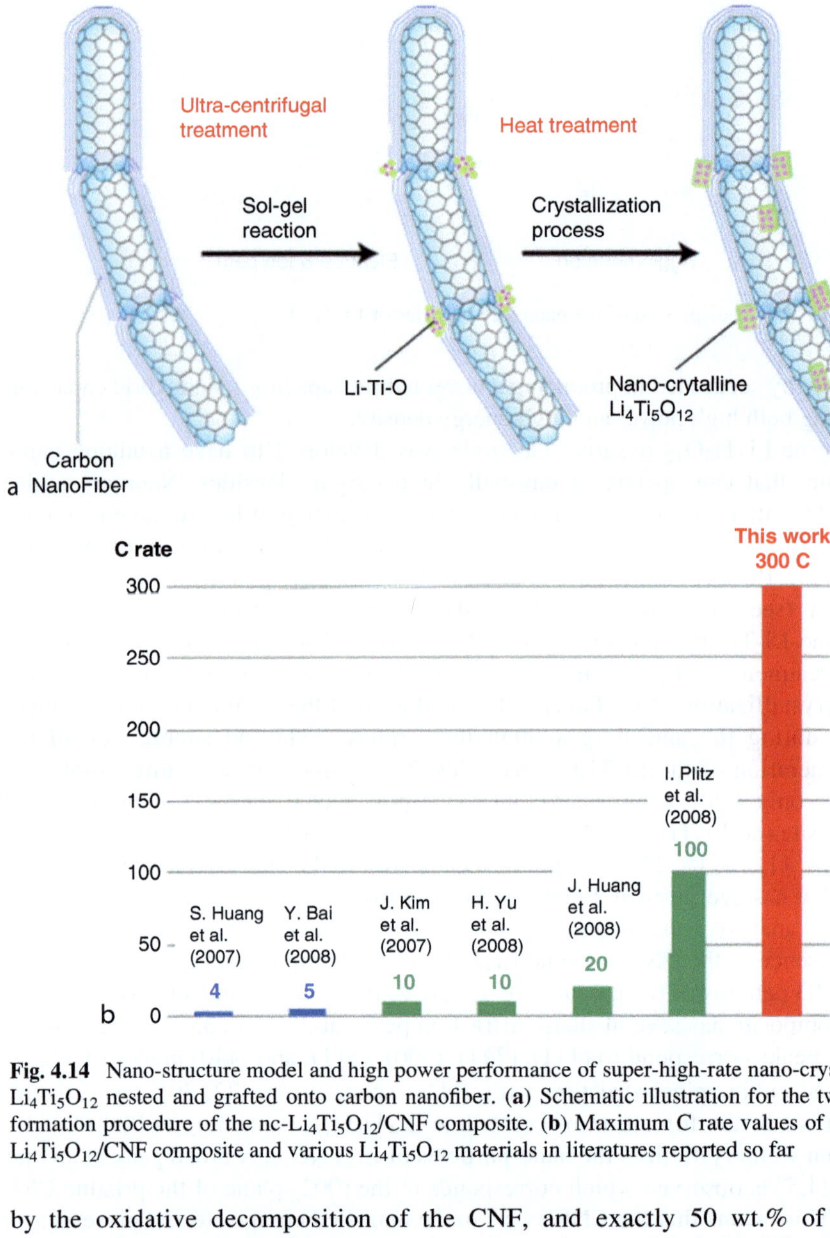

Fig. 4.14 Nano-structure model and high power performance of super-high-rate nano-crystalline $Li_4Ti_5O_{12}$ nested and grafted onto carbon nanofiber. (**a**) Schematic illustration for the two-step formation procedure of the nc-$Li_4Ti_5O_{12}$/CNF composite. (**b**) Maximum C rate values of the nc-$Li_4Ti_5O_{12}$/CNF composite and various $Li_4Ti_5O_{12}$ materials in literatures reported so far

by the oxidative decomposition of the CNF, and exactly 50 wt.% of the -nc-$Li_4Ti_5O_{12}$ remained. The 50 wt.% is well consistent with the weight ratio of the $Li_4Ti_5O_{12}$ to the CNF calculated on the basis of dosed Ti alkoxide weight before UC method. This fact implies that sol–gel reaction in the UC method stochiometrically proceeds, and the optimized (very short duration) annealing does not cause oxidative decomposition of the CNF. Such a stoichiometric preparation process (UC method and instantaneous annealing) is one of the important factors for cost-effectiveness of capacitor production.

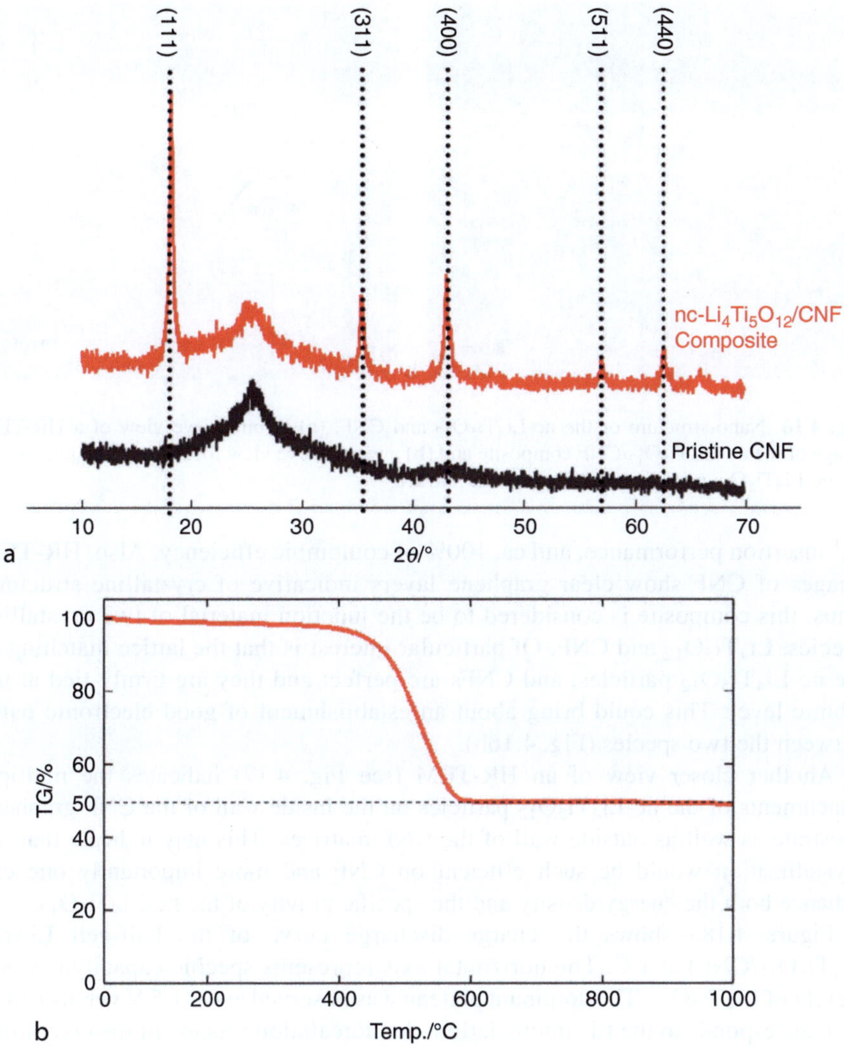

Fig. 4.15 Crystallinity and content of nc-$Li_4Ti_5O_{12}$ in the composite. (**a**) XRD patterns of the nc-$Li_4Ti_5O_{12}$/CNF composite and pristine CNF. (**b**) TG curve of the nc-$Li_4Ti_5O_{12}$/CNF composite at 1°C min^{-1} under air. The residual weight ratio corresponds to the content of nc-$Li_4Ti_5O_{12}$ in the composite

The nano-structures and crystallinity of nc-$Li_4Ti_5O_{12}$ and CNF in the composite is observed by HR-TEM (see Fig. 4.16). The image indicates that the edge or defect sites of CNF graphenes accommodate and graft nc-$Li_4Ti_5O_{12}$ particles. The results support that the $Li_4Ti_5O_{12}$ particles are formed through a nucleation process onto the edge and defect CNF sites. The clear facet of $Li_4Ti_5O_{12}$ reflects the high crystallinity that is consistent with the sharp XRD spectrum (see Fig. 4.16a), despite such a nanosize. Such a high crystallinity resulted in a reversible, smooth

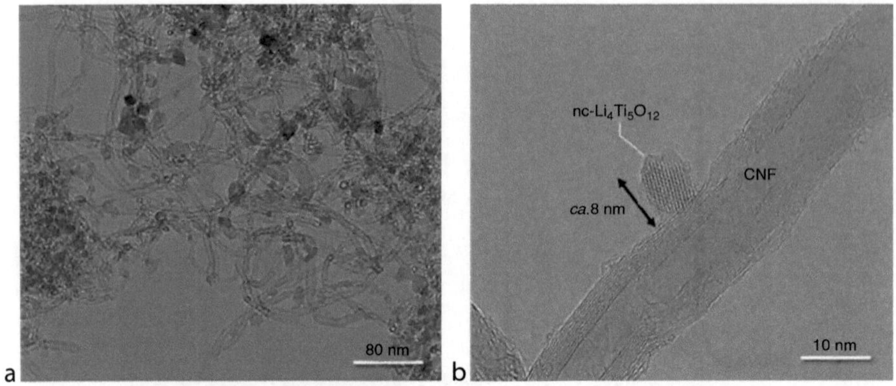

Fig. 4.16 Nano-structure of the nc-$Li_4Ti_5O_{12}$ and CNF. (**a**) A bird's eye view of a HR-TEM image of the nc-$Li_4Ti_5O_{12}$/CNF composite and (**b**) a worm's eye view focusing on the junction of the nc-$Li_4Ti_5O_{12}$ particle on CNF graphene surface

Li^+ insertion performance, and ca. 100% of coulombic efficiency. Also, HR-TEM images of CNF show clear graphene layers indicative of crystalline structure. Thus, this composite is considered to be the junction material of two crystalline species, $Li_4Ti_5O_{12}$ and CNF. Of particular interest is that the lattice matching of the nc-$Li_4Ti_5O_{12}$ particles, and CNFs are perfect and they are firmly tied at the atomic level. This could bring about an establishment of good electronic paths between the two species (Fig. 4.16b).

Another closer view of an HR-TEM (see Fig. 4.17) indicates the multiple attachments of the nc-$Li_4Ti_5O_{12}$ particles on the inside wall of the CNF graphene substrate as well as outside wall of the CNF matrices. This may indicate that the crystallization would be such efficient on CNF and more importantly one can enhance both the energy density and the specific gravity of the nc-$Li_4Ti_5O_{12}$.

Figure 4.18a shows the charge–discharge curve of the half-cell Li/(nc-$Li_4Ti_5O_{12}$/CNF) at 1 C. The horizontal axis represents specific capacity per unit weight of $Li_4Ti_5O_{12}$. The dominant plateau was observed at *ca.* 1.5 V versus Li/Li^+ that corresponds to the Li^+ intercalation–deintercalation process of the crystalline $Li_4Ti_5O_{12}$ [13, 24, 30], indicating that the capacity of the composite is determined by the redox capacity of nc-$Li_4Ti_5O_{12}$ in the composite. The obtained capacity was 167 mAh g^{-1} per $Li_4Ti_5O_{12}$, which is 95% of the theoretical capacity. It is noted here that the value of 167 mAh g^{-1} is obtained after subtracting the double-layer capacity of the CNF (8 mAh g^{-1}). This result indicates that almost all of the nc-$Li_4Ti_5O_{12}$ particles in the composite are electrochemically active, meaning that ionic and electric paths are fully established in the composite. The rate capability of the obtained composite is shown in Fig. 4.18b.

Even at a high rate of 300 C, the composite shows reversible capacity of 158 mAh g^{-1} per $Li_4Ti_5O_{12}$ which correspond to 95% of the capacity obtained at 1 C. Such an excellent rate-capability indicates that the optimized nano-structure of the nc-$Li_4Ti_5O_{12}$/CNF composites as observed in HR-TEM image well overcome the

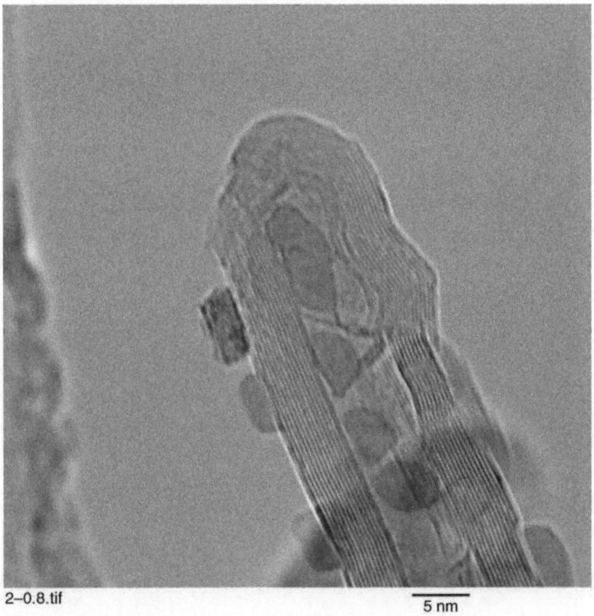

Fig. 4.17 Closer view of a HR-TEM image of multiple attachments of nc-$Li_4Ti_5O_{12}$ inside and outside wall of the CNF composite

inherent problems of $Li_4Ti_5O_{12}$ materials like poor Li^+ diffusivity and poor electronic conductivity. Probably, this is because the nano-crystallized $Li_4Ti_5O_{12}$ and $Li_4Ti_5O_{12}$/CNF junctions lead to facile ionic diffusion and electronic conduction, respectively. The cycleability of the composite is shown in Fig. 4.18c. Even after 9,000 cycles, 90% of the initial capacity is maintained, showing that the composite is electrochemically stable. The result strongly suggests that the aggregation and detachment of the nc-$Li_4Ti_5O_{12}$ particles hardly happen when they are operated at high-rate charge–discharge for a prolonged cycling.

Figure 4.19 shows Ragone plots obtained from charge–discharge measurements as laminate-type cell of the hybrid capacitor system ((nc-$Li_4Ti_5O_{12}$/CNF)/$LiBF_4$-PC/AC). The charge–discharge was performed between 1.5 and 3.0 V at various current densities ranging 0.2–30 mA cm^{-2} (0.18–26.8 A g^{-1}). For comparison, a conventional EDLC system (AC/$TEABF_4$-PC/AC) was also assembled and measured between 0 and 2.5 V. In a low power density range of 0.1–1 kW L^{-1}, the hybrid capacitor shows the energy density as high as 28–30 Wh L^{-1}, which is a value comparable to that of the Li-ion capacitors [31]. Even at a high power of 6 kW L^{-1}, the energy density of the hybrid capacitor remains at 15 Wh L^{-1} which is double that of the conventional EDLC system (AC/AC). The result reveals that the capacitor system can provide higher energy as compared with conventional EDLCs not only in the low power density region (0.1–1 kW L^{-1}) but also in the high power density region (1–6 kW L^{-1}). Accordingly, this configuration of capacitor system is anticipated as an energy device utilizable for both high energy and high power applications.

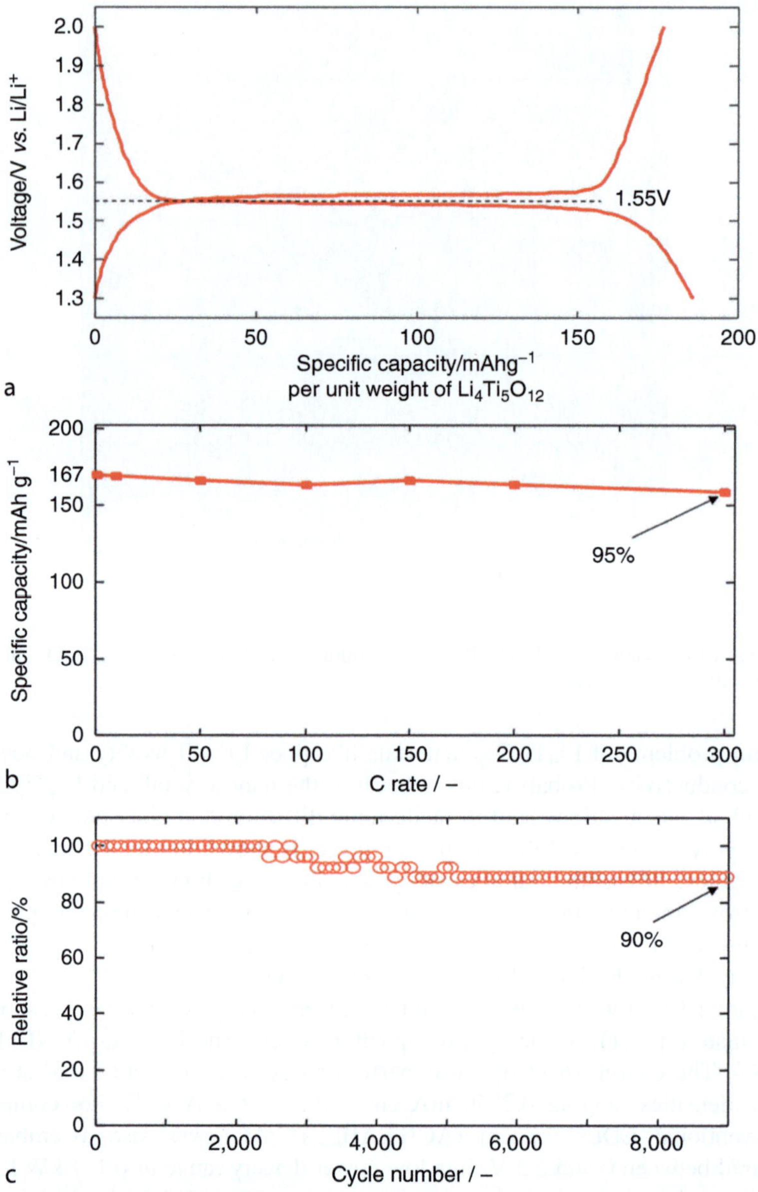

Fig. 4.18 Electrochemical properties of the nc-$Li_4Ti_5O_{12}$/CNF composite. (**a**) Charge–discharge curve of nc-$Li_4Ti_5O_{12}$/CNF composite between 1.0 and 3.0 V versus Li/Li^+ at 1 C. (**b**) Rate capability of the nc-$Li_4Ti_5O_{12}$/CNF composite in a C-rate ranges from 1 C to 300 C. (**c**) Cycleability of nc-$Li_4Ti_5O_{12}$/CNF composite between 1.0 and 3.0 V versus Li/Li^+ at 20 C. 1 M $LiBF_4$/EC + DEC(1:1) was used as the electrolyte

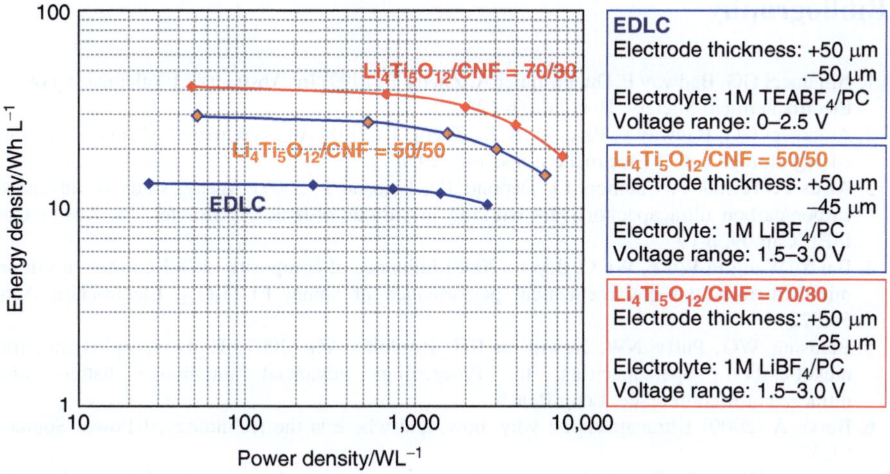

Fig. 4.19 Ragone plots of Nanohybrid capacitor systems ((nc-$Li_4Ti_5O_{12}$/CNF)/AC) with two different $Li_4Ti_5O_{12}$ loadings (50% and 70%) and conventional EDLC system (AC/AC) all in 1 M $LiBF_4$/PC

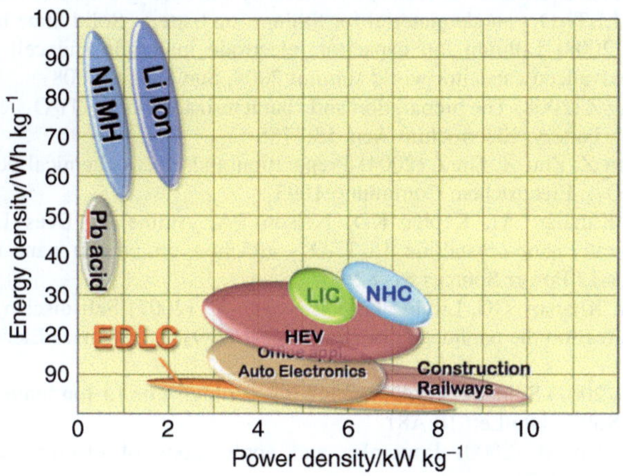

Fig. 4.20 Performances of lithium-ion capacitor (LIC) and nanohybrid capacitor (NHC) in Ragone plot

Future Perspectives

Hybrid capacitor systems are the promising approach to meet the goal to effectively increase the energy density. The investigation to develop hybrid capacitors has been initiated by "Li-ion capacitors (LIC)." And, now "Nanohybrid capacitor (NHC)" certainly achieves as high energy density as Li-ion capacitors with higher stability, higher safety, and higher productivity (see Fig. 4.20). Both LIC and NHC attain three-times higher energy density and are regarded as the next-generation supercapacitor systems.

Bibliography

1. Amatucci GG, Badway F, Du. Pasquier A, Zheng T (1999) In: Abstract of 196th meeting of the Electrochemical Society, Honolulu, p 122
2. Amatucci GG, Badway F, Pasquier ADu, Zheng T (2001) An asymmetric hybrid nonaqueous energy storage cell. J Electrochem Soc 148:A930
3. Azaïs P, Tetrais F, Caumont O, Depond JD, Lejosne J (2009) Ageing study of advanced carbon/carbon ultracapacitor cells working in various organic electrolytes. In: Abstract of ISEE'Cap 09, p 19
4. Bai Y, Wang F, Wu F, Wu C, Bao L (2008) Influence of composite LiCl–KCl molten salt on microstructure and electrochemical performance of spinel $Li_4Ti_5O_{12}$. Electrochim Acta 54:322
5. Baldsing WG, Puffy NW, Newnham RH, Pandolfo AG (2007) High-energy asymmetric nickel-carbon supercapacitors. In: Proceedings advanced automotive battery and ultracapacitor conference, Long Beach
6. Burke A (2000) Ultracapacitors: why, how, and where is the technology. J Power Sources 91:37
7. Burke A (2007) R&D considerations for the performance and application of electrochemical capacitors. Electrochim Acta 53:1083
8. Chen CH, Vaughey JT, Jansen AN, Dees DW, Kahaian AJ, Goacher T, Thackeray MM (2001) Studies of Mg-substituted $Li_{4-x}Mg_xTi_5O_{12}$ spinel electrodes (0x1) for lithium batteries. J Electrochem Soc 148:A102
9. Hao Y, Lai Q, Xu Z, Liu X, Ji X (2005) Synthesis by TEA sol–gel method and electrochemical properties of $Li_4Ti_5O_{12}$ anode material for lithium-ion battery. Solid State Ionics 176:1201
10. Hatozaki O (2008) Lithium ion capacitor: electrode materials and cell performance. In: Proceedings advanced capacitor world summit 2008, San Diego (2008)
11. Huang J, Jiang Z (2008) The preparation and characterization of $Li_4Ti_5O_{12}$/carbon nano-tubes for lithium ion battery. Electrochim Acta 53:7756
12. Huang S, Wen Z, Zhu X, Gu Z (2004) Preparation and electrochemical performance of Ag doped $Li_4Ti_5O_{12}$. Electrochem Commun 6:1093
13. Jansen AN, Kahaian AJ, Kepler KD, Nelson PA, Amine K, Dees DW, Vissers DR (1999) High-rate nano-crystalline $Li_4Ti_5O_{12}$ attached on carbon nano-fibers for hybrid supercapacitors. J Power Sources 81–82:902
14. Kazaryan SA, Kharsov GG, Litvinrnko SV, Kogan VI (2007) Self-discharge related to iron ions and its effect on the parameters of HES $PbO_2|H_2SO_4|C$ Systems. J Electrochem Soc 154: A751
15. Kim J, Cho J (2007) Spinel $Li_4Ti_5O_{12}$ nanowires for high-rate Li-Ion intercalation electrode. Electrochem Solid State Lett 10:A81
16. Koetz R, Carlen M (2000) Principles and applications of electrochemical capacitors. Electrochim Acta 45:2483
17. Laforgue A, Simon P, Fauvarque JF, Mastrangostino M, Soavi F, Sarrau JF, Lailler P, Conte M, Rossi E, Saguatti S (2003) Activated carbon/conducting polymer hybrid supercapacitors. J Electrochem Soc 150:A645
18. Machida K, Suematsu S, Ishimoto S, Tamamitsu K (2008) High-voltage asymmetric electrochemical capacitor based on polyfluorene nanocomposite and activated carbon. J Electrochem Soc 155:A970
19. Naoi K, Ishimoto S, Ogihara N, Nakagawa Y, Hatta S (2009) Encapsulation of nanodot ruthenium oxide into KB for electrochemical capacitors. J Electrochem Soc 156:A52
20. Naoi K, Simon P (2008) New materials and new configurations for advanced electrochemical capacitors. Interface 17:34
21. Nikkei Electronics (2008) LIC prevails EDLCs?! 991:77
22. Ohzuku T, Ueda A, Yamamoto N (1995) Zero-strain insertion material of $Li[Li_{1/3}Ti_{5/3}]O_4$ for rechargeable lithium cells. J Electrochem Soc 142:1431

23. Pandolfo AG, Hollenkamp AF (2006) Carbon properties and their role in supercapacitors. J Power Sources 157:11
24. Plitz I, Dupasquier A, Badway F, Gural J, Pereira N, Gmitter A, Amatucci GG (2006) Nanohybrid capacitor: The next generation electrochemical capacitors. Appl Phys A 82:615
25. Scharner S, Weppner W, Schmind-Beurmann P (1999) Evidence of two-phase formation upon lithium insertion into the $Li_{1.33}Ti_{1.67}O_4$ spinel. J Electrochem Soc 146:857
26. Shen CM, Zhang XG, Zhou YK, Li HL (2002) Preparation and characterization of nanocrystalline $Li_4Ti_5O_{12}$ by sol–gel method. Mater Chem Phys 78:437
27. Shu J (2008) Study of the interface between $Li_4Ti_5O_{12}$ electrodes and standard electrolyte solutions in 0.0–5.0 V. Electrochem Solid State Lett 11:A238
28. Simon P, Gogotsi Y (2008) Materials for electrochemical capacitors. Nat Mater 7:845
29. Takai S, Kamata M, Fujiine S, Yoneda K, Kanda K, Esaka T (1999) Diffusion coefficient measurement of lithium ion in sintered $Li_{1.33}Ti_{1.67}O_4$ by means of neutron radiography. Solid State Ionics 123:165
30. Thackeray MM (1995) Structural considerations of layered and spinel lithiated oxides for lithium ion batteries. J Electrochem Soc 142:2558
31. Yoshino A, Tsubata T, Shimoyamada M, Satake H, Okano Y, Mori S, Yata S (2004) Development of a lithium-type advanced energy storage device. J Electrochem Soc 151:A2180
32. Yu H, Zhang X, Jalbout AF, Yan X, Pan X, Xie H, Wang R (2008) High-rate characteristics of novel anode $Li_4Ti_5O_{12}$/polyacene materials for Li-ion secondary batteries. Electrochim Acta 53:4200
33. Zhou JH, Sui ZJ, Li P, Chen D, Dai YC, Yuan WK (2006) Structural characterization of carbon nanofibers formed from different carbon-containing gases. Carbon 44:3255

Chapter 5
Lead Acid Battery Systems and Technology for Sustainable Energy

Kathryn R. Bullock

Glossary

AGM	Absorptive glass mat, a battery separator material.
Ah	Ampere-hour: dc current multiplied by time of charge or discharge.
Monopolar cell	String of cells connected in series (+ − + − + −).
Bipolar plate	Conductive, nonporous substrate with negative active material on one side and positive active material on the opposite side.
e^-	Electron.
Float V	Voltage applied to sustain battery Ah capacity.
Flooded cell	Lead-acid cell saturated with aqueous sulfuric acid electrolyte.
Gel cell	Lead-acid cell with a gelling agent added to the electrolyte.
HEV	Hybrid electric vehicle.
KVA	Kilovolt amperes, unit of electrical energy in an ac circuit.
H_2SO_4	Reactant at both electrodes; electrolyte when aqueous.
MSDS	Material safety data sheet (for information on a battery product).
Pb	Lead metal, the main negative electro-active material.
PbO_2	Lead dioxide, the main positive electro-active material.
$PbSO_4$	Lead sulfate, the discharge product on both electrodes.
SLI	Starting, lighting and ignition automotive battery.
UPS	Uninterruptible power system.
VRLA cell	Lead-acid cell with one-way pressure-relief valve.

This chapter was originally published as part of the Encyclopedia of Sustainability Science and Technology edited by Robert A. Meyers. DOI:10.1007/978-1-4419-0851-3

K.R. Bullock (✉)
President, Coolohm, Inc., 980 Clover Court, Blue Bell, PA 19422-3015, USA
e-mail: coolohm@verizon.net

R.J. Brodd (ed.), *Batteries for Sustainability: Selected Entries from the Encyclopedia of Sustainability Science and Technology*, DOI 10.1007/978-1-4614-5791-6_5,
© Springer Science+Business Media New York 2013

Definition of the Subject

Work on optimizing battery designs to fit the needs of each emerging application has been an ongoing process since Gaston Planté first demonstrated the lead-acid battery in France in 1859 [1]. This article describes many different commercial lead-acid battery designs and electrical requirements in a wide range of applications. Commercial lead-acid batteries are increasingly used for sustainable energy storage and power system regulation. Their global availability and the low cost of their components, their reliability under many operating conditions and their established recycling industry are among the reasons that the technology is finding additional markets in sustainable energy systems.

Adapting battery designs to further optimize their performance in new applications has been a successful international effort for 150 years and continues to generate improved battery designs and power systems for the future.

Introduction

A wide range of designs and sizes of lead-acid batteries are manufactured for traditional markets. Examples of applications are automotive vehicle starting, lighting and ignition; stand-by power back-up for electrical and nuclear energy and safety systems; and vehicle propulsion. Growing use of lead-acid batteries for storing sustainable energy has led to new designs with improved performance and longevity. New lead-acid battery designs for hybrid electric vehicles are under development. A recent 144 V Ultra battery design with negative plates made of carbon and lead in a parallel configuration exceeded 100,000 miles when substituted for nickel metal hydride battery in a Honda Insight [2].

The electro-active ingredients of lead-acid batteries, based on lead, sulfuric acid, water and oxygen, are available around the globe. Small amounts of other common materials such as carbon, barium, and lignin from wood are also added. Plate separators, battery cases and other inert parts are generally made from glass and common plastics, such as polypropylene and polyethylene. Another environmental advantage is that a well-developed commercial battery recycling system has a proven record of recycling more than 97% of spent lead-acid cells and batteries in the United States and Europe. These advantages are major reasons why the lead-acid battery has remained the most widely used energy storage device for large-power sustainable energy systems.

Commercial designs range in size from single cylindrical 2-V "D" cells for portable equipment to large strings of prismatic battery modules for both stationary and motive power. A 6 V battery is achieved with a series connection of three 2-V cells in a string, in contrast to the four cells that are required for 6-V nickel-metal hydride or nickel-cadmium batteries. Higher lead-acid battery voltages in multiples of two are made by adding more cells to the string. Batteries for cars with gasoline engines or micro-hybrid systems typically have 6 cells connected in series to produce 12 V. DC standby-power systems that back-up telecommunication systems are usually 24 or 48 V modules. Medium hybrid electric vehicles (HEVs) with power-assist systems have a battery voltage in the range of 100–150 V. Some large stationary applications, as well as full hybrid electric vehicles, require a battery voltage over 200. The practice of balancing individual cells to maintain optimum electrical performance and long battery life increases in difficulty with the battery string voltage.

Lead-Acid Battery Markets

The largest and most familiar market for lead-acid batteries is vehicle starting, lighting and ignition (SLI). In addition to passenger cars and trucks, the SLI market includes batteries for motorcycles, off-road vehicles, motorboats, airplanes, racing cars, and other mobile devices. Only a small percentage of the total energy in the battery is used to start a car at moderate temperatures but the electrodes must have very high surface areas to yield the larger currents needed for cold engine starts.

SLI batteries must also provide higher amp hour (Ah) capacity at the lower currents needed to power the lighting, entertainment, safety, and other on-board electronic equipment. In addition to electrical drains from on-board equipment, batteries lose charge by internal chemical self-discharge processes. In cold weather these processes are slow, but when the vehicle is parked in hot environments, higher self-discharge rates can reduce the battery voltage and current needed for engine starting. When stored in hot environments, vehicles and boats that are not used regularly can fail to start because of these self-discharge processes.

Lead-acid batteries are also used for emergency power in uninterruptible power systems (UPS), telecommunications back-up systems, and safety equipment in critical applications such as hospitals and nuclear power plants. Stationary or "standby" battery designs and recharge profiles vary according to the specific system requirements. Large battery strings are connected in series to deliver up to several hundred volts.

In telecommunication back-up power systems, stationary batteries are typically charged continuously at low constant voltage, called the "float voltage" or at low constant current to maintain their energy levels. The batteries contain enough capacity to provide emergency power during long electrical outages of 8 h or longer or until a generator can be brought up to power the system. When power outages are short and infrequent and the battery electrical grid is restored, the float charging will continue. For longer electrical outages, a higher charging rate may be specified to recharge the battery as quickly as possible.

Large digital data centers that support online computers have been increasing rapidly in recent years. The market is driven by the very high cost penalty incurred when computers go offline due to a power outage in one of these centers. The batteries are designed to replace ac power and are therefore rated in KVA.

Deep-cycle batteries store and deliver energy for motive power applications, such as golf carts, forklift trucks, and airport service vehicles. Use of batteries to power robots for duty cycles in highly automated or hazardous applications is also growing. The number of charge–discharge cycles that they can deliver during their lifetime is a function both of the cell design and the battery cycle profile.

In a temperate climate zone, batteries used for solar energy storage generally do not undergo repetitive, deep-discharge cycling. Instead the energy storage system is designed to cycle at a daily partial-state-of-charge cycle in which the battery charges on sunny days and discharges when used at night or on overcast days. During the summer, the photovoltaic system will provide extra energy that can gradually increase the battery state of charge. But during the winter months the battery state of charge will gradually decrease. In temperate climates the use profile may, therefore, be shallow daily cycles within a single deep yearly cycle.

Energy from remote wind turbines can be stored in lead-acid batteries designed to accept high power peaks. These batteries may also be used to filter power peaks and maintain grid stability when wind energy is added to the electrical grid system.

Hybrid electric vehicles (HEVs) are a new and even more demanding partial state-of-charge applications. The battery powers the SLI functions but is also used to assist other functions in the vehicle propulsion system. A high current charges the on-board battery when the vehicle brakes are applied. When the vehicle accelerates, the battery discharges rapidly to provide the required power. If the car is parked for long time in a hot environment, the battery can also self-discharge. Although the battery-operating modes depend on the vehicle design and usage, many of these HEV batteries operate in a narrow range between about 30 and 70% state-of-charge [2].

Nickel metal hydride and lithium batteries are also under development for HEVs. However, safety, low cost and readily available materials give lead-acid batteries some important advantages for a consumer mass market. In addition, an industry is already in place and well capable of recycling nearly 100% of commercial lead-acid batteries.

In early hybrid electric vehicle testing, high currents needed for acceleration and regenerative braking and minimal recharge times limited the cycle life of VRLA batteries. Several modifications to the battery design have greatly improved the battery's high-rate performance in typical HEV driving cycles. For example, conductive particles have been added to the electro-active lead material in the negative plate to increase its electrical conductance and its electrolytic capacitance.

The new lead/carbon acid battery design, called the Ultra battery, shows promise for use in HEV and other partial-state-of-charge applications. Scientists at CSIRO in Australia invented the Ultra battery, and Furukawa in Japan has developed a manufacturing process that has been licensed in the United States and Europe.

The Ultra battery design [2] substitutes a carbon electrode for a portion of each negative lead electrode. The combination of the carbon negative and the lead-dioxide positive electrode provides a pseudo-capacitor in parallel with a normal lead-acid battery in the same cell. The pseudo-capacitor can operate at high currents for short periods of time to reduce the stress on the battery. Meanwhile, the battery can store chemical energy and supply electrical energy for a longer period of time at a more moderate rate.

Sources of Information on Lead-Acid Batteries

Battery manufacturers often classify lead-acid battery designs in their product specifications and data sheets according to the market application. Common terms are described below so that the reader can understand the type and appropriate use of each battery product.

Battery producers provide technical specifications and material safety data sheets (MSDS) for their products. The data sheets are often available on their websites. Section I of the MSDS shows the battery type and name, and the name and contact information of the manufacturer, including customer service and emergency telephone numbers. Chemicals in the battery are listed in Section 2 and physical and chemical characteristics are shown in Section 3. Section 4 lists fire and explosion hazards based on physical and chemical data. Section 5 provides important safety requirements during shipping and electrical installation and use to avoid shorting and electrical fires. Health hazards and precautions for safe handling, use and disposal are provided in Sections 6 and 7. Control measures, if needed, are provided in Section 8. Information on proper transportation, shipping and handling in Section 9 is essential to ensure good battery performance. Battery manufacturers also give additional information, including disclaimers, in Section 10.

Several large organizations have been instrumental in developing the battery industry in a responsible and innovative approach to meeting the needs of its customers. The Battery Council International [3] publishes manuals, specifications, test methods, and other literature for lead battery users and sponsors symposia on the battery industry. The International Lead Association (ILA) is a consortium of companies in mining, battery design and manufacturing, and recycling, their suppliers, contractors, and other organizations that have a direct interest in lead and its responsible use. ILA also manages an international consortium of companies that are doing cooperative research to develop and improve battery technologies for new applications, as well as environment, health, sustainable development and product-related research programs.

Battery Designs

Monopolar and Bipolar Electrodes

Commercial batteries are typically monopolar designs. Monopolar cells are made in a stack of a single positive electrode, a porous, nonconductive separator, and a single negative electrode. To increase the cell capacity, the series stacking process continues with separator/positive/separator/negative layers to produce a larger stack of plates that will deliver the required cell capacity. In each cell the negative electrodes are all connected in series by a conductive lead strap and terminal called the "top lead." The same method is used to connect the positive plates to the positive top lead.

The negative and positive terminals collect the current in each cell and conduct it to the external negative or positive post, respectively. The battery is sealed to avoid electrolyte leakage and air ingress, but one-way pressure relief valves are required to release small quantities of gas that may be generated during recharge. Most batteries have a pressure valve in each cell, but some designs use a single pressure relief valve for all cells.

Bipolar designs have been tested over many years, but are still under development. They are especially desirable for applications that require high voltage and power but can store only limited capacity per unit cell. One developing application is for hybrid electric vehicles. They require batteries with high current capability for short times to store regenerative braking energy and for vehicle acceleration. Each electrode in a bipolar cell has a positive side and a negative side with a solid partition between them that conducts electrons but not ions. The partition material must also be noncorrosive to prevent shorting through the partition walls.

Bipolar electrodes are stacked to produce a battery of cells connected at the partitions. If the cells in a bipolar battery are not rigorously sealed around the edges, electrolyte leaks can provide a shunt current path between cells, which reduces the battery performance. Researchers continue to look at new materials and designs for commercial bipolar batteries. Recent work to develop bipolar batteries is discussed further in Future Development at the end of this article (Fig. 5.1).

Flat and Tubular Electrodes

Lead-battery electrodes can be made as a flat plate with a lead grid as the current collector or as a tubular plate design with a lead rod current collector in the center of tubes. Monopolar electrode current collectors have a conductive lead grid that connects with the terminal. The current collector physically supports the electrode and also collects and carries the current to the electrical system.

Flat grids are often made by pouring molten lead into a mold and cooling it to form a grid with holes for the active materials. Other approaches to making grids are to punch holes in flat lead sheets or to slit the lead sheets and then expand the slits to make the holes for the active material paste.

Tubular positive plates are sometimes used in batteries for deep-discharge applications to reduce shedding of electro-active materials from the plate during cycling. The positive electrode has interconnected porous tubes that are filled with positive lead paste. Lead rods in the centers of tubes are connected in parallel by a conductive lead bar at the top, and the bar is connected to the positive terminal.

Tubular batteries have flat negative plates opposing the positive tubular plates. Recent approaches to preventing positive plate shedding with improved separator designs have limited the use of tubular positive plate batteries. They are primarily in applications that have deep-discharges or severe vibration.

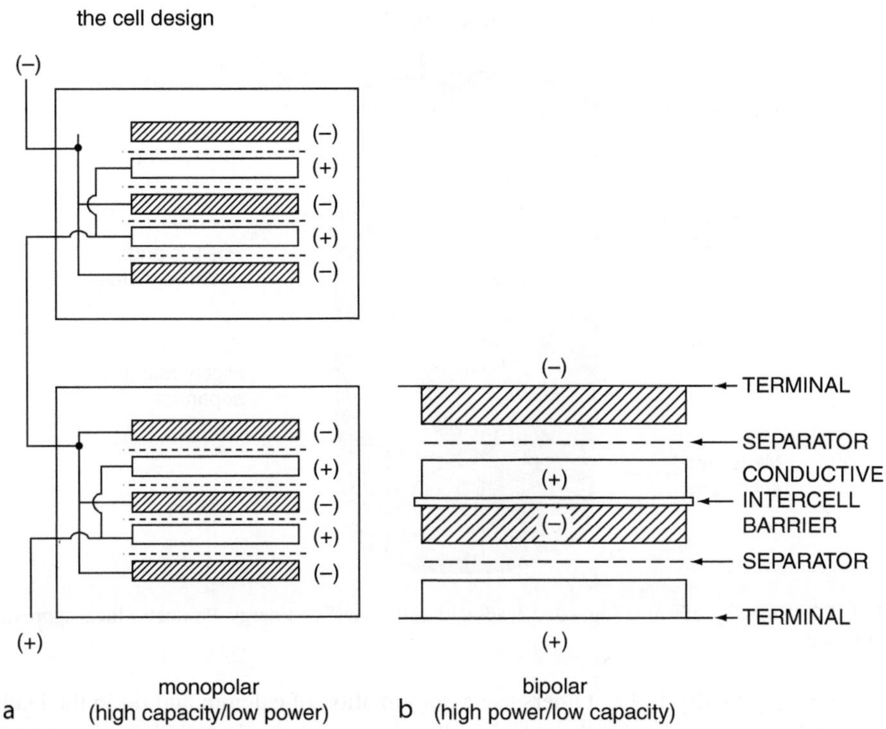

Fig. 5.1 Schematic of monopolar and bipolar battery designs [4] (Electrochemical Society approval obtained)

Lead Alloys in Grids and Gas Venting Systems

Because water electrolysis on pure lead is a very slow process, pure lead grids have good corrosion resistance in aqueous sulfuric acid. However, lead is a soft metal that is easily bent out of shape unless it is alloyed with stronger metals. This problem was solved between 1880 and 1890 by adding up to 12% antimony and a lesser amount of arsenic to the lead. These additives increase the mechanical strength of the grid, but unfortunately they also increase water loss and accelerate hydrogen evolution when the batteries are charged. Batteries with grids containing antimony are designed to vent the oxygen and hydrogen that water electrolysis produces, but periodic water additions are necessary to maintain life. Higher levels of room ventilation are also required to maintain safety.

Lead-alloys containing calcium were introduced in 1935 by Haring and Thomas of Bell Laboratories [5] to reduce water loss and ventilation requirements. These designs are often called low-maintenance batteries, because water additions are not required and now are often prevented by battery manufacturers to minimize electrolyte contamination. Some of the lead alloys in these designs have lower levels of

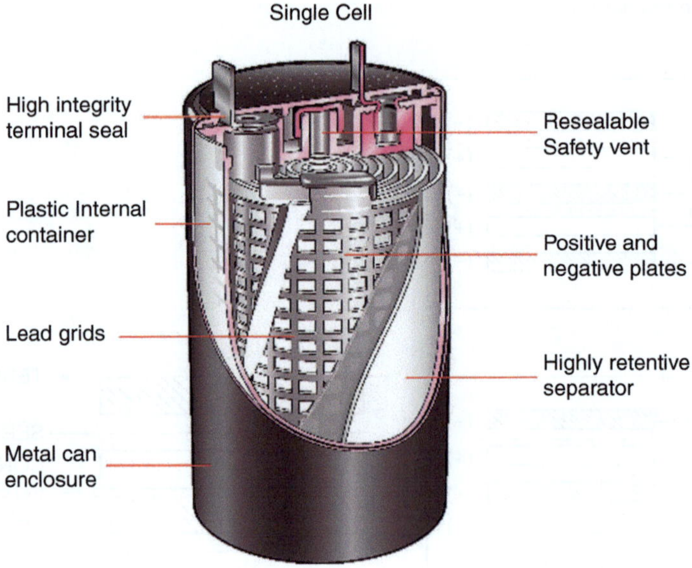

Fig. 5.2 Cutaway of valve-regulated lead-acid cell (EnerSys Energy Products, Inc., approval obtained)

antimony, typically 1–3%. Others use a combination of calcium and tin in the lead. Specially designed vent valves with a flame arrestor are installed to prevent an external spark from igniting hydrogen gas in the battery.

In about 1970, Sonnenschein in Germany [6] began using finely divided fumed silica to gel the sulfuric acid and reduce acid leakage. The technology was applied to portable or power equipment. At about the same time Don McClelland and John Devitt at Gates Rubber Company [7] invented a valve-regulated lead-acid (VRLA) battery with an absorptive glass mat (AGM). This valve reduces acid leakage while retaining the power capability of a liquid electrolyte system. The spirally wound battery design in Fig. 5.2, based on the Gates Rubber Company patents, is now manufactured by EnerSys Energy Products, Inc.

VRLA batteries have a one-way pressure relief valve that opens when the internal battery pressure increases to a specified valve design level between 1 and 30 psi. The valve closes when the internal battery pressure decreases. Some VRLA batteries have grids made of pure lead or pure lead with a little tin added. The design intent is to facilitate the recombination of the oxygen and hydrogen within the cell by re-forming the water during charge. This process, often called "the oxygen cycle," can reduce the amount of hydrogen vented over the life of the cell but will not entirely eliminate it. Room ventilation must be adequate to handle a worst-case condition, such as a high battery temperature, when the batteries are on charge. Water electrolysis and hydrogen venting accelerate with increasing battery temperatures during charge. Other options for controlling battery life and safety are monitoring and control of the environment, and smart battery charging using algorithms and electronic control methods.

Electrode Active Material Preparation

The electrodes in the first lead-acid battery were made by repeatedly charging and discharging lead plates in sulfuric acid mixed with forming agents to facilitate surface corrosion. Because the electrochemical formation process requires a large amount of time and energy to produce high battery capacity, electrochemically formed plates are now typically used mainly for heavy-duty industrial applications where very rugged plate designs are required.

Today, pasted plates are more common. They are made by mixing lead oxide powders with aqueous sulfuric acid and other minor additives to form a lead oxide paste that can be spread onto a lead grid. Since the negative plate surface area tends to shrink during battery cycling, a mixture of fine carbon particles, lignin derived from wood, and barium sulfate is typically added to maintain a high plate surface area. This additive is called an expander mixture. The expander mix is changing as new battery designs and applications emerge.

Electrolyte Systems: Flooded, Gel, and Absorptive Glass Mat (AGM)

The sulfuric acid used in lead-acid batteries is a combination of sulfuric acid (or dihydrogen sulfate, (H_2SO_4) and water (H_2O)). Acid concentrations in automotive batteries are about 35% H_2SO_4. The cells are "flooded" with excess electrolyte to prevent the battery from drying out during use.

Two types of electrolyte systems use methods to immobilize the liquid battery acid. The first design, which was developed in Germany, is called a "gel" battery because finely divided silica is added to "gel" or thicken the electrolyte. Gelling prevents the aqueous sulfuric acid from leaking or spilling if the battery is tipped. The other method, developed in the USA in about 1970, is absorbing a more highly concentrated acid in a porous glass mat (AGM). The acid concentration in the AGM battery was increased from about 35% in liquid and gel batteries to about 40% acid in AGM batteries. This reduces the water volume in each cell and makes the battery more sensitive to dry-out in hot environments. However, the AGM design has the advantage of higher discharge power compared to a gel battery design (Fig. 5.3).

A spirally wound, 2 V AGM battery was the first VRLA design used commercially. The valve was designed to open at 30–50 psi. The cell was made by stacking single strips of a positive plate, AGM separator, and a negative plate together, rolling the stack up like a jelly roll, and inserting it in a cylindrical case. The design was called a gas recombinant battery because it reduced the amount of water lost in the battery during charge by recombining the hydrogen and oxygen that are formed on overcharge.

Lower electrolyte volumes and higher acid concentrations of VRLA AGM batteries also allowed designers to lower the height of rectangular cases. Since typical rectangular battery containers cannot withstand high pressures, their

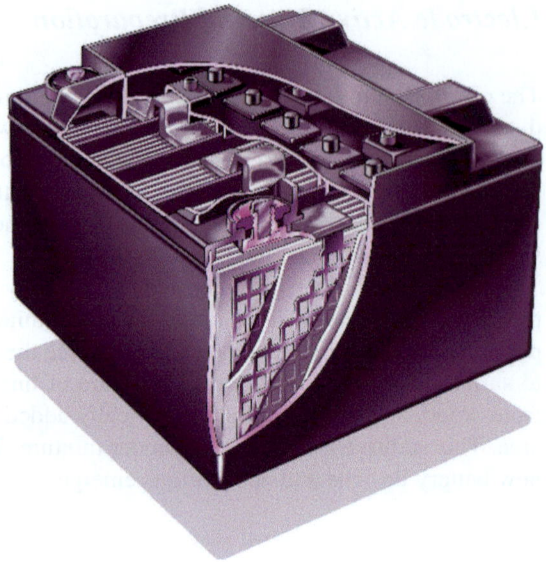

Fig. 5.3 Cut-away of VRLA AGM battery with low profile (EnerSys Inc., approval obtained)

valve-opening pressures may be only a few pounds per square inch (psi). Gel batteries became VRLA batteries when they adopted one-way pressure relief valves. Most of the Gel and AGM batteries are now rectangular. All of these designs are called VRLA batteries.

Gel batteries are finding a new market in storing energy from solar cell arrays. The acid concentration in gel batteries is similar to flooded lead-acid batteries. One-way pressure relief valves help to retain the water in hot environments. Compared to AGM batteries, gel batteries are less susceptible to drying out, because they contain more acid.

Battery Discharge Reactions

The lead-acid battery electrolyte (sulfuric acid) participates in the electrode reaction at both the positive and the negative plate when the battery discharges. Sulfuric acid dissolves in water (H_2O) and dissociates into ions in two steps. First, sulfuric acid forms hydrogen and bisulfate ions [1]

$$H_2SO_4 \Leftrightarrow H^+ + HSO_4^-.$$

These ions transport the electric charge between the electrodes. If the sulfuric acid in the electrolyte is very dilute, the bisulfate ion may dissociate further to hydrogen and sulfate ($SO_4^=$) ions: [2]

$$HSO_4^{=} \Leftrightarrow H^+ + SO_4^{=}.$$

The total reaction is [3]

$$H_2SO_4^{=} \Leftrightarrow 2H^+ + SO_4^{=}.$$

The standard negative electrode voltage on open circuit is $E_0 = -0.358$ V when measured versus a standard hydrogen electrode. During discharge, lead (Pb) oxidizes by reacting with the bisulfate ion and loses two electrons to form lead sulfate ($PbSO_4$). The reaction is [4]

$$Pb + HSO_4^- \Leftrightarrow PbSO_4 + H^+ + 2e^-.$$

The electrons from reaction [4] are transferred through the external electric circuit to the positive electrode, where lead dioxide (PbO_2) is reduced by reacting with two electrons (e^-), as well as hydrogen and bisulfate ions, forming lead sulfate and water. The reaction is [5]

$$PbO_2 + 3H^+ + HSO_4^- + 2e^- \Leftrightarrow PbSO_4 + 2H_2O.$$

The standard potential of this reaction [5] is +1.690 V versus a standard hydrogen electrode. (May want to use Parsons' book as reference in addition.)

The overall cell reaction is determined by adding reactions [4] and [5] to get the double sulfate reaction [6]

$$PbO_2 + Pb + 2H_2SO_4 \Leftrightarrow 2PbSO_4 + 2H_2O.$$

Reaction [6] is called the double sulfate reaction because sulfuric acid reacts at both electrodes. A resistive layer of lead sulfate crystals forms on both the positive and negative electrode surfaces. The equilibrium cell voltage is the sum of the two electrode reactions:

$$V^0 = +1.690 \text{ V} - (-0.358 \text{ V}) = 2.048 \text{ V}.$$

The voltage decreases as the sulfuric acid concentration decreases. A simple way of determining the relative state of charge of the battery is measuring the battery voltage on open circuit at room temperature. Temperature corrections can be made at higher or lower battery temperatures. Assuming that the battery is at room temperature is usually sufficient unless the batteries have been stored for hours in an uncontrolled hot or cold environment or are operated at high currents.

The theoretical specific energy of the lead-acid battery is calculated below using the molecular weights of the reactants and the chemical formulas. The number of electrons exchanged is an important factor in the energy density of a battery reaction, because it determines the total current that the reactants can deliver. A disadvantage of lead-acid batteries is the high molecular weight of lead. However, the water-based electrolyte is relatively light, and the reaction has the advantage of exchanging two electrons instead of only one.

The number of atoms in these reactions multiplied by the atomic weight of the atom determines the total molecular weight of the reactants:

$$\begin{aligned}&\text{O weighs } 16.0\ \mathrm{g/mole}(\times 10 \text{ moles weighs } 160.0\ \mathrm{g});\\&\text{H weighs } 1.0\ \mathrm{g/mole}(\times 4 \text{ moles weighs } 4.0\ \mathrm{g});\\&\text{Pb weighs } 207.2\ \mathrm{g/mole}(\times 2 \text{ moles weighs } 414.4\ \mathrm{g});\ \text{and}\\&S \text{ weighs } 32.1\ \mathrm{g/mole}(\times 2 \text{ moles weighs } 64.2\ \mathrm{g}).\end{aligned}$$

The total equivalent molecular weight of the reactants required to generate 2 moles of electrons in the discharge reaction is 646.6 g.

The calculation of the theoretical specific energy is shown below. The theoretical specific energy at 2.048 V is 170 W h/kg of material weight.

$$170\ \mathrm{Wh/kg\ of\ reactants} = \frac{2.048\ \mathrm{V} \times 2\ \mathrm{moles\ e^-} \times 96{,}500\ \mathrm{As/mole\ e^-}}{646.6\ \mathrm{g\ reactants} \times 3600\ \mathrm{s/h} \times 1\ \mathrm{kg}/1000\ \mathrm{g}}$$

This type of calculation can be useful as a means of comparing the theoretical specific energies of different batteries. Practical energy densities are design dependent and typically deliver no more than 20–40% of the theoretical specific energy.

Battery Discharge Rates

The reversible voltage of 2.048 V is for the lead-acid battery at a room temperature of 25°C and one atmosphere pressure in a dilute acid concentration. Lead-acid batteries have higher acid concentrations when fully charged, so the reversible voltage is higher. During discharge the reversible voltage drops rapidly at the beginning of the discharge until it reaches an initial voltage plateau. Then the voltage decreases more slowly to the discharge cut-off voltage. The initial drop is in part due to the internal resistance of the battery and in part due to the initial rapid depletion of the sulfuric acid concentration at the surface of both the positive and negative electrodes.

The actual discharge capacity is the current (I) multiplied by the discharge time (t) in ampere hours. The rate of discharge, or current, strongly affects the initial voltage drop and the slope of the plateau voltage. The higher the discharge rate, the shorter the discharge time and the smaller the capacity will be. This is because hydrogen and bisulfate ions must carry the charge between the electrodes through the porous separator and acid film at the electrode surfaces.

Table 5.1

Discharge amperes (I)	Discharge hours (t)	Ampere hour capacity (AH)[a]
5	9	45
12	3.6	43.2
25	1.4	35
$\ln C = n \ln I + \ln t$		
$4.078 = 1.15 \ln I + \ln t$		

[a]Based on three discharges of one battery

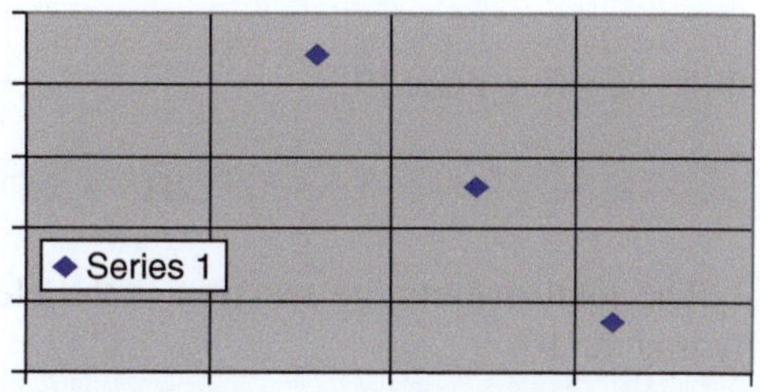

Fig. 5.4 Relationship between the discharge time and current in a lead-acid battery

The horizontal x axis is the log to the base 3 of the discharge current, and the vertical y axis is the log to the base e in discharge time.

The discharge time is a function of discharge current. $I^n\ t = C$ where I is current in amperes and t is the discharge time in hours for a specific battery design measured at three or more discharge currents and C is a constant. When these values are plotted on a log–log scale, the constant n is the slope of a straight line. It is typically 1.3–1.4 but approaches 1 at very small currents and 2 at high currents. The example in Table 5.1 and Fig. 5.4 is based on three discharges of a VRLA battery.

The discharge reactions are reversed when the cell is recharged. If each of these discharge reactions is reversed and the discharge reactions are added together, the reaction [7]

$$2PbSO_4 + 2H_2O \Leftrightarrow PbO_2 + Pb + 2H_2SO_4$$

is the reverse of the double sulfate reaction [6].

When the battery reaches a voltage level that corresponds to about 90% state of charge, the recharge reaction becomes less efficient as water in the electrolyte begins to form oxygen at the positive plate and hydrogen at the negative plate. This process accelerates as recharge currents and/or internal cell temperatures increase. To control water electrolysis, the charging current is generally decreased as the battery receives its finishing charge. Many lead-acid battery designs are able to recharge up to the gassing level relatively quickly, but the finishing charge will take longer to avoid rapid water loss and high gassing rates. The battery manufacturer's recommendations should be carefully followed, since charge methods and charging times are design-dependent.

Overcharge Reactions

Overcharge is defined as the amp hours of charge put into the battery in excess of the amount of Ah taken out of the battery during the previous discharge. Because charging is not 100% efficient, all lead-acid batteries must be overcharged to return them to full capacity.

The finishing charge and overcharge reactions that electrolyze water form a mixture of hydrogen (H_2) and oxygen (O_2) gases. At the positive plate, [8]

$$H_2O \Rightarrow 2H^+ + {}^1\!/\!_2\, O_2 + 2e^-$$

at the negative plate [9]

$$2H^+ + 2e^- \Rightarrow H_2.$$

The total overcharge reaction is the electrolysis of water to form oxygen and hydrogen [10]

$$H_2O \Rightarrow H_2 + {}^1\!/\!_2\, O_2.$$

Thermodynamic measurements of these reactions at platinum electrodes show that it takes less energy to electrolyze water in the electrolyte than to recharge lead sulfate. If this were true in the lead-acid battery, it could not be recharged. Lead-acid batteries recharge efficiently because of the low rate of water electrolysis on lead. The reason is that the hydrogen evolution reaction is impeded on the surface of the lead electrode. As a result, the lead-acid battery can deliver a higher voltage than other aqueous rechargeable batteries.

Impurities in the battery electrolyte lower the cell voltage and may increase water loss. When electrolyte purity is well controlled, the capacity can be returned more rapidly because the end of the recharge is more efficient. The final part of the recharge is controlled either by "floating" the battery at a voltage that is slightly above the open circuit or by applying a small "trickle" current to the battery. Floating or applying a trickle current near the end of charge minimizes the water loss.

The oxygen generated at the positive electrode is not very soluble in the sulfuric acid, so it vents out through the battery valve. However, in a VRLA battery, the oxygen cycle recombines the oxygen and prevents hydrogen from forming at the negative electrode by the following equations.

The positive plate reaction is [8] $H_2O \Rightarrow 2H^+ + {}^1\!/\!_2\, O_2 + 2e^-$ shown previously. The negative plate reactions may be a two-step process: [11] $Pb + {}^1\!/\!_2\, O_2 + H_2SO_4 \Rightarrow PbSO_4 + H_2O$ and [12] $PbSO_4 + 2H^+ + 2e^- \Leftrightarrow Pb + H_2SO_4$. When these equations are added up, the overall reaction is the double sulfate reaction [1] discussed above

$$2PbSO_4 + 2H_2O \Leftrightarrow PbO_2 + Pb + 2H_2SO_4$$

Since the oxygen generated at the positive electrode is not very soluble in the acid, it will vent out through the valve of a flooded lead-acid battery, and the opportunity to reform the water will be lost. A VRLA battery design allows the oxygen to travel to the negative plate through the AGM separator, because the separator pores are not fully saturated with acid. The acid volume is reduced in the separator by decreasing the percentage of water in the electrolyte. Because of the higher acid concentration, there is free space for oxygen to travel from positive to negative electrode as a gas instead of dissolving in the electrolyte. The oxygen reacts at the negative plate to reform water, because the negative electrode surface has a three-phase gas/liquid/solid interface.

Battery Maintenance and the Effect of Temperature

Lead-acid batteries have a variety of failure modes that depend on the design, the use mode, and the environment. Manufacturers generally provide information about the life of their battery designs in a particular application. Some factors that can shorten the useful battery life are, however, applicable to most designs and applications. One of the most important variables in predicting battery life is the ambient temperature.

Low ambient temperatures reduce the battery voltage and current while high temperatures reduce battery life. For example, automotive batteries often fail to start the engine in the winter for two reasons. First, the power (voltage times current) that is required to start a cold engine increases. And second the battery power is reduced because cold temperatures decrease both the voltage and current that the battery can deliver. High summer temperatures decrease battery life because of increased grid corrosion and other side reactions that lead to battery failure.

Although automotive batteries often fail to start a vehicle in the winter when the ambient temperature is low, battery storage at high temperatures is often the cause of the battery deterioration and failure. If the battery is subjected to high heat for a long time, reactions that wear out the battery, such as grid corrosion, water loss, and formation of hard lead sulfate on the plate surfaces, will be accelerated.

Application of batteries in other uncontrolled environments, such as outdoor telecommunications cabinets, photovoltaic systems, and marine uses also reduces battery life. When batteries are operated in hot environments, manufacturers often recommend reducing the charging current or float voltage to prevent the battery from overheating.

Battery life can also be reduced by storage on open circuit for a long time. Chemical reactions will self-discharge the battery gradually over its "shelf life." Since heat increases the rates of these chemical reactions, the battery shelf life will be longer if the battery is stored at a moderately cold temperature. The rates of most battery degradation mechanisms are reduced at low temperatures. The electrolyte in a fully charged battery does not freeze until the temperature reaches about minus 40°C. However, the battery freezing temperature will rise to 10–15°C when the battery is discharged.

Future Directions

Increasing use of renewable energy from wind turbines will increase the need for batteries with both higher power and increased energy storage. Wind is variable and tends to be higher at night when electricity is used less by energy consumers. When wind energy is fed into the ac electric grid, this variability frequently causes instability in the grid system. Several new approaches to developing higher-power lead-acid batteries for renewable energy storage are discussed below.

The thixotropic gel that is typically used in VRLA batteries is liquefied by agitation before being poured into the cell. It then solidifies around and within the plate separators. C&D Technologies developed a battery using a colloidal, poly-silica gel that does not have to be liquefied before it is poured into the cell. The gelling solution can be added after the AGM separator between the plates is soaked with acid, so less gel penetrates into plate area. The AGM separator maintains the high power capability of the AGM battery design, while the gel around the edges of the plates provides an additional source of aqueous electrolyte. This extends the cycle life by reducing the rate of water loss. This concept has been tested in a commercial 485 Ah C&D model 12SL 78-13 battery, which is designed for motive power and materials handling equipment [8].

Lead-acid batteries with high power capability may be required to stabilize the future grid system as more wind energy is fed into it. Use of a battery/capacitor combination may be necessary, but the physical space of the system must also be acceptable. The Ultra battery [2], which has both a super capacitor and a storage battery in the same module, is one very interesting way to address this need. In 2010, a 144 V Ultra battery design passed a 100,000-mile test in a Honda Insight at the Millbrook test track in the United Kingdom with no conditioning and in an excellent state of health. After passing that test, the battery was returned to Furukawa in Japan for continued use (Fig. 5.5) [2, 9].

If the history of lead-acid batteries is an indicator, the performance of a super capacitor and storage battery combination will likely be improved further as additional materials and designs are developed and field experience with the HEV and wind energy applications increases. Other sustainable energy systems that operate over a wide range of currents may also benefit from this work. Many types of carbons as well as other materials are being studied. The challenge is to find materials that are low cost and are also compatible with the lead-acid system.

High battery voltages and power capacities in a small footprint will also be needed to power equipment in off-grid applications. Bipolar lead-acid batteries are

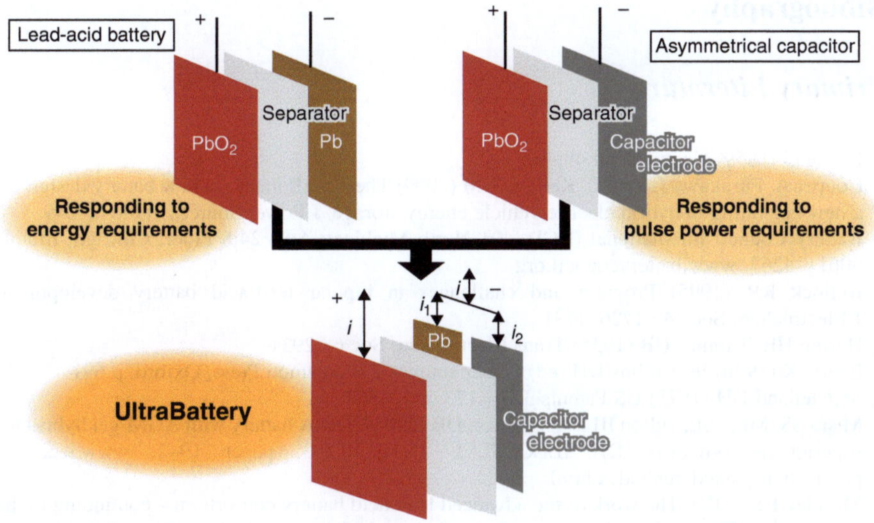

Fig. 5.5 Diagram of the construction of the Ultra Battery (Furukawa permission obtained)

most desirable for high voltage applications that do not require large amounts of energy. Researchers continue to look at new materials and designs that will improve the life and performance of bipolar batteries. Atraverda Ltd. in the U.K. uses Ebonex®, a ceramic material based on conductive titanium oxides, for its bipolar battery partitions and case.

Effpower® tested a lead infiltrated ceramic (LIC™) bipolar battery in a Honda Insight for 15,000 km in Gothenburg, Sweden [9]. Their conductive ceramic has pockets containing lead on the negative side and lead dioxide on the positive side of the partition. They have announced plans to start production of 24 V and 150 V lead-acid battery modules in 2011 in partnership with Banner Batterien in Austria. Both batteries are 6 Ah designs. The 24 V lead-acid battery module is rated at 5 KW/8.6 kg in a 90 × 253 × 203 mm module (0.58 KW/kg). The 150 V lead-acid battery module has 0.8 KW/kg. An Effpower® battery has also been installed in a 2007 Honda Civic Hybrid and is undergoing testing at the Millbrook testing grounds in the UK [9].

Bipolar batteries are susceptible to two failure modes that are not found in monopolar designs. (1) When seals around the outside edges of bipolar partitions fail, leakage currents between cells will cause a decline in cell performance. (2) An electrical short through a pinhole in the bipolar partition can cause sudden battery failure. Much progress has been made in bipolar module technologies. Finding the right materials, designs and construction technologies, and improving our understanding of application modes, will be ongoing as new bipolar battery designs are developed and applied.

Bibliography

Primary Literature

1. Planté G (1859) Compt Rend 49:402
2. Cooper A, Furakawa J, Lam L, Kellaway M (2009) The UltraBattery – a new battery design for a new beginning in hybrid electric vehicle energy storage. J Power Sources 188:642–649
3. Battery Council International (BCI), 401 North Michigan Ave, 24th Floor, Chicago, Illinois 60611–4267. www.batterycouncil.org
4. Bullock KR (1995) Progress and challenges in bipolar lead-acid battery development. J Electrochem Soc 142:1726–1731
5. Haring HE, Thomas UB (1935) Trans Electrochem Soc 68:293+
6. Eberts K (1970) In: Collins DH (ed) Power sources 2. Pergamon Press, Oxford, p 69+
7. McClelland DH (1975) US Patents 3,704,173 and 3,862,361
8. Misra SS, Mraz SL, Dillon III JD, Swanson DB (2003) VRLA battery with AGM-gel hybrid for superior performance. DB IEECE/IEEE INTELEC'03, paper 19–2. www.ieee.org/publications_standards/index.html
9. Moseley PT (2010) The work of the advanced lead-acid battery consortium – continuing in the footsteps of Gaston Planté, plenary lecture, Joseph Priestley Society, Chemical Heritage Foundation, Philadelphia

Books and Reviews

Bard AJ, Parsons R, Jordan J (1985) Standard potentials in aqueous solution. CRC Press, Boca Raton

Berndt D (2003) The lead-acid battery system. In: Ecclestone TL, Shepherd PA (eds) Maintenance-free batteries, 3rd edn. Research Press, Baddock

Bode H (1975) Lead acid batteries. Wiley, New York

Bullock KR, Salkind AJ (2011) Valve regulated lead-acid batteries. In: Reddy TB, Linden D (eds) Linden's handbook of batteries, 4th edn. McGraw Hill, New York (Chap 17)

Bullock KR, Vincent CA (1997) Secondary lead-acid cells. In: Vincent CA, Scrosati B (eds) Modern batteries, 2nd edn. University of Chicago Press, Chicago

Rand DAJ, Moseley PT, Garche J, Parker CD (2004) Valve-regulated lead-acid batteries. Elsevier, Amsterdam

Chapter 6
Rechargeable Batteries, Separators for

Shriram Santhanagopalan and Zhengming (John) Zhang

Glossary

Dry process	A process used to make a separator, which involves melting a polyolefin resin and extruding it into a film, thermal annealing to increase the size and amount of lamellar crystallites, and stretching to form tightly ordered micro pores.
Gurley	A measure of time taken for a predetermined quantity of air (or other specified fluid) to permeate across a porous membrane.
Nonwoven separators	Sheet, web, or matt of directionally or randomly oriented fibers, bonded by friction, and/or cohesion, and/or adhesion excluding paper and products which are woven, tufted, stitch-bonded incorporating binding yarns or filaments, or felted by wet milling.
Separator	A porous membrane placed between electrodes of opposite polarity, permeable to ionic flow but preventing electronic contact between the electrodes.
Shutdown separator	A separator that responds to increasing heat within a cell, by closure of the pores, such that it stops ionic flow between the anode and cathode, during a thermal event.
Wet process	A process used to make a separator that involves a solvent.

This chapter was originally published as part of the Encyclopedia of Sustainability Science and Technology edited by Robert A. Meyers. DOI:10.1007/978-1-4419-0851-3

S. Santhanagopalan (✉) • Z. (John) Zhang
Celgard, LLC, 13800 South Lakes Dr., Charlotte, NC 28273, USA
e-mail: shriram.santhanagopalan@nrel.gov; johnzhang@celgard.com

R.J. Brodd (ed.), *Batteries for Sustainability: Selected Entries from the Encyclopedia of Sustainability Science and Technology*, DOI 10.1007/978-1-4614-5791-6_6,
© Springer Science+Business Media New York 2013

Definition of the Subject

The battery industry has undergone tremendous growth over the last couple of years – both in terms of technological growth as well as in the variety of applications. The need to optimize battery materials to meet the increasing demand for energy as well as to extend the operating range continues to be a challenge – more so now than ever before. Correspondingly, the demand for novelty in separator membranes to match the newer battery chemistries and geometries continues to grow.

A separator is a porous membrane placed between electrodes of opposite polarity, permeable to ionic flow but preventing electronic contact between the electrodes [1, 2]. A variety of separators have been used in batteries over the years – from cedar shingles and sausage casing to present-day microporous flat sheet membranes made from polymeric materials. Their main function, however, continues to be the same – to keep the positive and negative electrodes apart. They should be very good electronic insulators and have the capability of conducting ions by either intrinsically being an ionic conductor or by soaking an electrolyte. They should minimize any processes that adversely affect the electrochemical energy efficiency of the batteries.

The separator is among the components that have sustained maximum growth in the battery industry with minimal changes to the material ingredients. Not much attention has been given to separators even in publications reviewing batteries [1–7]. The number of reviews on separators [8–17] is minimal compared to those on cell fabrication, their performance, and application in real life. Kinoshita et al. have reviewed different types of membranes/separators used in different electrochemical systems, including batteries [9]. Zhang presented a more recent review of the various separators used in liquid electrolyte systems [10]. This chapter is largely based on the work of Arora and Zhang [4].

The majority of the separators currently used in batteries were typically developed as spin-offs of existing technologies. They were usually not developed specifically for those batteries and thus were not completely optimized for battery systems in which they are currently used. One positive result of adapting existing technologies is that the separators are produced in high volumes at relatively low cost. The availability of low-cost separators is an important consideration in the commercialization of batteries, since the battery industry traditionally operates with thin profit margins and relatively small research budgets.

Introduction and Scope

The battery industry has seen enormous growth over the past few years in portable, rechargeable battery packs. The majority of this surge can be attributed to the revolution in the use of multimedia in mobile phones and personal digital assistants (PDAs), as well in laptop computers and other wireless electronics devices.

The introduction of vehicles implementing rechargeable batteries has increased the demand for batteries by several-fold. Batteries remained the mainstream source of power for systems ranging from mobile phones and personal digital assistants (PDAs) to electric and hybrid electric vehicles. The world market for batteries was approximately $41 billion in 2000, which included $16.2 billion primary and $24.9 billion secondary cells [20]. In 2010, the global demand was placed at $71 billion.

The Freedonia group estimates [21, 22] the aggregate US demand for primary and secondary batteries to be $16.8 billion in 2012 and that China will surpass the USA as the largest market with an estimated average annual growth at 7%. This growth will be driven by strong demand for battery-powered electronic devices like digital cameras and 3G wireless phones, increasing production of electrical and electronic equipment and the expansion in the automotive sector in the near future. The secondary battery demand has outpaced the primary battery market gains benefiting from strong growth in the use of high-drain portable electronic devices.

The tremendous progress in lithium-ion cells is clearly visible with as much as a threefold increase in the volumetric and gravimetric energy storage capability for both 18650 and prismatic cells between their initial introduction in the early 1990s and 2010. In last few years the lithium-ion production has expanded in South Korea (Samsung SDI, LG Chemical, etc.), China (ATL BYD, B&K, and Lishen, among others), and Japan. Several Japanese (Sanyo, Sony, MBI, NEC, etc.) and Korean (LG Chemical) manufacturers have also moved their manufacturing plants to China [23]. Japan, which controlled 94% of the global rechargeable battery market in 2000, has seen its market share drop to less than 50% of the global market [22, 24, 25]. The continued growth in lithium-ion battery market has led to a strong demand for battery separators. All the major separator manufacturers (Celgard, Asahi, and Tonen) have continually increased their capacity since 2003 [24, 26–29]. The industry has also witnessed the emergence of several new entrants [30–33].

There is not much information available on battery separator market in the literature. In 2009, it was estimated that about 50% of the rechargeable lithium battery market is the size of the components market, and separators constitute about 17% of this volume [22]. The Freedonia Group reported that the US demand for battery separators increased to $410 million in 2007 from $237 million in 1977, and $300 million in 2002, respectively [21, 22, 25].

The purpose of this chapter is to describe separators used in secondary batteries and characterization of their chemical, mechanical, and electrochemical properties, with particular emphasis on separators for lithium-ion batteries. The separator requirements, properties, and characterization techniques are described with respect to lithium-ion batteries. Despite the widespread use of separators, a need still exists for improving the performance, increasing its life, and extending the operating range.

Separator and Batteries

Batteries are built in many different shapes and configurations – button, flat, prismatic (rectangular), and cylindrical (AA, AAA, C, D, 18650 etc.). The cell components (including separators) are designed to accommodate a particular cell shape and design. The separators are either stacked between the electrodes or wound together with electrodes to form jellyrolls as shown in Fig. 6.1. Stacked cells are generally held together by pressure from the cell container. The lithium-ion gel polymer stacked cells are prepared by bonding/laminating layers of electrodes and separators together. The separator properties should not change significantly during the bonding process. In some cases, the separators are coated to help in bonding process and reduce the interfacial resistance [34, 35].

In the conventional way of making spirally wound cells, two layers of separators are wound along with the positive and negative electrodes, resulting in separator/negative/separator/positive configuration. They are wound as tightly as possible to ensure good interfacial contact. This requires the separators to be strong to prevent any contact between the electrodes through the separator. The separator also must not yield and reduce in width, or else the electrodes may contact each other. Once wound, the jellyroll is inserted into a can and filled with electrolyte. The separator must be wetted quickly by the electrolyte to reduce the electrolyte filling time. A header (cap) is then crimped onto the cell to cover the can top. In some prismatic cells, the electrode stack is pressed at high temperatures and pressures and then inserted into thin prismatic (rectangular) cans. A typical 18650 lithium-ion cell uses around 0.07–0.09 m^2 of separator, which is approximately 4–5% of the total cell weight [36].

A number of factors must be considered in selecting the best separator for a particular battery and application. The characteristics of each available separator must be weighed against the requirements and one selected that best fulfills these needs. A wide variety of properties are required of separators used in batteries. These include:

- Electronic insulator
- Minimal electrolyte (ionic) resistance
- Mechanical and dimensional stability
- Sufficient mechanical strength to allow easy handling
- Chemical resistance to degradation by electrolyte, impurities, and electrode reactants and products
- Effectiveness in preventing migration of particles between the two electrodes
- Readily wetted by electrolyte
- Uniformity in properties such as thickness, resistance, etc.

The above list presents a broad spectrum of requirements for separators in batteries. The order of importance of the various criteria varies, depending on the battery applications. In many applications, a compromise in requirements for the separator must be made to optimize performance, safety, cost etc. For example,

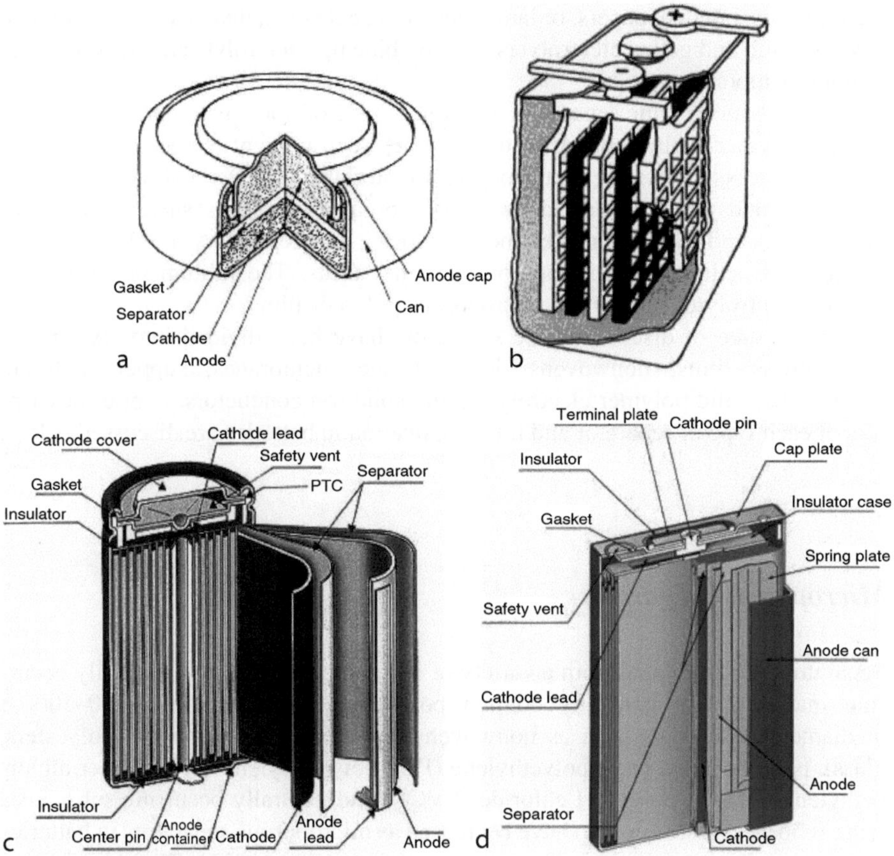

Fig. 6.1 Typical battery configurations (**a**) button cell, (**b**) stack lead acid, (**c**) spiral wound cylindrical lithium-ion; (**d**) spiral wound prismatic lithium-ion

batteries that are characterized by small internal resistance and consume little power require separators that are highly porous and thin; but the need for adequate physical strength may require that they be thick.

In addition to the above general requirements each battery type may have other specific requirements essential for good performance and/or safety. One typical example is that the separator used in sealed Nickel Metal Hydride (NiMH) batteries should be permeable to gas molecules for overcharge protection.

Separator Classification

Separators for batteries can be divided into different types, depending on their physical and chemical characteristics. They can be molded, woven, nonwoven, microporous, bonded, papers, or laminates. In recent years, there has been a trend to develop solid and gelled electrolytes that combine the electrolyte and separator into a single component.

In most batteries, the separators are either made of nonwoven fabrics or microporous polymeric films. Batteries that operate near ambient temperatures usually use separators fabricated from organic materials such as cellulosic papers, polymers, and other fabrics, as well as inorganic materials such as asbestos, glass-wool, and SiO_2. In alkaline batteries, the separators used are either regenerated cellulose or microporous polymer films. The lithium batteries with organic electrolytes mostly use microporous polymer films.

For the sake of discussion, the separators have been divided into six types – microporous films, nonwovens, ion-exchange membranes, supported liquid membranes, solid polymer electrolytes, and solid ion conductors. A brief description of each type of separator and their application in batteries are discussed below.

Microporous Separators

Separators are fabricated from a variety of inorganic, organic, and naturally occurring materials and generally contain pores that are greater than 50–100 Å in diameter. Materials such as nonwoven fibers (e.g., nylon, cotton, polyesters, glass), polymer films (e.g., polyethylene (PE), polypropylene (PP), polytetrafluoroethylene (PTFE), polyvinyl chloride (PVC)), and naturally occurring substances (e.g., rubber, asbestos, wood) have been used as microporous separators in batteries that operate at ambient and low temperatures (−40°C to 100°C). The microporous polyolefins (PP, PE, or laminates of PP and PE) are widely used in lithium-based nonaqueous batteries. More recently, other polymers have been employed, especially to enhance the window of temperature across which the polymer remains stable.

Nonwovens

Nonwovens are textile products that are manufactured directly from fibers. They are defined as a manufactured sheet, web, or matt of directionally or randomly oriented fibers, bonded by friction, and/or cohesion, and/or adhesion excluding paper and products which are woven, tufted, stitch-bonded incorporating binding yarns or filaments, or felted by wet milling whether or not needed. The fibers may be of natural or man-made origin. They may be staple or continuous filaments or maybe formed in situ [37].

The macroporous fibrous matrix is either dry laid, meltblown, or wet laid. The wet laid process is very similar to papermaking process. The fibers are bonded together by chemical or thermal bonding. The meltblown process is a binderless process and there the polymer fiber web is extruded. Typical pore size of fibrous matrix varies from 1 to 100 μm.

Nonwovens are widely utilized as separators for several types of batteries. Light-weight, wet laid nonwovens made from cellulosic, polyvinyl alcohol, and other fibers have achieved considerable success as separators for popular primary alkaline cells of various sizes. The key nonwoven attributes include consistently uniform basis weight, thickness, porosity, and resistance to degradation by electrolytes. Nonwovens are also successfully employed as separators in NiCd cells.

The materials used in nonwoven fabrics include a single polyolefin, or a combination of polyolefins, such as polyethylene (PE), polypropylene (PP), polyamide (PA), polytetrafluoroethylene (PTFE), polyvinylidine fluoride (PVdF), and polyvinyl chloride (PVC). Nonwoven fabrics have not been able to compete with microporous films in lithium-ion cells. This is primarily because of the inadequate pore-size structure and difficulty in making thin (<25 μm) nonwoven fabrics with acceptable physical properties. However, nonwoven separators have been used in button cells and bobbin cells when thicker separators and low discharge rates are acceptable.

Ion-Exchange Membranes

These membranes are generally fabricated from polymeric materials containing pores with diameters of less than 20 Å. The transport properties of ions in these membranes are characterized by strong interactions between the permeating species and the molecular structure of the polymer. This interaction is due to the presence of ion-exchange groups in the membrane, which allows the membrane to discriminate between permeating or migrating ions by virtue of their specific charge.

Radiation grafted membranes such as Permion® manufactured by RAI Research Corporation are ion-exchange membranes. Such membranes are used as battery separators in alkaline batteries. They are made from PE, PP, or Teflon-based films, which have excellent oxidation resistance and superior chemical resistance to alkali media. However, they are totally impervious to electrolyte flow, and therefore, have almost infinite resistance as a separator in this form. By using radiation grafting and cross-linking techniques, however, selected chemical species are grafted as pendant chains to the base structure of the linear polymer without altering the inert backbone. This modification imparts desirable hydrophilic properties to the films without materially impairing their excellent chemical resistance. This chapter provides a very limited discussion on ion-exchange membranes, as their application in batteries is very limited.

Supported Liquid Membranes

These types of separators consist of a solid matrix and a liquid phase, which is retained in the microporous structure by capillary forces. To be effective for batteries, the liquid in the microporous separator, which generally contains an organic phase, must be insoluble in the electrolyte, chemically stable, and still provide adequate ionic conductivity. Several types of polymers, such as polypropylene, polysulfone, polytetrafluoroethylene, and cellulose acetate, have been used for porous substrates for supported liquid membranes. The PVdF-coated polyolefin-based microporous membranes used in gel polymer lithium-ion battery fall into this category. Gel polymer electrolytes/membranes are only discussed briefly.

Polymer Electrolyte

Polymer electrolytes (e.g., poly(ethylene oxide), poly(propylene oxide)) have attracted considerable attention for batteries in recent years. These polymers form complexes with a variety of alkali metal salts to produce ionic conductors that serve as solid electrolytes. Its use in batteries is still limited due to poor electrode/electrolyte interface and poor room temperature ionic conductivity. Due to its rigid structure it can also serve as the separator. Polymer electrolytes are discussed briefly in the section Separators for Lithium-Ion Batteries.

Solid Ion Conductors

They serve as both separator and electrolyte. These are generally inorganic materials that are impervious barriers to gases and liquids. They allow one or more kinds of ions to migrate through their lattice when a potential gradient or a chemical gradient is present. These types of separators are beyond the scope of this article.

Separators for Lithium-Ion Batteries

All lithium-based batteries use nonaqueous electrolytes because of the reactivity of lithium in aqueous solution and because of electrolyte's stability at high voltage. The majority of these cells use microporous membranes made of polyolefins. In some cases, nonwovens made of polyolefins are either used alone or with microporous separators. This section will mainly focus on separators used in secondary lithium batteries.

Table 6.1 Types of separators used in different types of secondary lithium batteries

Battery system	Type of separator	Composition
Lithium-ion (liquid electrolyte)	Microporous	Polyolefins (PE, PP, PP/PE/PP)
Lithium-ion gel polymer	Microporous	PVdF
	Microporous	Polyolefins (PE, PP, PP/PE/PP) coated with PVdF or other gelling agents
Lithium polymer (e.g., Li-V_6O_{13})	Polymer electrolyte	Polyethylene oxide with lithium salt

Lithium secondary batteries can be classified into three types, a liquid-type battery using liquid electrolytes, a gel-type battery using gel electrolytes mixed with polymer and liquid, and a solid-type battery using polymer electrolytes. The types of separators used in different types of secondary lithium batteries are shown in Table 6.1. The liquid lithium-ion cell uses microporous polyolefin separators while the gel polymer lithium-ion cells either use PVdF separator (e.g., PLION® cells) or PVdF-coated microporous polyolefin separators. The PLION® cells use PVdF loaded with silica and plasticizer as separator. The microporous structure is formed by removing the plasticizer and then filling with liquid electrolyte. These are also characterized as plasticized electrolyte. In solid polymer lithium-ion cells, the solid electrolyte acts as both electrolyte and separator.

Sony's introduction of the rechargeable lithium-ion battery in the early 1990s precipitated a need for new separators that provided good mechanical and electrical properties. Since then, separators have played a significant role in improving the performance as well as safety of lithium battery technology. In 2010, 90% of the total rechargeable battery markets for mobile communication devices use lithium-based batteries whereas the nickel metal hydride batteries continue to dominate the automotive sector and a significant factor in the camera market as well as alternate to the alkaline primary cell.

The microporous polyolefin separator has been used extensively in lithium-ion batteries, since it is difficult for most other conventional separator materials to satisfy the characteristics required in lithium-ion batteries. In lithium-ion batteries two layers of separators are sandwiched between positive and negative electrodes and then spirally wound together in cylindrical and prismatic configurations. The pores of the separator are filled with an ionically conductive liquid electrolyte.

Microporous polyolefin membranes (see Fig. 6.2) in current use are thin (<30 μm) and are made of polyethylene (PE), polypropylene (PP), or laminates [38] of polyethylene and polypropylene. Polyolefin materials are preferred because they provide excellent mechanical properties, chemical stability, and acceptable cost [39, 40]. They have been found to be compatible with the cell chemistry and can be cycled for several hundred cycles without significant degradation in chemical or physical properties.

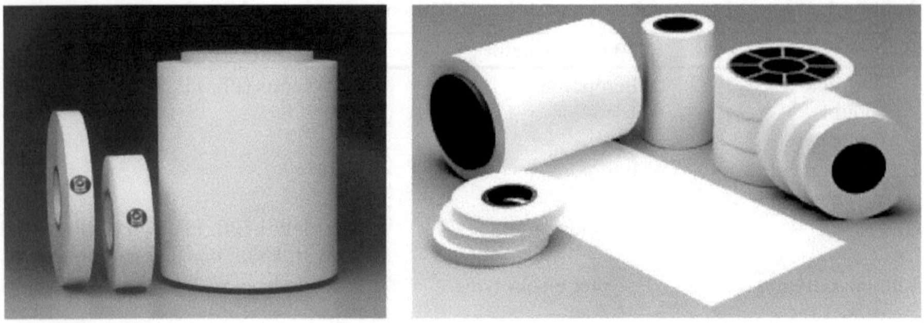

Fig. 6.2 Polyolefin separators used in lithium-ion batteries

Table 6.2 Major manufacturers of lithium-ion battery separators along with their typical products

Manufacturer	Structure	Composition	Process	Trade name
Asahi Kasai	Single layer	PE	Wet	HiPore
Celgard Inc	Single layer	PP, PE	Dry	Celgard
	Multilayer	PP/PE/PP	Dry	Celgard
	PVdF coated	PVdF, PP, PE, PP/PE/PP	Dry	Celgard
Entek Membranes	Single layer	PE	Wet	Teklon
Mitsui Chemical	Single layer	PE	Wet	
Nitto Denko	Single layer	PE	Wet	
DSM	Single layer	PE	Wet	Solupur
Tonen	Single layer	PE	Wet	Setela
Ube Industries	Multilayer	PP/PE/PP	Dry	U-Pore

Commercial membranes offer pore size in the range of 0.03–0.1 μm, and 30–50% porosity. The low melting point of PE enables their use as a thermal fuse. As the temperature approaches the melting point of the polymer, 135°C for PE and 165°C for PP, the porosity of the membrane is lost. The trilayer material (PP/PE/PP) [41] has been developed by Celgard® where a PP layer is designed to maintain the integrity of the film, while the low melting point of PE layer is intended to shutdown the cell if an over-temperature condition is reached [42]. Asahi Kasai's flat-film membrane "Hipore™" is available in thicknesses ranging from 20 μm to several hundred micrometers, and with highly uniform pore sizes ranging from 0.05 to 0.5 μm [43]. The major manufacturers of lithium-ion battery separators along with their typical products are listed in Table 6.2.

In recent years there has been a strong demand for higher-capacity lithium-ion cells because of the strong growth in portable electronics. One way to achieve higher capacity is by reducing the thickness of separators. At present, battery manufacturers routinely use separators 16 μm or thinner in higher-capacity (>2.6 Ah) cylindrical cells and 9 μm separators in lithium-ion gel polymer cells.

Separator Development

The process for making lithium-ion battery separators can be broadly divided into dry and wet processes. Both processes usually employ one or more orientation steps to impart porosity and/or increase tensile strength. Dry process involves melting a polyolefin resin, extruding it into a film, thermal annealing to increase the size and amount of lamellar crystallites, and precisely stretching to form tightly ordered micropores [44–48]. In this process, a row lamellar crystal structure is generated in the polymer in the initial extrusion step. This nonporous structure is highly oriented as a result of extrusion and annealing conditions. The films are then stretched to form micropores. This microporous structure is continuous throughout the bulk interior of the membrane [49].

Polypropylene and polyethylene microporous films obtained by this method are available from Celgard, [44, 46, 50, 51] and Ube [52]. The dry process is technologically convenient and environmentally benign because no solvents are required. However, biaxial stretching has been met with limited success to date and, as a result, the pores are slit-like in shape, and the mechanical properties of films are anisotropic. The tensile strength in the lateral direction is relatively low.

Wet process (phase inversion process) [53, 54] involves mixing of hydrocarbon liquid or some other low-molecular-weight substance generally with a polyolefin resin, heating and melting the mixture, extruding the melt into a sheet, orientating the sheet either in the machine direction (MD) or biaxially, and then extracting the liquid with a volatile solvent [55, 56]. Separators made by wet process are available from Asahi Kasei [57], Tonen [58–60], Mitsui Chemicals [57], and more recently from Polypore/Membrana and Entek [31]. The structure and properties of the membranes can be changed by controlling the composition of the solutions and the evaporation or removal of solvents in the gelation and solidification processes. The separators made by wet process use ultrahigh-molecular-weight polyethylene (UHMWPE). The use of UHMWPE gives good mechanical properties as well as some degree of melt integrity.

Ihm et al. have given an overview of the wet process by preparing a separator with polymer blends of high-density polyethylene (HDPE) and ultrahigh-molecular-weight polyethylene (UHMWPE) [54]. They showed that the mechanical strength and drawing characteristics are influenced by the content and the molecular weight of the UHMWPE contained in a polyolefin blending solution. The manufacturing process of typical microporous film by dry and wet process is compared in Table 6.3.

A simplified flowchart for separator manufacturing process is shown in Fig. 6.3 [61]. The virgin polymer is prepared and mixed with processing aids (e.g., antioxidants, plasticizer, etc.) and then extruded. The extruded polymer then goes through different steps, which vary from process to process. For dry process, it can involve film annealing and stretching, while for wet process, it can involve solvent extraction and stretching. The finished film is then slit into required widths and packed into boxes and shipped to the battery manufacturers. With the advent of thinner separators, the film handling during manufacturing steps has become very

Table 6.3 Manufacturing process of typical microporous film

Process	Mechanism	Raw material	Properties	Typical membranes	Manufacturers
Dry process	Drawing	Polymer	Simple process Anisotropic film	PP, PE, PP/PE/PP	Celgard, Ube
Wet process	Phase separation	Polymer + Solvent	Isotropic film	PE	Asahi, Tonen
		Polymer + Solvent + Filler	Large pore size High porosity	PE	Asahi

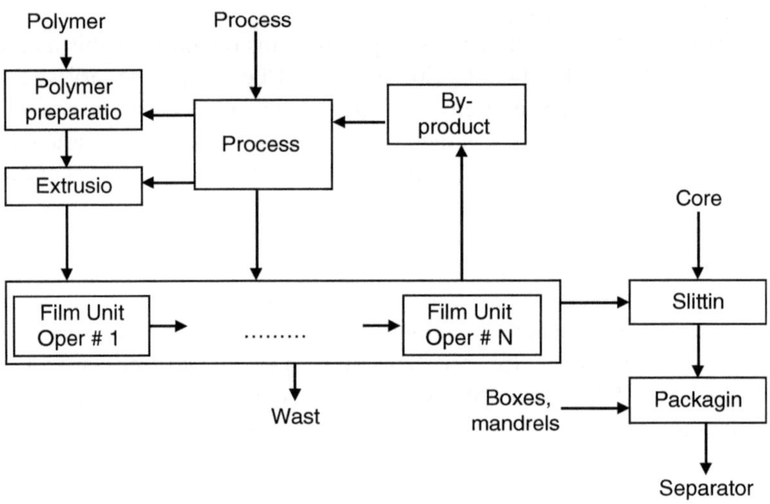

Fig. 6.3 Generalized process for lithium-ion separator manufacturing [61]. Each step of the separator manufacturing process has online detection systems to monitor the quality of the separator

important for the final quality of the film. Each step of the separator manufacturing process has online detection systems to monitor the quality of the separators.

Uniaxially oriented films generally have high strength in only one direction, whereas biaxially oriented films are relatively strong in both machine direction (MD) and transverse direction (TD). However, biaxial orientation often tends to introduce TD shrinkage. This shrinkage, at elevated temperatures, can allow electrodes to contact each other. The separator must have sufficient strength in the machine direction so that it does not decrease in width or break under the stress of winding. The typical requirement for the mechanical strength in a 25-μm separator is 1,000 kg/mm^2 [54].

The typical properties of some commercial microporous membranes are summarized in Table 6.4. Celgard 2730 and Celgard 2400 are single-layer PE and PP separators, respectively, while Celgard 2320 and 2325 are Trilayer separators of

Table 6.4 Typical properties of some commercial microporous membranes

Separator/ properties	Celgard 2730	Celgard 2400	Celgard 2320	Celgard 2325	Asahi Hipore	Tonen Setela
Structure	Single Layer	Single layer	Trilayer	Trilayer	Single layer	Single layer
Composition	PE	PP	PP/PE/PP	PP/PE/PP	PE	PE
Thickness (um)	20	25	20	25	25	25
Gurley (s)	22	24	20	23	21	26
Ionic resistivity[a] (Ω-cm^2)	2.23	2.55	1.36	1.85	2.66	2.56
Porosity (%)	43	40	42	42	40	41
Melt temperature (°C)	135	165	135/165	135/165	138	137

[a]In 1 M $LiPF_6$ EC:EMC (30:70 by volume)

20 and 25 μm thickness. Asahi and Tonen separators are single-layer PE separators made by wet process. Basic properties such as thickness, Gurley, porosity, melt temperature, and ionic resistivity are reported in Table 6.4. These properties are defined in a subsequent section.

Efforts have been made to find a new route for dry process using biaxial stretching techniques for preparing polypropylene microporous films, which may have submicron pore size and narrow size distribution, high permeability to gasses and liquids combined with good mechanical properties. The biaxially stretched polypropylene microporous films (Micpor®) were made by using nonporous polypropylene films of high beta-crystal content [62]. The porosity of these films can be as high as 30–40%, with an average pore size of approximately 0.05 μm. The pores on the surface were almost circular in shape compared to slit-like pores observed in uniaxial stretched samples and exhibited high permeability to fluids with good mechanical properties and almost circular pore shape with narrow pore-size distribution [63–65].

The PP/PE bilayer [38] and PP/PE/PP trilayer separators were developed by Celgard. Multilayer separators offer advantages of strength and combine the lower melting temperature of PE with the high-temperature strength of PP. Nitto Denko has also patented a single-layer separator made from a blend of PE/PP by the dry stretch process [66]. According to the patent, the separator has microporous regions of PE and PP. On heating in an oven, the impedance of the separator increases near the melting point of PE and the impedance remains high until beyond the melting point of PP. However, battery performance data have not been presented.

Microporous polyethylene separator material composed of a combination of randomly oriented thick and thin fibrils of ultrahigh-molecular-weight polyethylene (UHMWPE), Solupur®, manufactured by DSM Solutech, is also an interesting separator material for lithium-ion batteries. Solupur® is fabricated in standard grades with base weights ranging from 7 to 16 g/m^2 and mean pore size ranging from 0.1 to 2.0 μm and a porosity of 80–90% [67]. Ooms et al. carried out a study on a series of DSM Solupur materials with different permeability. Rate capability and cycling tests of these materials were compared with commercial available separators in

CR2320-type coin cells. Solupur® materials showed low tortuosity, high strength and puncture resistance, excellent wettability, good high rate capability and low-temperature performance because of its high porosity and UHMWPE structure [68].

Nitto Denko has also developed a battery separator made by a wet process that had high puncture strength and high heat rupture resistance [69]. They used a polyolefin resin with a high-molecular-weight rubber as its main component material and cross-linked through oxidation in air. The melt rupture temperature, as measured by thermomechanical analysis, was over 200°C in this material. They also tried cross-linking ultrahigh-molecular-weight polyethylene with electron-beam and ultraviolet irradiation, but this had the side effect of causing deterioration in the polyolefin including rupture of the main chains and therefore resulted in reduced strength.

ENTEK Membranes LLC has developed Teklon™ – a highly porous, ultrahigh-molecular-weight polyethylene separator for lithium-ion batteries. At the writing of this publication, the separator is available in small quantities. Pekala et al. characterized Celgard™, Setela™, and Teklon™ separators in terms of their physical, mechanical, and electrical properties [70].

Celgard's separators are the best-characterized battery separators in literature as they have been widely used in numerous battery systems. Bierenbam et al. [55] has described the process, physical, and chemical properties, and end-use applications. Fleming and Taskier [71] described the use of Celgard microporous membranes as battery separators. Hoffman et al. [72] presented a comparison of PP and PE Celgard microporous materials. Callahan discussed a number of novel uses of Celgard membranes. Callahan and coworkers [73] also characterized Celgard membranes by SEM image analysis, mercury porosimetry, air permeability, and electrical resistivity, and later characterize the puncture strength and temperature/impedance data for Celgard membranes [38]. Spotnitz et al. reported short-circuit behavior in simulated, spirally wound cells, as well as impedance/temperature behavior and thermomechanical properties [42]. Yu [74] found that a trilayer structure of PP/PE/PP Celgard™ microporous membranes provided exceptional puncture strength.

Nonwoven materials such as cellulosic fibers have never been successfully used in lithium batteries. This lack of interest is related to the hygroscopic nature of cellulosic papers and films, their tendency to degrade in contact with lithium metal, and their susceptibility to pinhole formation at thickness of less than 100 μm.

Asahi Chemical Industry carried out an exploratory investigation to determine the requirements for cellulose-based separators for lithium-ion batteries [75]. In an attempt to obtain an acceptable balance of lithium-ion conductivity, mechanical strength, and resistance to pinhole formation, they fabricated a composite separator (39–85 μm) that consists of fibrilliform cellulosic fibers (diameter 0.5–5.0 μm) embedded in a microporous cellulosic (pore diameter: 10–200 nm) film. The fibers are intended to reduce the possibility of separator meltdown under exposure to heat generated by overcharging or internal short circuiting. The resistance of these films was equal to, or lower than, the conventional polyolefin-based microporous separators. The long-term cycling performance was also very comparable.

Pasquier et al. [76] used paper-based separators in flat pouch-type lithium-ion batteries and compared the performance with cells made with Celgard-type polyolefin-based separators. The paper separators had good wetting properties, good mechanical properties, but did not provide the shutdown effect essential for large lithium-ion batteries. Their resistance was similar to polyolefin separators and when all water traces were removed from paper, their cycling performance was similar to Celgard separators. The paper-based separators can be used in small flat pouch-type cells where high strength and shutdown behavior is not required. For larger spherically wound cells, which require strong separators with shutdown feature, paper-based separators cannot be used.

Recently Degussa announced that they have developed Separion® separators for lithium batteries by combining the characteristics of flexible polymeric separators with the advantages of chemical and thermally resistant and hydrophilic ceramic materials. Separion® is produced in a continuous coating process. Ceramic materials, e.g., alumina, silica, and/or zirconia, are slip coated and hardened onto a support [77, 78]. According to Degussa, Separion separators have an excellent high temperature stability, superior chemical resistance, and good wettability, especially at low temperatures. They tested the performance and safety behavior of Separion separator in 18,650 cells and found the performance to be comparable to polyolefin-based separators [79].

The potential use of polymeric ion-exchange membranes in the next generation single-ion secondary lithium polymer batteries was shown by Sachan et al. [80, 81]. Conductivities exceeding 10^{-4} S/cm with transference numbers of unity were achieved for Nafion converted to Li^+ salt form. However, little work has continued to introduce ion-exchange membranes in lithium batteries. The cost associated with the manufacturing of such membranes has been a significant barrier to their commercial viability.

To obtain a thin (less than 15 μm) separator for lithium batteries, Optodot has taken a different approach of high-speed coating of a metal oxide sol gel coating on a smooth surface followed by a delamination step to provide the free-standing separator. Using this approach, separator with thicknesses from 6 to 11 μm was made on large-scale production coating equipment [82]. They found that the sol gel separators with a thickness in the middle of this range of 8–9 μm have the preferred combination of thinness and strength. The metal oxide sol gel coating is water based with no organic solvents present. The coating formulations include a polymer and a surfactant. The polymer provides improved coating rheology, mechanical strength, and other properties. The surfactant provides improved wetting properties on the substrate. The films prepared were around 11 μm thick, with 45% porosity, completely wettable in nonaqueous electrolyte, and melt temperature greater than 180°C. While these films are relatively thin and should help in increasing the capacity, they may not be strong enough for use in tightly wound cells. Moreover, the shutdown temperature of the separator seems to be very high and thus not suitable for existing lithium-ion chemistries.

Gineste et al. carried out the grafting of hydrophilic monomers onto PP or PE separators to improve the wettability of separators used in secondary lithium batteries with a lower content of wetting agents [83, 84]. They used a PP film (Celgard 2505) of 50 μm thickness after irradiating in air by electron beams with a dose ranging from 0.5 to 4 Mrad. The irradiated film was grafted by a monofunctional monomer (acrylic acid, AA), in the presence of difunctional cross-linking agent (diethyleneglycol dimethacrylate, DEGDM). The separators start losing mechanical properties, when the grafting ratio is higher than 50%.

Separator Requirements

In lithium-based cells, the essential function of battery separator is to prevent electronic contact, while enabling ionic transport, between the positive and negative electrodes. It should be usable on high-speed winding machines and possess good shutdown properties. The most commonly used separators for primary lithium batteries are microporous polypropylene membranes. Microporous polyethylene and laminates of polypropylene and polyethylene are widely used in lithium-ion batteries [85]. These materials are chemically and electrochemically stable in secondary lithium batteries. A key requirement for the separators for lithium primary batteries is that their pore size be small enough to prevent dendritic lithium penetration through them.

The general requirements [86] for lithium-ion battery separators are given below:

1. *Thickness* – The lithium-ion cells used in consumer applications use thin microporous separators (<25 μm). The separators being developed for EV/HEV applications will require thicker (~40 μm) separators. The thicker the separator, the greater the mechanical strength and the lower the probability of punctures during cell assembly, but reduce the amount of active materials that can be placed in the same size can. The thinner separators lower the internal resistance, take up less space, and permit the use of longer electrodes for increased capacity and rate capability.
2. *Permeability* – The separators should not limit the electrical performance of the battery under normal conditions. Typically the presence of separator increases the effective resistivity of the electrolyte by a factor of 6–7. The ratio of the resistivity of the separator filled with electrolyte divided by the resistivity of the electrolyte itself is called MacMullin number. The rate capability of the battery is inversely proportional to the MacMullin number; however, it has been shown that increase in resistance across the electrode material has a larger impact on the cell performance for comparable changes made to the separator [87].
3. *Gurley* (*air permeability*) – Air permeability is proportional to electrical resistivity, for a given separator morphology. It is used in place of electrical resistance (ER) measurements once the relationship between Gurley and ER is established. The separator should have low Gurley values for good electrical performance.

4. *Porosity* – It is implicit in the permeability requirement; typically lithium-ion battery separators have a porosity of 40%. Control of porosity is very important for battery separators. Specification of percent porosity is commonly an integral part of separator acceptance criteria. The porosity of separators used in alkaline zinc MnO_2 cells is typically around 80–90%.
5. *Wettability* – The separators should wet out quickly and completely in typical battery electrolytes. The lack of wetting results in localized spots of high resistance.
6. *Electrolyte absorption and retention* – A separator should be able to absorb and retain electrolyte. Electrolyte absorption is needed for ion transport. The microporous membranes usually do not swell on electrolyte absorption.
7. *Chemical stability* – The separators should be stable in battery for a long period of time. It should be inert to both strong reducing and strong oxidizing conditions and should not degrade or loose mechanical strength or produce impurities, which can interfere with the function of the battery. The separator must be able to withstand the strong oxidizing positive electrode and the corrosive nature of the electrolyte at temperatures as high as 75°C. The greater the oxidation resistance, the longer the separator will survive in a cell. Polyolefins (e.g., polypropylene, polyethylene) exhibit high resistance to most of the conventional chemicals, good mechanical properties, and moderate temperature range for application making it one of the ideal polymers for lithium-ion battery separators. Polypropylene separators exhibit better mechanical properties and minimal oxidation when in contact with the positive electrode in a lithium-ion cell. Thus, the performance of trilayer (PP/PE/PP) separators with PP as the outside layer and PE as inner layer are superior.
8. *Dimensional stability* – The separator should lay flat and should not curl at the edges when unrolled as this can greatly complicate cell assembly. The separator should also not shrink when exposed to electrolyte. The cell winding should not affect the porous structure in any adverse way.
9. *Puncture strength* – The separators used in wound cells require high puncture strength to avoid penetration of electrode material through the separator. If particulate material from the electrodes penetrates the separator, an electrical short will result and the battery will be rejected. The separators used in lithium-ion batteries require more strength then the one used in lithium primary batteries. The primary lithium batteries have only one rough electrode and thus it requires less strength. As empirically observed, for most applications, the puncture strength should be at least 400 g/mil for separators used in lithium-ion cells. Mix penetration strength is a better measure of separator strength in a battery compared to puncture strength.
10. *Mix penetration strength* – The susceptibility of separators to particle penetration is characterized by mix penetration strength [45]. During the winding of the spiral wrap construction considerable mechanical pressure is applied to the cathode–separator–anode interface. Any loose particle could be forced through the separator and short the cell. The mix penetration strength should be at least 100 kgf/mil for separators used in lithium-ion cells.

11. *Thermal stability* – Lithium-ion batteries can be poisoned by water and so materials going into the cell are typically dried at 80°C under vacuum. Under these conditions, the separator must not shrink significantly and definitely must not wrinkle. Each battery manufacturer has specific drying procedures. The requirement of less than 5% shrinkage after 60 min at 90°C (in vacuum) in both MD and TD direction is a reasonable generalization.
12. *Pore size* – A key requirement of separators for lithium batteries is that their pores be small enough to prevent dendritic lithium penetration through them. Membranes with submicron pore sizes have proven adequate for lithium batteries.
13. *Tensile strength* – The separator is wound with the electrodes under tension. The separator must not elongate significantly under tension in order to avoid contraction of the width. A tensile strength specification is sometimes given, but the key parameter is Young's Modulus in the machine direction. Since Young's Modulus is difficult to measure, 2% offset yield is a good measure; less than 2% offset at 1,000 psi is acceptable for most winding machines.
14. *Camber* – Ideally, when a strip of separator is laid out, the separator should be straight and not bow or skew. In practice, however, some camber is often observed. If sufficiently extreme, this can cause misalignment between the electrodes and separator. Camber can be measured by laying the separator flat on a table parallel with a straight meter stick. The camber should be less than 0.2 mm/m of separator.
15. *Shutdown* – Lithium-ion battery separators provide some margin of protection against short circuit and overcharge in lithium-ion cells. The separators exhibit a large increase in impedance at temperature about 130°C that effectively stops ionic transport between the electrodes [88, 89]. The greater the mechanical integrity of the separator above 130°C, the greater the margin of safety the separator can provide. If the separator looses mechanical integrity, then the electrodes can come into direct contact, react chemically, and result in thermal runaway. The shutdown behavior of a separator can be characterized by heating the separator (saturated with electrolyte) to high temperatures and simultaneously monitoring the electrical resistance of the separator [42, 89].
16. *High temperature stability* – A separator might provide an extra margin of safety if it can prevent the electrodes from contacting one another at high temperatures. Separators with good mechanical integrity at high temperatures can provide a greater margin of safety for lithium-ion cells. Thermal Mechanical Analysis (TMA) can be used to characterize the high temperature stability of separators. Utilizing TMA, the separator is held under constant load and the degree of elongation versus temperature is measured; the temperature at which the separator loses mechanical integrity, the elongation increases dramatically.
17. *Electrode interface* – The separator should form a good interface with the electrodes to provide sufficient electrolyte flow.

Table 6.5 General requirements for lithium-ion battery separator [86]

Parameter	Goal
Thickness[a,b] (μm)	<25
Electrical resistance (MacMullin No.[c], dimensionless)	<8
Electrical resistance (ohms-cm^2)	<2 Ω-cm^2
Gurley[d] (s)	~25 ml
Pore size[e] (μm)	<1
Porosity (%)	~40
Puncture strength[f] (g)	>300 g/mil
Mix penetration strength (kgf)	>100 kgf/mil
Shrinkage[g] (%)	<5% in both MD and TD
Tensile strength[h]	<2% offset at 1,000 psi
Shutdown temperature (°C)	~130
High temperature melt integrity (°C)	>150
Wettability	Complete wet out in typical battery electrolytes
Chemical stability	Stable in battery for long period of time
Dimensional stability	Separator should lay flat; be stable in electrolyte
Skew	<0.2 mm/m

[a]ASTM D5947-96, "Standard test methods for physical dimensions of solid plastics specimens," ASTM International.
[b]ASTM D2103, "Standard specification for polyethylene film and sheeting," ASTM International.
[c]Caldwell DL, Poush KA (1984) US Patent 4,464,238
[d]ASTM D726, "Standard test methods for identification of fibers in textiles," ASTM International.
[e]ASTM E128-99, "Standard test method for maximum pore diameter and permeability of rigid porous filters for laboratory use," ASTM International.
[f]ASTM D3763, "Standard test method for high-speed puncture properties of plastics using load and displacement sensors," ASTM International.
[g]ASTM D1204, "Standard test methods for linear dimensional changes of nonrigid hermoplastic sheeting or film at elevated temperatures," ASTM International.
[h]ASTM D882, "Standard test method for tensile properties of thin plastic sheeting," ASTM International.

In addition to the above properties, the separator must be essentially free of any type of defects (pinholes, gels, wrinkles, contaminants, etc.). All of the above properties have to be optimized before a membrane qualifies as a separator for a lithium-ion battery. The general requirements for lithium-ion battery separators are also summarized in Table 6.5.

Separator Properties/Characterization

Separators are characterized by structural and functional properties; the former describes what they are and the latter how they perform. The structural properties include chemical (molecular) and microcrystalline nature, thickness, pore size, pore-size distribution, porosity, and various chemical and physical properties such as chemical stability, and electrolyte uptake. The functional properties of interest are electrical resistivity, permeability, and transport number. It is useful to characterize separator materials in terms of their structural and functional properties, and to establish a correlation of these properties with their performance in batteries. A variety of techniques are used to evaluate separators. Some of these techniques are discussed in this section.

Gurley Number

Separator permeability is usually characterized by air permeability. The Gurley number expresses the time required for a specific amount of air to pass through a specific area of separator under a specific pressure. The standard test method is described in ASTM-D726 (B).

The Gurley number is used to characterize separators because the measurement is accurate and easy to make, and deviations from specific values are a good indication of problems. Air permeability (Gurley) is proportional to electrical resistance (ER), for a given separator morphology [90]. Gurley can be used in place of ER measurements once the relationship between Gurley and ER is established. A lower Gurley value means higher porosity, lower tortuosity, and accordingly lower ER.

Electrical Resistance

The measurement of separator resistance is very important to the art of battery manufacture because of the influence the separator has on electrical performance. Electrical resistance is a more comprehensive measure of permeability then the Gurley number, in that the measurement is carried out in the actual electrolyte solution. The ionic resistivity of the porous membrane is essentially the resistivity of the electrolyte that is embedded in the pores of the separator. Typically, a microporous separator, immersed in an electrolyte, has an electrical resistivity about six to seven times that of a comparable volume of electrolyte, which it displaces. It is a function of the membrane's porosity, the tortuosity, the resistivity of the electrolyte, the thickness of the membrane, and the extent to which the electrolyte wets the pores of the membrane [83]. The electrical resistance of the separator is the true performance indicator of the cell. It describes a predictable voltage loss within the cell during discharge and allows one to estimate rate limitations.

Classical techniques for measuring electrical resistivity of microporous separators have been described by Falk and Salkind [5] and Robinson and Walker [42]. The resistivity of an electrolyte is more accurately determined by AC methods since DC can polarize the electrodes and cause electrolysis of the solution. Modern AC impedance measuring systems allow rapid measurements of cell resistance over a wide range or frequencies from which resistance can be calculated free of capacitance effects. Compared to the DC techniques, the equipment required and the theory necessary to interpret the AC techniques are more complex; however, AC measurements yield information about long-range migration of ions and polarization phenomena occurring within the cell. In an AC measurement, a sinusoidal voltage is applied to a cell, and the sinusoidal current passing through the cell as a result of this perturbation is determined. A four-electrode cell is usually used for resistivity measurements. The outer two electrodes serve to apply a sinusoidal potential, and the resulting current passing through the inner two electrodes is measured. This technique is employed to avoid the complications arising from a nonuniform potential field near the outer two electrodes. An excellent review of experimental techniques for measuring electrical resistivity in aqueous solution is available [91, 92].

The separator resistance is usually characterized by cutting small pieces of separators from the finished material and then placing them between two blocking electrodes. The separators are completely saturated with the electrolyte. The resistance (Ω) of the separator is measured at a certain frequency by AC impedance techniques. The frequency is chosen so that the separator impedance is equal to the separator resistance. In order to reduce the measurement error, it is best to do multiple measurements by adding extra layers. The average resistance of single layer is determined from multiple measurements. The specific resistivity, ρ_s (Ω-cm), of the separator saturated with electrolyte is given by

$$\rho_s = \frac{R_s A}{l} \tag{6.1}$$

where R_s is the measured resistance of separator in Ω, A is the electrode area in square centimeters and l is the thickness of membrane in centimeters. Similarly, the specific resistivity of the electrolyte, ρ_e (Ω-cm), is given by

$$\rho_e = \frac{R_e A}{l} \tag{6.2}$$

where R_e is the measured resistance of electrolyte in ohms. The ratio of the resistivity of a separator membrane to that of the electrolyte is called MacMullin number, N_m, which can be used to predict the influence of the separator on battery performance [93].

$$N_m = \frac{\rho_s}{\rho_e} = \frac{\tau^2}{\varepsilon} \tag{6.3}$$

where τ is the tortuosity and ε is the porosity of the separator. The MacMullin number describes the relative contribution of a separator to cell resistance. It is almost independent of electrolyte used and also factors out the thickness of the material. It assumes that the separator wets completely in the electrolyte used for the test. From Eqs. 6.1 and 6.3 the electrical resistance of a microporous membrane is given by the following [5, 94]:

$$R_m = \rho_e\left(\frac{\tau^2 l}{\varepsilon A}\right) \tag{6.4}$$

It has been shown for Celgard membranes that the membrane resistance can be related to the Gurley number by Eq. 6.5. where R_m is the membrane resistance (Ω), A is the membrane area (cm^2), ρ_e is the specific electrolyte resistance (Ω-cm), t_{gur} is the Gurley number (10 cm^3 air, 2.3 mmHg), d is the pore size, and 5.18×10^{-3} a scaling factor [90].

$$R_m A = \frac{\rho_e}{5.18 \times 10^{-3}} t_{\text{gur}} d \tag{6.5}$$

The usual procedure for characterizing battery separators is to cut several test samples from the finished material. Thus, only a small portion of the separator is actually examined. Ionov et al. has proposed an alternative technique to measure the resistance of a separator over a large separator area [95]. In this technique the separator material is passed through an electrolyte bath between electrical resistance measuring transducers. The set of transducers installed in the bath transverse to the moving sheet of separator material examines the whole surface of the material. If the production process ensures good uniformity in the physicochemical properties of the separator material over the whole surface, the transducer outputs will be close to one another. A nonuniform separator will cause significant deviations from the average value at various sections of the material examined. In this case, the sections having lower or higher resistance compared with the average value should be regarded as flawed.

Porosity

The porosity is important for high permeability and also for providing a reservoir of electrolyte in the cell. A higher and uniform porosity is desirable for unhindered ionic current flow. Nonuniform porosity leads to nonuniform current density and can further lead to reduced activity of the electrodes. Cell failure can result, if during discharge, some areas of the electrodes work harder than other.

Porosity of a separator is defined as the ratio of void volume to apparent geometric volume. It is usually calculated (Eq. 6.6) from the skeletal density, basis weight, and dimensions of the material and so may not reflect the accessible porosity of the material.

$$\text{Porosity}(\%) = 1 - \frac{\left(\frac{\text{Sample weight}}{\text{Sample volume}}\right)}{\text{Polymer density}} \times 100\% \tag{6.6}$$

The standard test method is described in ASTM D-2873. The actual or accessible porosity can also be determined by the weight of liquid (e.g., Hexadecane) absorbed in the pores of the separator. In this method, the separator weight is measured before and after dipping in Hexadecane solvent and the porosity is calculated (Eq. 6.7) by assuming that volume occupied by Hexadecane is equal to the porous volume of the separator.

$$\text{Porosity}(\%) = \frac{\text{Volume occupied by Hexadecane}}{\text{Volume of Polymer Volume occupied by Hexadecane}} \times 100 \tag{6.7}$$

Tortuosity

Tortuosity is the ratio of mean effective capillary length to separator thickness. The tortuosity factor τ of a separator can be expressed by

$$\tau = \frac{l_s}{d} \tag{6.8}$$

where l_s is the ion path through the separator and d is the thickness of the separating layer.

Tortuosity is a long-range property of a porous medium, which qualitatively describes the average pore conductivity of the solid. It is usual to define τ by electrical conductivity measurements. With knowledge of the specific resistance of the electrolyte and from a measurement of the sample membrane resistance, thickness, area, and porosity, the membrane tortuosity can be calculated from Eq. 6.3.

This parameter is widely used to describe the ionic transport by providing information on the effect of pore blockage. A tortuosity factor $\tau = 1$, therefore, describes an ideal porous body with cylindrical and parallel pores, whereas values of $\tau > 1$ refer to more hindered systems. Higher tortuosity is good for dendrite resistance but can lead to higher separator resistance.

Pore Size and Pore-Size Distribution

For any battery applications, the separator should have uniform pore distribution to avoid performance losses arising from nonuniform current densities. The submicron pore dimensions are critical for preventing internal shorts between the anode and the cathode of the lithium-ion cell, particularly since these separators tend to be as thin as 25 μm or less. This feature will be increasingly important as battery manufacturers continue to increase the cell capacity with thinner separators. The pore structure is usually influenced by polymer composition, and stretching conditions, such as drawing temperature, drawing speed, and draw ratio. In the wet process, the separators produced by the process of drawing after extraction (as claimed by Asahi Chemical and Mitsui Chemical) are found to have much larger pore size (0.24–0.34 μm) and wider pore-size distribution than those produced by the process of extraction (0.1–0.13 μm) after drawing (as claimed by Tonen) [54].

The testing of battery separators and control of their pore characteristics are important requirements for proper functioning of batteries. Mercury porosimetry has been historically used to characterize the separators in terms of percentage porosity, mean pore size, and pore-size distribution [96]. In this method, the size and volume of pores in a material are measured by determining the quantity of mercury, which can be forced into the pores at increasing pressure. Mercury does not wet most materials and a force must be applied to overcome the surface tension forces opposing entry into the pores.

The hydrophobic (e.g., polyolefins) separators are also characterized with Aquapore (nonmercury porosimetry) technique, where water is used in place of mercury. This is a very useful technique for characterizing polyolefin-based separators used in lithium batteries [97]. Porosimetry gives pore volume, surface area, mean pore diameter, and pore-size distribution. In a typical experiment, the sample is placed in the instrument and evacuated. As the pressure increases, the quantity of water forced into the pores increases in proportion to the differential pore volume, the size of the pores corresponding to the instantaneous pressure. Thus, increasing the pressure on a membrane having a given pore-size distribution results in a unique volume versus pressure or pore diameter curve. The pressure required for intrusion of water into a pore of diameter D is given by following equation

$$D = \frac{4\gamma \cos \theta}{p} \tag{6.9}$$

where D is the diameter of the pore assuming the pore to be cylindrical, p is the differential pressure, γ is the surface tension of the nonwetting liquid, water, and θ is the contact angle of water. The pores generally are not of spherical shape or a constant diameter. They usually vary in their form and size. Thus, statements of any pore diameter are always to be viewed with the above in mind.

Another technique, Capillary Flow Porometry has been developed by Porous Materials Inc. [98] to characterize battery separators [99, 100]. The instrument can measure a number of characteristics of battery separators such as size of the pore at its most constricted part, the largest pore size, pore-size distribution, permeability, and envelope surface area [101].

Scanning electron microscopy (SEM) is also used to examine separator morphology. SEM pictures of some commercial membranes are shown in Figs. 6.4–6.6. The surface SEM of Celgard 2400, 2500, and 2730 are shown in Fig. 6.4. It is clear from the images that the pores are uniformly distributed. Both Celgard 2400 and 2500 are single-layer PP separators, but the pore size of Celgard 2500 is substantially larger than Celgard 2400. Thus, it has lower resistance and is more suited for high rate applications. Figure 6.5 shows the surface SEM and cross-section SEM of Celgard 2325. The surface SEM only shows the PP pores while the PE pores are

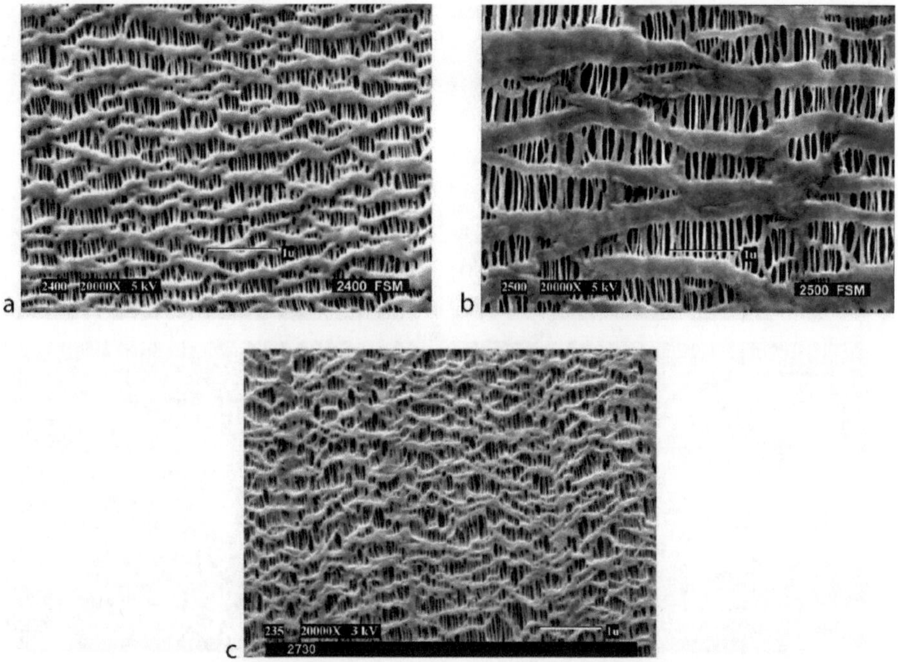

Fig. 6.4 Scanning electron micrographs of surface of single-layer Celgard separators used in lithium batteries (**a**) 2400 (PP), (**b**) 2500 (PP), and (**c**) 2730 (PE)

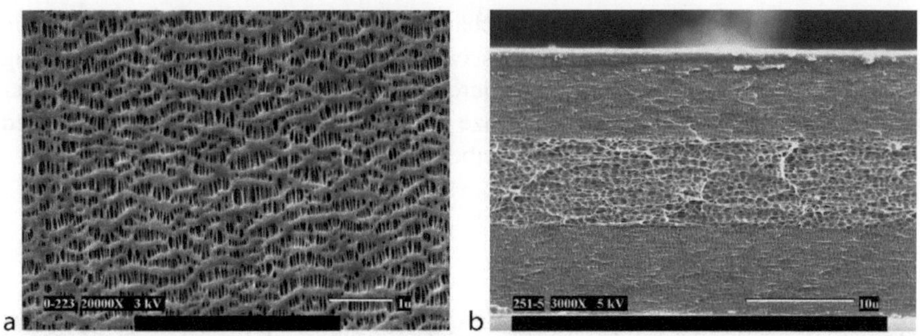

Fig. 6.5 Scanning electron micrographs of Celgard trilayer separator (2325) used in lithium-ion batteries (**a**) surface SEM; (**b**) cross-section SEM

visible in the cross-section. It is clear from the image that all three layers are of equal thickness. The SEM pictures of separators made by wet process are shown in Fig. 6.6. The pore structure of all of these membranes is very similar. Asahi-1 (Fig. 6.6b) separator has significantly larger pores compared to the other membranes.

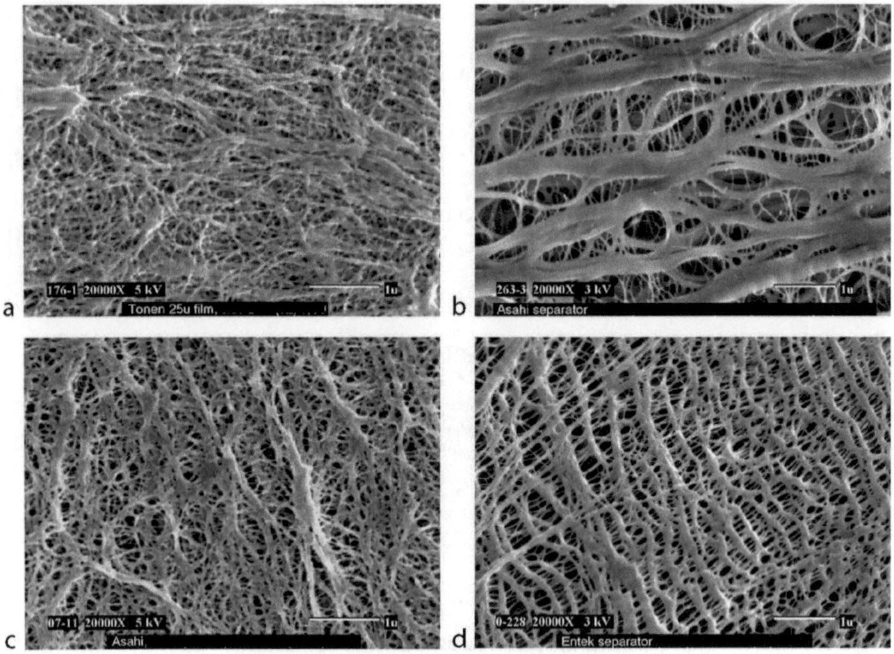

Fig. 6.6 Scanning electron micrographs of separators made by wet process and used in lithium-ion batteries (**a**) Setela (Tonen), (**b**) Hipore–1 (Asahi), (**c**) Hipore-2 (Asahi), and (**d**) Teklon (Entek)

Image analysis has been used to characterize the pore structure of synthetic membrane materials [102]. The films can also been characterized by scanning tunneling microscopy, atomic force microscopy, and field emission scanning electron microscopy [49, 103]. The pore size of the membranes can also be calculated from Eq. 6.5, once the MacMullin number and Gurley values are known.

Puncture Strength

A separator is required to have sufficient physical strength to endure the rigors of cell assembly and day-to-day charge–discharge cycling. Physical strength is required to withstand basic handling, cell blocking/assembly, physical shock, punctures, abrasion, and compression.

The Puncture Strength (PS) is the weight that must be applied to a needle to force it completely through a separator [55, 104]. It has been used to indicate the tendency of separators to allow short circuits in a cell that may occur due to holes generated in the separator by the rough surface of an electrode during the battery assembly and charge–discharge cycle. The PS requirement for lithium-ion batteries is higher them lithium-foil batteries, because the separator must contend with two rough surfaces. Commercially available puncture strength machines made for textiles tend to give meaningless results when testing battery separator membranes. More reproducible results can be obtained with a load frame (such as an Instron Machine). The mix penetration strength is a better measure of mechanical strength for battery separators as it measures the force required to create a short through the separator when electrode mix is pushed through it.

The strength of the separator depends greatly on the materials used and the manufacturing method. The wet-biaxial method simultaneously stretches in the MD and TD directions and thus achieves a material that has tensile modulus and rupture strength in both directions. Both high polymer entanglement and stretching help increase the physical strength of the separator.

Mix Penetration Strength

The force required to create a short through a separator due to mix (electrode material) penetration defines mix penetration strength. In this test, force (with a ½-inch diameter ball) is applied on the positive electrode/separator/negative electrode sandwich and the force at which the mix penetrates through the separator and creates an electronic short is called mix penetration force. Mix penetration strength is used to indicate the tendency of separators to allow short circuits during battery assembly. The mix penetration resistance test is more closely related to particle penetration resistance compared to puncture resistance [45].

Tensile Strength

The tensile strength measurements (e.g., Young's Modulus, Percent offset strength, elongation at break, stress at break) can be made by utilizing widely known standard procedures. These tests are carried out in both MD and TD directions. The tensile properties are dependent on the manufacturing process. The Uniaxially oriented films have high strength in only one direction, whereas biaxially oriented films are more uniformly strong in both MD and TD directions. ASTM test method D882-00 "Standard test method for tensile properties of thin plastic sheeting" is an appropriate test.

The separator should be strong enough to withstand mechanical handling during cell winding and assembly. It should be dimensionally stable and should not neck down during winding. The decrease in width will allow the electrodes to touch each other and create a short. Thus, the tensile property of the separator should be very strong in MD direction compared to TD direction.

Shrinkage

Shrinkage test is carried out on both MD and TD directions. In this test, the dimensions of separators are measured and then stored at 90°C for a fixed time. The shrinkage is then calculated from the change in dimensions as shown in Eq. 6.10,

$$\text{Shrinkage } (\%) = \frac{L_i - L_f}{L_i} \times 100 \tag{6.10}$$

where L_i is the initial length and L_f is the final length of separator after high temperature storage. The uniaxially stretched separators tend to shrink in MD direction only, while the biaxially stretched separators shrink in both MD and TD directions. The shrinkage of separators can also be compared by carrying out Thermal Mechanical Analysis (TMA) test at a constant load and rate.

Shutdown

Separator shutdown is a useful and essential mechanism for limiting temperature and preventing venting in short-circuited cells [42]. It usually takes place close to the melting temperature of the polymer when the pores collapse turning the porous ionically conductive polymer film into a nonporous insulating layer between the electrodes. At this temperature, a significant increase in cell impedance occurs and passage of current through the cell is restricted. This prevents further electrochemical activity in the cell, thereby shutting the cell down before an explosion can occur.

The ability of the PE-based separator to shutdown the battery is determined by its molecular weight, percent crystallinity (density), and process history. Material properties and processing methods might need to be tailored so that the shutdown response is spontaneous and complete. The optimization needs to be done without affecting the mechanical properties of the material in the temperature range of interest. This process is easier to do with the trilayer separators, since one material is utilized for the shutdown response and another for the mechanical properties. Polyethylene containing separators, in particular trilayer laminates of polypropylene, polyethylene, and polypropylene, appear to have the most attractive properties for preventing thermal runaway in lithium-ion cells [101–105]. The shutdown temperature of 130°C is usually sufficient to control the cell heating and avoid thermal runaway in lithium-ion cells. A lower temperature shutdown will be desirable if it does not affect the separator mechanical properties or high temperature cell performance in any adverse way.

The shutdown property of separators is determined by measuring the impedance of a separator while the temperature is linearly increased [42, 89]. Figure 6.7 shows actual measurement for Celgard® 2325 membrane. The heating rate was around 60°C/min and the impedance was measured at 1 kHz. The rise in impedance corresponds to a collapse in pore structure due to melting of the separator. A 1,000-fold increase in impedance is necessary for the separator to stop thermal runaway in the battery. The drop in impedance corresponds to opening of the separator due to coalescence of the polymer, and/or to penetration of the separator by the electrodes; this phenomenon is referred to as a loss in "melt integrity." This test is fairly reliable in indicating the temperature at which the impedance rises, but some variability in characterizing the subsequent drop in impedance may occur.

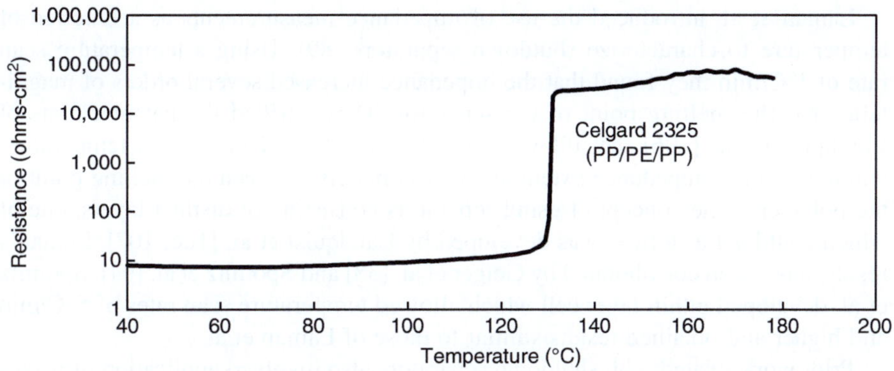

Fig. 6.7 Internal impedance (at 1 kHz) of Celgard 2325 (PP/PE/PP) separator as a function of temperature. Heating rate: 60°C/min

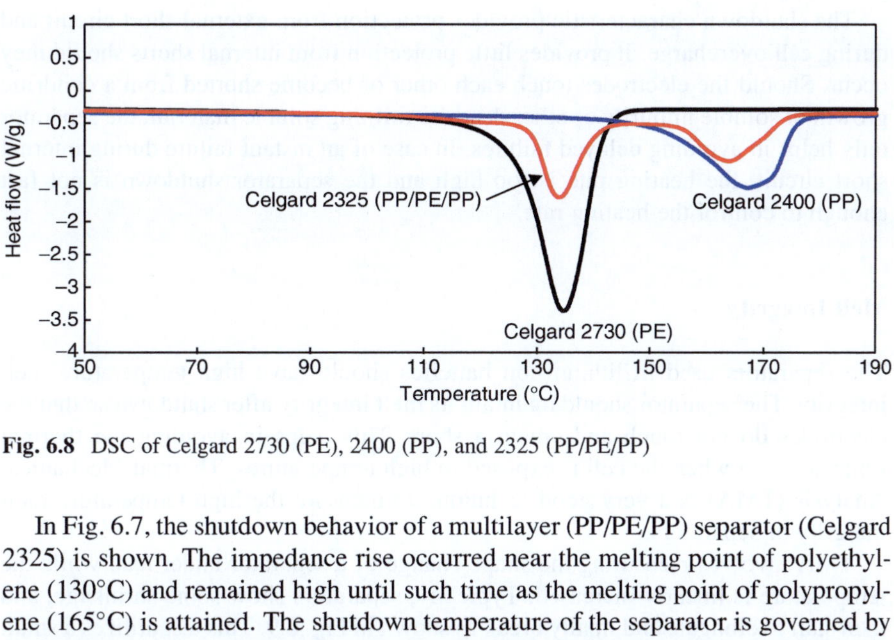

Fig. 6.8 DSC of Celgard 2730 (PE), 2400 (PP), and 2325 (PP/PE/PP)

In Fig. 6.7, the shutdown behavior of a multilayer (PP/PE/PP) separator (Celgard 2325) is shown. The impedance rise occurred near the melting point of polyethylene (130°C) and remained high until such time as the melting point of polypropylene (165°C) is attained. The shutdown temperature of the separator is governed by the melting point of the separator material. At the melting point, the pores in the separator collapse to the form a relatively nonporous film between the anode and the cathode. This was confirmed by DSC. The DSC scan in Fig. 6.8 gives a peak melting temperature of 135°C for Celgard 2730, 168°C for Celgard 2400, and 135/165°C for Celgard 2325. The shutdown behavior of thinner separators (<20 μm) is very similar to thicker separators. The battery manufacturers have been very successful in using the thinner separators without compromising on the shutdown behavior of the separators.

Laman et al. introduced the use of impedance measurements as a function of temperature to characterize shutdown separators [89]. Using a temperature scan rate of 1°C/min they found that the impedance increased several orders of magnitude near the melting point of the separator. They verified the patent claims of Lundquist et al. [106] that bilayer separators of PE and PP gave a temperature window of high impedance extending approximately between the melting point of the polymers. The concept of using separators consisting of distinct layers, one of which could act as a fuse, was developed by Lundquist et al. [106, 107]. Laman's results have been corroborated by Geiger et al. [38] and Spotnitz et al. [94]. Spotnitz et al. developed a thin layer cell which allowed temperature scan rates of 5°C/min and higher and obtained results similar to those of Laman et al.

Prior work related with shutdown separators also involved application of waxes on membranes [108, 109]. In these cases, the wax or low melting polymers were coated on the polyolefin separator. The disadvantage of this technique is that the coating can block the pores of the separator and thus can affect the performance by increasing separator resistance. Moreover, the coating level has to be very high to get complete shutdown.

The shutdown characteristic provides protection from external short circuit and during cell overcharge. It provides little protection from internal shorts should they occur. Should the electrodes touch each other or become shorted from a dendritic growth of soluble impurity or other dendrite forming soluble material, the separator only helps in avoiding delayed failures. In case of an instant failure during internal short circuit, the heating rate is too high and the separator shutdown is not fast enough to control the heating rate.

Melt Integrity

The separators used in lithium-ion batteries should have high temperature melt integrity. The separator should maintain its melt integrity after shutdown so that the electrodes do not touch and create a short. This helps in avoiding the thermal runaway even when the cell is exposed to high temperatures. Thermal Mechanical Analysis (TMA) is a very good technique to measure the high temperature melt integrity of separators.

TMA involves measuring the shape change of a separator under load while the temperature is linearly increased. Typically, separators show some shrinkage, and then start to elongate and finally break as shown in Fig. 6.9. This test utilizes a small separator sample (about 5–10 mm length (MD) and about 5 mm width), which is held in mini-instron-type grips. The sample is held with a constant 2 g load while the temperature is ramped at 5°C/min past the melting point until the tension ruptures the film. Three parameters are reported from TMA test – shrinkage onset temperature, melt temperature, and melt rupture temperature. TMA has proven to be a more reproducible measure of melt integrity of the separator [107].

Figure 6.9 shows the TMA data for two different Celgard membranes. The shrinkage onset temperature, deformation temperature, and rupture temperature are

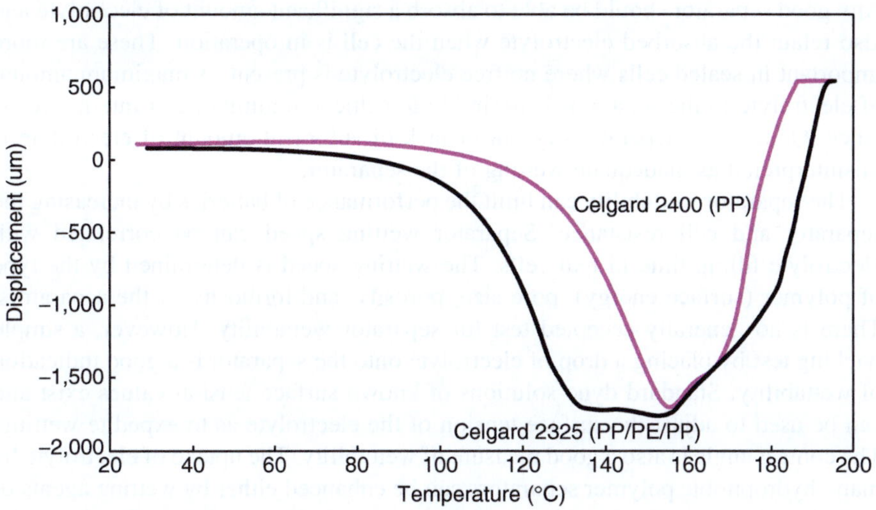

Fig. 6.9 TMA of Celgard 2400 (PP) and 2325 (PP/PE/PP). A constant load (2 g) is applied while the temperature is ramped at 5°C/min

Table 6.6 TMA data for typical Celgard separators

	Celgard 2730	Celgard 2400	Celgard 2325
Shrinkage onset temperature (°C)	100	121	106
Deformation temperature (°C)	125	156	135, 154
Rupture temperature (°C)	140	183	192

summarized in Table 6.6. The single-layer PP membrane (Celgard 2400) showed a higher softening temperature (~121°C), a deformation temperature around 160°C, and a very high rupture temperature around 180°C. The multilayer polypropylene/polyethylene/polypropylene separator (Celgard 2325) combined the low-temperature shutdown property of polyethylene with the high temperature melt integrity of polypropylene, resulting in a separator with softening (~105°C) and melt temperature (~135°C) very similar to PE and rupture temperature (~190°C) very similar to PP.

Separators with melt integrity greater than 150°C are desirable for lithium-ion cells. The Trilayer separators with polypropylene on the outside helps in maintaining the melt integrity of the separators at higher temperatures compared to single-layer PE separators. The choice of a shutdown separator for bigger lithium-ion cells being developed for hybrid and electric vehicles is highly specific to the design of the cell.

Wettability and Wetting Speed

Two physical properties of separators, which are important to the operating characteristics of a battery, are electrolyte absorption and electrolyte retention.

Any good separator should be able to absorb a significant amount of electrolyte and also retain the absorbed electrolyte when the cell is in operation. These are more important in sealed cells where no free electrolyte is present. A maximum amount of electrolyte in the separator is desirable to achieve minimum cell internal resistance. Quite often separator dry-out or lack of sufficient amount of electrolyte is misinterpreted as inadequate wetting of the separator.

The separator wettability can limit the performance of batteries by increasing the separator and cell resistance. Separator wetting speed can be correlated with electrolyte filling time in real cells. The wetting speed is determined by the type of polymer (surface energy), pore size, porosity, and tortuosity of the separators. There is no generally accepted test for separator wettability. However, a simple wicking test by placing a drop of electrolyte onto the separator is a good indication of wettability. Standard dyne-solutions of known surface tension values exist and can be used to adjust the surface tension of the electrolyte as to expedite wetting. The contact angle is also a good measure of wettability. The uptake of electrolyte by many hydrophobic polymer separators can be enhanced either by wetting agents or ionic-functional groups (e.g., ion-exchange membranes).

Effect of Separator on Cell Performance and Safety

Although the material of a battery separator is inert and does not influence electrical energy storage or output, its physical properties greatly influence performance and safety of the battery. This is especially true for lithium-ion cells and, thus, the battery manufacturers have started paying more attention to separators while designing the cells. The cells are designed in such a way that separators do not limit the performance, but if the separator properties are not uniform, or if there are other issues, it can affect the performance and safety of cells. This section will focus on the effect of the separator properties on cell performance and safety. Table 6.7 shows different types of safety and performance tests for lithium-ion batteries and the corresponding important separator property and how it affects performance and/or safety.

To achieve good performance of lithium-ion cells, the separators should have low resistance and low shrinkage and strength across the thickness. The separator with high resistance will perform poorly during high rate discharge and will also increase the cell charging time. Low shrinkage is a very important characteristic for separators, especially for higher capacity cells. These cells are used in high-speed laptop computers, which can experience higher temperatures (~70–75°C) under certain conditions [110]. This can lead to shrinkage of separators and ultimately higher cell resistance and poor long-term cycling. The shrinkage in TD direction can lead to safety issues because of an internal short between the electrodes. Larger pores can lead to shorts during cell manufacturing or can fail during Hipot testing. Larger pores will allow more soft shorts and higher self-discharge, especially during high temperature storage. Very small pore size can lead to higher resistance

Table 6.7 Safety and performance tests for lithium-ion batteries and the corresponding important separator property and its affect on the cell performance and/or safety

Cell property	Separator property	Comments
Cell capacity	Thickness	Cell capacity can be increased by making the separator thinner
Cell internal resistance	Resistance	Separator resistance is a function of thickness, pore size, porosity, and tortuosity
High rate performance	Resistance	Separator resistance is a function of thickness, pore size, porosity, and tortuosity
Fast charging	Resistance	Low separator resistance will aid in overall faster charging by allowing higher and/or longer constant current charging
High temperature storage	Oxidation resistance	Oxidation of separators can lead to poor storage performance and reduce performance life
High temperature cycling	Oxidation resistance	Oxidation of separators can lead to poor cycling performance
Self-discharge	Weak areas, pinholes	Soft shorts during cell formation and testing can lead to internal current leakage
Long-term cycling	Resistance, shrinkage, pore size	High resistance, high shrinkage, and very small pore size can lead to poor cycling performance
Overcharge	Shutdown behavior; high temperature melt integrity	Separator should completely shutdown and then maintain its melt integrity at high temperatures
External short circuit	Shutdown behavior	Separator shutdown stops the cells from overheating
Hotbox	High temperature melt integrity	Separator should be able to keep the two electrodes apart at high temperatures
Nail crush	Shutdown (to stop delayed failure)	In case of internal shorts, the separator may be the only safety device to stop the cell from overheating.
Bar crush	Shutdown (to stop delayed failure)	In case of internal shorts, the separator may be the only safety device to stops the cell from overheating.

and poor cycle life during high temperature cycling and storage. Thus, the pore size of the separator should be optimized to achieve good strength and performance.

One of the ways to increase cell capacity is by decreasing the thickness of separators. The newer high capacity cells (>2.6 Ah) generally use 16 and 12 μm separators as compared to 20–25 μm separators used in cells with 2.2–2.4 Ah capacity. The thinner separators offer lower resistance and help in increasing the capacity; but the amount of electrolyte they can hold is less and their mechanical strength is often not as high as thicker separators. Thus, appropriate changes should be made in cell design to keep the cell safe. The handling and manufacturing of thinner separators is also a challenge for the separator manufacturers. They are required to maintain the same electrical and mechanical properties and better quality for thinner separators. The separator manufacturers have installed better controls and quality standards. Many battery experts are of the opinion that the

16 μm is the thinnest that can be used while still maintaining the stringent performance and safety requirements of lithium-ion cells. The use of thinner separators has often resulted in voluminous battery recalls [111].

The separators inside the lithium-ion batteries experience extreme oxidizing environment on the side facing the positive electrode and extreme reducing environment on the side facing the negative electrode. The separators should be stable in these conditions during long-term cycling especially at high temperatures. Separators with poor oxidation resistance can lead to poor high temperature storage performance and poor long-term cycling behavior. The oxidation resistance properties of trilayer (PP/PE/PP) separators with PP as the outside layer and PE as inner layer is superior compared to polyethylene separators. This is because of the better oxidation resistance properties of polypropylene in contact with the positive electrode in a lithium-ion cell.

The products formed by the decomposition of the electrolyte can also block the pores of the separator, leading to increase in cell resistance. The separators with lower resistance also help in better low-temperature performance. At very low temperatures, the resistance of the electrolytes is very high and thus smaller contribution from separator helps in keeping the cell resistance low.

The lithium-ion cells have demonstrated power loss when aged and/or cycled at high temperatures. Norin et al. [112] demonstrated that the separator is at least partly responsible for the power loss due to the intrinsic increase in its ionic resistance. They showed that impedance increased significantly upon cycling and/or aging of lithium-ion cells at elevated temperatures and that separators account for about 15% of the total cell impedance rise. They later reported that the loss in ionic conductivity of the separator was due to blocking of the separator pores with the products formed due to electrolyte decomposition, which was significantly accelerated at elevated temperatures [113].

There are several groups that regulate, or provide testing, to verify safe operation of lithium-ion cells under abuse conditions. The US Department of Transportation (DOT) classifies all lithium-ion batteries as hazardous materials for shipping in the same category as lithium metal primary batteries [114]. The DOT grants exceptions based on the cell capacity and ability of the cells to pass specified tests. In addition, the UL Laboratories [115, 116], the International Electrotechnic Commission [117], and the United Nations (UN) [118] have developed standardized safety testing procedures. These tests are designed to assure that cells are safe to ship and are resistant to typical abuse conditions such as internal shorting, overcharge, overdischarge, vibration, shock, and temperature variations that may be encountered in normal transportation environments.

Underwriters Laboratories (UL) requires that consumer batteries pass a number of safety tests (UL 1642 [119] and UL-2054 [120]). There are similar recommendations from UN for transport of dangerous goods, [121] the International Electrotechnical Commission (IEC), and the Japan Battery Association [122]. An abnormal increase in cell temperature can occur from internal heating caused by either electrical abuse – overcharge or short circuit – or mechanical abuse – nail penetration or crush. Higher cell temperature could also be a result of external

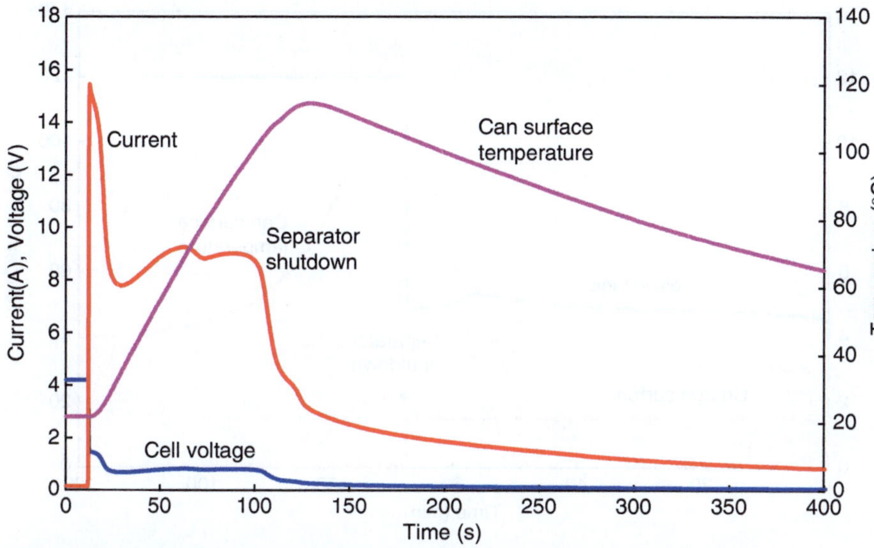

Fig. 6.10 Typical short-circuit behavior of a 18650 lithium-ion cell with shutdown separator and without PTC (positive temperature coefficient) and CID (current interrupt device). This test simulates external short circuit of cell

heating. For this reason, lithium-ion cells used in battery packs are designed with safety control circuits that have redundant safety features (PTC, CID, vent, thermal fuse, etc.). Shutdown separators are one of the safety devices inside the cell and act as a last line of defense. The separator shutdown is irreversible, which is fine for polyethylene-based separators, which melt around 130°C.

The impedance of the separator increases by two to three orders of magnitude due to an increase in cell temperature, resulting from cell abuse (e.g., short circuit, overcharge). The separator should not only shutdown around 130°C, but it should also maintain its mechanical integrity at higher temperatures, preferably at temperatures as high as 200°C. If the separator does not shutdown properly then the cell will continue to heat during an overcharge test and can lead to thermal runaway. The high temperature melt integrity of separators is also a very important property to keep the cell safe during extended overcharge or during extended exposure to higher temperatures.

Figure 6.10 shows a typical short-circuit curve for an 18650 lithium-ion cell with shutdown separator, $LiCoO_2$ positive electrode, and MCMB carbon negative electrode. For the results shown in Fig. 6.10, the cell tested did not have other safety devices (e.g., CID, PTC), which usually work before separator shutdown. As soon as the cell is short circuited externally through a shunt resistor, the cell starts heating because of the large current drained through the cell. The shutdown of the separator, which occurs around 130°C, stops the cell from heating further. The current decrease is caused by increase of battery internal resistance due to separator shutdown. The separator shutdown helps in avoiding the thermal runaway of the cell.

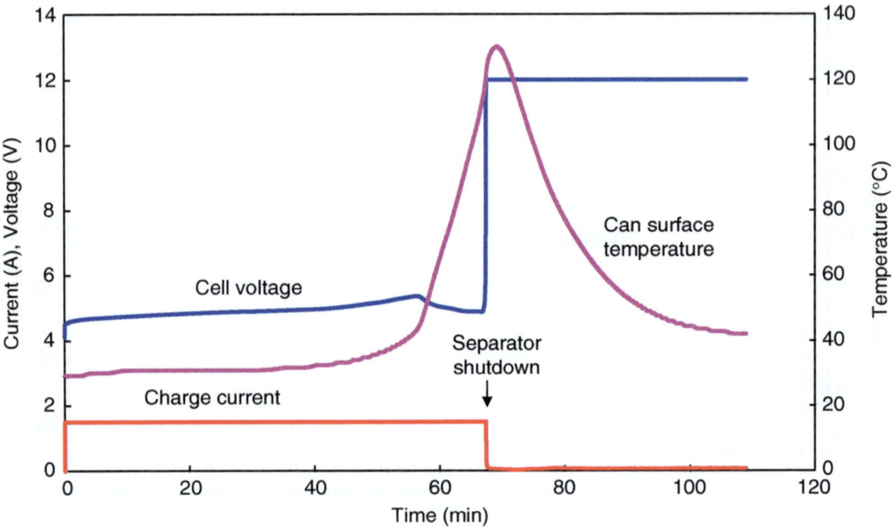

Fig. 6.11 Typical overcharge behavior of a 18650 lithium-ion cell with shutdown separator. The PTC (Positive Temperature Coefficient) and CID (Current Interrupt Device) were removed from the cell header

Cells can be overcharged when the cell voltage is incorrectly detected by the charging control system, or when the charger breaks down. When this happens, the lithium ions remaining in the cathode are removed and more lithium ions are inserted into the anode then under standard charging conditions. If the lithium insertion ability of the carbon anode is limited, lithium metal in the form of dendrites may be deposited on the carbon and cause a drastic reduction in thermal stability. At higher charging rates, the heat output increases greatly because the joule heat output is proportional to I^2R. Several exothermic reactions (e.g., reaction between lithium and electrolyte, thermal decomposition of anode and cathode, thermal decomposition of electrolyte, etc.) occur inside the cell as its temperature increases. Separator shutdown happens when cell temperature reaches melting point of polyethylene as shown in Fig. 6.11. The CID and PTC of the 18650 cells were removed, to identify the role of the separator. The current decrease is caused by increase of battery internal resistance due to separator shutdown. Once the pores of the separator have closed due to softening, the battery cannot continue to be charged or discharged, and thus thermal runaway is prevented. During continued overcharge, the separator should maintain its shutdown feature and should not allow the cell to heat again. It should also maintain its melt integrity and should not allow the two electrodes to touch each other.

The separator should also not allow any dendrite to penetrate through the separator to avoid internal shorts. During an internal short, separator is the only safety device, which can stop the thermal runaway. If the heating rate is not too high then the separator shutdown can help in controlling the heating rate and stop

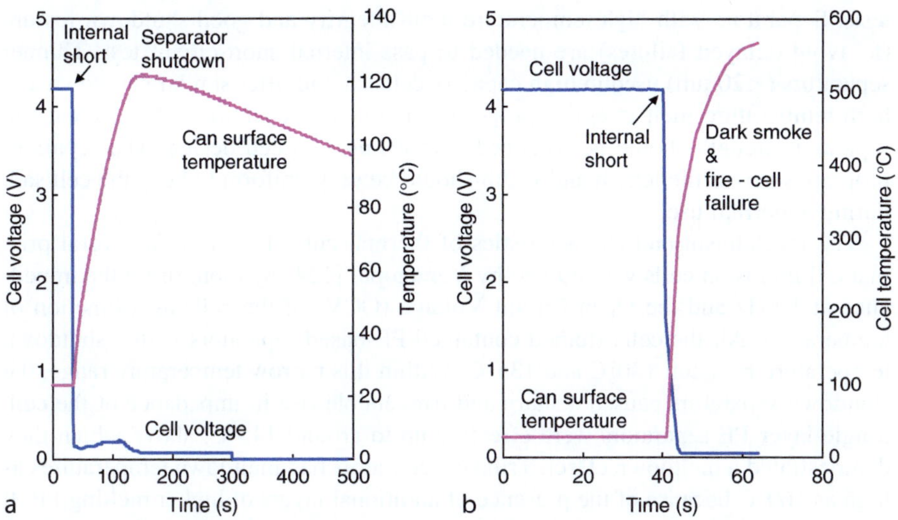

Fig. 6.12 Typical nail penetration behavior of an 18650 lithium-ion cell with shutdown separator. This test simulates internal short circuit of a cell. (**a**) Cell passed nail penetration test; (**b**) cell failed nail penetration test

thermal runaway. In a nail penetration test, an instantaneous internal short results the moment the nail penetrates into battery. Enormous heat is produced from current flow (double-layer discharge and electrochemical reactions) in the circuit by the metal nail and electrodes. Contact area varies according to depth of penetration. In general, the shallower the penetration depth, the smaller the contact area and therefore the greater are the local current density and heat production. Thermal runaway is likely to take place as local heat generation induces electrolyte and electrode materials to decompose. On the other hand, if the battery is fully penetrated, the increased contact area would lower the current density, and the cell could pass the nail penetration test. A detailed investigation of various internal short-circuit scenarios is presented by Santhanagopalan et al. [229]. Internal short-circuit tests are more difficult to pass than the external short-circuit tests described earlier, because the nature of the short cannot be determined a priori.

Figure 6.12 shows the typical nail penetration behavior of an 18650 lithium-ion cell with shutdown separator, $LiCoO_2$ positive electrode, and MCMB carbon negative electrode. Clearly, there was a voltage drop from 4.2 to 0.0 V, instantaneously, as the nail penetrated through (when internal short circuit occur), and temperature rose. When the heating rate is low the cell stops heating when the temperature is close to separator shutdown temperature as shown in Fig. 6.12a. If the heating rate is very high, then the cell continues to heat and fails the nail penetration test as shown in Fig. 6.12b. In this case, the separator shutdown is not fast enough to prevent thermal runaway. Thus, separator only helps in avoiding delayed failures in case of internal short circuit as simulated by nail and bar crush tests. Separators with high temperature melt integrity and good shutdown feature (to avoid delayed failures) are needed to pass internal short-circuit test. Thinner separators (<20 μm) used in high capacity cells should offer similar shutdown and high temperature melt integrity properties as thicker separators. The decrease in separator strength should be balanced with changes in cell design. The separator properties across the length and width should be very uniform to keep the cell safe during abnormal use.

The mechanism and characteristics of thermal cut-off devices in several prismatic lithium-ion cells was studied by Venugopal [124] by monitoring the impedance at 1 kHz and the Open Circuit Voltage (OCV) of the cells as a function of temperature. All the cells studied contained PE-based separators with a shutdown temperature between 130°C and 135°C. Within this narrow temperature range, the shutdown separators caused a sharp and irreversible rise in impedance of the cell. Single-layer PE separators were effective up to around 145°C, above which they demonstrated a meltdown effect. Trilayer separators had meltdown temperatures as high as 160°C because of the presence of additional layers of higher melting PP. It was found that the separators, alone, are not able to shutdown the cell completely. In case of an overcharged test, the cell could continue to charge at lower currents even after the shutdown event, rendering the cell a potential hazard if not disposed of immediately and safely. This usually does not become an issue in commercial cells because the cell manufacturers have addressed this issue by including multiple cut-off devices within a single cell. The use of inorganic coatings and fillers has also drawn considerable attention in order to prevent cell failure in such cases [78, 125].

Development efforts are under way to displace the use of microporous membranes as battery separators and instead use gel or polymer electrolytes. Polymer electrolytes, in particular, promise enhanced safety by eliminating organic volatile solvents. The next two sections are devoted to solid polymer and gel-polymer-type lithium-ion cells with focus on their separator/electrolyte requirements.

Separator for Lithium Polymer Batteries

Due to their high theoretical capacity, lithium polymer batteries have long been identified as a very promising technology to meet the requirements of upcoming applications such as standby power and electric vehicles. Research and development of polymer electrolytes for ambient-temperature rechargeable lithium batteries has always been very active. Rapid progress for the past two decades in this field has led to numerous monographs and reviews [126–131]. These polymers are generally polyethers, poly(ethylene oxide) (PEO), or poly(propylene oxide) (PPO).

Solid polymer electrolytes serve two principal roles in rechargeable lithium batteries. Not only do they function as the traditional electrolyte, i.e., the medium for ionic transport, but also as the separator which insulates the cathode from the anode. Consequently, the polymer electrolyte must have sufficient mechanical integrity to withstand electrode stack pressure and stresses caused by dimensional changes, which the rechargeable electrodes undergo during charge/discharge cycling.

Lithium polymer electrolytes formed by dissolving a lithium salt LiX (where X is preferably a large soft anion) in poly(ethylene oxide) PEO can find useful application as separators in lithium rechargeable polymer batteries [132–134]. Thin films must be used due to the relatively high ionic resistivity of these polymers. For example, the lithium-ion conductivity of PEO-Li salt complexes at 100°C is still only about 1/100 the conductivity of a typical aqueous solution.

A polymer electrolyte with acceptable conductivity, mechanical properties, and electrochemical stability has yet to be developed and commercialized on a large scale. The main issues that must be resolved for a completely successful operation of these materials are the reactivity of their interface with the lithium metal electrode and the decay of their conductivity at temperatures below 70°C. Croce et al. found an effective approach for reaching both of these goals by dispersing low-particle-size ceramic powders in the polymer electrolyte bulk [135, 136]. They claimed that this new "nanocomposite polymer electrolytes" had a very stable lithium electrode interface and an enhanced ionic conductivity at low temperature, combined with good mechanical properties. Fan et al. [137] has also developed a new type of composite electrolyte by dispersing fumed silica into low to moderate molecular weight PEO.

The gel-type polymer electrolyte prepared by dispersing ceramic powders (e.g., Al_2O_3) into a matrix formed by a lithium salt solution contained in a poly(acrylonitrile) (PAN) network was reported by Appetecchi et al. [138] These new types of composite gel electrolytes had high ionic conductivity, wide electrochemical stability, and particularly, high chemical integrity even at temperatures above ambient. Kim et al. [139] used a blend of PVdF-HFP and PAN as a matrix polymer to attain high ionic conductivity and good mechanical strength. The PAN can give mechanical integrity and structural rigidity to a porous membrane without inorganic fillers. The high ionic conductivity was due to the high volume of pores and a high affinity of the membrane for electrolyte solution [140].

Separator for Lithium-Ion Gel Polymer Batteries

The solid polymer electrolyte approach provides enhanced safety, but the poor ambient-temperature conductivity excludes their use for battery applications, which requires good ambient-temperature performance. In contrast, the liquid lithium-ion technology provides better performance over a wider temperature range, but electrolyte leakage remains a constant risk. Midway between the solid polymer electrolyte and the liquid electrolyte is the "hybrid polymer" electrolyte concept leading to the so-called gel polymer lithium-ion batteries. Gel electrolyte is a two-component system, namely, a polymer matrix swollen with a liquid electrolyte. The gel polymer electrolyte approach to the lithium-ion technology combines the positive attributes of both the liquid (high ionic conductivity) and solid polymer electrolytes (elimination of leakage problems).

Gel polymer lithium-ion batteries replace the conventional liquid electrolytes with an advanced polymer electrolyte membrane. These cells can be packed in light-weight plastic packages as they do not have any free electrolyte and they can be fabricated in any desired shape and size. They are now increasingly becoming an alternative to liquid electrolyte lithium-ion batteries, and several battery manufacturers, such as Sanyo, Sony, and Panasonic, have started commercial production [141, 142]. Song et al. [143] have recently reviewed the present state of gel-type polymer electrolyte technology for lithium-ion batteries. They focused on four plasticized systems, which have received particular attention from a practical viewpoint, i.e., poly (ethylene oxide) (PEO), poly (acrylonitrile) (PAN) [144], poly (methyl methacrylate) (PMMA) [145, 146], and poly (vinylidene fluoride) (PVdF) based electrolytes [147–150].

One particular version of the lithium-ion gel polymer cells, also known as plastic lithium-ion cell (PLION™), was developed by Bellcore [151–153]. In this case, Gozdz et al. developed a microporous plasticized PVdF-HFP-based polymer electrolyte that served both as separator and electrolyte. In PLION™ cells, the anode and cathode are laminated onto either side of the gellable membrane. Good adhesion between the electrodes and the membranes is possible because all three sheets contain significant amounts of a PVdF copolymer that can be melted and bonded during the lamination step.

The PVdF-HFP separators used in PLION™ cells were around 3 ml thick, and had poor mechanical properties. It has been reported that the major source of rate limitation in PLION™ cells was the separator thickness [154]. The rate capability of these cells can be significantly improved by decreasing the separator thickness to that typically used in liquid electrolyte system. Moreover, in the absence of shutdown function, the separator does not contribute to cell safety in any way. Park et al. reported that the HFP content in separators did not have any significant impact on cell performance [155]. The Bellcore process has proven to be an elegant laboratory process but is difficult to implement in large-scale production.

To overcome the poor mechanical properties of polymer- and gel-polymer-type electrolytes, microporous membranes impregnated with gel polymer electrolytes, such as PVdF, PVdF-HFP, and other gelling agents, have been developed as an electrolyte material for lithium batteries [156–166]. Gel-coated and/or gel-filled separators have some characteristics that may be harder to achieve in the separator-free gel electrolytes. For example, they can offer much better protection against internal shorts when compared to gel electrolytes and can therefore help in reducing the overall thickness of the electrolyte layer. In addition the ability of some separators to shutdown at a particular temperature allows safe deactivation of the cell under overcharge conditions.

The shutdown behavior of PVdF-coated Celgard a trilayer membrane is shown in Fig. 6.13. The shutdown is defined by the sharp increase in resistance around 130°C.

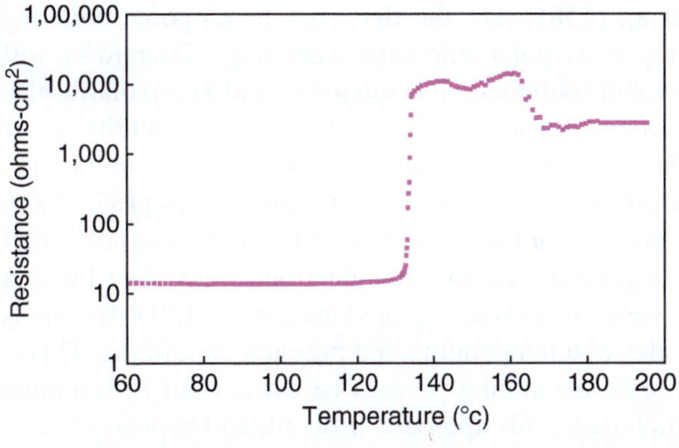

Fig. 6.13 Internal impedance (at 1 kHz) of PVdF-coated Celgard trilayer separators as a function of temperature. Heating rate: 60°C/min

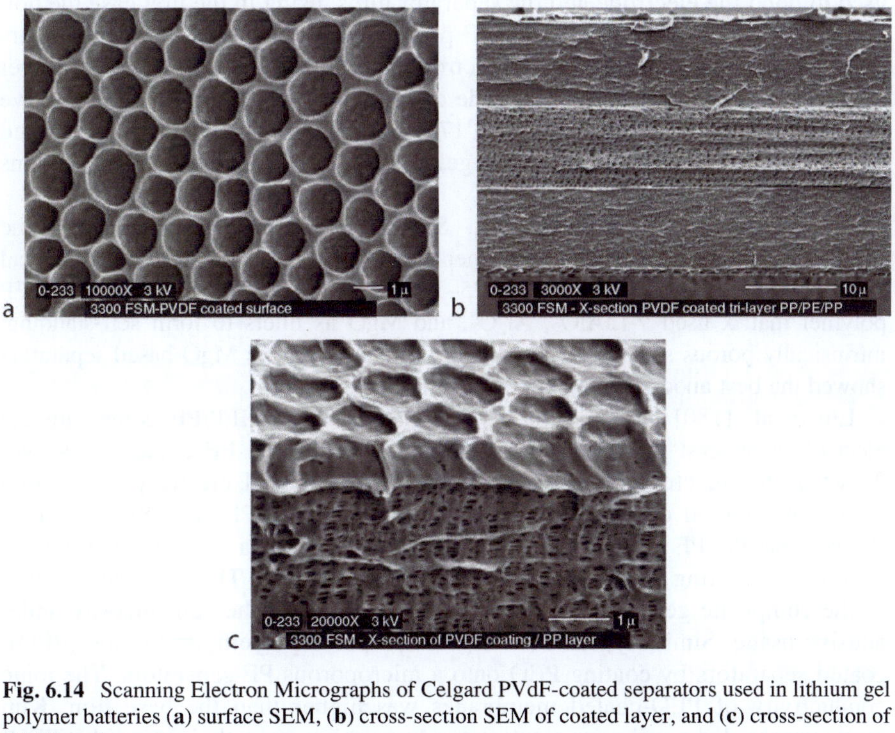

Fig. 6.14 Scanning Electron Micrographs of Celgard PVdF-coated separators used in lithium gel polymer batteries (**a**) surface SEM, (**b**) cross-section SEM of coated layer, and (**c**) cross-section of PVdF coating

The PVdF coating should be porous and should not block the pores to maintain similar ionic conductivity. The scanning electron micrographs of PVdF-coated membrane is shown in Fig. 6.14. The cross-section SEM of Celgard 3300 provides visual evidence that the coating is porous and is not blocking the pores of the top PP layer.

Abraham et al. [158] were the first ones to propose saturating commercially available microporous polyolefin separators (e.g., Celgard®) with a solution of lithium salt in a photopolymerizable monomer and a nonvolatile electrolyte solvent. The resulting batteries exhibited low discharge rate capability due to the significant occlusion of the pores with the polymer binder and the low ionic conductivity of this plasticized electrolyte system. Dasgupta and Jacobs [157, 168] patented several variants of the process for the fabrication of bonded-electrode lithium-ion batteries, in which a microporous separator and electrode were coated with a liquid electrolyte solution, such as ethylene-propylene-diene (EPDM) copolymer and then bonded under elevated temperature and pressure conditions. This method required that the whole cell assembling process be carried out in scrupulously anhydrous conditions, which make this approach difficult, and expensive.

The later methods, proposed by Motorola [159, 170] and Mitsubishi Electric [171] researchers, differ in implementation details, but they share a common feature in that a separate adhesive layer (PVdF) is applied to the separator and used to bond the electrode and the separator films, using in the first case the hot, liquid electrolyte as an in situ PVdF plasticizer. Sony [172, 173] researchers described the use of thin, liquid electrolyte-plasticized polyacrylonitrile layer directly applied either to the electrode or the separator surfaces as an effective ion-conductive adhesive. Sanyo [174, 175] investigators, on the other hand, used thermally polymerizable additives to gel, or solidify, liquid electrolyte solutions in a wound, packaged battery.

The ceramic fillers (e.g., Al_2O_3, SiO_2, TiO_2) can greatly influence the characteristics and properties of polymer electrolyte by enhancing the mechanical stability and the conductivity [135, 175–178]. Prosini et al. [179] in a PVdF-HFP polymer matrix used γ-$LiAlO_2$, Al_2O_3, and MgO as fillers to form self-standing, intrinsically porous separators for lithium-ion batteries. The MgO-based separators showed the best anode and cathode compatibilities.

Liu et al. [180] has successfully prepared a PVdF-HFP/PE composite gel electrolyte by cast method. They showed that when the PE content was over 23 wt.%, the electrical impedance of the composite gel electrolyte increased rapidly by several orders, around the melting point of PE. The SEM pictures showed that the PE particles were fused and formed into a continuous film at or near the PE melting point, which cuts off the ion diffusion. This shutdown feature of the composite gel electrolyte can help in preventing the cell runaway under abusive usage. Similarly, Kim et al. [181] prepared polyethylene oxide (PEO)-coated separators by coating PEO onto a microporous PE separators. The ionic conductivity of PEO-coated membranes was higher than the base film. Kim et al. prepared the polymer electrolytes by coating polyethylene oxide (PEO) and polyethylene glycol dimethacrylate (PEGDMA) onto a microporous polyethylene membrane (Asahi Kasei, 25 um, 40% porosity) [182]. They showed that the relative weight ratio of PEO and PEGDMA coated onto the microporous membrane played a critical role in determining the uptake of electrolyte solution and ionic conductivity.

Separator for Aqueous Batteries

The aqueous batteries use water-based electrolytes (e.g., KOH electrolyte for NiCd, NiMH, and H_2SO_4 electrolyte for Lead acid), which are less resistive than non-aqueous electrolytes. Polyolefin materials are generally suitable for use in the manufacture of separators for these batteries, but they are not inherently wettable by aqueous electrolytes. Such electrolytes are therefore unable to penetrate the pores of a separator formed from such a material, so that ion migration through the pores in solution will not occur without modification. This problem is sometimes overcome by treating the polyolefin material with a surfactant, which allows an aqueous electrolyte to wet the material. However, such surfactant can be removed from the surfaces of the polyolefin material when electrolyte is lost from the device, for example during charging and discharging cycles, and it is not subsequently replaced on the material when the electrolyte is replenished.

This problem has also been addressed by modifying the surface properties of the polyolefin materials used to form polymeric sheets, by graft copolymerizing a monomeric substance to its surface, which, after copolymerization, confers hydrophilic properties, and, in some cases ion-exchange properties. This technique has been found to be practical when the porous substrate is formed from PE, which lends itself well to a graft-copolymerization reaction of this kind. However, it has been found that, when such a reaction is attempted using polyolefin materials other than PE, the rate of the grafting reaction is reduced significantly.

Graft polymerization is a convenient method for the modification of the physical and chemical properties of polymer materials and is of particular interest for synthesis of the hydrophilic membranes. Graft copolymerization can be achieved by various methods such as an exposure to ionizing radiation or ultraviolet light and the use of chemical initiators. Ionizing radiation is one of the most promising methods because of its rapid and uniform formation of active sites for initiating grafting throughout the matrix. Under appropriate experimental conditions, modifications of polymer properties can be accomplished not only on the surface but also throughout the polymer.

There have been several reports on radiation grafting of acrylic and methacrylic acid onto various substrates. These include both the direct grafting method and the pre-irradiation method to synthesize ion-exchange membranes. Two cation exchange membranes modified with the carboxylic acid group for battery separator was prepared by radiation-induced grafting of acrylic acid (AA) and methacrylic acid (MA) onto a polyethylene film by Choi et al. [183]. They found that KOH diffusion flux of AA-grafted PE membrane and MA-grafted PE membrane increased with an increase in the degree of grafting. AA-grafted PE membrane had a higher diffusion flux then MA-grafted PE membrane. Electrical resistance of both membranes decreased rapidly with an increase in the degree of grafting up to 120%, and then leveled off.

Battery separators having carboxylic acid group were prepared by radiation-induced grafting of acrylic acid onto a polyolefin nonwoven fabric (PNF). The PNF comprised of approximately 60% polyethylene and 40% polypropylene. It was found that the wetting speed, electrolyte retention, thickness, and ion-exchange capacity increased, whereas the electrical resistance decreased with increasing grafting yield [184]. The surface characteristics of the separators can also be modified by plasma discharge.

The subsequent subsections discuss separators used in lead acid and nickel metal hydride batteries.

Separators for Lead Acid Batteries

It has been a long time since the invention of the lead acid battery, but it still represents the most important secondary chemical power source – both in number of types and diversity of application. The lead acid battery has maintained its leading role for so many decades due to its competitive electrical characteristics and price, and due to its adaptability to new applications. It is manufactured in a variety of sizes and designs, ranging from less than 1 Ah to over 10,000 Ah [185].

Lead acid batteries can be classified into three major types or categories, namely, automotive (SLI), stationary, and motive power (industrial). In addition, there are many special batteries that cannot be easily categorized as either of the above types. These types of batteries are constructed with different materials and designed to meet the requirements of their intended end uses, each with a particular separator requirement with specific material composition, mechanical design, and physical, chemical, and electrochemical properties, tailored for the battery and its relevant specific uses. These batteries are generally available in flooded electrolyte or valve regulated (sealed) versions. In this section the types and properties of separators used for lead acid batteries are reviewed. The reader is referred to recent reviews published by Boehnstedt [12, 186, 187] and others [188–190] for detailed descriptions of lead acid separators.

Flooded Electrolyte Lead Acid

Separators currently used in lead acid batteries can be classified based on their materials of construction into four major types: plastic (PE/silica, PVC/silica, Sintered PVC), paper (phenolic resin impregnated cellulose), glass (glass fiber mat), and rubber (hard rubber/silica, flexible rubber/silica, coated rubber/silica) separators. Table 6.8 shows the different types of separators used in batteries along with their manufacturing process and main features. Glass, paper, and sintered PVC separators can be classified as macroporous separators having an average pore diameter greater than 10 μm while all other separators can be classified as microporous separators having an average pore diameter smaller than 1 μm. All of these separators can be utilized as leaf separators in battery construction.

Table 6.8 Typical separators used in lead acid battery systems

Separator	Class	Manufacturing process	Properties
Wood	Paper	Cellulosic separators are made from cotton linters or craft pulp and generally coated with phenolic resin for acid resistance and strength	Comparatively large pore size and relatively high electrical resistance
Hard rubber	Rubber	Made by mixing natural rubber, rehydrated precipitated silica, and sulfur. This is then extruded and calendared, vulcanized under water, and dried.	Finer pore diameter (0.2 μm average), relatively lower electrical resistance, excellent oxidation resistance, retards antimony transfer
Flexible rubber	Rubber	Made by mixing natural rubber, rehydrated precipitated silica. This is then extruded and calendared, irradiated with an ionizing electron beam and dried.	Flexible, fine pore structure (0.06 μm average), retards antimony transfer
Glass mat rubber	Rubber/ glass mat	Made by mixing polymeric emulsion, precipitated silica, and rubber. This is then coated on a fiberglass mat and finally cured and dried.	Finer pore diameter (<0.2 μm average), high porosity, excellent thermal dimensional stability
Sintered PVC	Plastic	Made by sintering PVC powder of a particle size ranging between 10 and 20 μm	Medium pore size (10–20 μm), generally good chemical resistance
Synthetic PVC	Plastic	Made from mixture of PVC, silica fine powder, and a solvent, and then extruded, calendared, and extracted	Small to medium average pore size and relatively low electrical resistance
Synthetic pulp with glass mat	Plastic/ glass	Made from blending PE synthetic pulp, synthetic fiber, and fine silica powder, and then heat treated	Medium pore size, low electrical resistance, and long service life at high temperatures; more difficult to process and assemble
Polyethylene (PE)	Plastic	Made from a mixture of UHMW PE powder, fine silica powder, and mineral oil. The mixture is extruded as a film, calendered, and made porous by extraction	Fine pore size, low electrical resistance, high puncture resistance, and strongly resistant to oxidation
Glass fiber mat	Glass	Deposition on a single sheet, a mixture of fibers dispersed in an aqueous solution	Excellent wettability, durable in an acid environment, good resiliency, high temperature stability, more difficult to process and assemble

Polyethylene can be used also as enveloped separators around either the positive or the negative plate. The use of "envelope" separators is popular in small, sealed cells, SLI, motive power, and standby batteries to facilitate production and to control lead contamination during manufacturing.

The environment of the lead acid battery (e.g., automotive battery) has been increasing in severity in recent years. The improvements and development of the separators have proceeded in accordance with the changes in the specifications for the batteries which were first made with wooden separators (preferred wood was Oregon Ceder as it contained small amounts of Lignin that enhances the performance of lead negative), then progressed through microporous rubber separators, cellulose separators and synthetic pulp separators (SPG) with glass mats, PVC separators, and now polyethylene separators have evolved. This sequential change in separator technology has provided continuous improvements in the charge and discharge efficiency of batteries and has given high vehicle-starting capability and reliability. Moreover, short circuits (caused by particles of active material dislodged from the battery plates) are prevented due to the smaller pores and excellent electrochemical oxidation resistance of the PE separator. These features contributed greatly to the improvement in battery life.

Rubber separators have good voltage characteristics, ability to retard antimony transfer, properties to retard dendrite growth, and good electrochemical compatibility [191]. Due to the hydrophilic properties of the rubber composition, the separators are highly wettable and renewable for the dry-charging process. Paik et al. showed that ACE-SIL (sulfur cured, hard rubber) separators performed well in industrial stationary or traction batteries, FLEX-SIL (electron-beam-cured, flexible rubber separator) separators are suited for deep-cycling batteries, and MICROPOR-SIL (a coated, glass mat, rubber separator) separators have been found to be a good choice for high rate discharging or cranking applications and for various types of gel cells [192]. Recently Daramic® DC UHMW PE has demonstrated excellent performance in these applications, as well.

Polyvinylchloride (PVC) and polyethylene (PE) separators have been the most commonly used separators in automotive batteries for the last 20 years. Polyethylene separators have a narrow pore-size distribution. The PVC separator is built up by sintering PVC powder in general of a particle size ranging between 10 and 20 μm. The decrease of particle size in the sintered product is negligible compared to the particle size of the raw materials. The pores are dispersed homogenously with a medium size ranging between 10 and 20 μm. Since a PVC separator exclusively consists of PVC, it exhibits advantageously good chemical resistance against acid and alkaline solutions. Unlike PE, PVC is disadvantaged due to its brittleness. The decline in PVC separators in recent years is in part due to their tendency to yield chloride ions from chemical attack.

The battery separators currently used by most flooded-cell-type lead acid battery manufacturers are of the microporous PE type. It was invented in the late 1960s by W. R. Grace & Co. [193]. The term "polyethylene separators" is somewhat misleading, since such a separator consists mainly of agglomerates of precipitated silica, being held within a network of extremely long-chain UHMWPE [194]. A typical PE

Table 6.9 Comparison of properties of different separators used in lead acid batteries

Property	Rubber	Cellulose	PVC	PE	Glass fiber
Year available	1930	1945	1950	1970	1980
Electrical resistance	Very poor	Poor	Poor	Very good	Very good
Porosity	Sufficient	Good	Poor	Good	Very good
Battery performance (cold crank)	Poor	Sufficient	Sufficient	Very good	Very good
Maximum pore size	Good	Poor	Sufficient	Very good	Poor
Mean pore diameter	Good	Poor	Poor	Very good	Poor
Purity	Good	Fair	Good	Good	Good
Resistance to shorting	Good	Poor	Poor	Very good	Poor
Corrosion resistance	Very good	Poor	Good	Very good	Good
Oxidation resistance	Fair	Poor	Good	Very good	Very good
Envelopable (sealability)	Very poor	Very poor	Sufficient	Very good	Very poor
Flexibility	Brittle	Brittle	Brittle	Excellent	Good

separator formulation comprises precipitated silica (~60 wt.%), UHMW PE (~20 wt.%), mineral process oil (~15 wt.%), as well as some processing aids, like antioxidants and/or proprietary surface tension modifiers [195, 196].

The microporous PE separator is commercially manufactured by passing the ingredients through a heated extruder, passing the extrudate generated by the extruder through a die and into the nip formed by two heated calendar rolls to form a continuous web, extracting a substantial amount of the processing oil from the web by use of a solvent, drying the extracted web, slitting the web into lanes of predetermined width, and winding the lanes into rolls [195].

The PE separators have excellent microporous structure for electrolyte flow with minimal lead particle deposits; excellent ductility, strength, and toughness for envelopability and plate puncture resistance; excellent oxidation, chemical, and thermal resistance to resist premature deterioration; good manufacturability with high production efficiency and relatively low raw material cost which reduces overall manufacturing costs [196]. The PE pocket separation is in almost all aspects significantly superior to leaf separation. Only PE separators can be enveloped and can develop good sealability. These have low electrical resistance, sufficient porosity, small pore size, and great resistance to both shorting and corrosion. The PE separator, by virtue of its low electrical resistance, generally provides better cold cranking performance. They are very flexible and offer excellent oxidation resistance if the residual oil content is controlled and/or proprietary chemical modifiers have been incorporated. A comparison of the properties of different types of separators is given in Table 6.9.

PE separators have contributed to improved battery specific energy and specific power, increased battery cycle life, and higher temperature operating capabilities. PE separators have gained in popularity and have generally replaced PVC, cellulose, glass fiber, and other conventional separators. The transition to microporous PE envelope separators started in the USA in the 1970s, followed by Europe in 1980s. Today, PE separators have captured almost 100% of the US market and more than 70% of the remaining worldwide automotive markets [197].

In a flooded-cell-type lead acid battery, the battery separator typically has "ribs" or protrusions extending from at least one planer face of the separator. Such ribs are either formed integrally with the backweb of the separator, or they can be subsequently applied to the backweb as a bead of the same or different material as the backweb, or they can be formed by embossing the backweb. The ribs function is to provide proper spacing between the plates and to provide a space wherein free electrolyte resides. The ribs also provide pressure to hold the electrodes in contact with the separator. This reduces the need for precise dimensional control on the cell components. Microporous PE separators typically have a configuration comprising a backweb having a predetermined thickness, and a plurality of parallel or patterned ribs spaced apart by a predetermined distance and extending outwardly from one planar surface of the backweb. The ribs extend continuously in a longitudinal direction parallel to the edges of the separator material. The thickness of the backweb and height and spacing of the ribs is specified to the separator manufacturer by the battery manufacturer; based on specifications designed to maximize certain battery characteristics desired by the battery manufacturer. SLI batteries tend to have separators that are thinner than "industrial" lead acid batteries used for standby power sources and traction devices.

Endoh [198] has reported that one reason for the occasionally shortened life of batteries assembled with PE pocket separators is the development of internal short circuits at the bottom part of the PE separator due to anodic corrosion causing active material to shed from the positive plates and leak through the separators. He also found that when synthetic pulp (SP) separators with glass mats are used, it is possible to not only restrain the shedding from positive plates, but also to protect the separators against intensive oxidation so that no internal short circuits develop on charge. He concluded that the use of SP separators with glass mat is required to produce long service-life batteries, especially in tropical regions.

Higashi et al. [199] carried out endurance test under high temperature conditions on automotive batteries made with three different types of separators. One group was assembled with PE pocket separators for the negative plates, another with PE pocket separators with glass mats for the positive plates, and a third with leaf-type synthetic pulp separators with glass mats. They concluded that battery assembly with PE pocket separators with glass mat is an effective way to achieve good endurance (i.e., life extension at high temperature) and leaf-type synthetic pulp separators with glass mats are the best approach for hot climatic conditions.

Valve Regulated Lead Acid (VRLA)

The valve regulated lead acid battery is an important development in lead acid battery technology. These batteries operate on the principle of oxygen recombination, using a "starved" or immobilized electrolyte. The oxygen generated at the positive electrode during charge can, in these battery designs, diffuse to the negative electrode, where it can react, in the presence of sulfuric acid, with freshly formed lead. The separator material should provide innumerable gas channels between the plates through which oxygen can flow from the positive to the negative electrode. These batteries differ from its flooded electrolyte precursor in a number of important ways [200]. They have been manufactured for many years with microfiber glass separators, also called absorptive glass mat (AGM). They are inherently resistant to acid stratification and have the additional important advantage of being essentially maintenance free. The separator is a crucial component in determining the useful life of a VRLA cell. While a prime function of the VRLA cell separator is to hold the cell's electrolyte in place, it must also offer characteristics that prevent major failure mechanisms occurring in the cell's positive and negative plates.

The microglass separator, since its discovery by McClelland and Devitt, has been the material of choice for VRLA designs [201, 202]. It is a wet laid nonwoven (glass fiber) "paper" and is manufactured on a paper machine. The type of paper machine used by the manufacturer can influence the separator properties. Three properties – porosity, uniformity, fiber directionality – are important attributes that can be influenced by the type of fiber used. The glass fiber, which has a zero contact angle with the acid, is durable in the acid environment, and the fine fiber structure also has good resiliency to allow for a sustained pressure against the plate. The microglass separator has a porosity in the 90–95% range and is very conformable. It can adapt to imperfections in the plate surface. The separator also has high temperature stability. Recent studies have shown that higher levels of fine fiber and higher separator compression provide improved cycle performance in VRLA batteries [203–206].

On the other hand, AGM separators offer little control over the oxygen transport rate or the recombination process. The arrival of too much oxygen to the negative plate could result in overheating, hindrance of the battery's ability to recharge, or even a loss of capacity. Furthermore, AGM separators exhibit low puncture resistance.

Nickel Systems

The nickel-based systems have traditionally included the following systems – nickel-iron (Ni/Fe), nickel-cadmium (NiCd), nickel metal hydrides (NiMH), nickel hydrogen (Ni/H_2), and nickel-zinc (Ni/Zn). Of these, the metal hydride chemistry has been the most successful in the secondary battery market. All nickel systems are based on the use of a nickel oxide active material (undergoing one valence change from charge to discharge or vice-versa). The electrodes can be pocket type, sintered type, fibrous type, foam type, pasted type, or plastic roll-bonded type. All systems use an alkaline electrolyte, KOH.

The sealed nickel metal hydride battery uses hydrogen, absorbed in a metal alloy, for the active negative material. The NiMH batteries have a higher energy density and are considered environmentally friendly than the NiCd battery. However, the sealed NiMH battery has limited rate capability and is less tolerant of overcharge. The self-discharge rate is generally higher when conventional nylon separators are used [207]. The presence of oxygen and hydrogen gases cause the polyamide materials to decompose, producing corrosion products which poison the nickel hydroxide, promoting premature oxygen evolution and also forming compounds capable of a redox shuttle between the two electrodes which further increases the rate of self-discharge [208]. Ikoma et al. carried out a detailed investigation to study the self-discharge mechanism and contribution of separators [209]. They used nonwoven fabric made of conventional polyamide (PA), PP (with surfactant), and a nonwoven fabric whose main material was sulfonated-PP (hydrophilic) as separators. When nonwoven fabric made of chemically stable sulfonated-PP is used as a separator instead of a conventional polyamide separator, the self-discharge rate of the NiMH battery was strongly depressed, to the same level as that for the NiCd battery [208, 210, 211].

Nagarajan et al. [212] used differential scanning calorimeter (DSC) to study the materials used as separators in commercial AA cells. They found that Sanyo and Matsushita cells containing nonwoven fabrics fabricated from conventional polyamide as separators exhibited substantially higher self-discharge due to the shuttle reaction of the ammonia and amine. Scimat Ltd. has shown that acrylic acid grafted nonwoven polyolefin separators have the ability to absorb chemical impurities, for example ammonia, from the alkaline environments. It has been shown that by using a grafted polyolefin separator the free ammonia present inside a NiMH cell is trapped by the separator resulting in a reduction in self-discharge to levels normally associated with NiCd cells [213]. In October 2002, Scimat Ltd. announced the launch of the next generation of separators for NiMH and NiCd cells using its second-generation grafting technology [214].

The commonly used separator material now is the surface-treated polypropylene. The surface treatment helps in making the polypropylene permanently wettable. Surface treatments involve the grafting of a chemical such as acrylic acid to the base fibers to impart wettability and are accomplished using a variety of techniques such as UV or cobalt radiation. Another method of imparting wettability to the polypropylene is a sulfonation treatment where the base fiber material is exposed to fuming sulfuric acid. The separator surface is designed to be made hydrophilic to the electrolyte.

Cheng et al. [215] carried out the impedance study on a foam-type NiMH battery with nonwoven PP separator to determine the main causes of early cycle deterioration. Their data indicated that the decrease in the voltage characteristic of the battery was due to drying out of the separator that increases the ohmic resistance of the battery, and that decay of the total discharge capacity is due to an inactive surface that increases the charge-transfer resistance of the battery.

Mathematical Modeling of Separators

Computer simulations have been used as an important tool for understanding and optimizing battery performance since the 1970s [216–218]. Continued progress in computational tools has enabled ever-increasing sophistication in battery modeling and a steady increase in the number of systems to which modeling has been applied. Today, it is possible to obtain simulation codes for all of the major rechargeable batteries, some of which are available in the public domain [219].

The mathematical models of different types of batteries (lead acid [220, 221], NiMH [222], lithium-ion [223, 224]) have been developed during the last few years [219]. This has led to a better understanding of those systems. The present models consider usually, the thickness and porosity of the separators. Very little has been done in incorporating the effect of physical and chemical properties of separators on the performance and safety of batteries. This is also because the microstructure of separators and their effect on transport properties in batteries are generally known only qualitatively.

A thorough understanding of the microstructure of separators would be beneficial for modeling studies and optimization of electrochemical systems. This will help in making the battery model predictions more practical and reliable. The separator pore structure is usually very complex. It consists of a porous network of interconnected pores, which are filled with liquid electrolyte. A complete description of the pore structure would require a very intricate model. Simulations are only practically possible if a simplified quasi-continuum model involving a few parameters represents the structure. In such an approach, the "effective" electrolytic conductivity, σ_{eff}, is often defined by [93]

$$\sigma_{\mathrm{eff}} = \varepsilon^{\alpha}\sigma_0, \alpha \approx 1.5 \tag{6.11}$$

where σ_0 is the bulk ionic conductivity of the electrolyte, ε is the void volume fraction of separator filled with electrolyte, and α is the Bruggeman exponent. The general applicability of alpha ~1.5 appears questionable because separator pores are never of an ideal shape. Fan and White [224] chose a α value of 2.5 for separators in NiCd batteries and Doyle et al. [225] used 3.3 for lithium-ion batteries. Arora et al. [226] measured the value as 2.4 for PVdF-based separators by measuring the separator and electrolyte conductivity at different salt concentration. Doyle et al. used an even higher Bruggeman exponent of 4.5 for quantifying the ionic conductivity of their plasticized electrolyte membrane [86].

Patel et al. showed that a Bruggeman exponent of 1.5 is often not valid for real separator materials, which do not have uniform spherical shape [227]. Porous networks based on other morphologies such as oblate (disk-type) ellipsoids or lamella increase the tortuous path for ionic conductivity and result either in a significant increase of the exponent α, or in a complete deviation from the power law. They showed that spherical or slightly prolate ellipsoidal pores should be preferred for separators, as they lead to higher ionic conductivity separators.

Tye [122] explained that separator tortuosity is a key property determining transient response of a separator and steady-state electrical measurements do not reflect the influence of tortuosity. He recommended that the distribution of tortuosity in separators be considered; some pores may have less tortuous paths than others. He showed mathematically that separators with identical average tortuosity and porosities could be distinguished by their non-steady-state behavior if they have different distributions of tortuosity.

Doyle et al. [86] used a mathematical model to examine the effect of separator thickness for the PVDF:HFP gel electrolyte system and found that decreasing separator thickness below 52 μm caused only a minor decrease in ohmic drop across the cell. The voltage drop in the electrodes was much more significant. Mao and White [228] developed a mathematical model for discharge of a Li/TiS_2 cell. Their model predicted that increasing the thickness of the separator from 25 to 100 μm decreased the discharge capacity from 95% to about 90%; further increasing separator thickness to 200 μm reduced discharge capacity to 75%. These theoretical results indicate that conventional separators (25–37-μm thick) do not significantly limit mass transfer of lithium. Santhanagopalan et al. [229] studied the influence of the separator on several internal short-circuit scenarios.

The use of electroactive polymers for overcharge protection has been recently reported for lithium-ion batteries [230, 231]. The electroactive polymer incorporated into a battery's separator is an attractive new option for overcharge protection. Thomas et al. [233] developed a mathematical model to explain how electroactive polymers such as polythiophene can be used to provide overcharge protection for lithium-ion batteries. The model shows that, as the cell potential exceeds the oxidation potential of the polymer, the cell is transformed, over a time scale of a few minutes, from a battery into a resistor, after which a steady-state overcharge condition is attained.

Summary

The ideal battery separator would be infinitesimally thin, offer no resistance to ionic transport in electrolytes, provide infinite resistance to electronic conductivity for isolation of electrodes, be highly tortuous to prevent dendritic growths, and be inert to chemical reactions. Unfortunately, such a product is not commercially feasible. Actual separators are electronically insulating membranes whose ionic resistivity is brought to the desired range by manipulating the membranes thickness and porosity.

It is clear that no single separator satisfies all the needs of battery designers, and often, optimization is performed on a case-by-case basis. It is ultimately the application that decides which separator is most suitable. This chapter is intended to be a useful tool to help battery manufacturers in selecting the most appropriate separators for their batteries and respective applications. The information provided is purely technical and does not include other very important parameters, such as cost of production, availability, etc.

There has been a continued demand for thinner battery separators to increase battery power and capacity. This has been especially true for lithium-ion batteries used in portable electronics. However, it is very important to ensure the continued safety of batteries, and this is where the role of the separator is greatest. Thus, it is essential to optimize all the components of battery to improve the performance while maintaining the safety of these cells. Separator manufacturers continue to work along with the battery manufacturers to create the next generation of batteries with increased reliability and performance, but always keeping safety in mind.

This chapter has attempted to present a comprehensive review of literature on separators used in various batteries. It is evident that a wide variety of separators are available, and that they are critical components in batteries. In many cases, the separator is one of the major factors limiting the life and/or performance of batteries. Consequently, development of new improved separators would be very beneficial for the advanced high capacity batteries.

Future Directions

Up until the last few years, most of the separators and membranes historically used had not been customized for specific battery applications. Thus, future research should be aimed at developing custom-separators that are specifically tailored for the individual battery applications. For example, the form factor of batteries has drastically changed since 2005 – with the advent of sleek consumer electronic devices, and large format batteries and packs for automotive applications. Efforts toward better heat dissipation, extended operating windows, and more stringent requirements on long-term durability have significantly expanded. The general objectives of separator research should be: (a) to find new and cost-effective separators, (b) to understand the separator properties in batteries, and (c) to optimize separator properties related to specific cell performance, life, and safety. The battery separators for tomorrow will demand more than just good insulation and mechanical filtration; they will require unique electrochemical properties. More work focusing on precise control of the membrane properties guided by fundamental insight into the membrane behavior will set the trend for the next generation separators.

Acknowledgments The authors would like to acknowledge numerous colleagues and staff members at Celgard, LLC and Polypore, International Inc. for their help with the preparation and publication of this chapter.

Bibliography

Primary Literature

1. Linden D, Reddy TB (2002) Handbook of batteries, 3rd edn. McGraw Hill, New York
2. Besenhard JO (ed) (1999) Handbook of battery materials. Weimheim, Wiley-VCH
3. Berndt D (2003) Maintenance free batteries, 3rd edn. Research Studies Press, Taunton, Somerset
4. Arora P, Zhang Z (2004) Chem Rev 104:4419
5. Brodd RJ, Friend HM, Nardi JC (eds) (1995) Lithium ion battery technology. ITE-JEC Press, Brunswick
6. Wakihara M, Yamamoto O (eds) (1998) Lithium ion batteries, fundamentals and performance. Wiley-VCH, New York
7. Yoshino A (1995) Chem Ind 146:870
8. Schalkwijk WAV (ed) (2002) Advances in lithium ion batteries. Kluwer, New York
9. Kinoshita K, Yeo R (1985) Survey on separators for electrochemical systems. LBNL, January 1985
10. Zhang SS (2007) J Power Sources 164 (1):351
11. Benett J, Choi WM (1995) Developments in small cell separators. In: Proceedings of the 10th annual battery conference on applications & advances, IEEE, New York, 10–13 Jan 1995, p 265
12. Boehnstedt W (1999) In: Besenhard JO (ed) Handbook of battery materials. VCH Wiley, Amsterdam
13. Spotnitz R (1999) In: Besenhard JO (ed) Handbook of battery materials. VCH Wiley, Amsterdam
14. Shirai H, Spotnitz R (1996) In: Kozawa A, Yoshio M (eds) Lithium ion secondary battery – materials and applications. Nikkan Kogyo Shin-bun, Tokyo, p 91 (in Japanese)
15. Shirai H, Spotnitz R, Atsushi A (1997) Characterization of separators for lithium ion batteries – a review. Chem Ind 48:47 (in Japanese)
16. Hiroshi T, In: Ogumi Z (ed) (1997) The latest technologies of the new secondary battery materials. CMC, Tokyo, p 99 (in Japanese)
17. Hiroshi T, In: Oyama N (ed) (1998) Advanced technologies for polymer battery. CMC, Tokyo, p 165 (in Japanese)
18. Koichi K, In: Oyama N (ed) (1998) Advanced technologies for polymer battery. CMC, Tokyo, p 174 (in Japanese)
19. Kiyoshi K, In: Kanamura K (ed) (2004) Lithium secondary battery technology for the 21st century. CMC, Tokyo, p 116
20. Brodd RJ, Bullock KR, Leising RA, Middaugh RL, Miller JR, Takeuchi EJ (2004) Electrochem Soc 151:K1
21. http://www.freedoniagroup.com/Batteries.html. Accessed Feb 2010
22. About Edison batteries, Inc. http://www.optodot.com/sys-tmpl/htmlpage/. Accessed Feb 2010
23. Industry News: Looking Back, Looking Forward. Battery & EV Technology, January 2004, 28:2
24. Advanced rechargeable battery industry 2001/2002. Nomura Research Institute Limited, Japan
25. Battery and fuel cell components (2009) The Fredonia Group, Cleveland, OH
26. Pilot C (2004) The worldwide rechargeable battery market. Presented at Batteries 2004, 6th edn, Paris, 2–4 June 2004
27. Celgard Inc. http://www.celgard.com. Accessed Feb 2010
28. http://www.celgard.com/news-room/press-releases/2010/celgard-expansion.asp. Accessed Feb 2010

29. http://www.asahi-kasei.co.jp/asahi/en/ir/supplement/1003 transcript.pdf. Accessed Feb 2010
30. Advanced Battery Technology, February 2004, 40:22
31. Pekala RW, Khavari M (2003) US Patent 6,586,138
32. www.porouspower.com. Accessed Feb 2010
33. http://www.sumitomo-chem.co.jp/english/division/itrc.html. Accessed Feb 2010
34. Hamano K, Yoshida Y, Shiota H, Shiraga S. Aihara S, Murai M, Inuzuka T (2003) US Patent 6,664,007 B2
35. Sun L (2003) US Patent 2003/0152828A1
36. Johnson BA, White RE (1998) J Power Sources 70:48
37. Hoffman HG (1995) In: Proceedings of the tenth annual battery conference on applications and advances, LongBeach, 10–13 Jan 1995, IEEE, New York, p 253
38. Geiger M, Callahan RW, Diwiggins CF, Fisher HM, Hoffman DK, Yu WC, Abraham KM, Jillson MH, Nguyen TH (1994) The eleventh international seminar on primary and secondary battery technology and application, Fort Lauderdale, FL, 28 Feb–3 Mar 1994, Florida Educational Seminars, Boca Raton
39. Tanba H (1999) Molding Process 11:759
40. Adachi A, Spotnitz RM et al (1997) Osaka chemical marketing center. Osaka, Japan, p 69
41. Callahan RW, Nguyen KV, McLean JG, Propost J, Hoffman DK (1993) In: Proceedings of the 10th international seminar on primary and secondary battery technology and application, Fort Lauderdale, 1–4 Mar 1993. Florida Educational Seminars, Boca Raton
42. Yu WC, Hux SE (1999) US Patent 5,952,120
43. Hipore. http://www.asahikasai.co.jp/memrbane/english/tradenm/t07.html. Accessed Feb 2010
44. Druin ML, Loft JT, Plovan SG (1974) US Patent 3,801,404
45. Schell WJ, Zhang Z (1999) In: The fourteenth annual battery conference on applications and advances, Long Beach, 12–15 Jan 1999. IEEE, New York, p 161
46. Isaacson RB, Bierenbaum HS (1971) US Patent 3,558,764
47. Kamei E, Shimomura Y (1986) US Patent 4,563,317
48. Yu TH (1996) Processing and structure-property behavior of microporous polyethylene – from resin to final film. Ph.D. Dissertation, Virginia Polytechnic Institute and State University, Virginia
49. Sarada T, Sawyer LC, Ostler MI (1983) J Membr Sci 15:97
50. Bierenbaum HS, Daley LR, Zimmerman D, Hay IL (1974) US Patent 3,843,761
51. Hamer EAG (1986) US Patent 4,620,956
52. Hiroshi K, Tetuo A, Akira K (1997) US Patent 5,691,047
53. Kesting RE (1985) Synthetic polymeric membranes, 2nd edn. Wiley, New York
54. Ihm DW, Noh JG, Kim JY (2002) J Power Sources 109:388
55. Bierenbaum HS, Isaacson RB, Druin ML, Plovan SG (1974) Ind Eng Chem Prod Res Dev 13:2
56. Norimitsu K, Kotaru T, Koichi K, Hidehiko F (2000) US Patent 6,153,133
57. Michiyuki A (1996) Japan Patent 8064194
58. Kotaro T, Koichi K, Tatsuya T, Kenkichi O (1991) US Patent 5,051,183
59. Koichi K, Kotaro T, Mamoru T, Tatsuya T (1996) Japan Patent 8012799
60. Akinao H, Kazuo Y, Hitoshi M (2000) US Patent 6,048,607
61. Userguide (2003) FreedomCar separator costing document, February 2003
62. Xu M, Hu S, Guan J, Sun X, Wu W, Zhu W, Zhang X, Ma Z, Han Q, Liu S (1992) US Patent 5,134,174
63. Fisher HM, Wensley CG (2002) US Patent 6,368,742
64. Zhu W, Zhang X, Zhao C, Zu W, Hou J, Xu M (1996) Polym Adv Tech 7:743
65. Sadamitsu K, Ikeda N, Hoki M, Nagata K, Ogino K (2002) World Patent Application 02066233A1
66. Higuchi H, Matsushita K, Ezoe M, Shinomura T (1995) US Patent 5,385,777
67. Calis GHM, Daemen APM, Gerrits NSJA, Smedinga JT (1997) J Power Sources 65:275

68. Ooms FGB, Kelder EM, Schoonman J, Gerrits N, Smedinga J, Calis G (2001) J Power Sources 97–98:598
69. Yamamura Y, Ooizumi S, Yamamoto K (2001) Separator for rechargeable lithium-ion batteries with high puncture strength and high melt rupture temperature, Nitto Denko Technical Report, vol 39, p 39. http://www.nitto.com/rd/rd6_1.html. Accessed Feb 2010
70. Pekala RW, Khavari M, Dobbie G, Lee D, Fraser-Bell G (2000) In: 17th international seminar & exhibit on primary and secondary batteries, Fort Lauderdale, 6–9 Mar 2000. Florida Educational Seminars, Boca Raton
71. Fleming R, Taskier H (1990) Prog Batt Sol Cells 9:58
72. Hoffman D, Fisher H, Langford E, Diwiggins C (1990) Prog Batt Sol Cells 9:48
73. Spotnitz R, Ferebee M, Callahan RW, Nguyen K, Yu WC, Geiger M, Dwiggens C, Fischer H, Hoffman D (1995) In: Proceedings of the 12th international seminar on primary and secondary battery technology and applications, Fort Lauderdale, 6–9 Mar 1995. Florida Educational Seminars, Boca Raton
74. Yu WC, Callahan RW, Diwiggins CF, Fischer HM, Geiger MW, Schell WJ (1994) In: North America membrane society conference, Breckenridge, CO
75. Kuribayashi I (1996) J Power Sources 63:87
76. Pasquier AD, Gozdz A, Plitz I, Shelburne J (2002) In: 201st meeting, The Electrochemical Society, Philadelphia, 12–17 May 2002
77. Augustin S, Volker H, Gerhard H, Christian H (2002) Desalination 146:23
78. http://www.separion.com. Accessed Feb 2010
79. Hying C (2004) Separion separators for lithium batteries – safety & performance. In: Batteries 2004, 6th edn. Paris, 2–4 June 2004
80. Sachan S, Ray CA, Perusich SA (2002) Polym Eng Sci 42:1469
81. Sachan S, Perusich S (1999) In: Electrochemical Society Meeting, Seattle
82. Carlson SA (2004) Membrane & separation technology news. 22, 8
83. Abraham KM (1993) Electrochim Acta 38:1233
84. Gineste JL, Pourcell G (1995) J Membr Sci 107:155
85. Hoffman DK, Abraham KM (1991) In: Proceedings of the fifth international seminar on lithium battery technology and applications, Deerfield Beach. Florida Educational Seminars, Boca Raton, FL
86. Fuller TF, Doyle M, Newman J (1994) J Electrochem Soc 141:1
87. USABC (2001) Development of low cost separators for lithium-ion batteries, RFPI 2001
88. Laman FC, Sakutai Y, Hirai T, Yamaki J, Tobishima S (1992) Extended abstract. In: 6th International meeting lithium batteries, Münster, Germany, 10–15 May 1992, p 298
89. Laman FC, Gee MA, Denovan J (1993) J Electrochem Soc 140:L51
90. Robinson RG, Walker RL (1963) In: Collins DH (ed) Batteries. MacMillan, New York, p 15
91. Lander JJ, Weaver RD, Salkind AJ, Kelley JJ (1964) In: Cooper JE, Fleischer A (eds) Characteristics of separators for alkaline silver oxide zinc secondary batteries. Screening methods. NASA Technical Report NAS 5–2860
92. Kilroy WP, Moynihan CT (1978) J Electrochem Soc 125:520
93. MacMullin RB, Muccini GA (1956) AIChE J 2:393
94. Spotnitz R, Ferebee MW (1996) Meeting abstracts, 6–11 Oct 1996. The Electrochemical Society Inc., vol 96-2, Fall Meeting, San Antonio, TX
95. Ionov VV, Isakevitch VV, Katalevsky EE, Chernokoz AJ (1990) J Power Sources 30:321
96. Lowell S, Shields E (1991) Powder surface area and porosity, 3rd edn. Chapman and Hill, New York
97. PMI conference 2000 proceedings (2000) PMI short course, Ithaca, NY, 16–19 Oct 2000
98. Porous materials Inc. http://www.pmiapp.com. Accessed Feb 2010
99. Jena AK, Gupta KM (1999) J Power Sources 80:46
100. Jena AK, Gupta KM (2001) J Power Sources 96:214
101. Venugopal G, Moore J, Howard J, Pendalwar S (1999) J Power Sources 77:34
102. Zeman L, Denault L (1992) J Membr Sci 71:221

103. Chen RT, Saw CK, Jamieson MG, Aversa TR, Callahan RW (1994) J Appl Polym Sci 53:471
104. Fujii T, Mochizuki T (1998) US Patent, 5,759,678
105. Venugopal G (1997) The role of plastics in lithium-ion batteries. In: Proceedings of the 3rd annual conference on plastics for portable and wireless electronics, Philadelphia, 14–15 Oct 1997, p 11
106. Lundquist JT, Lundsager CB, Palmer NL, Troffkin HJ, Howard J (1998) US Patent 4,731,304
107. Lundquist JT, Lundsager CB, Palmer NI, Troffkin HJ (1987) US Patents 4,650,730
108. Faust MA Suchanski MR, Osterhoudt HW (1988) US Patent No. 4,741,979
109. Matthias U, Dieter B, Heinrich R, Thomas B (2003) US Patent 6,511,517
110. Maleki H, Shamsuri AK (2003) J Power Sources 115:131
111. http://www.cpsc.gov/cpscpub/prerel/prhtml07/07011.html. Accessed Feb 2010
112. Norin L, Kostecki R, McLarnon F (2002) Electrochem Solid State Lett 5:A67
113. Kostecki R, Norin L, Song X, McLarnon F (2004) J Electrochem Soc 151:A522
114. Hazardous materials regulations, Code of Federal Regulations, CFR49 173.185
115. UL1640, Lithium batteries, Underwriters Laboratories, Illinois
116. UL2054, Household and commercial batteries, Underwriter Laboratories, Illinois
117. Secondary lithium cells and batteries for portable applications, International Electrotechnic Commission, IEC 61960–1 and IEC 61960–2
118. Recommendations on the transport of dangerous goods (1999) Manual of Tests and Criteria, United Nations, New York
119. Safety Standard for Lithium Batteries (1995) UL 1642, 3rd edn. Underwriters Laboratories Inc, Illinois
120. Standard for Household and Commercial Batteries (1993) UL 2054. Underwriter Laboratories, Illinois
121. UN Recommendations on the Transport of Dangerous Goods, December 2000
122. Tye FL (1983) J Power Sources 9:89
123. Japan Battery Association (1997) A guideline for the safety evaluation of secondary lithium cells. Japan Battery Association, Tokyo
124. Venugopal G (2001) J Power Sources 101:231
125. Kim MH (2008) LG Li-Ion technology for automotive applications. Presented at the AABC 2008, Tampa, FL
126. Alamgir M, Abraham KM (1994) Chapter 3. In: Pistoia G (ed) Lithium batteries: new materials, developments and perspectives, vol 5, Industrial Chemistry Library. Elsevier, New York
127. Gray FM (1997) Polymer electrolytes, RSC materials monograph. The Royal Society of Chemistry, Cambridge
128. Fauteux D, Massucco A, McLin M, Van Buren M, Shi J (1995) Electrochim Acta 40:2185
129. North M, Markin TL, Hooper A, Tofield BC (1984) In: Second international meeting on lithium batteries, Extended Abstracts #19, Paris, France, 25–27 Apr 1984
130. Appetecchi GB, Dautzenberg G, Scrosati BJ (1996) J Electrochem Soc 143:6
131. Armand M (1983) Solid State Ionics 9/10:745
132. Lightfoot P, Mehta MA, Bruce PG (1993) Science 262:883
133. Vincent CA, Scrosati B (1993) Modern batteries. An introduction to electrochemical power sources. Arnold, London
134. Appetecchi GB, Passerini S (2002) J Electrochem Soc 149:A891
135. Croce F, Appetecchi GB, Persi L, Scrosati B (1998) Nature 394:4496
136. Croce F, Persi L, Ronci F, Scrosati B (2000) Solid State Ionics 135:47
137. Fan J, Fedkiw PS (1997) J Electrochem Soc 144:399
138. Appetecchi GB, Romagnoli P, Scrosati B (2001) Electrochem Commun 3:281
139. Kim DW, Sun YK (2001) J Power Sources 102:41
140. Chojnacka J, Acosta JL, Morales E (2001) J Power Sources 97–98:819
141. Nishi Y (2002) In: van Schalkwijk W, Scrosati B (eds) Advances in lithium ion batteries. Kluwer/Plenum, New York

142. EE Times.com. http://www.eet.com/story/OEG19990121S0013. Accessed Feb 2010
143. Song JY, Wang YY, Wan CC (1999) J Power Sources 77:183
144. Min HS, Ko JM, Kim DW (2003) J Power Sources 119–121:461
145. Jo SI, Sohn HJ, Kang DW, Kim DW (2003) J Power Sources 119–121:478
146. Kim HS, Kum KS, Cho WI, Cho BW, Rhee HW (2003) J Power Sources 124:221
147. Abraham KM, Alamgir M (1994) Solid State Ionics 70–71:20
148. Schmutz C, Tarascon JM, Gozdz AS, Schumtz CN, Warren PC, Shokoohi FK (1995) In: Proceeding electrochemical society, 1995. Rechargeable lithium and lithium ion batteries, vol 94–28, pp 330–335
149. Murata K, Izuchi S, Yoshihisa Y (2000) Electrochim Acta 45:1501
150. Jiang Z, Carroll B, Abraham KM (1997) Electrochim Acta 42:2667
151. Song JY, Cheng CL, Wang YY, Wan CC (2002) J Electrochem Soc 149:A1230
152. Gozdz AS, Schmutz CN, Tarascon JM, Warren PC (1995) US Patent 5,456,000
153. Gozdz AS, Tarascon JM, Schmutz CN, Warren PC, Gebizlioglu OS, Shokoohi F (1995) In: Tenth annual battery conference on advances and applications, Long Beach, 10–13 Jan 1995. IEEE, New York, p 301
154. Tarascon JM, Gozdz AS, Schumtz CN, Shokoohi FK, Warren PC (1996) Solid State Ionics 86–88:49
155. Pasquier AD, Waren PC, Culver D, Gozdz AS, Amatucci G, Tarascon JM (1999) Electrochem Soc Proc 99(24):360
156. Park CK, Kakirde A, Ebner W, Manivannan V, Chai C, Ihm DJ, Shim YJ (2001) J Power Sources 97–98:775
157. Dasgupta S, Jacobs JK (1995) US Patent 5,437,692
158. Abraham KM, Alamgir M, Hoffman DK (1995) J Electrochem Soc 142:683
159. Pendalwar SL, Howard JN, Venugopal G, Oliver M (1998) US Patent 5,716,421
160. Gozdz AS, Plitz I, Du Pasquier A, Zheng T (2001) In: Proceedings of the 200th ECS meeting, Phoenix, AZ, vol 2000–2001, Fall 2001, p 336
161. Gozdz AS (2001) US Patent 6,328,770
162. Kim DW, Oh B, Park JH, Sun YK (2000) Solid State Ionics 138:41
163. Wang Y, Sejdic JT, Steiner R (2002) Solid State Ionics 148:443
164. Gozdz AS, Plitz I, DuPasquier A, Zheng T (2007). Presented at the 198th meeting of the electrochemical society, Phoenix, AZ, 22–27 Oct 2000
165. Spotnitz RM, Wensley CG (2000) US Patent 6,322,923
166. Fabrice C, Bradford R (2002) WO 02/50929 A2
167. Jeong YB, Kim DW (2004) J Power Sources 128:256
168. Dasgupta S, Jacobs JK (1996) US Patent 5,498,489
169. Balsara N (2010) Block copolymer separators for lithium batteries. In: Presented at the 2010 annual merit review of the vehicle technologies program. http://www1.eere.energy.gov/vehiclesandfuels/pdfs/merit_review_2010/electrochemical_storage/es088_balsara_2010_p.pdf. Accessed Feb 2010
170. Eschbach FO, Oliver M (1997) US Patent 5,681,357
171. Hamano K, Shiota H, Shiraga S, Aihara S, Yoshida, Y, Murai M, Inuzuka T (1999) US Patent 5,981,107
172. Akashi H (1997) US Patent 5,658,686
173. Akashi H (1996) Paper presented at the international symposium on polymer electrolytes, ISPE-5, Uppsala, Sweden, 11–16 Aug 1996
174. Fujii T (2000) In: Proceeding of 17th international seminar & exhibit on primary and secondary batteries, Boca Raton, 6–9 Mar 2000. Florida Educational Seminars, Boca Raton
175. Nakane I, Narukawa S (2000) Power 2000. In: The 8th annual international conference on power requirements for mobile computing and wireless communications, San Diego, CA, 24–27 Sept 2000
176. Kim KM, Ryu KS, Kang SG, Chang SH, Chung IJ (2001) Macromol Chem Phys 202:866
177. Kim KM, Park NG, Ryu KS, Chang SH (2002) Polymer 43:3951

178. Kim KM, Ko JM, Park NG, Ryu KS, Chang SH (2003) Solid State Ionics 161:121
179. Prosini PP, Villano P, Carewska M (2002) Electrochim Acta 48:227
180. Liu X, Kusawake H, Kuwajima S (2001) J Power Sources 97–98:661
181. Kim DW, Ko JM, Chun JH, Kim SH, Park JK (2001) Electrochem Comm 3:535
182. Kim DW, Noh A, Chun JH, Kim SH, Ko JM (2001) Solid State Ionics 144:329
183. Choi SH, Park SY, Nho YC (2000) Radiat Phys Chem 57:179
184. Choi SH, Kang HJ, Ryu EN, Lee KP (2001) Radiat Phys Chem 60:495
185. Mardegain SB (2003) Battery Power Prod Technol 1:12
186. Boehnstedt W (1996) J Power Sources 59:45
187. Boehnstedt WJ (2004) J Power Sources 133:59
188. Lander JJ (1974) In: Proceedings of the symposium on battery separators. The Electrochemical Society, Columbus, OH, p 4
189. Prout L (1993) J Power Sources 46:117
190. Vinal GW (1945) Storage batteries. Wiley, New York
191. Butherus AD, Lindenberger WS, Vaccaro FJ (1970) Lead-acid battery: electrochemical compatibility of plastics. Bell Syst Tech J 49(7):1377–1392
192. Paik SL, Terzaghi G (1995) J Power Sources 53:283
193. Kung J (1994) J Power Sources 48:129
194. Wang LC, Harvey MK, Stein HL, Scheunemann U (1997) The role of UHMW-PE in microporous PE separators. In: Proceedings of the 12th annual battery conference on applications & advancesIEEE, New York, 14–17 Jan 1997, p 69
195. Larsen DW, Kehr CL (1996) US Patent 3,351,495
196. Boehnstedt W (2001) J Power Sources 95:234
197. Wang LC, Harvey MK, Ng JC, Scheunemann U (1998) J Power Sources 73:74
198. Endoh H (1996) J Power Sources 599:51
199. Higashi T, Endoh H (1998) J Power Sources 73:110
200. Rand DAJ, Woods R, Dell RM (1998) Batteries for electric vehicles. Research Studies Press, Taunton, UK. ISBN. ISBN 0-86380-205-0
201. McClelland DH, Devitt JL (1975) US Patent 3,862,861
202. Zguris GC (1998) J Power Sources 73:60
203. Zguris GC (1997) J Power Sources 67:307
204. Zguris GC (1996) J Power Sources 59:131
205. Pavlov D, Ruevski S, Naidenov V, Sheytanov G (2000) J Power Sources 85:164
206. Ferreira AL (1999) J Power Sources 78:41
207. Fetcenko MA, Venkatesan S, Ovshinsky S (1992) In: Proceedings of the symposium on hydrogen storage materials, batteries & electrochemistry. Electrochemical Society, Pennington, NJ, p 141
208. Wada M (1994) Polymer role in advanced battery technology. Polym Adv Tech 5:645–652
209. Ikoma M, Hoshina Y, Matsumoto L, Iwakura C (1996) J Electrochem Soc 143:1904
210. Ikoma M, Takahashi O, Tsuboi R, Matsumoto L (1993) Denki Kagaku 61:997
211. Furukawa N (1994) J Power Sources 51:45
212. Nagarajan GS, Van Zee JW (1998) J Power Sources 70:173
213. Cook JA, Lancaster IM (1998) Electrochem Soc Proc 98–15:55
214. Scimat's latest separators (2002) Batteries International, October 2002
215. Cheng S, Zhang J, Liu H, Leng Y, Yuan A, Cao C (1998) J Power Sources 74:155
216. Newman J, Tiedemann W (1975) AIChE J 21:25
217. Newman JS (1991) Electrochemical systems, 2nd edn. Prentice-Hall, Englewood Cliffs
218. Vidts PD, White RE (1997) J Electrochem Soc 144:1343
219. Ceder G, Doyle M, Arora P, Fuentes Y (2002) Computational modeling and simulation for rechargeable batteries. MRS Bull 27(8):619–623
220. Newman J, Tiedemann W (1997) J Electrochem Soc 144:3081
221. Gu H, Nguyen TV, White RE (1987) J Electrochem Soc 134:2953
222. Vidts PD, Delgado J, White RE (1995) J Electrochem Soc 142:4006

223. Doyle M, Fuller TF, Newman J (1993) J Electrochem Soc 140:1526
224. Fan D, White RE (1991) J Electrochem Soc 138:17
225. Doyle M, Newman J, Gozdz AS, Schumtz CN, Tarascon JM (1996) J Electrochem Soc 143:1890
226. Arora P, Doyle M, Gozdz AS, White RE, Newman J (2000) J Power Sources 88:219
227. Patel KK, Paulsen JM, Desilvestro J (2003) J Power Sources 122:144
228. Mao Z, White RE (1993) J Power Sources 43–44:181
229. Santhanagopalan S, Ramadass P, (Zhengming) Zhang J (2009) Analysis of internal short-circuit in a lithium ion cell. J Power Sources 194(1):550–557
230. Denton F, Howard JN, Anani AA, Fernandez JM (2001) US Patent 6,228,516
231. Takita K, kono K, Takashima T, Okamoto K (1991) US Patent 5,051,183
232. Mao H, Wainwright DS (1990) US Patent 6,074,766
233. Thomas-Alyea KE, Newman J, Chen G, Richardson TJ (2004) A concentrated solution theory model of transport in solid-polymer-electrolyte fuel cells. J Electrochem Soc 151: A509

Chapter 7
Lithium Battery Electrolyte Stability and Performance from Molecular Modeling and Simulations

Grant D. Smith and Oleg Borodin

Glossary

Interfacial impedance	The resistance to lithium motion at the interface between the electrolyte and an electrode due to structure imposed on the electrolyte by the electrode, barriers to intercalation into the electrode, lithium desolvation energy, and/or the SEI layer.
Ionic liquid electrolyte	An electrolyte comprised of a lithium salt dissolved in a room-temperature ionic liquid.
One-electron reduction	The reduction of an electrolyte molecule by a single electron resulting in formation of species believed to be important contributors to the outer part of the SEI.
Organic liquid electrolyte	An electrolyte comprised of a lithium salt dissolved in liquid carbonates, esters, ethers, or a mixture thereof.
Oxidative stability	The ability of an electrolyte to remain electrochemically stable against oxidation at the surface of the cathode.

This chapter was originally published as part of the Encyclopedia of Sustainability Science and Technology edited by Robert A. Meyers. DOI:10.1007/978-1-4419-0851-3

G.D. Smith (✉)
Department of Materials Science and Engineering, University of Utah,
Salt Lake City, UT 84112, USA
e-mail: gdsmithwmi@gmail.com

O. Borodin
Department of Materials Science and Engineering, University of Utah,
Salt Lake City, UT 84112, USA

Electrochemistry Branch, SEDD, AMSRD–ARL–RDRL–SED–C, 2800 Powder Mill Rd,
Adelphi, MD 20783, USA
e-mail: Oleg.Borodin@utah.edu

R.J. Brodd (ed.), *Batteries for Sustainability: Selected Entries from the Encyclopedia of Sustainability Science and Technology*, DOI 10.1007/978-1-4614-5791-6_7,
© Springer Science+Business Media New York 2013

SEI layer	The layer of electrolyte decomposition products that forms at the interface between an electrode (primarily the anode) and the electrolyte, called the solid electrolyte interphase or solid electrolyte interface. The SEI is believed to be comprised of an inner SEI layer and an outer SEI layer.
Solid polymer electrolyte	An electrolyte comprised of a lithium salt dissolved in a polymer matrix.
Two-electron reduction	The reduction of an electrolyte molecule by two electrons resulting in formation of species believed to be important contributors to the inner part of the SEI.

Definition of the Subject

Lithium batteries are complex devices whose performance optimization necessitates that they be well understood on multiple timescales and length scales, ranging from systems level to molecular. Optimization of the electrolyte, in particular, requires detailed, fundamental, molecular-level understanding of the chemical features that lead to stable electrolytes with large electrochemical windows – what electrolyte solvents and additives facilitate the formation of stable solid electrolyte interface layers and which features of the electrolyte result in good bulk and interfacial lithium transport properties as well as thermal stability, low-temperature transport, and low volatility/high safety [1, 2]. Molecular modeling has contributed significantly to this fundamental understanding of electrolyte properties and will continue to contribute to it in the future.

Introduction

Electrolytes and electrode/electrolyte interfaces are well suited for molecular modeling studies because important timescales and length scales are accessible to these powerful methodologies, and because gleaning molecular-level information from experiment is intrinsically challenging. While molecular modeling methods have not yet reached the point where they can be used exclusively in the design of electrolytes for improved battery performance, they have provided important insight into electrolyte stability, transport mechanisms, and the nature of the interface between the electrolyte and electrodes. These insights, or more precisely, a subset thereof, are the subject of this chapter.

The primary molecular modeling methods that have been extensively applied to lithium battery electrolytes and electrode/electrolyte interfaces are molecular orbital calculations and molecular dynamics simulations. The former involves ab initio and density functional methods and will be referred to quantum chemistry or QC calculations. In QC methods, atomic centers and electrons are treated explicitly and the optimized molecular orbitals (with corresponding energies and molecular/complex geometries) are obtained. QC methods are extremely powerful and accurate given adequate levels of theory (e.g., electron correlation treatment or density functional used) and adequate atomic basis sets, but quickly become computationally prohibitive with increasing system size. Hence, QC methods are applied to individual molecules or small clusters of molecules and as such do not typically allow for direct inclusion of condensed phase/solvent effects. Furthermore, while it is possible to perform molecular dynamics simulations using QC methods, the computational cost of this approach is such that the size of the system, the length of the trajectory, and the level of theory that can be employed are typically severely compromised. Despite their computational limitations, QC methods provide invaluable information relevant to the stability and performance of lithium battery electrolytes, including properties such as the electron affinity and ionization potential of a molecule or molecular cluster, solvation/complexation energies, and the energetics of various reactions. QC methods are also invaluable in providing information needed in the development of atomistic potential functions, or force fields, for molecular dynamics (MD) simulations.

MD simulations of electrolytes for lithium batteries retain the atomistic representation of the electrolyte molecules but do not treat electrons explicitly. Instead the influence of electrons on intermolecular interactions is subsumed into the description of the interatomic interactions that constitute the atomistic potential or force field. The interatomic potential used in MD simulations is made up of dispersion/repulsion terms, Coulomb interactions described by partial atomic charges, and in some cases, dipole polarizability described by atom-based polarizabilities. The importance of explicit inclusion of polarization effects is considered below. In the most accurate force fields, interatomic potentials are informed by high-level QC calculations. Specifically, QC calculations provide molecular geometries, conformational energetic, binding energies, electrostatic potential distributions, and dipole polarizabilities that can be used to parameterize atomic force fields.

Because MD simulations use atomic force fields (i.e., do not explicitly include electrons), they are computationally orders of magnitude more efficient than QC methods. Hence, MD simulations can be run to predict bulk liquid properties, understand solvation/coordination structure, elucidate transport mechanisms, and study the interface between liquids and solid substrates, e.g., electrodes and electrolytes. Typical MD simulations involve hundreds of molecules and thousands of atoms, and trajectories of many nanoseconds can be routinely generated. Unlike QC methods, however, most classical MD simulation methodologies do not allow for chemical reactions, specifically the making and breaking of covalent bonds. An exception is MD simulations that involve reactive force fields, such as the ReaxFF, that allow for bond breaking and formation through use of bond order potentials and partial atomic charges that depend upon the bonding environment of an atom. Reactive force fields are computationally more expensive than nonreactive atomistic force fields, resulting in typically smaller simulation systems and shorter trajectories. Furthermore, reactive force fields, in an effort to minimize complexity and number of parameters, are less accurate in the prediction of thermophysical and transport properties than the better nonreactive force fields.

Quantum Chemistry Studies of Oxidative Stability of Electrolyte Compounds

Electrolyte oxidative stability is one of the key parameters for designing electrolytes compatible with a cathode. An electrolyte should be either anodically stable toward the cathode or form a stable passivation layer on the cathode that is capable of conducting lithium ions and accommodating electrode dimensional changes during cycling. QC calculations were used to estimate the oxidative stability of electrolyte solvents [3, 4], additives [5], anions [6–8], and redox shuttle molecules [9]. A free-energy cycle shown in Fig. 7.1 is used to relate the gas-phase adiabatic ionization potential (IP) to the absolute oxidation potential using the free energy of solvation $\Delta G^0{}_S(M)$ and $\Delta G^0{}_S(M^+)$ for the reduced and oxidized forms of a molecule M. IP is defined as the enthalpy changes of the oxidation reaction in the gas phase at 298 K. QC calculations yield the standard oxidative potential $E^0{}_{abs}(M)$ at the absolute electrochemical scale [10, 11] with the reference vacuum state corresponding to an electron at rest in a field-free vacuum.

Using the free-energy cycle shown in Fig. 7.1, $E^0{}_{abs}(M)$ is expressed by Eq. 7.1

$$E^0{}_{abs}(M) = IP + \left[-T\Delta S + \Delta G^0{}_s(M^+) - \Delta G^0{}_s(M)\right]/F \tag{7.1}$$

where F is Faraday constant 23.061 kcal per volt gram equivalent and $-T\Delta S$ is the entropy difference in the gas phase between M and M^+. Most electrochemical experiments, however, report oxidation potential versus standard hydrogen electrode (SHE) or Li^+/Li in the lithium battery literature instead of using absolute oxidation potential. The SHE at room temperature (25°C) is estimated [11] to have the absolute potential of 4.42 V in aqueous solutions. Older International Union of Pure and Applied Chemistry (IUPAC) recommendation [10] of 4.44 V is often used. The IUPAC recommended values for the absolute SHE potential in nonaqueous solvents are within a few tens of volts of the value of the absolute SHE potential in aqueous solutions. For example, it was recommended by IUPAC to use the following absolute potential values for SHE in nonaqueous solvents: 4.13 ± 0.06 V for acetone and 4.60 ± 0.10 for acetonitrile. The Li^+/Li standard absolute electrode potential is -3.04 V versus SHE. Most commonly [4, 5, 7, 9], the aqueous value of $4.44 - 3.04 = 1.4$ V is used to convert from the absolute electrochemical scale used in QC calculations to the experimentally measured potential versus Li^+/Li.

An extensive QC study [4] of 160 structurally unrelated organic molecules reported prediction of IPs with a standard deviation of 0.14 V at B3LYP/6-311++G (2df,2p)//B3LYP/6-31+G(d). Such good agreement was achieved after correcting IPs obtained from QC calculations for the systematic underestimation of the experimental data by adding 0.28 V to all QC IP values. At the next step, $E^0{}_{abs}(M)$ was calculated from IP using Eq. 7.1. Widely available in QC codes, polarized continuum models (PCM) are commonly used to estimate free energy of solvation of the reduced and oxidized forms of M (see Eq. 7.1 and Fig. 7.1). The standard redox potentials of 270 structurally unrelated organic molecules in acetonitrile were predicted with the

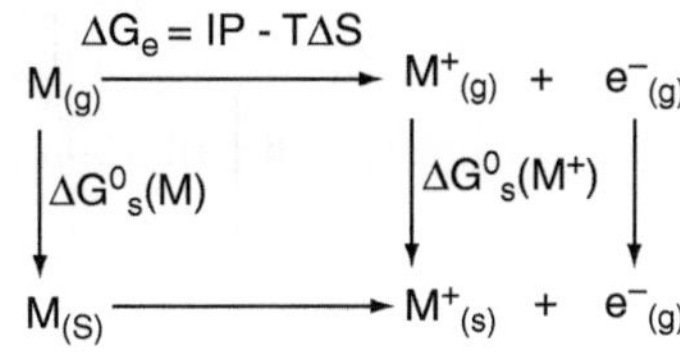

Fig. 7.1 Free-energy cycle for the redox reaction $M \rightarrow M^+ + e^-$, $M_{(g)}$ denotes molecule M in gas phase, while $M_{(S)}$ denotes the solvated molecule

standard deviation from experiments of 0.17 V [4]. This accuracy is sufficient for guiding the electrolyte solvent development. One caveat, however, should be mentioned. For a number of alkene molecules, the deviation between the experiment and the theory was about 0.5–0.8 V (i.e., 11–18 kcal/mol). No explanation was given for this discrepancy [4].

The other QC studies [5, 9] focused on the redox additives that could be reversibly reduced and oxidized for the overcharge protection. The calculated oxidative potential using B3PW91/6-31 + G(d,p) density functional theory (DFT) level were found in close agreement with the experimental data with the root mean-square deviation is 0.08 V and the maximum deviation is 0.15 V, indicating high fidelity of QC calculations of the solvent oxidation potential using PCMs to account for the solvation effects [5]. The experimental measurements were performed in 1 M lithium hexafluorophosphate ($LiPF_6$) ethylene carbonate (EC)/ethylmethyl carbonate (EMC) (1/2, v/v) supporting electrolyte on Pt wire. The correlation coefficients (R^2) of IP (calc. QC), E_{ox}(calc. QC) with respect to E_{ox}(exp.) were 0.84 and 0.98, respectively. Similarly, Wang, Buhrmester, and Dahn [9] reported an excellent agreement between the calculated values for 17 redox shuttle additives with the root-mean-square deviation between the calculated and measured oxidation potentials being 0.15 V with the maximum deviation of 0.25 V, indicating that DFT calculations at B3LYP/6-31 G(d,p) could be effectively used for screening redox additives.

Comparison of the common electrolyte solvents (EC, propylene carbonate [PC], dimethyl carbonate [DMC], EMC, vinylene carbonate [VC], dimethoxyethane [DME]) oxidative stability with experiments was reported by Zhang et al. [3]. While trends of the oxidative stability were reasonably captured in this study, typical deviations between experiments and simulations were reported to be around 0.5–1.0 V. Note that Zhang et al. [3] did not use the value of 1.4 V to convert from the absolute to Li^+/Li potential scale, instead they used the Li/Li^+ and M/M^+ cycles with a number of calculated/estimated quantities, resulting in the absolute potential versus Li^+/Li being around 2.2 V. Application of the value of 1.4 or 1.54 V derived from SHE potential in water and acetonitrile and using the standard Li^+/Li vs. SHE potential will result in an improved agreement between QC-based values reported by Zhang et al. [3] and experiments.

Ue et al. [6] and Johansson [7, 8] have reported anion oxidative potentials. They found a significant correlation between the predicted single molecule gas-phase QC values and experimental oxidation potential measurements performed on numerous solvents and salt concentrations. Figure 7.2 and Table 7.1 shows the vertical (e.g., anion (charge=−1e) geometry was used in the oxidized molecule (charge = 0) energy calculation) ionization energy (ΔE_v) and free energy of anion ionization

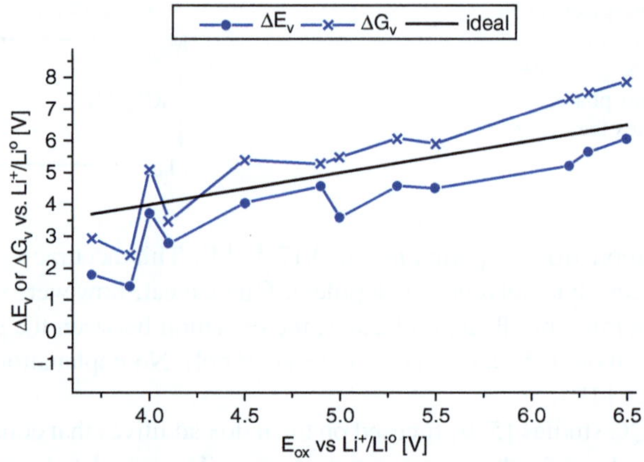

Fig. 7.2 Experimental anion oxidation potential compared with the vertical electron transition energy (ΔE_v) and the vertical free-energy difference (ΔG_v) from Johansson [7]. The highest experimental values from Table 7.1 are shown in this figure (Reproduced with permission of ACS)

Table 7.1 Anion oxidation potentials (in V) from experiments and vertical electron transition energy from DFT calculations by Johansson [7, 8]

Anion	Exp	QC	Anion	Exp	QC
AsF_6^-	6.5–6.8	6.05	$[1,3\text{-}BF_3\text{-}(N_2C_3H_3)]^-$; "imidazolide"$^-$	4.9	4.58
PF_6^-	6.3–6.8	5.65	$[(C_2O_4)_2B]$; BOB^-	>4.5	4.05
BF_4^-	6.2–6.6	5.22	$[(1,2\text{-}C_6F_4O_2)_2B]^-$ 4 F-BBB^-,	4.1	2.79
$N(SO_2CF_3)_2^-$,$TFSI^-$	6.1–6.3, 5.3	4.52	$N_5C_4^-$; $TADC^-$	>4.0	3.72
$(CF_3SO_2)_3C^-$, methide$^-$	6.1, 5.5	4.57	$(2,3\text{-}C_{10}H_6O_2)_2B]^-$; BNB^-	3.9	1.43
$CF_3SO_3^-$	5.9–6.0, 5.0	3.6	$[(1,2\text{-}C_6H_4O_2)_2B]^-$ BBB^-	3.7	1.79

(ΔG_v) from the Johansson work [8] that employed VSXC DFT functional and 6-311 + G(2df,p) basis set. Johansson [8] used vertical instead of adiabatic energies in calculations to lower the computational cost. QC calculations generally reproduced the order of oxidation potentials with the exceptions of the large anions that have low oxidative stability where deviations up to a few volts were observed. Another case where QC calculations fail to reproduce oxidative potential trends was the comparison of highest occupied molecular orbital (HOMO) calculations and measured oxidative potential of propargyl methanesulfonate (PMS) and EC [12]. Calculations indicated higher HOMO absolute value implying higher oxidative stability of EC compared to PMS by 0.8–1.0 V, while oxidative stability of PMS on Pt was higher compared to EC by ~0.2 V in the $LiPF_6$-based electrolytes [12]. It is important to note that anodic stability of electrolytes on active cathode surfaces could be significantly (up to a few volts) lower than the electrolyte stability on Pt or glassy carbon; therefore, one would need to include surfaces in the electrolyte calculations in order predict oxidative stability of electrolytes on active materials [1].

While the focus of this section was on the prediction of oxidation potentials, reduction potentials of EC, PC, DMC, VC, and vinylethylene carbonate (VEC) were calculated using quantum chemistry methods and compared with experiments [13–15].

Modeling of Reductive Decomposition of Electrolytes

The properties of the solid-electrolyte interface (SEI) that forms as a result of reduction of the electrolyte solvent and lithium salt are key elements in determining the performance of lithium ion batteries [1, 16]. The electrolyte solvent is typically comprised of a mixture of linear and cyclic carbonates, with EC being a major component of most electrolytes. The solvent and lithium salt anion undergo one- and two-electron reduction at the anode, typically Li-intercalated graphite. During the first cycle, the reaction products of electrolyte reduction form a thin SEI film, which in the ideal case protects the electrolyte from further reductive decomposition during subsequent cycles while at the same time contributing significantly to the resistance for Li^+ exchange between the electrolyte and the electrode. There have been number of experimental studies focusing on understanding of mechanisms of the SEI formation, a detail discussion of which can be found in Ref. [1, 17]. In general, the SEI is thought to consist of an inner and an outer layer. The inner layer is believed to be primarily comprised of doubly (two electron) reduced compounds, such as Li_2CO_3, while the outer SEI layer is thought to be comprised largely of reaction products of singly (one electron) reduced electrolyte species [1, 18–21], the nature of which remains controversial.

Given the importance of the SEI layer in lithium batteries, it is clear that improved understanding of the mechanisms of electrolyte decomposition and the formation of the SEI layer would facilitate optimization of battery performance. The reductive decomposition of organic solvents at the anode–electrolyte interface is largely responsible for the formation of the SEI. The electrochemical reactions leading to the formation of the SEI are complex and not well understood. Fortunately both QC and molecular dynamics simulations have provided some insight into reaction pathways that are believed to be important in the formation of the SEI. Reduction pathways for EC [14, 15, 22–24], PC [15, 23, 25], and VC [15, 23, 24], among others, have been most extensively studied extensively computationally, with the greatest focus on EC. Two schemes have been proposed in the literature for reductive decomposition of solvents at the anode interface. The first involves decomposition after electron transfer to Li^+-solvent complexes, while the latter involves direct decomposition on the electrode surface. Computational insights into both of these schemes for EC are considered below.

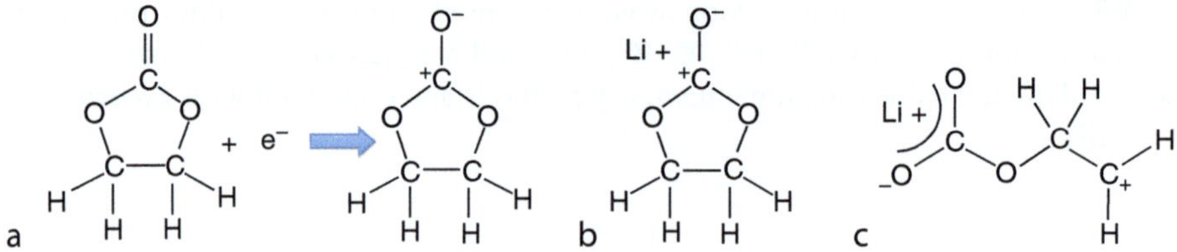

Fig. 7.3 (**a**) Reduction of EC by one electron to form the closed radical anion c-EC^{-}; (**b**) Li^{+}/c-EC^{-} complex; (**c**) Li^{+}/c-EC^{-} complex

Reduction of Li^{+}–EC Complexes

One-Electron Reduction of EC

The most important component of the current generation of electrolytes is EC. The initial step in the commonly accepted single-electron reduction mechanism is shown in Fig. 7.3a. QC calculations [14] have shown that this reaction is energetically unfavorable in the gas phase and energetically favorable when solvent effects are taken into account through the polarized continuum model approach. However, the reaction becomes dramatically more favorable when Li^{+} is complexed to the carbonate molecule, resulting in the cation/anion complex shown in Fig. 7.3b. The next step in the one-electron reduction process is commonly believed to be the opening of the closed EC radical anion (complexed to Li^{+}), denoted as c-EC^{-}, to form an open radical, shown in Fig. 7.3c. This process involves a substantial energy barrier leading ultimately to a substantially lower energy radical anion [14].

The fate of the o-EC^{-} radical anion formed from one-electron reduction of EC is a matter of central importance in understanding the formation of the SEI in EC-based electrolytes, and is a matter of controversy. Dilithium ethylene dicarbonate $(CH_2OCO_2Li)_2$ has been proposed as a major component of the outer SEI layer [1], resulting from combination reactions of pairs of o-EC^{-}, as shown in Scheme I in Fig. 7.4. Fourier transform infrared (FTIR) and x-ray photoelectron spectroscopy (XPS) studies of SEI films formed on lithium metal in EC-based electrolytes reveal the presence of alkyl carbonates [19, 26], while FTIR shows the presence of alkyl carbonates in the SEI formed from EC/diethyl carbonate electrolytes on graphite [20]. Hydrolysis of alkyl carbonate species formed from reduction of EC on noble metal electrolytes leads to the formation of ethylene glycol, which has been taken as evidence for the formation of $(CH_2OCO_2Li)_2$ [27]. Nuclear magnetic resonance (NMR) studies of the surface species formed on graphite in EC electrolytes resulted in spectra consist with that of alkyl carbonates, and in particular, the spectrum of $(CH_2OCO_2Li)_2$ [28]. A combined transmission electron microscopy (TEM)/FTIR study of the SEI formed on graphite in a EC/lithium perchlorate ($LiClO_4$) electrolyte revealed the formation of alkyl carbonates, one of whose chemical composition (O/C ratio = 1.5) is consistent with $(CH_2OCO_2Li)_2$ [29].

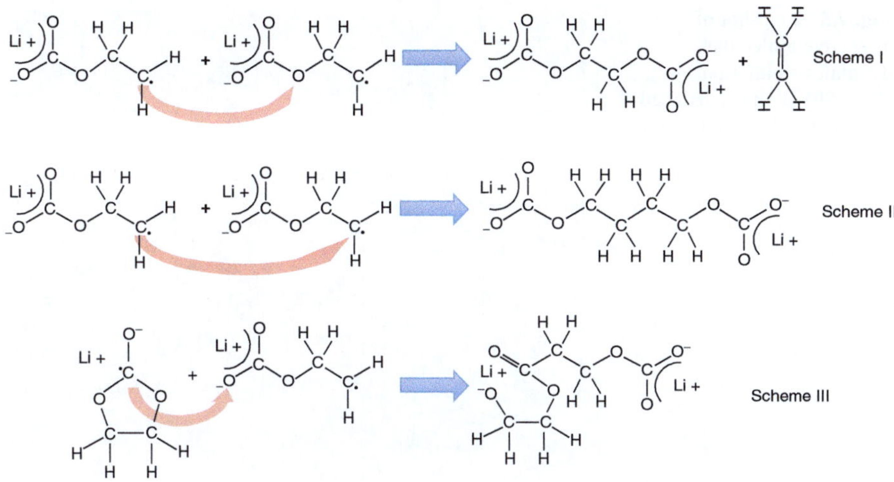

Fig. 7.4 Proposed reaction schemes involving singly reduced EC

In addition to experimental investigations, QC studies have provided important insight into single reduction reactions of EC. In one study, DFT studies at B3PW91/6-311++G(d,p) level of theory were performed in order to investigate the energetics of radical termination reactions of EC^- [14]. Both $(CH_2OCO_2Li)_2$ and dilithium butylene dicarbonate $(CH_2CH_2OCO_2Li)_2$ (Scheme II, Fig. 7.4) are believed to form by the reaction of two o-EC^-, while an ester-carbonate compound forms from the reaction of o-EC^- with c-EC^- as shown in Fig. 7.4 Scheme III. In terms of energy and free energy, $(CH_2CH_2OCO_2Li)_2 < (CH_2OCO_2Li)_2 + CH_2 = CH_2 <$ ester-carbonate compound [14]. Furthermore, the DFT study revealed that while Li^+/o-EC^- is much lower in energy than Li^+/c-EC^- (by about 27 kcal/mol), there is a substantial energy barrier for opening c-EC^- to form o-EC^-, around 13 kcal/mol in energy and 11–12 kcal/mol in free energy. For the elementary reaction $A \leftrightarrow A^*$, where A^* represents an activated complex, the Eyring rate equation is given as [30]

$$k = \frac{k_b T}{h} \exp\left[-\frac{\Delta G^*}{k_b T}\right] \tag{7.2}$$

At room temperature, the estimated lifetime c-EC^-, given as $1/k$, is around 100 μs. Hence, reactions involving the high-energy but long-lived c-EC^- form of the EC radical anion to form the ester-carbonate compound shown in Fig. 7.4 should be considered as a possible contributor to the outer SEI in EC-based electrolytes.

While the presence of $(CH_2OCO_2Li)_2$ in the SEI formed from EC-based electrolytes is not disputed, there is limited evidence for the presence of $(CH_2CH_2OCO_2Li)_2$ and the ester-carbonate compound. The combined TEM/FTIR study that claimed evidence for the presence of $(CH_2OCO_2Li)_2$ also found evidence for an alkyl carbonate compound with O/C ratio = 1, which the

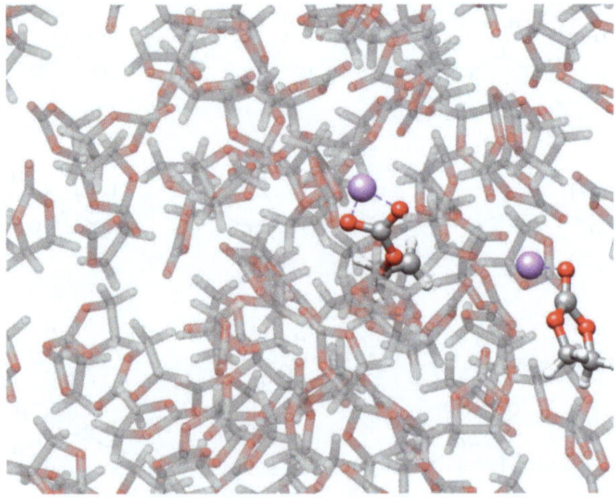

Fig. 7.5 Snapshot of a reactive molecular dynamics simulation (ReaxFF) of singly reduced EC species in a solution of EC molecules [31]

investigators associated with $(CH_2CH_2OCO_2Li)_2$ [29]. This O/C ratio is also consistent with the ester-carbonate compound. However, in their NMR analysis of surface compounds that was consistent with the presence of $(CH_2OCO_2Li)_2$, Xu et al. ruled out the possibility of these compounds [28]. In general, however, studies (e.g., FTIR and XPS) that reveal the presence of alkyl carbonates in the SEI formed in EC-based electrolytes do not allow for the differentiation between $(CH_2OCO_2Li)_2$ and $(CH_2CH_2OCO_2Li)_2$ and the ester-carbonate compound.

In addition to QC studies, reactive molecular dynamics (RMD) simulations using the reactive force field ReaxFF have been used to gain insight into reactions of singly reduced EC in the condensed (solution) phase [31]. In this study the reaction of Li^+/o-EC^- with both Li^+/o-EC^- and Li^+/c-EC^- has been studied in a solution of EC molecules. A snapshot of the system is shown in Fig. 7.5. RMD simulations were used to determine the free energy as a function of reaction coordinate (see below) and to examine the propensity of various radical combination reactions to occur in the condensed phase of an EC solvent.

The free energy in the condensed phase was determined for the reactions of Li^+/o-EC^- with Li^+/o-EC^- according to Scheme I and Scheme II in Fig. 7.4. ReaxFF simulations predict essentially no barrier for formation of $(CH_2CH_2OCO_2Li)_2$ while a barrier of around 12 kcal/mol is seen for the formation of $(CH_2OCO_2Li)_2 + CH_2 = CH_2$. ReaxFF simulations also predict $(CH_2CH_2OCO_2Li)_2$ to be of lower energy than $(CH_2OCO_2Li)_2 + CH_2 = CH_2$, in qualitative agreement with ab initio predictions [14]. The only reaction observed in an o-EC^-/EC system containing many Li^+/o-EC^- complexes was the rapid combination of o-EC^- radical pairs to form $(CH_2CH_2OCO_2Li)_2$ (Scheme II, Fig. 7.4). No formation of $(CH_2OCO_2Li)_2 + CH_2 = CH_2$ was observed in simulations of o-EC^-/EC system, indicating the presence of an appreciable barrier to this reaction.

The free energy of the condensed phase reaction path for Li^+/o-EC^- with Li^+/c-EC^-, Scheme III, Fig. 7.4, as a function of the reaction coordinate (C-O distance) reveals a barrier for this reaction of around 6 kcal/mol, but it is significantly less than that for ring opening reaction as estimated from QC of 11–12 kcal/mol [14]. This barrier is sufficiently large such that no reactions were observed in simulations of the o-EC^-/c-EC^-/EC system at 313 K. At elevated temperature (500 K), the reaction shown in Scheme III, Fig. 7.4, was observed in ReaxFF simulations on the timescale of a hundred picoseconds. At this temperature, the estimated lifetime (mean time to ring opening) of c-EC^- is, according to Eq. 7.2, much greater, around 10 ns. At lower temperatures, the separation in timescale between the reaction of Li^+/o-EC^- with Li^+/c-EC^- and the ring opening reaction will become even greater due to the much higher activation energy for the latter reaction. These results imply that the reaction of Li^+/o-EC^- with Li^+/c-EC^- occurs rapidly enough that this reaction could be important in the formation of the outer SEI in EC-based electrolytes.

In summary, a review of experimental and QC studies of compounds present or likely to be present in the outer SEI formed at the anode in EC-based electrolytes leads to the conclusion that these studies cannot be easily reconciled regarding the importance/presence of $(CH_2CH_2OCO_2Li)_2$, $(CH_2OCO_2Li)_2$, and the ester-carbonate compound. Condensed phase ReaxFF simulations have demonstrated that while $(CH_2CH_2OCO_2Li)_2$ and the ester-carbonate compound can easily form through the mechanisms shown in Fig. 7.4, $(CH_2OCO_2Li)_2$, often suggested as the most important compound of the outer SEI, does not form through the commonly held mechanism of combination of two o-EC^- radicals (Scheme I, Fig. 7.4). Furthermore, the large barrier for the ring opening reaction leading to formation of o-EC^- from c-EC^-, revealed by previous QC studies, combined with the relatively low barrier for formation of the ester-carbonate compound from reaction of c-EC^- with o-EC^- implies that the presence of the ester-carbonate compound as a component of the SEI in EC-based electrolytes should be given serious consideration. Furthermore, reactions of the relatively long-lived c-EC^- with electrolyte additives such as VC may also be important.

Two-Electron Reduction of EC

In addition to participating in the formation of the outer SEI through one-electron reduction, EC can undergo two-electron reduction. The resulting compounds are believed to be important components of the inner part of the SEI. QC reveals that this is a two-step process involving one-electron reduction to the o-EC radical followed by reduction by the second electron [14]. QC reveals that this an energetically unfavorable reaction in the gas phase but energetically favorable in the presence of Li^+, resulting in the formation of ethylene gas and $LiCO_3{}^-$. This doubly reduced form of EC can react with an additional Li^+ form lithium carbonate (Li_2CO_3) or with a Li^+/EC complex to form dilithium ethylene dicarbonate, also shown in Fig. 7.4. Both of these compounds are believed to be present in the inner SEI [16, 32, 33].

Decomposition at Electrode Surfaces

Computational investigation of the reduction of carbonate compounds occurring directly on the anode (e.g., lithium metal) surface has received less attention than reactions in the gas phase or bulk electrolyte. In one example, a QC study of a number of important electrolyte compounds, including EC, reacting with lithium clusters, has been conducted [14]. While the level of theory is modest, qualitative trends were clear. Reaction of EC with the lithium cluster results in the formation of ethylene plus Li_2CO_3, a known major component of the inner SEI. The energetics of this reaction is highly favorable, while there was found to be an appreciable activation energy for the reaction. This high activation energy is somewhat surprising due to the known reactivity of lithium metal with carbonate solvents, and may reflect the computational approach and modest level of theory employed. ReaxFF simulations of EC in contact with lithium metal show fast reaction on a timescale of picoseconds (ps) resulting in formation primarily of Li_2CO_3 and ethylene gas [31]. These initial RMD simulations on lithium metal show no signs of formation of dicarbonate compounds or the presence of an appreciable barrier for the formation of Li_2CO_3.

The decomposition of EC on graphitic anode with C = O, C–OH, and C–H terminations was studied at 450 K in ab initio simulations using the Perdew–Burke–Ernzerhof (PBE) functional, with a more accurate PBE0 functional used as for spot checks [34]. Fast EC decomposition on ps timescale occurred upon two-electron reduction at the graphite edge rich in oxidized sites (C = O and C-OH), yielding CO_3^{2-} and $OC_2H_4O^{2-}$. A potential reaction between $OC_2H_4O^{2-}$, or HOC_2H_4OH in the proton rich environment, with CO_2 or CO_3^{2-} was suggested as a possible route to forming $(CH_2CO_2Li)_2$ SEI product but was not observed presumably because of short simulation timescales [34]. The C = O termination was significantly more active in decomposing EC than C–H termination.

Molecular Dynamics Simulations of Bulk Electrolyte Properties

Organic Liquid Electrolytes

Molecular dynamics simulations were used to study a number of electrolytes with potential interest to lithium battery applications: EC:DMC/$LiPF_6$[35], EC/LiTFSI [36, 37], DMC/LiTFSI [38], GBL/LiTFSI [38], oligoethers/Li salts [39–41], acetamide/LiTFSI [42], EC/$LiBF_4$ [43], PC/$LiBF_4$ [43, 44], PC/$LiPF_6$ [44], DMC/$LiBF_4$ [43], oligoethers/$LiPF_6$ [45–47], and PC/LiTFSI [37]. The lithium cation coordination by solvent molecules, cation–anion aggregation, and transport properties were derived from MD simulations. It is important to pay attention to the reported simulation time because some of the earlier simulations by Li et al. [48]

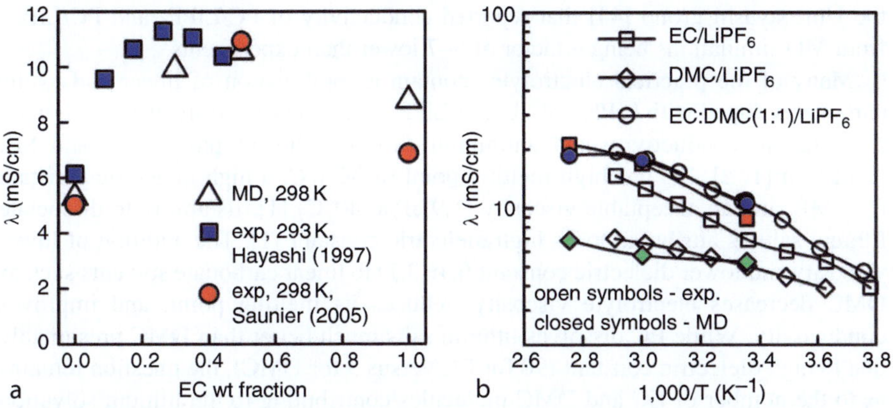

Fig. 7.6 Solvent composition (*left*) and temperature (*right*) dependence of ion conductivity from MD simulations [35] and experiments [51, 52]. Solvent: Li ratios were the following: EC:Li = 13, DMC:Li = 10, (EC + DMC): Li =11.8 for EC:DMC mol/mol = 1/1 and (EC + DMC): Li =10.7 for EC:DMC mol/mol = 3/7

and Soeten [43] were only 200 ps and 100 ps long [43, 48], which is less than one EC-lithium residence time (~0.5 ns) [35] and significantly less than the lithium–anion residence time of 2–3 ns found for similar electrolytes at 298 K [35]. Experience indicates that MD simulation trajectories should be longer than the cation–anion or lithium-solvent residence time in order to obtain an equilibrated ion aggregation and the composition of the ion solvation shell. Recent MD simulations [35, 38] were performed for 7–20 ns to ensure a complete equilibration and extraction of the accurate conductivity values.

Early study of one $LiBF_4$ in PC, EC, and DMC by Soeten [43] provided initial insight into the structure of the lithium solvation shell. Four strongly bound solvent molecules in a tetrahedral arrangement around lithium cation were reported [43], which is qualitatively consistent with the recent MD studies [35, 36, 38] and ab initio simulations of Li^+ in 32 EC [34]. MD simulations with polarizable force fields predicted ion diffusion coefficients [35, 36, 38, 49] within 40% of experiments for DMC/LiTFSI, EC/LiTFSI, PC/LiTFSI, DMC/LiTFSI, (EO_n)/LiTFSI, where EO is ethylene oxide repeat unit, n = 2,5,12 and EC:DMC/$LiPF_6$. MD simulations employing nonpolarizable force field [37], on the other hand, predicted ion diffusion a factor of 2–4 slower for EC/LiTFSI and PC/LiTFSI and 5–7 times slower for PC/$LiBF_4$ and PC/$LiPF_6$ [44]. Conductivity was predicted a factor of 5 lower for PC/$LiPF_6$ [50] and a factor of 2 lower for PC/LiTFSI [37] and EC/LiTFSI [37] in MD simulations using nonpolarizable force fields. MD simulations using polarizable force fields predicted electrolyte conductivity within 20–40%. Figure 7.6 shows conductivity of EC: DMC/$LiPF_6$ as a function of solvent composition and temperature. MD simulations slightly overestimate EC/$LiPF_6$ conductivity but overall quality of the conductivity predictions is satisfactory, especially in light of recent results from the Umebayashi group [44] that reported conductivity of PC/$LiPF_6$ and PC/$LiBF_4$ from MD simulations being a factor of 5–7 lower than experiments.

Many of the practical electrolytes contain a combination of linear and cyclic carbonates doped with $LiPF_6$ salt. Using $LiPF_6$ salt yields electrolytes with high ion dissociation, conductivity and aluminum current collector passivation and SEI formation [1, 2]. EC has high melting point of 36.4°C, a high dielectric constant ($\varepsilon \approx 90$), and an acceptable viscosity (1.9 cP at 40°C) [1]. Its ability to dissociate lithium salts is attributed to its high dielectric constant [1]. The addition of lower viscosity and lower dielectric constant ($\varepsilon \approx 3.1$) to linear carbonate solvents such as DMC decreases electrolyte viscosity, reduces its melting point, and improves conductivity. While EC dissolves lithium salts much better than DMC presumably due to high dielectric constant (90 for EC versus 3 for DMC), the question remains as to the number of EC and DMC molecules contributing to the lithium solvation shell in the mixed EC:DMC electrolyte. Intuitively, one expects EC contribution to dominate over DMC because of the difference in dielectric constants [1]. The issue of the competitive lithium cation solvation in mixed linear and cyclic carbonate solvents was addressed in a recent MD simulation and QC study [35]. Both EC and DMC were found to participate in the Li^+ solvation in mixed EC:DMC (1:1) electrolytes despite a huge difference between their dielectric constants (90 for EC versus 3 for DMC). These results were found in agreement with the reanalyzed [53] Raman spectroscopy experiments. A more detailed examination of the mixed electrolyte MD data revealed that the free lithium cations slightly favor EC solvation, while Li^+ contact ion pairs with PF_6^- anions leads to a preference for DMC in the Li^+ solvation shell. These conclusions are in accord with the quantum chemistry studies that indicated that $LiPF_6$ ion pair binds slightly stronger to DMC compared to EC, while binding energy of Li^+-EC is larger than Li^+-DMC. Analysis of the lithium cation coordinated by four solvent molecules (EC and DMC as shown in Fig. 7.7) indicated that the most stable configuration is not $Li^+(EC)_4$, but $Li^+(EC_3$, DMC) and energy of the $Li^+(EC_2, DMC_2)$ cluster is only 0.8–1.6 kcal/mol higher than $Li^+(EC_3$, DMC) depending on the level of theory used.

MD simulations offer a unique opportunity to investigate the degree of ion correlation from both structural and dynamic points of view. The degree of ion dissociation α_s (static) defined as the fraction of Li^+ and PF_6^- with no counterion in their first solvation shell, i.e., r(Li-P) < 4.4 Å for all pairs, is shown in Fig. 7.8. The fraction of free ions increases near linearly with increasing EC fraction in electrolyte and that increasing temperature slightly decreases the fraction of free ions. Surprisingly, the DMC/$LiPF_6$ 1 M electrolyte, which exhibits a reasonable DC conductivity shown in Fig. 7.6, has a very low fraction of free ions (1.7% at 298 K). The degree of dynamic correlation (α_d) was extracted from MD simulations using ion self-diffusion coefficients and conductivity (λ) as given by Eqs. 7.3–7.4 in order to compare with α_s (static),

$$\alpha_d = \frac{\lambda}{\lambda^{N-E}} \tag{7.3}$$

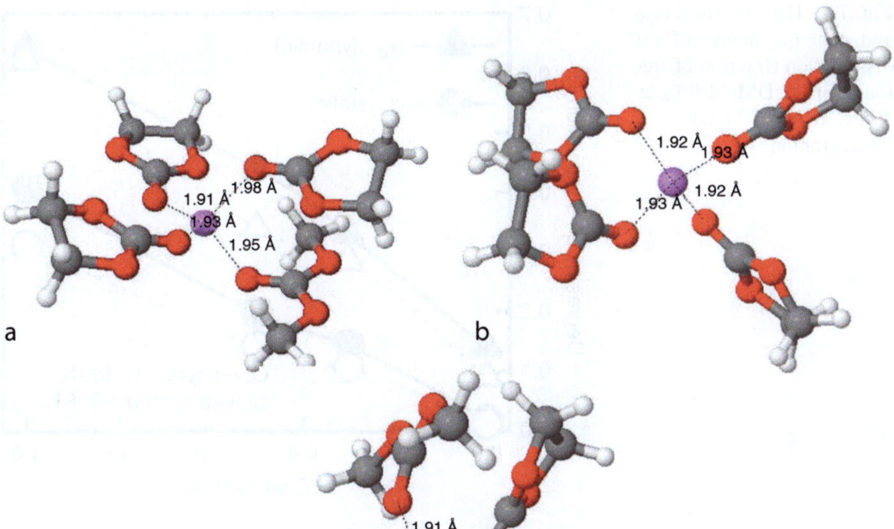

Fig. 7.7 Geometry of the three most stable clusters of Li^{4+} coordinated by EC and DMC solvent molecules from M05-2X/6-31 G* geometry optimization

$$\lambda^{N-E} = \frac{e^2}{Vk_BT}(n_+D_+ + n_-D_-) \tag{7.4}$$

k_B is the Boltzmann constant, T is temperature, V is the volume of the simulation box, and D_- and D_+ are self-diffusion coefficients of the anion and cation, respectively. The degree of ion dynamic correlation α_d is plotted versus electrolyte composition and temperature in Fig. 7.8 together with the static fraction of free ions from the structural analysis α_s. It was found that α_d is always higher than the fraction of free ions from the structural analysis (α_s), indicating an important contribution from the charged ion clusters transport to ion conductivity discussed in details elsewhere [35]. Joint pulsed gradient spin echo (pgse)-NMR and conductivity analysis of LiTFSI in multiple electrolytes at solvent:Li ratio of 20 at 298 K for PC/LiTFSI and DMC/LiTFSI yielded α_d values 0.62 and 0.11 in accord with MD predictions for EC:DMC/$LiPF_6$ shown in Fig. 7.8 provided $LiPF_6$ and LiTFSI yield

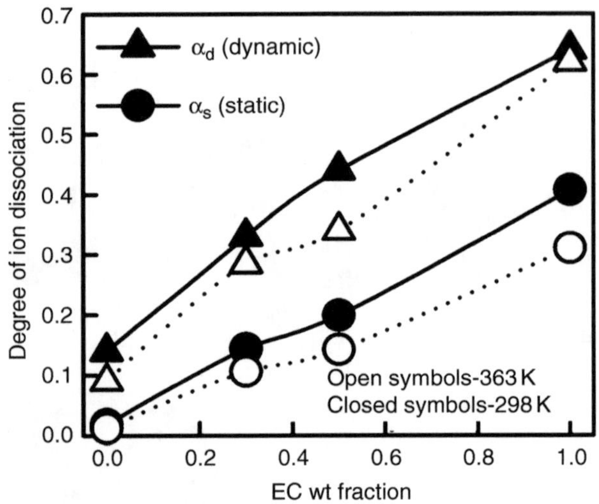

Fig. 7.8 The dynamics (α_d) and static (α_s) degree of ion dissociation (fraction of free ions) for EC:DMC/$LiPF_6$ at 298 K at an ~1 M salt concentration

similar α_d [54]. This indeed seems to be the case as similar α_d were found for LiTFSI and $LiPF_6$ salts in γ-butyrolactone (GBL) electrolyte from a combination of pgse-NMR and conductivity analysis [55].

Analysis of the solvent residence times relative to the diffusion coefficients allows one to estimate the frequency of the solvent exchange during the time it takes for a solvent to diffuse its own size [49]. For EC:DMC/$LiPF_6$ electrolytes [49], an approximately equal contribution from vehicular motion with the first solvation shell and solvent exchange for Li solvated by EC was found. In contract to the lithium motion in carbonates, in oligoethers, Li^+ moves 3–4 solvent molecules before its solvation shell is renewed, indicating that motion or the Li^+ together with its solvation shell dominates over solvent exchange mode of the lithium transport in oligoethers.

Polymeric Electrolytes

Solid polymer electrolytes (SPEs) are formed by doping a polymer with a lithium salt. SPEs have several advantages over traditional organic solvent–based electrolytes that include negligible volatility, leading to increased safety and reduced environmental impact, and intrinsic mechanical integrity, which could eliminate the need for a separator [56]. However, SPEs suffer from a major disadvantage, namely, dramatically reduced lithium ion conductivity compared to organic solvent-based electrolytes. Consequently, SPEs can currently only be employed at elevated temperatures, typically 60–80°C [56]. Simulation and modeling studies of SPEs have concentrated on elucidating the mechanism of ion motion in SPEs, and in understanding how polymer structure influences ion motion.

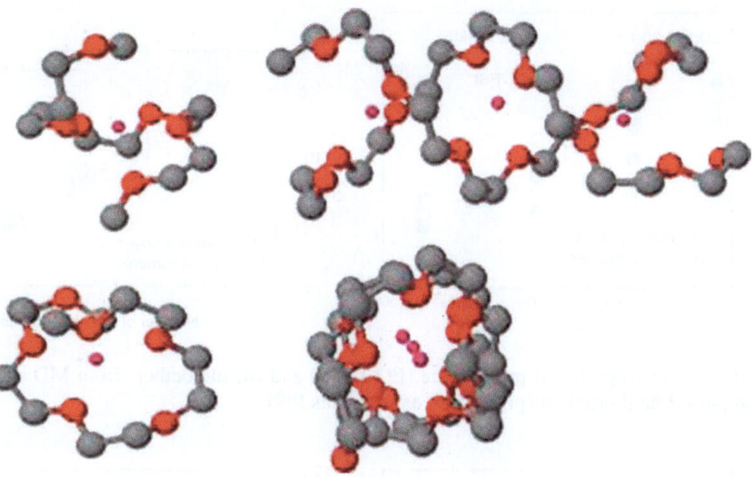

Fig. 7.9 A representative configuration of Li^+ cation coordination by an oligoether chain from MD simulations

Polyethylene oxide (PEO)-based polymers have been investigated most extensively as SPEs both experimentally [56] and in simulations [40, 57–67]. These polymers are very effective at dissolving lithium salts due to strong Li^+-ether oxygen binding. QC [68] and MD simulation studies [58] of oligoethers (low molecular weight PEO) as well as PEO have revealed that Li^+ cations are very strongly coordinated by an average of 5–6 ether oxygen atoms in configurations similar to that shown in Fig. 7.9. Simulations reveal that coordination of Li^+ usually involves a single oligoether or two segments of a single PEO chain as shown in Fig. 7.9 [58].

The nature of ether-Li^+ binding has a very strong influence on the mechanism for Li^+ conduction in PEO-based SPEs. This can be understood by first comparing the mechanism of Li^+ diffusion in an oligoether versus that in organic solvents. As discussed in Organic Liquid Electrolytes, Li^+ motion in organic liquid electrolytes occurs by a mixture of vehicular (with solvent shell) and structural (exchange of solvent shell) diffusion [49]. In contrast, simulations reveal that Li^+ diffusion in pentaglyme (six repeat unit PEO) + LiTFSI, compared with experiment in Fig. 7.10, occurs almost entirely by a vehicular mechanism [49]. In other words, a pentaglyme/Li^+ complex diffuses on average distances many times larger than the size of the pentaglyme before Li^+ changes complexation to a different pentaglyme molecule. The residence time of Li^+ with a pentaglyme molecule is around 50 ns, compared to around 1 ns in propylene carbonate [49].

While the lack of an efficient structural diffusion mechanism has relatively little influence on Li^+ motion in pentaglyme because of the fast center-of-mass diffusion of pentaglyme, the situation is quite different in higher molecular weight PEO. Because the center-of-mass diffusion of PEO is negligible, a Li^+ cation necessarily *must* change coordinating PEO chains, i.e., undergo "jumps" between

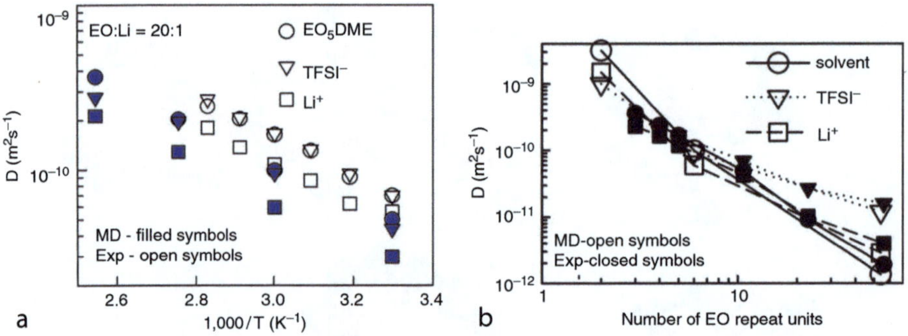

Fig. 7.10 Ion diffusion in (**a**) pentaglyme (EO_5DME) and (**b**) oligoethers from MD simulations [49] and pulsed field gradient (pfg)-NMR experiments [69]

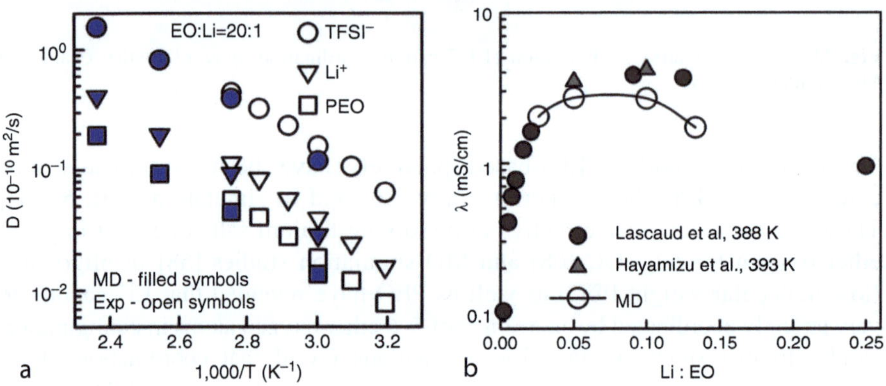

Fig. 7.11 (**a**) Diffusion in a PEO/LiTFSI electrolyte from simulation and experiment. (**b**) Ionic conductivity in a PEO/LiTFSI electrolyte from simulation and experiment

PEO chains, in order to move distances greater than the size of the PEO chain. Figure 7.11a shows ion and polymer diffusion in a PEO-LiTFSI SPE as a function of inverse temperature from simulation and experiment [38], while Fig. 7.11b shows conductivity as a function of salt concentration [58]. In both cases, excellent agreement between simulation and experiment can be seen. Compared with pentaglyme (Fig. 7.10), it can be seen that Li^+ transport is considerably slower in the SPE, while the diffusion of TFSI is reduced relatively little, indicative of the relative independence of TFSI motion from that of the solvent/polymer. This leads to a significantly lower transference number in the SPE compared to the liquid electrolyte as a significantly greater fraction of charge is carried by TFSI in the former due to the very slow Li^+ motion. Furthermore, while pentaglyme motion is faster than that of Li^+ in the oligoether electrolyte, PEO center-of-mass motion is considerably slower than that of cation, supporting the supposition that the vehicular mechanism, which is so important in the oligoether, does not contribute significantly to Li^+ in the SPE. Note that Li^+ motion in the PEO

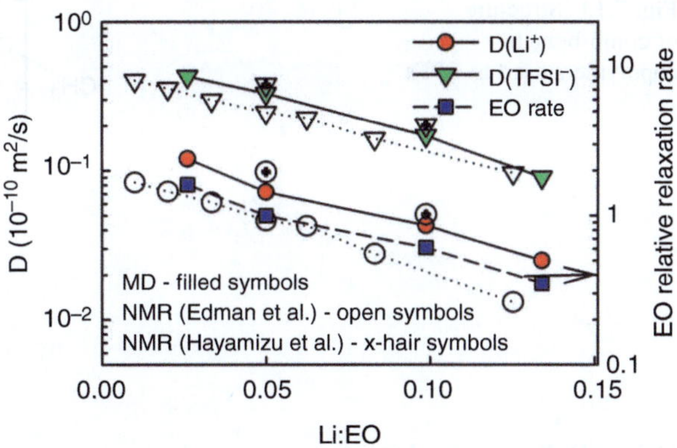

Fig. 7.12 Dependence of Li^+ and $TFSI^-$ motion on salt concentration from both experiment and simulation

SPE can be well reproduced by a model that allows for 1-D diffusion along Gaussian chains, motion with a segment of the chain with occasional interchain jumps [59, 66].

Coupling of Ion Motion and Polymer Segmental Motion

Figure 7.12 shows the dependence of Li^+ and $TFSI^-$ motion on salt concentration from both experiment and simulation [58]. Also shown is the time (relative to that at for Li:EO = 0.05) from simulations at which the mean-square displacement of the ether oxygen atoms reaches 50 $Å^2$, the length scale of the cage imposed by neighbor packing. This timescale is directly related to the timescale of conformational transitions. Figure 7.12 reveals that while TFSI motion remains much faster than that of Li^+ for all salt concentrations, the translational dynamics of both ions follow the same dependence on salt concentration as polymer segmental/conformational dynamics, i.e., a dramatic decrease in mobility with increasing salt concentration. These results show that both cation and anion dynamics are strongly coupled with local polymer relaxation. The slowing of segmental/conformational dynamics of the polymer with increasing salt concentration is due to the strong binding between ether oxygen atoms and Li^+, which greatly restricts conformational motion. The correlation of both cation and anion dynamics with local polymer relaxation is further illustrated by simulations in which the barrier for dihedral/conformational transitions were artificially increased by 1.3 kcal/mol [58]. Simulations of the PEO/LiTFSI SPE at EO:Li = 20:1 at 393 K revealed a decrease in the diffusion coefficient of both $TFSI^-$ and Li^+ by a factor of 3, and an increase in the polymer segmental relaxation time by about the same factor.

An interesting feature of the PEO SPE is the maximum in conductivity for Li:EO ratio around 0.05, as can be seen in Fig. 7.11b. Simulations reveal that this is due to two counter effects. On the one hand, ionic conductivity increases with increasing salt concentration due to the increase in the number of charge carriers per unit volume. On the other hand, increasing salt concentration decreases the rate of polymer dynamics dramatically decreasing Li^+ transport as well (to a lesser extent) TFSI transport.

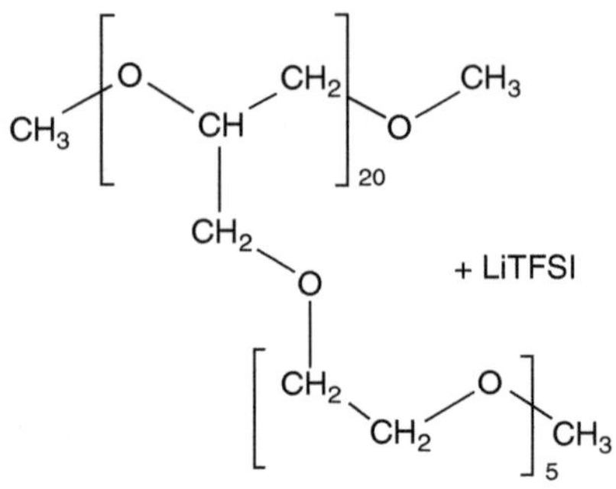

Fig. 7.13 Structure of comb-branch copolymers used as SPEs

Other Polymer Architectures

In summary, simulation studies of SPEs based on linear PEO have revealed that strong binding of ether oxygen to Li^+, while facilitating dissolution of lithium salts, reduces the mobility of Li^+ by suppressing structural diffusion. Furthermore, the vehicular mechanism is eliminated in the polymer due to negligible center-of-mass motion of the coordinating polymer chain, necessitating interchain hopping for transport, which is found to be a rare event. While both anion and cation motion are found to be coupled to local polymer relaxation, the anion mobility is found to be significantly greater than that of the strongly bound Li^+ in the PEO-based SPEs, leading to very low transference numbers. Finally PEO crystallizes at around 60°C, precluding its use at lower temperatures due to very poor transport of Li^+ in the crystalline phase. On the other hand, about 60°C (in the molten state), PEO does not provide any mechanical properties that would allow for elimination of a separator, a major potential advantage of SPEs

Comb-Branched Polyethers

Comb-branch polyethers with architectures similar to that shown in Fig. 7.13 have been investigated in efforts to develop SPEs that take advantage of the ability of oligoethers to coordinate Li^+ while at the same time (1) preventing crystallinity due to use of short side chains and (2) allowing in principle for separate optimization of the backbone properties from those of the side chains. The ideal comb-branch SPE might, for example, use a glassy backbone polymer, thereby resulting in good mechanical properties, while Li^+ transport is carried out by flexible ether side chains. The use of relatively short side chains not only reduced/prevents crystallinity, but may also facilitate interchain hopping, which is necessary for large-scale Li^+ transport in SPEs, due to sharing of Li^+ cations between side chains. The optimization of the properties/chemical structure of comb-branch SPEs would be greatly facilitated by an improved understanding of Li^+ transport mechanisms in these polymers, particularly the roles played by the polymer backbone and side chains in ion transport.

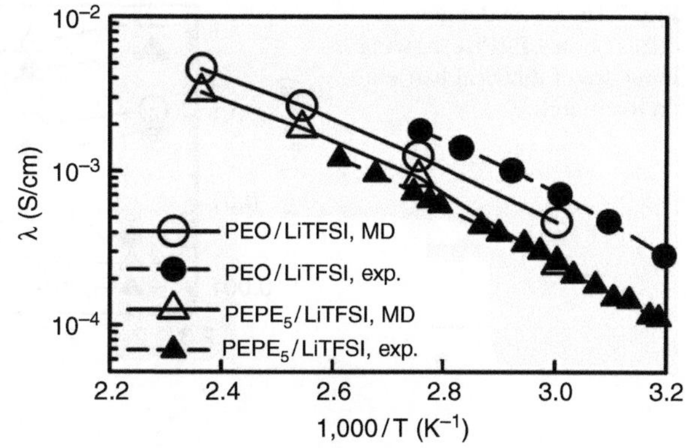

Fig. 7.14 Ionic conductivity in comb-branch SPEs from simulation and experiment

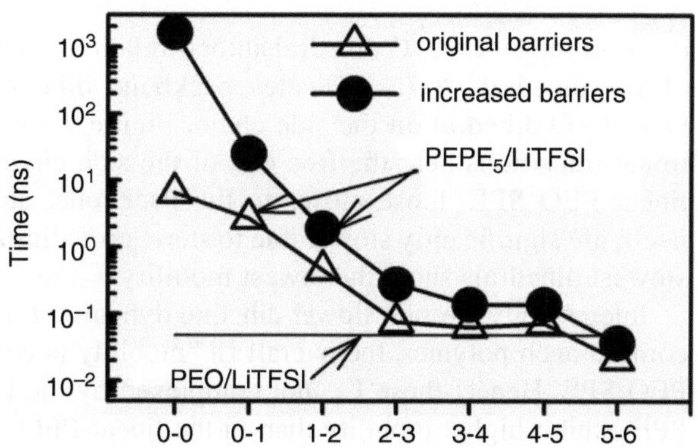

Fig. 7.15 Conformational relaxation time as a function of position of the dihedral in comb-branch polymer electrolytes from MD simulation

MD simulations of comb-branch polyethers of the structure shown in Fig. 7.14 ($PEPE_5$) have been performed [60]. SPEs formed from this comb-branch polymer have also been studied experimentally [70–72]. A comparison of the ionic conductivity of SPEs with LiTFSI and this comb-branch polymer from simulation and experiment, along with the conductivity of a linear PEO SPE with same salt concentration is shown in Fig. 7.14 [60]. Good agreement between experiment and simulation is apparent. Furthermore, the conductivity of the comb-branch SPE is very similar to that of the linear PEO SPE. The slightly higher conductivity in the linear PEO SPE is facilitated by the lower molecular weight (2,380 Da) of the linear polymer compared to the comb-branch (around 6,000 Da).

Analysis of the MD simulations of the $PEPE_5$ reveal that Li^+ occurs primarily by hopping of the cation from one side chain to another [60]. The six oxygen atoms of the side chains facilitated coordination of Li^+, as anticipated. However, a fraction of Li^+ cations are partially coordinated by the polyether backbone. These cations have very slow dynamics and do not contribute appreciably to Li^+ motion. The slow dynamics of cations partially coordinated by the chain backbone can be understood by considering conformational dynamics in the comb-branch polymer. Figure 7.15 shows conformational relaxation time, based on

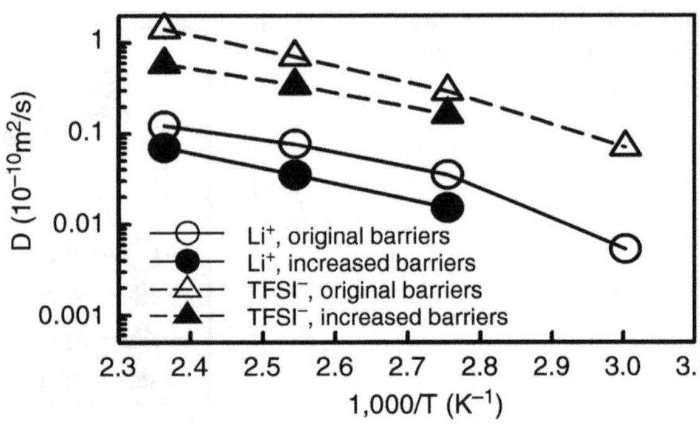

Fig. 7.16 Li$^+$ and anion diffusion in a $PEPE_5$ showing influence of dihedral barriers on ion mobility

decay of the O–C–C–O dihedral autocorrelation function, as a function of position of the dihedral. "0-0" indicates backbone dihedrals, while "5-6" is the last O–C–C–O dihedral on the side chain. Figure 7.15 reveals that while conformational transitions near the free end of the side chains are similar to those in the linear PEO SPE, those closer to the backbone, and particularly the backbone itself, are significantly slower due to steric crowding. Those Li$^+$ complexed by the slowest dihedrals show the lowest mobility.

Interestingly, despite slower dihedral dynamics for the polymer backbone on the comb-branch polymer, the overall Li$^+$ mobility is comparable to that in the linear PEO SPE. Hence, those Li$^+$ not complexed by the backbone in the comb-branch SPE exhibit higher mobility than in the linear PEO SPE, indicating that the short side chain architecture does promote interchain hopping, as hoped for. Furthermore, it appears that Li$^+$ mobility, at least for those not complexed by the backbone, is largely independent of backbone motion [60]. This independence is further demonstrated in Fig. 7.16, which shows the influence of backbone dihedral barriers in the $PEPE_5$ SPE on ion mobility. When the backbone dihedral barriers are increased, the polymer backbone becomes glassy on the timescale of the simulations. Despite the elimination of backbone conformational dynamics, the mobility of the ions is reduced by less than a factor of 2, less than that observed in the linear PEO SPE when the barriers was increased by only 1.3 kcal/mol (see previous discussion). Furthermore, Li$^+$ diffusion increased by about 30% when a repulsive interaction between the backbone ether oxygen atoms and Li$^+$ was introduced, thereby eliminating the role of the backbone in cation coordination. In summary, simulations support the supposition that a comb-branch copolymer with ether side groups and a glassy, noncoordinating backbone would show conductivity comparable to or even superior to ether-based polymers, with improved mechanical properties.

Finally, it may seem reasonable that shortening the ether side chains (to less than six ether oxygen atoms) would support interchain hoping by promoting multiple side-chain coordination of Li$^+$. Combined with a noncoordinating backbone, however, shorter side chains could be effective in increasing Li$^+$ mobility.

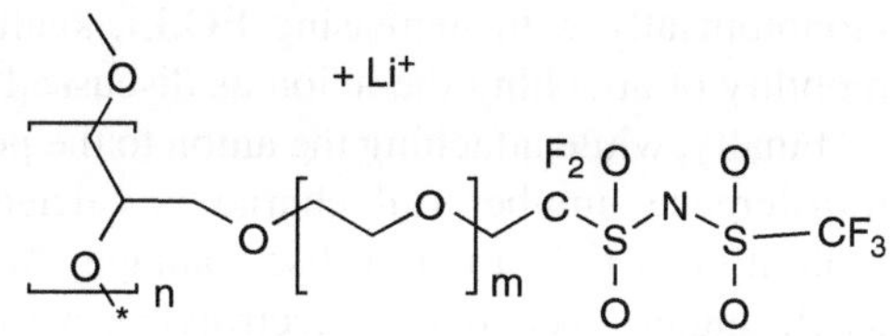

Fig. 7.17 Structure of a PEO-based single ion conductor

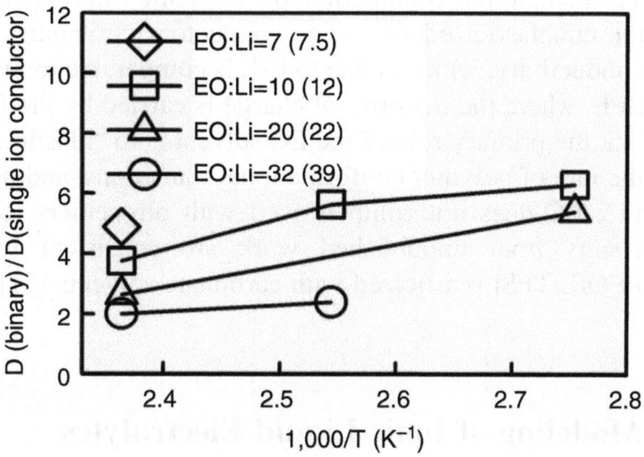

Fig. 7.18 A comparison of the lithium diffusion coefficient in PEO-based single ion conductors (Fig. 7.14) and PEO-based binary comb-branch polymer electrolytes (Fig. 7.10) from simulations

Single-Ion Conductors

As discussed above, one of the disadvantages of PEO-based SPEs is their low transference number due to relatively high anion mobility. Large-scale anion motion in comb-branch SPEs can be eliminated by attaching the anions to the polymer, such as shown in Fig. 7.17. Here, the TFSI anion of LiTFSI has been attached to the oligoether side chain to form a polyanion with essentially no large-scale mobility, resulting in a single-ion conductor. Here, the ether oxygen to Li^+ ratio is controlled by the length of the side chains. Unfortunately, simulations show that reducing the mobility of the TFSI anion has a deleterious effect on Li^+ mobility, particularly for low temperatures and short side chains/small EO:Li^+ ratios. Simulations reveal that anions play an important role in the ability of Li^+ cations to undergo interchain jumps in PEO-based electrolytes, particularly at higher ion concentrations. Attaching TFSI to the side chain reduces the mobility of the anion and apparently its ability to assist in Li^+ hops. The effect is reduced for longer side chains, due both to decreased importance of anions for the lower salt concentration, and possibility increased anion mobility on the longer tether.

Figure 7.18 shows the lithium diffusion coefficient in a PEPE binary (free anion) SPE (Fig. 7.13) and single ion conductors (Fig. 7.17) from unpublished simulations for various EO:Li ratios. Simulation reveals that the lithium diffusion coefficient in the single ion conductor is about a factor of 2–6 lower than in the binary PEO SPEs. Note that unlike the binary PEO SPEs, the PEPE-based single ion conductor does not show a maximum in conductivity around EO:Li = 20, but increases monotonically with increasing EO:Li, again due to the deleterious effect on Li^+ mobility of attaching the anion as discussed above.

Finally, while attaching the anion to the polymer has the advantage of improving transference number (all charge is carried by Li^+), reducing the already low mobility of Li^+ is a major disadvantage. Addition of a solvent/plasticizer to form a gel, single-ion conductor electrolyte is a possible solution. Simulations reveal that gel electrolytes formed from 37 wt.% ethylene carbonate plasticizer with a $PEPE_5$: TFSI single ion conductor show dramatic improvement in ionic conductivity over the nonplasticized single ion conductor, particularly at lower temperatures. Here, conductivity, which is due to Li^+, is comparable to that for the comb-branch binary SPE, where the majority of charge is carried by the free anion. Simulations reveal that the primary role of the EC solvent is to "plasticize" the polymer, i.e., increase the rate of polymer conformational transitions, and not to coordinate and transport Li^+; EC does not compete well with oligoethers for coordination of Li^+. These results from unpublished work are consistent with the recent findings for PEO/LiTFSI plasticized with carbonate solvents [37].

Modeling of Ionic Liquid Electrolytes

Room-Temperature Ionic Liquid-Based Binary Electrolytes

Two types of ionic liquid electrolytes have been extensively studied: (1) ionic liquid solvents comprised of ions shown in Fig. 7.19 doped with lithium salts and called binary room temperature ionic liquid (RTIL) electrolytes; and (2) molten salts, in which the lithium solvating host such as oligoether is attached to an anion such as $TFSI^-$, thus forming an ionic liquid (also called molten salt) without a free solvent. Binary ionic liquid electrolytes have attracted a significant attention because of their negligible vapor pressure; good thermal and electrochemical stability in the range of 5.3 V [73]; good dissolution properties with many organic and inorganic compounds and lithium salts; and low flammability [74]. A wide variety of possible anions and cations offer a potential to tailor RTIL properties to a particular application. The solvent free molten salt electrolytes (type b) are attractive due to their potential to eliminate ion concentration polarization during battery operation. However, low oxidative stability of oligoether solvating host attached to anion and low conductivity counterbalance this advantage.

Hundreds of MD simulation studies of pure RTIL solvents have been reported [75, 79, 80]. Accurate prediction of density, heat of vaporization, conductivity, viscosity, self-diffusion coefficients, structure factors, and crystal structure was reported [75, 80–83] for RTIL comprised of ions listed in Fig. 7.19, while other MD studies [84] reported ion transport an order of magnitude slower than experiments. An order of magnitude slower prediction of RTIL transport cannot be fully attributed to the absence of polarization in the force field as turning off

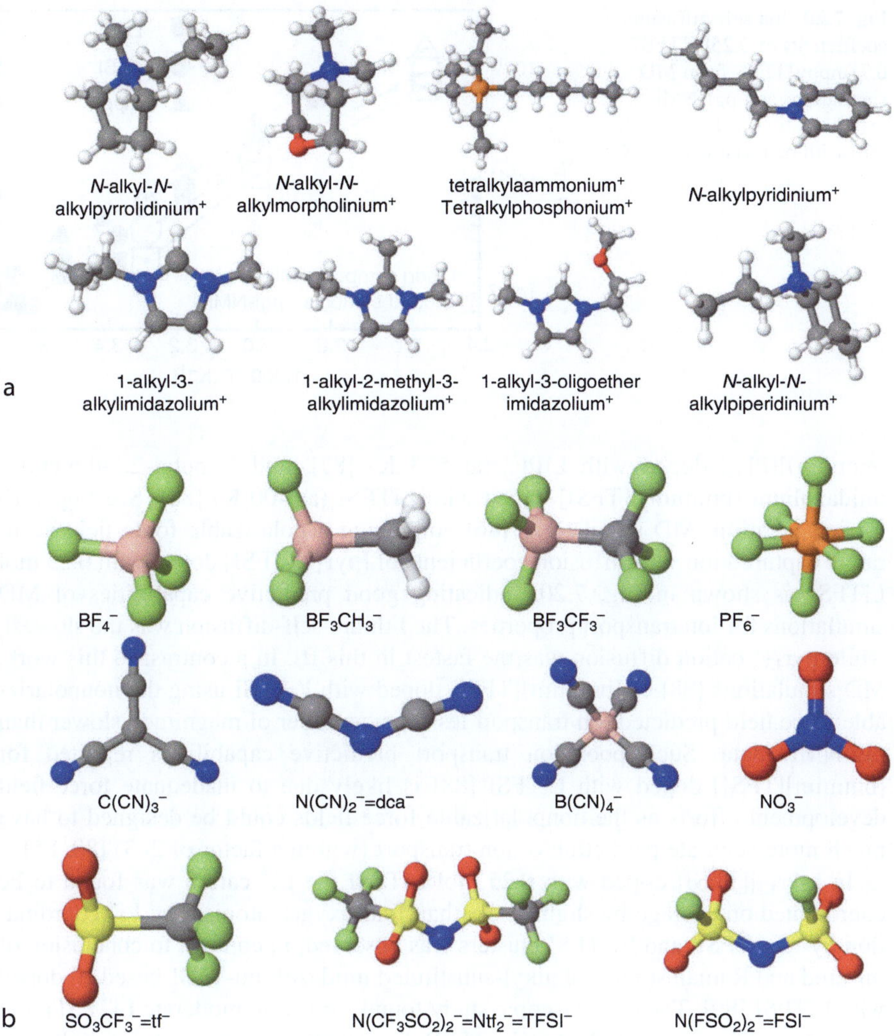

Fig. 7.19 A list of representative anion and cations in ionic liquid solvents used in MD simulations [75–78]

polarization slowed down ion transport only by a factor of 1.3–3 and not an order of magnitude in published simulations [75, 85]. The relation between ion transport, ion volume, and a combination of the heat of vaporization and the cation–anion binding energy has been reported [76].

Unlike the tremendous interest in simulating pure RTIL solvents, only a very limited number of MD simulations studies of ILs doped with lithium salts have been reported: [pyr_{11}][TFSI] doped with LiTFSI and [pyr_{13}][TFSI] doped with LiTFSI (at 303, 333, 393, and 500 K) [86], [1-ethyl-2,3-dimethyl-imidazolium

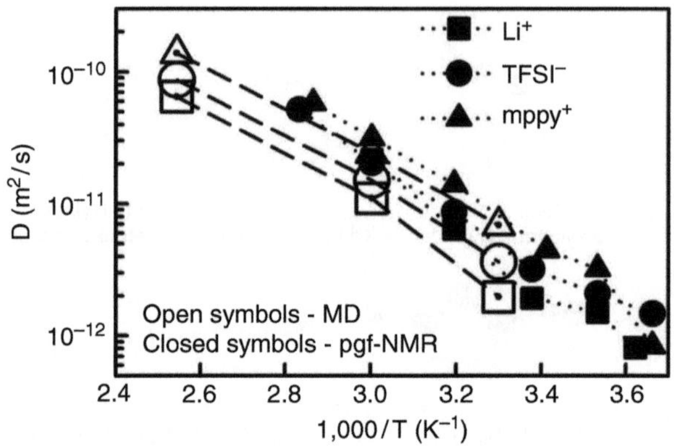

Fig. 7.20 Ion self-diffusion coefficients of $0.25Li^{+}TFSI^{-}$-$0.75mppy^{+}TFSI^{-}$ from MD simulations and pgf-NMR experiments (Used from Ref. [86] with permission of ACS)

(emmim)][PF_6] doped with $LiPF_6$ (at 523 K) [87], and [1-butyl-2,3-dimethyl-imidazolium (bmmim)][TFSI] doped with LiTFSI (at 400 K) [88]. See Fig. 7.19 for ion notation. MD simulations [86] employing a polarizable force field accurately captured ion self-diffusion coefficients of [pyr_{13}][TFSI] doped with 0.25 mol LiTFSI as shown in Fig. 7.20, indicating good predictive capabilities of MD simulations for ion transport properties. The lithium self-diffusion was the slowest, while pyr_{13}^{+} cation diffusion was the fastest in this IL. In a contrast to this work, MD simulations [88] of [bmmim][TFSI] doped with LiTFSI using the nonpolarizable force field predicted ion transport less than an order of magnitude slower than pfg-NMR data. Such poor ion transport predictive capabilities reported for [bmmim][TFSI] doped with LiTFSI [88] is likely due to inadequate force field development efforts as the nonpolarizable force fields could be designed to have much more accurate prediction of ion transport (within a factor of 2–3) [80, 85].

In [pyr_{13}][TFSI] doped with 0.25 mol LiTFSI the Li^{+} cation was found to be coordinated on average by slightly less than four oxygen atoms. The Li^{+} coordination by four $TFSI^{-}$ and by TFSI clusters was observed, in contrast to conclusion of infrared and Raman studies of alkyl-substituted imidazolium-TFSI-based IL doped with LiTFSI [89]. The spectroscopic study found for low to moderate LiTFSI mole fractions, $0.08 < \times < 0.2$, the $[Li(TFSI)_2]^{-}$ solvating cage involving bidentate coordinations of Li^{+} with two oxygen atoms of one anion in the trans (C_2) conformation and two oxygen atoms of the other anion in the cis (C_1) conformation. It is not clear how to explain the difference between MD results and analysis of the spectroscopic study. Turning off ion polarization in MD simulations dramatically reduced probability of forming extended $(Li^{+})_n(TFSI^{-})_m$ aggregates but did not result in the Li^{+} cation solvation by only two $TFSI^{-}$.

The mechanism of lithium transport was also studied in [pyr_{13}][TFSI] + 0.25 LiTFSI [86]. The Li^{+} cation transport was found to occur primarily by exchanging $TFSI^{-}$ anions in the first coordination shell of a Li^{+} with a smaller (~30%) contribution also due to Li^{+} cations diffusing together with their first coordination shell. It is important to note that despite high conductivity of IL-based electrolytes (~10^{-2} S/cm) the lithium contribution to the charge transport remains quite low.

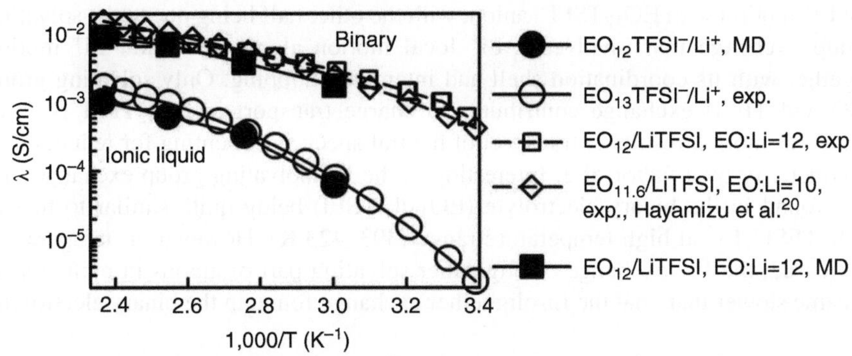

Fig. 7.21 Conductivity of binary electrolytes EO_{12}/LiTFSI and EO_nTFSI$^-$/Li^+ IL from MD simulations ($n = 12$) and experiments ($n = 13$). Hayamizy data for binary electrolyte from Ref. [69] are also shown (Reproduced with permission of ACS from Ref. [39])

In fact, conductivity due to the Li^+ cation transport only in [pyr$_{13}$][TFSI] + 0.25 LiTFSI was found to be somewhat greater than that for a model poly(ethylene oxide)(PEO-based)/LiTFSI polymer electrolyte but 1–2 orders of magnitude lower than conductivity of ethylene carbonate/LiTFSI liquid electrolyte depending on temperature [86].

Solvent-Free Molten Salts Electrolytes

MD simulations provided insight into the lithium coordination and transport mechanism in the solvent-free molten salt electrolyte comprised of TFSI$^-$ tethered to oligoethylene oxide (EO) (EO_{12}TFSI$^-$/Li^+) [39]. Behavior of EO_{12}TFSI$^-$/Li^+ was contrasted with the ion transport and aggregation in a binary solution of EO_{12} doped with LiTFSI salt (EO_{12}/LiTFSI). It was found from MD simulations and experiments that attaching TFSI$^-$ anion to the chain end of the oligoether resulted in reduction of conductivity by one order of magnitude at 423 K and two orders of magnitude at room temperature, as shown in Fig. 7.21. The lithium cations were separated from TFSI$^-$ anion by oligoether in most instances (75–90%) for both EO_{12}/LiTFSI and EO_{12}TFSI$^-$/Li^+. For the EO_{12}/LiTFSI binary electrolyte, the fraction of separated Li^+/TFSI$^-$ ions was similar to the degree of ion uncorrelated motion (α_d, see Eqs. 7.3–7.4) indicating mostly uncorrelated Li^+ and TFSI$^-$ ion motion in agreement with conclusions the coordinated pfg-NMR and conductivity study [69]. In contrast to this, a significant correlation between lithium and anion motion was found in EO_{12}TFSI$^-$/Li^+ molten salt with a degree of ion uncorrelated motion (α_d) being about (~45–55%).

The Li^+ cation self-diffusion coefficient in a binary EO_{12}/LiTFSI electrolyte was dominated (90%) by Li^+ vehicular diffusion of the Li^+ with an EO_{12} solvent. In EO_{12}TFSI$^-$/Li^+ molten salt electrolyte, half of the Li^+ motion was attributed to the Li^+ motion with $EO_{12}TSFI^-$ anion, with the other half being due to the solvating group exchange that involves a Li^+ local motion along the chain, Li^+ motion together with its coordination shell and interchain hopping. Only solvating group (EO and TFSI) exchange contributes to charge transport in $EO_{12}TFSI^-/Li^+$, as vehicular motion results in transport of neutral species accounting for reduction of conductivity by a factor of 2. Interestingly, the Li^+ solvating group exchange rate was found in the binary electrolyte (EO_{12}/LiTSFI) being quite similar to that in $EO_{12}TFSI^-/Li^+$ at high-temperature range (393–423 K). However, as temperature decreased, the Li^+ exchange of oligoether solvating part of anions in molten salts became slower than that the Li-oligoether exchange found in the binary electrolyte.

SEI Model Compounds

The ability of an electrolyte to form a stable and conductive SEI on electrodes is paramount for efficient battery operation as discussed previously. Solubility of SEI model compounds by DMC and EC reduction on anode was studied in simulations by Tasaki et al. [90]. They suggested the following order for the SEI forming salts to be soluble in DMC and EC: lithium dilithium ethylene dicarbonate (Li_2EDC) > lithium methyl carbonate (LiMC, $LiOCO_2CH_3$) > LiOH > $LiOCO_2C_2H_5$ > $LiOCH_3$ > LiF > $(LiCO_2)_2$ > Li_2CO_3 > Li_2O. This order agreed with experiments performed in DMC quite well [90]. MD simulations [91] have been used to investigate the structure and transport of two SEI components: Li_2EDC and LiMC. These SEI components are commonly reported in the outer part of SEI [92, 93] and probably exist in a disordered state in it because of the high fraction of other components present, thus inhibiting formation of the crystalline phase.

Ionic conductivity of Li_2EDC and LiMC predicted from MD simulations is shown in Fig. 7.22. At high temperature, conductivity of alkyl carbonates is comparable to that of liquid electrolytes such as EC/LiTFSI [36]. However, the much stronger temperature dependence of alkyl carbonate conductivity compared to typical liquid electrolytes (e.g., EC/LiTFSI) leads to significantly lower conductivity of alkyl carbonates at lower temperature compared to liquid electrolytes. The extrapolated conductivity of Li_2EDC at $-30°C$ is on the order of 10^{-10} S/cm, suggesting that an SEI comprised primarily of Li_2EDC with typical width of $\sim$10 nm would give rise to significant SEI layer resistance. Conductivity of LiMC is somewhat greater than that for Li_2EDC but also would result in a quite low conductivity of SEI. The extrapolated Li_2EDC conductivity from MD simulations is around 2×10^{-8} S/cm at room temperature in a good agreement with the value 10^{-9} S/cm measured by Phillip Ross group from LBNL [94]. These results are in stark contrast with a dramatically faster ion transport in Li_2EDC predicted by Tasaki et al. [90] at room temperature. The lithium diffusion in Li_2EDC

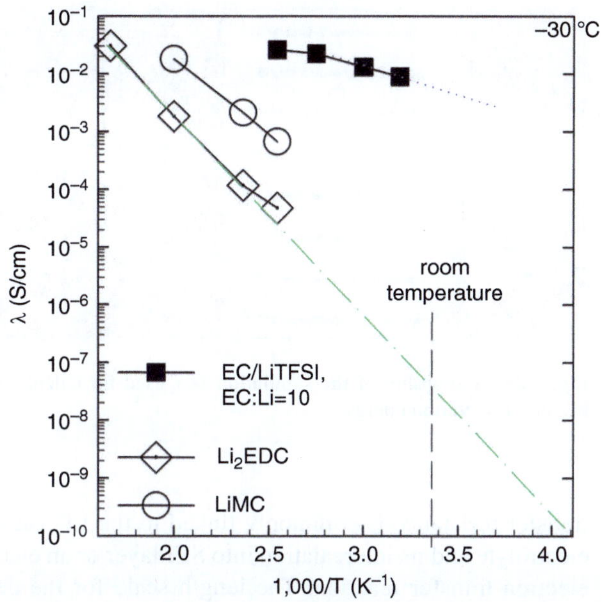

Fig. 7.22 Conductivity of lithium dilithium ethylene dicarbonate (Li_2EDC) > lithium methyl carbonate (LiMC) from MD simulations [91] compared to conductivity of EC/LiTFSI from MD simulations [36]

of 8×10^{-12} m^2/s predicted from Tasaki et al. [90] is approximately four and a half orders of magnitude higher than the extrapolated from high temperature $D(Li^+)$ of 2×10^{-17} m^2/s self-diffusion from Borodin et al. [91].

Molecular Dynamics Simulations of Electrode/Electrolyte Interfaces

Graphite Interface

Experimental studies [95–98] of the secondary cells at low temperature (less than −20°C) have consistently found that $R_{ct} > R_{SEI} > R_{bulk}$ for mixed carbonate solvents using conventional $LiBF_4$, $LiPF_6$, and LiBOB salts for a number of tested cells. Thus, the charge transfer resistance (R_{ct}) resistance is the dominant contribution to the secondary lithium cell resistance at low temperature, followed by R_{SEI} (SEI resistance) and R_{bulk} (bulk electrolyte resistance). Similar results have been reported for ionic liquid [99] and polymeric electrolytes [100]. Therefore, the conventional matrix for choosing liquid electrolytes based upon bulk electrolyte conductivity and its ability to form a stable SEI layer needs to be augmented with understanding and optimization of the electrolyte interfacial properties such as interfacial resistance. Intriguingly, the interfacial resistance does not always correlate with the bulk electrolyte resistance creating both opportunities and additional challenges for optimization of electrolyte materials. Importantly, the charge

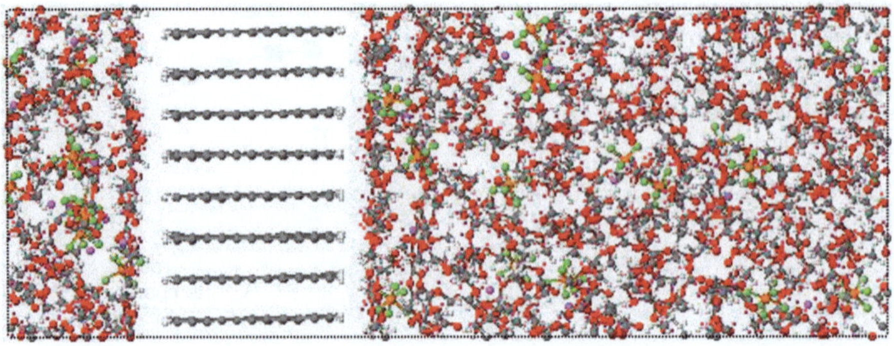

Fig. 7.23 A snapshot of the simulation box used for calculating the free-energy barrier of the lithium desolvation energy

transfer resistance is commonly linked to the Li^+ cation desolvation process from electrolyte and its intercalation into SEI layer or an electrode followed by the actual electron transfer reaction. The length scale for the desolvation process is on the scale of nanometers and, therefore, could be accessed via molecular simulations together with the fundamental electrolyte properties such as conductivity, ion self-diffusion, and the degree of ion aggregation.

MD simulations were performed on EC/$LiPF_6$ and EC:DMC(3:7)/$LiPF_6$ 1 M electrolytes next to the material blocking for anion and solvent but nonblocking to lithium cation and measure the interfacial resistance and free energy due to lithium desolvation from electrolyte. At the initial stage, graphite was chosen as the electrode material and did not include an SEI or controlled voltage between electrodes. MD simulations have been performed at 298 K for two sets of EC/$LiPF_6$ and EC:DMC(3:7)/$LiPF_6$ at 1 M salt concentration next to graphite as shown in Fig. 7.23 for EC:DMC(3:7)/$LiPF_6$ | Graphite. The electron transfer process has not being included in the simulated model.

The free-energy profile for the lithium desolvation ($\Delta G(z)$) in the range of $k_B T$ could be obtained from equilibrium MD simulations using Eq. 7.5:

$$\Delta G(z) = -k_b T \cdot \log(\rho(z)) \tag{7.5}$$

where k_B is the Boltzmann constant, T is temperature, and $\rho(z)$ is the normalized lithium probability profile calculated perpendicular to graphite (z-direction). Figure 7.24 shows the lithium free-energy profile $\Delta G(z)$ calculated from Eq. 7.2 using the lithium cation probability profile obtained from 20 ns MD simulations with the first 10 ns omitted as equilibration. The free-energy profile $\Delta G(z)$ shows a local minimum of −0.5 kcal/mol at ~5 Å from the graphite indicative of a preferential lithium position in the Helmholtz plane. The free energy sharply increases as the lithium approaches graphite and is forced to desolvate from electrolytes at distances closer than 4 Å from graphite. Note, that even long (10 ns) simulation trajectory

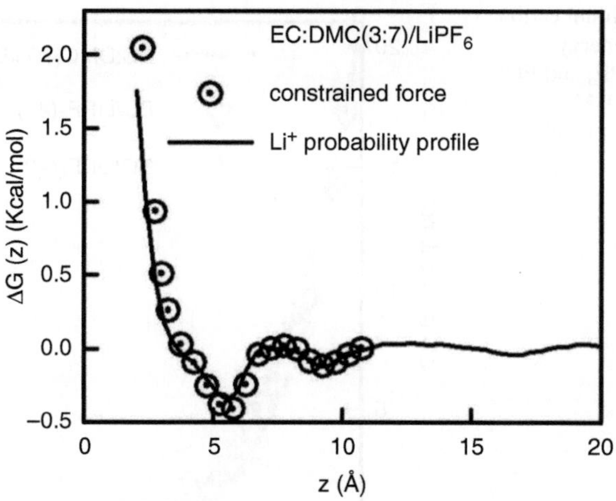

Fig. 7.24 The free-energy profile for the lithium cation desolvation from EC:DMC(3:7)/$LiPF_6$ at 1 M electrolyte at 298 K calculated using the Li^+ probability profile from equilibrium MD simulations and the integration of the constrained force method. $Z = 0$ at the position of hydrogen atoms of graphite

allowed us to sample the free-energy profile only for energies lower than 1.5 kcal/mol, indicating that an advanced sampling technique needs to be used.

The $\Delta G(z)$ was also calculated using the integration of the constrain force technique, in which a Li^+ is constrained at a certain position relative to the surface and the constraining force is calculated (for all points shown in Fig. 7.24). This methodology was previously used to investigate small molecule permeation through a lipid bilayer [101]. The simulations were 6–12 ns for each constrained distance. The integral of the constrained force gives a free-energy profile that is also shown in Fig. 7.24. An excellent agreement between two methods of calculating the free-energy profile is observed for free energies below 1 kcal/mol, where the equilibrium MD is expected to provide an adequate sampling.

Two independent sets of simulations of EC/$LiPF_6$ 1 M | graphite for calculating $\Delta G(z)$ have been performed in order to ensure convergence of simulation results and their independence from the state of the lithium coordination shell. In the first simulation, the test-lithium was initially solvated only by EC molecules in the first coordination shell, while in the second set of simulations the lithium (with the constrained distance to graphite) had one PF_6^- anion in its first coordination shell. The free-energy profiles for the lithium desolvation from these electrolytes were calculated by the integration of the constrain force and are shown in Fig. 7.25.

A similar barrier of ΔG(z) is observed for EC/$LiPF_6$ and EC:DMC(3:7)/$LiPF_6$ 1 M electrolytes next to graphite is in accord with the experimental evidence. If one assumes that the charge-transfer reaction occurs when a Li^+ is located between outer layer carbon and hydrogen atoms then a value of $\Delta G(z)$ around 12–19 kcal/mol is obtained, which is consistent with the experimental [102] activation energies of 16.4–17 kcal/mol.

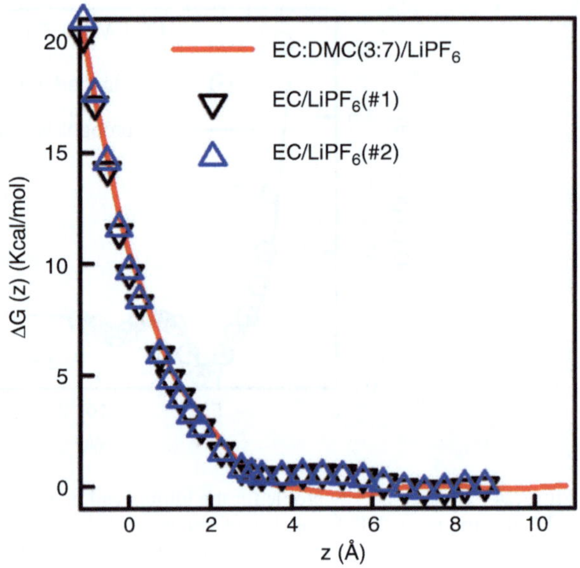

Fig. 7.25 The lithium cation desolvation free-energy profile for EC/$LiPF_6$ and EC:DMC(3:7)/$LiPF_6$ 1 M electrolytes next to graphite at 298 K

$LiFePO_4$/Electrolyte Interface

Lithium iron phosphate ($LiFePO_4$) is an attractive cathode material for lithium-ion batteries for a number of reasons, including good thermal stability, low fabrication costs (from abundant starting materials), nontoxicity, a reasonably high theoretical specific capacity and an inherently flat voltage – time discharge profile [103]. Interfacial impedance for $LiFePO_4$ cells, which includes both anode/electrolyte and cathode/electrolyte contributions, can be an important source of overall cell impedance at rates and temperatures of practical interest, with both conventional solvent–based electrolytes and ionic liquid–based electrolytes [99]. It has long been held that resistance to Li^+ transport through the anode/electrolyte interface contributes significantly to cell impedance due to the presence of a poorly conducting SEI that forms on most anode materials due to reduction of the electrolyte [104]. Recently, more attention has been focused on contributions of Li^+ transport across the cathode/electrolyte interface to cell impedance. Interfacial effects, particularly those associated with Li^+ desolvation during intercalation, are thought to play an important role in determining the impedance of $LiFePO_4$-based cathodes [105, 106].

An atomistic simulation MD simulation of a common carbonate-based organic electrolyte, ethylene carbonate:dimethyl carbonate (EC:DMC = 3:7) with approximately 1 mol/kg $LiPF_6$, referred to as the organic liquid electrolyte or OLE, and an ionic liquid–based electrolyte (ILE), 1-ethyl 3-methyl-imidazolium: bis(fluorosulfonyl)imide ($EMIM^+$:FSI^-) with 1 mol/kg LiFSI, in contact with $LiFePO_4$ has been carried out [107]. Simulations were carried out using quantum chemistry–based polarizable force at 363 K on a 3-D periodic orthorhombic

$LiFePO_4$ film + electrolyte systems such as shown in Fig. 7.26. POMF studies analogous to those for graphite were carried out in order to investigate the free-energy profile of bringing a Li^+ cation from the bulk electrolyte to the $LiFePO_4$ surface.

Poisson Potential Profile and Double Layer Structure for the ILE

The Poisson potential for the ILE is shown in Fig. 7.27. In addition to the potential for the $LiFePO_4$/ILE system, the potential for the $LiFePO_4$ film alone (in vacuo) is shown. For the $LiFePO_4$ in vacuo, the potential at the outer edge of the $LiFePO_4$ surface, i.e., $y \approx 0.5$ Å ($y = 0$ at the surface lithium layer, with y being the direction of the surface normal) reaches a constant value which is arbitrarily assigned a value of 0 V. For $LiFePO_4$ in contact with the ILE system, the potential exhibits structure with a period of 2.5–3 Å, oscillating between negative and positive values, which reflects the atomic-scale surface-induced ordering in the electrolyte. The surface-induced local structure in the ILE reflected in the potential is clear in the snapshot of the system (Fig. 7.26) FSI^- and $EMIM^+$ are strongly segregated in the surface layer, with the former concentrated near surface Li^+ and the latter near surface oxygen atoms. The electrostatic potential reaches the bulk value (arbitrarily set to 0 V) at about 1 nm from the surface, indicating a relatively narrow double layer. The potential at the outer edge of the $LiFePO_4$ surface (the electric surface potential) is about −0.25 V with respect to the bulk electrolyte. Hence, the presence of the ILE results in a negative potential at the $LiFePO_4$ surface. This negative potential is due to the excess negative charge in the electrolyte between 1.5 and 2.5 Å, resulting from excess FSI^- in the interfacial layer. Between the potential minimum near the surface and the next minimum at around 4 Å, there is a large positive potential of about 0.6 V with respect to the bulk electrolyte.

Poisson Potential Profile and Double Layer Structure for the OLE

The Poisson electrostatic potential for the $LiFePO_4$/OLE is shown in Fig. 7.28. Also shown is the potential for the $LiFePO_4$ film alone (in vacuo). The potential at the outer edge of the $LiFePO_4$ surface is negative with respect to the bulk (around −0.8 V). This electric surface potential is much more negative than observed for the ILE because of the propensity of negatively charged carbonyl oxygen atoms from the carbonate molecules to approach the surface and coordinate Li^+ and the lack of a strong positive charge (such as the hydrogen and nitrogen atoms in $EMIM^+$) to offset the resulting negative charge density. Between the negative potential minimum near the surface (around 1 Å) and the next minimum at around 5 Å, there is a positive potential comparable in magnitude to that observed in the ILE. The electrostatic potential reaches

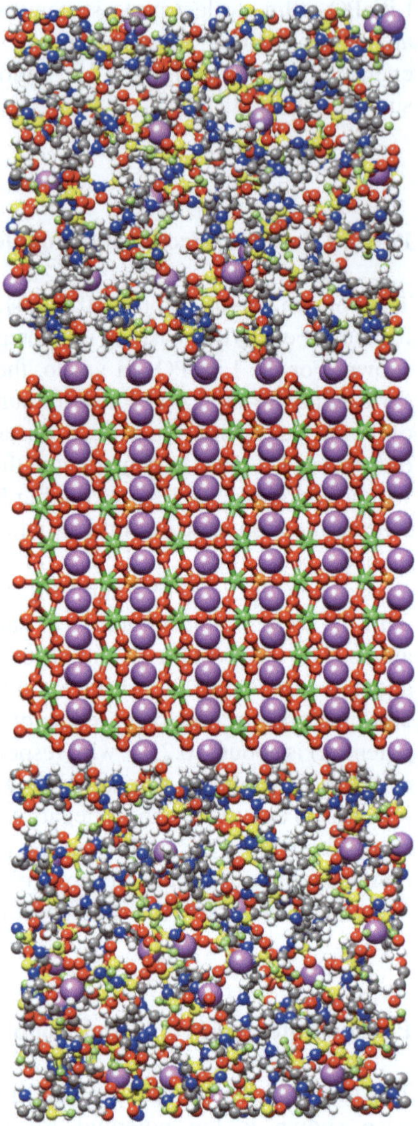

Fig. 7.26 Snapshot from an MD simulation of an ionic liquid electrolyte in contact with $LiFePO_4$

a constant (bulk) value at about 1 nm from the surface as was observed for the ILE electrolyte, indicating a narrow double layer.

The carbonyl oxygen atoms of the carbonate molecules, which have an appreciable negative charge (−0.4e to −0.5e) and are principally responsible for solvating Li^+ in bulk EC/DMC electrolytes [35], play the same role for the surface Li^+. Unlike the ILE, however, there are no exposed highly positively charged atoms in the OLE available to solvate the $LiFePO_4$ surface oxygen atoms. The CH_2/CH_3 groups in EC/DMC have a small positive charge in the range of +0.1e to +0.15e and

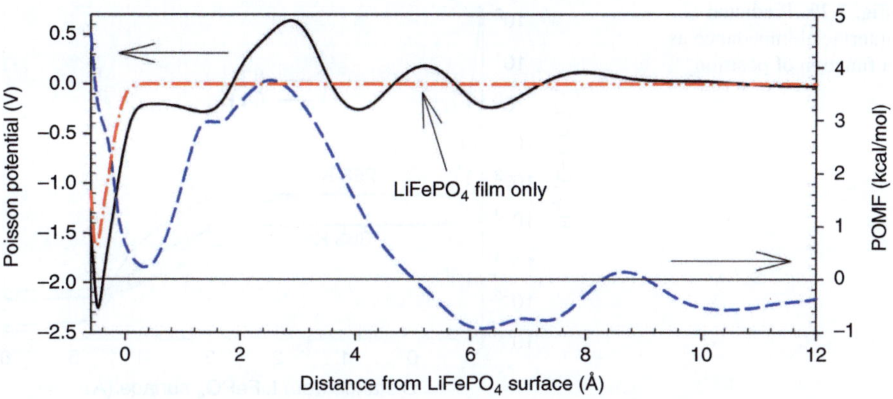

Fig. 7.27 Poisson potential and Li^+ free energy as a function of position with respect to the interface with $LiFePO_4$ in an ionic liquid electrolyte

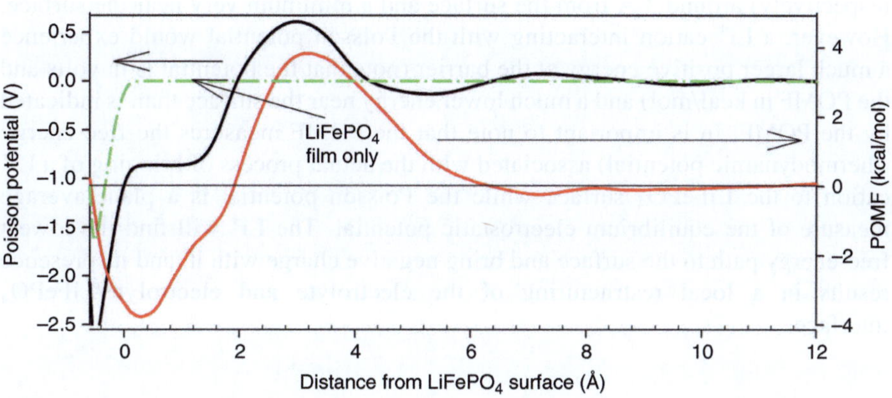

Fig. 7.28 Poisson potential and Li^+ free energy as a function of position with respect to the interface with $LiFePO_4$ in an organic liquid electrolyte

the hydrogen atoms from these groups are the main coordinating atoms for the surface O atoms. This results in a dense layer of EC and DMC centered 3–4 Å from the surface as indicated by atomic and molecular density profiles.

Free Energy of Li^+ as a Function of Position Relative to the $LiFePO_4$ Surface

The potential of mean force (POMF), i.e., the free energy as a function of the position of a Li^+ relative to the $LiFePO_4$ surface, is shown in Fig. 7.27 for the ILE/$LiFePO_4$ system and in Fig. 7.28 for the OLE/$LiFePO_4$ system. For both the OLE and ILE systems the POMF is qualitatively similar to the Poisson potential. First,

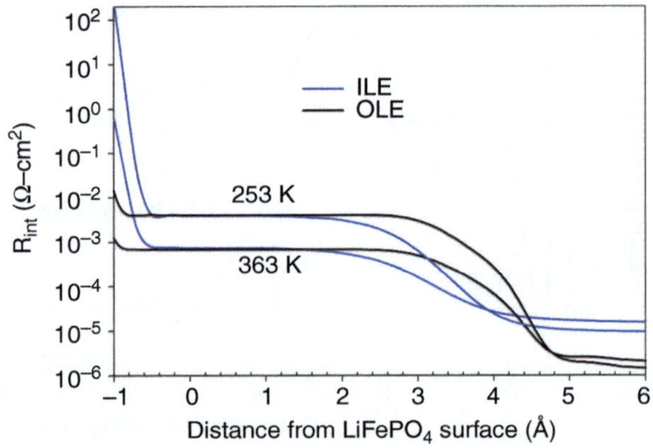

Fig. 7.29 Predicted interfacial impedance as a function of position

there is little structure in the POMF beyond 1 nm from the surface. Similarly, both the POMF and Poisson potential exhibit a maximum (in energy and potential, respectively) around 3 Å from the surface and a minimum very near the surface. However, a Li^+ cation interacting with the Poisson potential would experience a much larger positive energy at the barrier (note that the potential is in volts and the POMF in kcal/mol) and a much lower energy near the surface than is indicated by the POMF. In is important to note that the POMF measures the free energy (thermodynamic potential) associated with the actual process of bringing of a Li^+ cation to the $LiFePO_4$ surface while the Poisson potential is a plane-average measure of the equilibrium electrostatic potential. The Li^+ will find the lowest free-energy path to the surface and bring negative charge with it, and its presence results in a local restructuring of the electrolyte and electrolyte/$LiFePO_4$ interface.

Interfacial Impedance

The total interfacial impedance from the bulk electrolyte to a point y relative the $LiFePO_4$ surface can be estimated (in units of Ω-cm^2) as the integral of the local resistivity:

$$R_{\text{int}}(y) = \int_{y=14}^{y} -2.71 \times 10^{-9} \frac{kT \exp\left[\frac{\Delta G(y)}{kT}\right]}{D_{Li^+}(y) N_{\text{bulk}} e^2} dy \qquad (7.6)$$

where ΔG is Li^+ free energy, *D* is the Li^+ diffusion in the *y* direction, N_{bulk} is the concentration of Li^+ in the bulk electrolyte, and *T* is temperature. Using values of these quantities as determined from simulation, $R_{int}(y)$ for the ILE and the OLE systems was estimated using Eq. 7.2 [107]. Results are shown in Fig. 7.29, where it can be seen that the interfacial impedance for transporting Li^+ from $y = 14$ Å to about $y = 3$ Å is very small. In this region, the greater interfacial impedance for the ILE is due to the bulk lower Li^+ diffusion compared to that in the OLE. Beginning at around $y = 3$ Å, the decrease in Li^+ concentration due to the free-energy barrier observed for both the OLE and the ILE systems results in an increase in resistivity. The free energy of almost 4 kcal/mol, relative to the bulk and the corresponding decrease in Li^+ concentration in this region, does not lead to an appreciable contribution to R_{int}. An important conclusion from this work is that the higher bulk viscosity/lower bulk Li^+ mobility for the ILE compared to the OLE does not appear lead to a significant barrier for transporting Li^+ to the $LiFePO_4$ surface.

If/when the resistance for Li^+ transport through the electrolyte/$LiFePO_4$ interface is to make a significant contribution to overall cell impedance (in the absence of an SEI layer at the cathode/electrolyte interface), simulations indicate that it is likely due to the final desolvation process associated with intercalation of Li^+ into $LiFePO_4$. The rapid increase in free energy associated with this process leads to large resistivity according to Eq. 7.2 and a rapid increase in R_{int}, as shown in Fig. 7.29. Unfortunately, the range of positions (for $y < 0$) over which the free-energy profile used to estimate $N_{Li^+}(y)$ is accurate is unknown. Furthermore, use of Eq. 7.2 to estimate the equilibrium $N_{Li^+}(y)$ is questionable for $y < 0$ due to adsorption of Li^+ on the surface and the localization of Li^+ into specific channels (galleries). Nevertheless, assuming that the free-energy profile from simulation is valid to $y = -1$ Å, Fig. 7.29 indicates that interfacial impedance associated with the initial stages of intercalation from the surface into the $LiFePO_4$ gallery is significantly greater for the ILE than the OLE, but in either case is estimated to be quite small. The greater interfacial impedance for the ILE is not due to differences in the energetics of the desolvation process but rather to the much lower propensity of Li^+ to populate the $LiFePO_4$ surface in the ILE.

While interfacial impedance associated with bringing the cation to the $LiFePO_4$ surface and the initial stages of intercalation appears to be small at 363 K, the situation may be quite different at lower temperatures. Equation 7.2 can be applied at any temperature of interest assuming that $D_{Li} + (y)$ and $\Delta G_{Li} + (y)$ are independent of temperature. The estimated interfacial impedance at 253 K so calculated is shown in Fig. 7.29. Certainly $D_{Li} + (y)$ will decrease with temperature, particularly for the ILE, which, combined with increased interfacial structure with decreasing temperature, makes the estimated interfacial impedance at 253 K shown in Fig. 7.29 lower than would be obtained from direct determination of $D_{Li} + (y)$ and $\Delta G_{Li} + (y)$ at lower temperatures. The estimated contributions of the desolvation process associated with the initial stages of Li^+ intercalation are much greater at the lower temperature than at 363 K, particularly for the ILE. This result is intriguing in light of recent experimental studies that claim that interfacial impedance at the cathode electrode is important at sub-ambient temperatures for OLE-based systems, albeit, desolvation process on the anode was found to have a higher activation energy [108, 109]. Simulation results indicate that this could be exacerbated by the use of ILEs.

Summary

In summary, MD simulations of an OLE and an ILE in contact with $LiFePO_4$ reveal surface-induced structure in the electrolytes that extends about 1 nm from the $LiFePO_4$ surface. The Poisson electrostatic potential was found to be large and negative within about 1 Å of the surface and large and positive in the region of 2–4 Å from the $LiFePO_4$ surface. The negative potential at the $LiFePO_4$ surface results from an excess of negative charge from the electrolyte near the surface that participates in the coordination of the surface Li^+ atoms. The positive potential slightly further from the surface results from the orientation of the electrolyte molecules at the $LiFePO_4$ surface. This positive electrostatic potential is qualitatively consistent with a free-energy barrier for bringing Li^+ cations from the bulk electrolyte to the surface found to be 4 kcal/mol. However, this barrier and the corresponding decrease in Li^+ concentration due to lack of coordinating atoms (carbonyl oxygen atom) in the OLE in this range of distances from the interface, large local positive charge in the ILE, and the highly structured nature of the interfacial electrolyte, which reduces its ability to solvate Li^+, do not contribute significantly to interfacial impedance. The latter, if and when important, appears to be dominated by the desolvation process associated with intercalation of Li^+ into the galleries of $LiFePO_4$.

Future Directions

The future will see increased reliance on modeling for understanding of electrolyte electrochemical stability, electrolyte/electrode interactions, and the structure and properties of the SEI layer. Particularly valuable will be simulation methods that accurately reproduce reactions between the electrolyte, salts, and electrolyte additives with the electrodes. These methods, which will include QM and reactive molecular dynamics methods, will be invaluable in providing fundamental insight into the mechanisms of electrolyte decomposition, reactions leading to the formation of the SEI as well as in predicting the structure of the SEI. As efforts toward developing high-voltage cathodes increase, simulations will play an important role in the development of electrolytes that are oxidatively stable at the higher voltages as well as in development of strategies for passivating the cathode surface.

Acknowledgments The authors are grateful to the Department of Energy under Contract grant DE-SC0001912 to University of Utah and Air Force Office of Scientific Research contract number FA9550-09-C-0110 for the financial support of this work. We would also like to thank T. Richard Jow and Phil Ross for their comments and suggestions. This work was partially supported by an Interagency Agreement between the U.S. Department of Energy and the U.S. Army Research Laboratory under DE-IAOI-11EE003413 for the office of vehicle technologies programs including Batteries for Advanced Transporation Technologies (BATT) program.

Bibliography

1. Xu K (2004) Nonaqueous liquid electrolytes for lithium-based rechargeable batteries. Chem Rev 104:4303
2. Armand M, Tarascon JM (2008) Build better batteries. Nature 451:652
3. Zhang XR, Pugh JK, Ross PN (2001) Computation of thermodynamic oxidation potentials of organic solvents using density functional theory. J Electrochem Soc 148:E183
4. Fu Y, Liu L, Yu HZ, Wang YM, Guo QX (2005) Quantum-chemical predictions of absolute standard redox potentials of diverse organic molecules and free radicals in acetonitrile. J Am Chem Soc 127:7227
5. Han YK, Jung J, Yu S, Lee H (2009) Understanding the characteristics of high-voltage additives in Li-ion batteries: solvent effects. J Power Sour 187:581
6. Ue M, Murakami A, Nakamura S (2002) Anodic stability of several anions examined by Ab initio molecular orbital and density functional theories. J Electrochem Soc 149:A1572
7. Johansson P (2006) Intrinsic anion oxidation potentials. J Phys Chem A 110:12077
8. Johansson P (2007) Intrinsic anion oxidation potentials (vol 110, pg 12077, 2006). J Phys Chem A 111:1378
9. Wang RL, Buhrmester C, Dahn JR (2006) Calculations of oxidation potentials of redox shuttle additives for Li-ion cells. J Electrochem Soc 153:A445
10. Trasatti S (1986) The absolute electrode potential – an explanatory note (recommendations 1986). Pure Appl Chem 58:955–966
11. Isse AA, Gennaro A (2010) Absolute potential of the standard hydrogen electrode and the problem of interconversion of potentials in different solvents. J Phys Chem B 114:7894
12. Abe K, Hattori T, Kawabe K, Ushigoe Y, Yoshitake H (2007) Functional electrolytes. J Electrochem Soc 154:A810
13. Zhang XR, Kostecki R, Richardson TJ, Pugh JK, Ross PN (2001) Electrochemical and infrared studies of the reduction of organic carbonates. J Electrochem Soc 148:A1341
14. Wang YX, Nakamura S, Ue M, Balbuena PB (2001) Theoretical studies to understand surface chemistry on carbon anodes for lithium-ion batteries: reduction mechanisms of ethylene carbonate. J Am Chem Soc 123:11708
15. Vollmer JM, Curtiss LA, Vissers DR, Amine K (2004) Reduction mechanisms of ethylene, propylene, and vinylethylene carbonates. J Electrochem Soc 151:A178
16. Verma P, Maire P, Novák P (2010) A review of the features and analyses of the solid electrolyte interphase in Li-ion batteries. Electrochimica Acta 55(22):6332–6341
17. Aurbach D (2000) Review of selected electrode-solution interactions which determine the performance of Li and Li ion batteries. Power Sour 89:206
18. Aurbach D et al (1994) The Correlation between the surface-chemistry and the performance of Li-carbon intercalation anodes for rechargeable rocking-chair type batteries. J Electrochem Soc 141:603
19. Aurbach D, Einely Y, Zaban A (1994) The surface-chemistry of lithium electrodes in alkyl carbonate solutions. J Electrochem Soc 141:L1
20. Aurbach D et al (1995) The study of electrolyte-solutions based on ethylene and diethyl carbonates for rechargeable Li batteries. 1. Li metal anodes. J Electrochem Soc 142:2873–2882
21. Aurbach D et al (1995) The study of electrolyte-solutions based on ethylene and diethyl carbonates for rechargeable Li batteries. 2. graphite-electrodes. J Electrochem Soc 142:2882–2890
22. Li T, Balbuena PB (2000) Theoretical studies of the reduction of ethylene carbonate. Chem Phys Lett 317:421
23. Tasaki K, Kanda K, Kobayashi T, Nakamura S, Ue M (2006) Theoretical studies on the reductive decompositions of solvents and additives for lithium-ion batteries near lithium anodes. J Electrochem Soc 153:A2192

24. Wang YX, Nakamura S, Tasaki K, Balbuena PB (2002) Theoretical studies to understand surface chemistry on carbon anodes for lithium-ion batteries: how does vinylene carbonate play its role as an electrolyte additive? J Am Chem Soc 124:4408
25. Wang YX, Balbuena PB (2002) Theoretical insights into the reductive decompositions of propylene carbonate and vinylene carbonate: density functional theory studies. J Phys Chem B 106:4486
26. Aurbach D, Weissman I, Schechter A, Cohen H (1996) X-ray photoelectron spectroscopy studies of lithium surfaces prepared in several important electrolyte solutions. A comparison with previous studies by Fourier transform infrared spectroscopy. Langmuir 12:3991
27. Aurbach D, Gofer Y, Benzion M, Aped P (1992) The behavior of lithium electrodes in propylene and ethylene carbonate – the major factors that influence Li cycling efficiency. J Electroanal Chem 339:451
28. Xu K, Lam YF, Zhang SS, Jow TR, Curtis TB (2007) Solvation sheath of Li + in nonaqueous electrolytes and its implication of graphite/electrolyte interface chemistry. J Phys Chem C 111:7411
29. Naji A, Ghanbaja J, Humbert B, Willmann P, Billaud D (1996) Electroreduction of graphite in $LiClO_4$-ethylene carbonate electrolyte. Characterization of the passivating layer by transmission electron microscopy and Fourier-transform infrared spectroscopy. J Power Sour 63:33–39
30. Eyring H (1935) The activated complex in chemical reactions. J Chem Phys 3:107
31. Bedrov D, Smith GD, van Duin A (2010) Reactions of singly-reduced ethylene carbonate in lithium battery electrolytes. A molecular dynamics simulation study using the reaxFF. J Phys Chem B (accepted)
32. Aurbach D, Levi MD, Levi E, Schechter A (1997) Fail stabilization mechanisms graphite electrodes. J Phys Chem B 101:2195
33. Aurbach D, Moshkovich M, Cohen Y, Schechter A (1999) The study of surface film formation on noble-metal electrodes in alkyl carbonates/Li salt solutions, using simultaneous in situ AFM, EQCM, FTIR, and EIS. Langmuir 15:2947
34. Leung K, Budzien JL (2010) Ab initio molecular dynamics simulations of the initial stages of solid-electrolyte interphase formation on lithium ion battery graphitic anodes. Phys Chem Chem Phys 12:6583
35. Borodin O, Smith GD (2009) Quantum chemistry and molecular dynamics simulation study of dimethyl carbonate: ethylene carbonate electrolytes doped with $LiPF_6$. J Phys Chem B 113:1763
36. Borodin O, Smith GD (2006) LiTFSI structure and transport in ethylene carbonate from molecular dynamics simulations. J Phys Chem B 110:4971
37. Wu H, Wick CD (2010) Computational Investigation on the role of plasticizers on ion conductivity in poly(ethylene oxide) LiTFSI electrolytes. Macromolecules 43:3502–3510
38. Borodin O, Smith GD (2006) Development of many-body polarizable force fields for Li-battery applications: 2. LiTFSI-doped oligoether, polyether, and carbonate-based electrolytes. J Phys Chem B 110:6293–6299
39. Borodin O et al (2006) Li^+ Transport in lithium sulfonylimide-oligo(ethylene oxide) ionic liquids and oligo(ethylene oxide) doped with LiTFSI. J Phys Chem B 110:24266
40. Siqueira LJA, Ribeiro MCC (2006) Molecular dynamics simulation of the polymer electrolyte poly(ethylene oxide)/$LiClO_4$. II. Dynamical properties. J Chem Phys 125:214903
41. Borodin O, Smith GD (2000) Molecular dynamics simulation study of LiI-doped diglyme and poly(ethylene oxide) solutions. J Phys Chem B 104:8017–8022
42. Li S et al (2008) Molecular dynamics simulation of LiTFSI-acetamide electrolytes: structural properties. J Phys Chem B 112:6398
43. Soetens JC, Millot C, Maigret B (1998) Molecular dynamics simulation of $Li^+BF_4^-$ in ethylene carbonate, propylene carbonate, and dimethyl carbonate solvents. J Phys Chem A 102:1055

44. Takeuchi M et al (2009) Ion-ion interactions of $LiPF_6$ and $LiBF_4$ in propylene carbonate solutions. J Molec Liq 148:99
45. Brandell D, Liivat A, Aabloo A, Thomas JO (2005) Molecular dynamics simulation of the crystalline short-chain polymer system $LiPF_6$ center dot PEO_6 (M-w similar to 1000). J Mater Chem 15:4338–4345
46. Brandell D, Liivat A, Kasemagi H, Aabloo A, Thomas JO (2005) Molecular dynamics simulation of the $LiPF_6$ center dot PEO_6 structure. J Mater Chem 15:1422–1428
47. Borodin O, Smith GD, Jaffe RL (2001) Ab initio quantum chemistry and molecular dynamics simulations studies of $LiPF_6$/poly(ethylene oxide) interactions. J Comput Chem 22:641
48. Li T, Balbuena PB (1999) Theoretical studies of lithium perchlorate in ethylene carbonate, propylene carbonate, and their mixtures. J Electrochem Soc 146:3613
49. Borodin O, Smith GD (2007) Li^+ transport mechanism in oligo(ethylene oxide)s compared to carbonates. J Solution Chem 36:803
50. Newman J, Thomas KE, Hafezi H, Wheeler DR (2003) Modeling of lithium-ion batteries. J Power Sour 119:838
51. Hayashi K, Nemoto Y, Tobishima S, Yamaki J (1997) Electrolyte for high voltage Li/LiMn1.9Co0.1O4 cells. J Power Sour 68:316–319
52. Saunier J, Gorecki W, Alloin F, Sanchez JY (2005) NMR study of cation, anion, and solvent mobilities in macroporous poly(vinylidene fluoride). J Phys Chem B 109:2487–2492
53. Morita M, Asai Y, Yoshimoto N, Ishikawa M (1998) A Raman spectroscopic study of organic electrolyte solutions based on binary solvent systems of ethylene carbonate with low viscosity solvents which dissolve different lithium salts. J Chem Soc Faraday Trans 94:3451–3456
54. Hayamizu K, Aihara Y, Arai S, Martinez CG (1999) Pulse-gradient spin-echo H-1, Li-7, and F-19 NMR diffusion and ionic conductivity measurements of 14 organic electrolytes containing $LiN(SO_2CF_3)_2$. J Phys Chem B 103:519–524
55. Aihara Y et al (2004) Ion transport properties of six lithium salts dissolved in gamma-butyrolactone studied by self-diffusion and ionic conductivity measurements. J Electrochem Soc 151:A119
56. Tarascon JM, Armand M (2001) Issues and challenges facing rechargeable lithium batteries. Nature 414:359
57. Snyder JF, Ratner MA, Shriver DF (2002) Polymer electrolytes and polyelectrolytes: Monte Carlo simulations of thermal effects on conduction. Solid State Ionics 147:249
58. Borodin O, Smith GD (2006) Mechanism of ion transport in amorphous poly(ethylene oxide)/LiTFSI from molecular dynamics simulations. Macromolecules 39:1620–1629
59. Diddens D, Heuer A, Borodin O (2010) Understanding the lithium transport within a Rouse-based model for a PEO/LiTFSI polymer electrolyte. Macromolecules 43:2028
60. Borodin O, Smith GD (2007) Molecular dynamics simulations of comb-branched poly (epoxide ether)-based polymer electrolytes. Macromolecules 40:1252–1258
61. Borodin O, Smith GD, Bandyopadhyaya R, Redfern P, Curtiss LA (2004) Molecular dynamics study of nanocomposite polymer electrolyte based on poly(ethylene oxide)/$LiBF_4$. Model Sim Mater Sci Eng 12:S73
62. Londono JD et al (1997) Cation environment in molten lithium iodide doped poly(ethylene oxide). Macromolecules 30:7151–7157
63. Annis BK, Kim MH, Wignall GD, Borodin O, Smith GD (2000) A study of the influence of LiI on the chain conformations of poly(ethylene oxide) in the melt by small-angle neutron scattering and molecular dynamics simulations. Macromolecules 33:7544–7548
64. Saboungi ML et al (2002) Coherent neutron scattering from PEO and a PEO-based polymer electrolyte. Solid State Ionics 147:225
65. Muller-Plathe F, van Gunsteren WF (1995) Computer simulation of a polymer electrolyte: lithium iodide in amorphous poly(ethylene oxide). J Chem Phys 103:4745–4756
66. Maitra A, Heuer A (2007) Cation transport in polymer electrolytes: a microscopic approach. Phys Rev Lett 98:227802

67. Maitra A, Heuer A (2007) Understanding segmental dynamics in polymer electrolytes: a computer study. Macromol Chem Phys 208:2215
68. Sutjianto A, Curtiss LA (1998) Li+—diglyme complexes: barriers to lithium cation migration. J Phys Chem A 102:968
69. Hayamizu K, Akiba E, Bando T, Aihara Y (2002) H-1, Li-7, and F-19 nuclear magnetic resonance and ionic conductivity studies for liquid electrolytes composed of glymes and polyetheneglycol dimethyl ethers of $CH_3O(CH_2CH_2O)n\ CH_3$ (n = 3-50) doped with LiN $(SO_2CF_3)_2$. J Chem Phys 117:5929
70. Buriez O et al (2000) Performance limitations of polymer electrolytes based on ethylene oxide polymers. J Power Sour 89:149
71. Kerr JB et al (2002) From molecular models to system analysis for lithium battery electrolytes. J Power Sour 110:389
72. Kerr JB (2004) Nazri GA, Pistoia G (eds) Edsin Lithium batteries: sience and technology. Kluwer, Boston
73. Paillard E et al (2009) Electrochemical and physicochemical properties of $PY_{14}FSI$-based electrolytes with LiFSI. J Electrochem Soc 156:A891
74. Fox DM et al (2003) Flammability, thermal stability, and phase change characteristics of several trialkylimidazolium salts. Green Chem 5:724–727
75. Borodin O (2009) Polarizable force field development and molecular dynamics simulations of ionic liquids. J Phys Chem B 113:11463
76. Borodin O (2009) Relation between heat of vaporization, ion transport, molar volume, and cation and anion binding energy for ionic liquids. J Phys Chem B 113:12353
77. Smith GD et al (2010) A comparison of fluoroalkyl-derivatized imidazolium:TFSI and alkyl-derivatized imidazolium:TFSI ionic liquids: a molecular dynamics simulation study. Phys Chem Chem Phys 12:7064
78. Smith GD et al (2008) A comparison of ether- and alkyl-derivatized imidazolium-based room-temperature ionic liquids: a molecular dynamics simulation study. Phys Chem Chem Phys 10:6301
79. Maginn EJ (2009) Molecular simulation of ionic liquids: current status and future opportunities. J Phys Condens Matter 21:17
80. Koddermann T, Paschek D, Ludwig R (2007) Molecular dynamic simulations of ionic liquids: a reliable description of structure, thermodynamics and dynamics. Chemphyschem 8:2464–2470
81. Hooper JB, Borodin O (2010) Molecular dynamics simulations of N, N, N, N-tetramethylammonium dicyanamide plastic crystal and liquid using a polarizable force field. Phys Chem Chem Phys 12:4635
82. Borodin O, Gorecki W, Smith GD, Armand M (2010) Molecular dynamics simulation and pulsed-field gradient NMR studies of Bis(fluorosulfonyl)imide (FSI) and Bis [(trifluoromethyl)sulfonyl]imide (TFSI)-based ionic liquids. J Phys Chem B 114:6786
83. Kelkar MS, Shi W, Maginn EJ (2008) Determining the accuracy of classical force fields for ionic liquids: atomistic simulation of the thermodynamic and transport properties of 1-ethyl-3-methylimidazolium ethylsulfate ([emim][EtSO4]) and its mixtures with water. Ind Eng Chem Res 47:9115–9126
84. Tsuzuki S et al (2009) Molecular dynamics simulations of ionic liquids: cation and anion dependence of self-diffusion coefficients of ions. J Phys Chem B 113:10641
85. Bedrov D, Borodin O, Li Z, Smith GD (2010) Influence of polarization on structural, thermodynamic, and dynamic properties of ionic liquids obtained from molecular dynamics simulations. J Phys Chem B 114:4984
86. Borodin O, Smith GD, Henderson W (2006) Li^+ cation environment, transport and mechanical properties of the LiTFSI doped *N*-Methyl-*N*-alkylpyrrolidinium$^+$TFSI$^-$ ionic liquids. J Phys Chem B 110:16879

87. Niu SA, Cao Z, Li S, Yan TY (2010) Structure and transport properties of the LiPF6 doped 1-ethyl-2,3-dimethyl-imidazolium hexaftuorophosphate ionic liquids: a molecular dynamics study. J Phys Chem B 114:877
88. Monteiro MJ, Bazito FFC, Siqueira LJA, Ribeiro MCC, Torresi RM (2008) Transport coefficients, Raman spectroscopy, and computer simulation of lithium salt solutions in an ionic liquid. J Phys Chem B 112:2102
89. Lassegues JC, Grondin J, Aupetit C, Johansson P (2009) Spectroscopic identification of the lithium ion transporting species in LiTFSI-doped ionic liquids. J Phys Chem A 113:305
90. Tasaki K et al (2009) Solubility of lithium salts formed on the lithium-ion battery negative electrode surface in organic solvents. J Electrochem Soc 156:A1019
91. Borodin O, Smith GD, Fan P (2006) Molecular dynamics simulations of lithium alkyl carbonates. J Phys Chem B 110:22773
92. Xu K et al (2006) Syntheses and Characterization of lithium alkyl mono- and dicarbonates as components of surface films in Li-ion batteries. J Phys Chem B 110:7708
93. Zhuang GV, Xu K, Yang H, Jow TR, Ross PN (2005) Lithium ethylene dicarbonate identified as the primary product of chemical and electrochemical reduction of EC in 1.2 M $LiPF_6$EC: EMC electrolyte. J Phys Chem B 109:17567–17573
94. Ross P (2006), in BATT report
95. Zhang SS, Xu K, Jow TR (2006) EIS study on the formation of solid electrolyte. Electrochimica Acta 51:1636
96. Zhang SS, Xu K, Jow TR (2006) Charge and discharge characteristics of a commercial $LiCoO_2$-based 18650 Li-ion battery. J Power Sour 160:1403–1409
97. Xu K (2007) "Charge-transfer" process at graphite/electrolyte interface and the solvation sheath structure of Li + in nonaqueous electrolytes. J Electrochem Soc 154:A162
98. Abraham DP, Heaton JR, Kang SH, Dees DW, Jansen AN (2008) Investigating the low-temperature impedance increase of lithium-ion cells. J Electrochem Soc 155:A41
99. Kobayashi Y et al (2007) Comparative study of lithium secondary batteries using nonvolatile safety electrolytes. J Electrochem Soc 154:A677
100. Seki S et al (2005) Degradation mechanism analysis of all-solid-state lithium polymer secondary batteries by using the impedance measurement. J Power Sour 146:741
101. Bemporad D, Essex JW, Luttmann C (2004) Permeation of small molecules through a lipid bilayer: a computer simulation study. J Phys Chem B 108:4875
102. Xu K, Lam YF, Zhang SS, Jow TR, Curtis TB (2007) Solvation sheath of Li + in nonaqueous electrolytes and its implication of graphite/electrolyte interface chemistry. J Phys Chem C111:7411–7421
103. Padhi AK, Nanjundaswamy KS, Goodenough JB (1997) Phospho-olivines as positive-electrode materials for rechargeable lithium batteries. J Electrochem Soc 144:1188
104. Aurbach D (2003) Electrode-solution interactions in Li-ion batteries: a short summary and new insights. J Power Sour 119:497
105. He P, Zhang X, Wang YG, Cheng L, Xia YY (2008) Lithium-ion intercalation behavior of $LiFePO_4$ in aqueous and nonaqueous electrolyte solutions. J Electrochem Soc 155:A144
106. Yu DYW et al (2007) Study of $LiFePO_4$ by cyclic voltammetry. J Electrochem Soc 154:A253
107. Smith GD, Borodin O, Russo SP, Rees RJ, Hollenkamp AF (2009) A molecular dynamics simulation study of $LiFePO_4$/electrolyte interfaces: structure and Li + transport in carbonate and ionic liquid electrolytes. Phys Chem Chem Phys 11:9884
108. Jow TR et al (2010) Electrolytes SEI and charge discharge kinetics in Li-ion batteries. ECS Trans 25:3
109. Ogumi Z (2010) Interfacial reactions of lithium-ion batteries. Electrochemistry 78:319

Chapter 8
Lithium Ion Batteries, Electrochemical Reactions in

Paul J. Sideris and Steve G. Greenbaum

Glossary

Diffraction	A phenomenon which occurs when a propagating wave encounters or interacts with an object. Diffraction techniques have become a standard method for the investigation of the atomic structure of matter.
Nuclear magnetic resonance (NMR)	A condition in which magnetic nuclei in the presence of an external magnetic field absorb and reemit electromagnetic radiation (in the radiofrequency regime). The energy absorbed depends on the strength of the magnetic field and a number of chemical and structural properties of the matter under investigation.
Photoelectron	The electrons ejected from matter after having absorbed electromagnetic radiation of a particular wavelength.
Relaxation	The process by which a system returns to equilibrium after a perturbation, usually characterized by a specific time *t*.
Solid electrolyte interphase (SEI)	Electrolyte decomposition products, both organic and inorganic, that form a protective layer on the electrodes (predominantly the anode) of lithium-ion batteries which is necessary for optimal performance and cell longevity.

This chapter was originally published as part of the Encyclopedia of Sustainability Science and Technology edited by Robert A. Meyers. DOI:10.1007/978-1-4419-0851-3

P.J. Sideris (✉) • S.G. Greenbaum
Department of Physics and Astronomy, Hunter College of CUNY, 695 Park Ave, New York, NY 10065-5024, USA
e-mail: psideris@hunter.cuny.edu; steve.greenbaum@hunter.cuny.edu

R.J. Brodd (ed.), *Batteries for Sustainability: Selected Entries from the Encyclopedia of Sustainability Science and Technology*, DOI 10.1007/978-1-4614-5791-6_8,
© Springer Science+Business Media New York 2013

Definition of the Subject and Its Importance

Despite their spectacular success in portable electronics applications, continued technical advances of lithium-ion batteries are crucial to establishing large-scale storage applications such as electric vehicles and enabling development of renewable intermittent energy sources, i.e., wind and solar. Paramount considerations in realizing scaled-up battery systems are safety, cost, energy density, and service lifetime. Some of these applications also require rapid charge and discharge capability. To move beyond the current generation of lithium-ion batteries, it is necessary to understand some of the outstanding materials issues of the individual components (i.e., electrodes and electrolytes) as well as the battery system as a whole where the components interact under conditions of elevated temperature and electric current flow.

The rational design of compounds, complexes, and systems consisting of them for particular applications is the basis of modern materials science. The advent of powerful computing hardware and software in the last decade has made great strides in predicting material properties without doing a single experiment. On the other hand, the need for experimental verification has often resulted in the discovery of new scientific principles and behavior, and driven the development of not only the materials themselves but also ever more potent and versatile experimental tools. The purpose of this article is to describe several analytical methods with examples of their use in investigating material properties of rechargeable lithium battery cell components and systems, especially in the context of the outstanding technical challenges to further development. In addition, a few examples detailing how computational studies can aid in the interpretation of experimental results and in the development of new materials will be presented.

Introduction

Since their successful commercialization in the 1990s, extensive research has been carried out on lithium-ion batteries. The main material challenges have been recently summarized by Goodenough and Kim [1]. A lithium-ion battery cell consists of three main components: two electrodes – an anode (the negative terminal), and a cathode (positive terminal), as well as a non-aqueous electrolyte which physically separates the two electrodes, but allows conduction of ions between them during cycling (discharging and charging) of the battery. The work of Whittingham on titanium(II) sulfide as a cathode material was a breakthrough in the field and generated a lot of interest in utilizing intercalation chemistry for energy storage applications [2]. Since then, lithium metal oxides –$LiCoO_2$, $LiNiO_2$, $LiMn_2O_4$– have been employed as cathode materials. Due to the high reactivity of lithium metal, safety concerns have prompted the use of lithiated graphitic carbon as a replacement anode. Typical liquid electrolytes include lithium salts dissolved in organic carbonates.

The advancement of battery technology ultimately resides in a fundamental understanding of the structural, chemical, and electronic changes in the component materials that occur upon cycling. Experimental techniques commonly used in battery science can be broadly categorized as probing either the long-range or local structure. For a significant advancement in the field, both sets of techniques need to be utilized effectively to fully understand the chemistry and structure-property relationships of materials.

Electrochemistry

The overwhelming majority of commercially available anode and cathode materials for Li-ion batteries undergo an insertion reaction. The reaction mechanism involves the migration of Li^+ cations into (insertion, intercalation) and out of (extraction, deintercalation) cathode and anode materials, both of which often have a layered or tunnel structural motif. Since Li metal anodes have been largely replaced by graphitic carbon due to safety concerns, the positive electrodes in Li-ion batteries act as a source of lithium. During discharge, Li^+ migrates from the anode, through the electrolyte, to the cathode. Concomitantly, an electron is released for every Li^+ ion involved in the migration to an external circuit where it can power a device and perform work. The half-reactions, or one component of a redox (reduction-oxidation) reaction, of the two electrodes composed of lithium metal oxides of the form $LiMO_2$ (where M is a transition metal) and graphite can be written as,

$$\mathrm{Li}MO_2 \leftrightarrow \mathrm{Li}_{1-x}M\mathrm{O}_2 + x\mathrm{Li}^+ + xe^-, \tag{8.1}$$

for the cathode and

$$x\mathrm{Li}^+ + xe^- + 6\mathrm{C} \leftrightarrow \mathrm{Li}_x\mathrm{C}_6, \tag{8.2}$$

for the anode, respectively. In the $LiMO_2$ case, the transition metal gets oxidized from M^{3+} to M^{4+} during charging, with a subsequent reduction from M^{4+} to M^{3+} during discharge. Initially, layered $LiNiO_2$ was considered a promising cathode material considering it is cheaper and less toxic than cobalt [3], but the delithiated compounds were recognized as being too reactive with the organic electrolyte, posing a significant safety hazard. Moreover, Li–Ni exchange in the material over repeated cycles drastically hindered performance [4]. Issues stemming from environmental concerns and the cost of cobalt have motivated the study of similar layered compounds containing little to no Co as replacement cathode materials, such as $LiNi_{1-y}Co_yO_2$ [5], $LiMnO_2$ [6], $LiNi_{1-y}Mn_yO_2$ [7], and $LiNi_{1-y-z}Mn_yCo_zO_2$ [8, 9]. Structural instability of the electrodes, in conjunction with unavoidable side reactions which occur in the cell, limits the amount of charge and discharge cycles a battery can undergo. The rate of cycling also has a pronounced effect on the structure of many electrode materials and is an important factor in the depletion of performance.

A few classes of materials, such as binary copper compounds (copper sulfide CuS and copper oxide CuO) and metal fluorides ($M^{n+}F_n$), undergo a conversion reaction during cycling. These compounds have generated interest as potential electrode materials due to their high specific capacities (~600–1,000 mAh/g) and excellent cycling reversibility [10]. In this reaction mechanism, lithium compounds (lithium sulfide Li_2S, lithium oxide Li_2O, lithium fluoride LiF) are formed along with metal nano-particles during discharging. The overall reaction can be written as

$$M_xX_y + ny\mathrm{L}_i \leftrightarrow y\mathrm{Li}_nX + xM^0, \tag{8.3}$$

where X is either O or F, and M^0 denotes a transition metal in the metallic state. The reaction typically involves micron-sized M_xX_y particles which react to form composite materials composed of nano-sized metallic particles (on the order of 1–10 nm) and Li_nX which occurs on the first cycle. In subsequent redox cycles, the size of these particles remains virtually unchanged. The confinement of these metal particles in the lithium salt matrix is thought to play a crucial role in the electrochemical activity in the material. Of the M_xX_y compounds, only metal fluorides are considered as alternative positive electrodes in lithium-ion batteries. This is due to their ionic character, which in turn necessitates higher operating voltages [11, 12]. Unfortunately, fluorides tend to be large bandgap insulators, prohibiting their use in batteries. To circumvent the inherent issues of the fluorides, Amatucci and coworkers have prepared a mixed conductive matrix (MCM) composed of nanocarbon-metal-fluoride composite materials [13, 14]. In a few instances, a conductive oxide such as MoO_3 can also be used to enhance conductivity [15]. The drastic reduction of the metal fluoride particles, along with the presence of the highly conductive matrix in the nanocomposites results in enhanced electrochemical activity and cycling properties. The smaller particle size decreases the tunneling lengths of the electrons during the redox reaction in addition to increasing the available surface area, facilitating the diffusion of Li. The presence of a conducting matrix between the nanograins of the materials allows the reaction to progress.

The choice of electrolytes highly depends on the operating voltages and nature of the cell, i.e., a polymer or liquid-based battery. Employing a polymer electrolyte imposes further criteria, as electrochemical stability of the backbone and sidechains becomes relevant [16, 17]. Many liquid solvents are available to use as electrolytes, each with a respective viscosity and dielectric constant that can be selected to facilitate the ionic conductivity. Typically, ethylene carbonate and dimethyl carbonate are used. In contrast, only a handful of lithium salts – lithium hexafluorophosphate ($LiPF_6$), lithium triflate ($LiSO_3CF_3$), lithium tetrafluoroborate ($LiBF_4$), lithium perfluorosulfonimide ($Li^+[CF_3SO_2NSO_2CF_3]^-$), among others – have been used for the electrolyte. Additionally, polyethylene oxide (PEO) has been the primary focus of polymer electrolytes. Despite this, details regarding the nature of salt-polymer interactions as well as the local structure of polymer electrolytes remain elusive. Ongoing research is focused on finding new, highly conductive salts with a large electrochemical operating window which dissolves in PEO at low temperature [18].

During charging and discharging, a small amount of the electrolyte separating the two electrodes decomposes, producing a film across the surface of the electrode materials. This film is referred to as the solid electrolyte interphase (SEI). The formation of an SEI is critical in the function and longevity of the battery–it is electrically insulating, which protects the remaining electrolyte from further decomposition, and allows for the diffusion of lithium during the discharging and charging processes. Ethylene carbonate (EC) is present in the overwhelming majority of commercially available electrolyte compositions, due to its ability to form this protective layer on the surface of graphite. In the event that the SEI cracks or flakes off, lithium dendrites could form which could lead to a "chemical short circuit." The end result would be the cell overheating causing flammability and, in the worst case scenario explosions. A thin passivating layer which has some degree of flexibility to expand and contract along with the anticipated volume changes of the electrodes upon intercalation and deintercalation is also desirable. While the capacity of a battery cell is primarily determined by the structure and electronic properties of the electrode materials, the longevity is intricately tied to the frequency of sidereactions that occur between the electrolyte and electrode. Control of the chemistry of this interface is of crucial importance to the longevity and performance of the battery [19].

To date, many of the decomposition products are not fully known. To complicate matters, SEI products depend on several variables, such as degree and rate of cycling, as well as temperature. However, it is well established that achieving a desired thickness of the insulating layers is essential to prolonging cell life and acquiring stable performance. As the thickness increases, the rate-determining step of the charge transfer reaction becomes the migration of the ion across the interphase. Irreversible capacity loss occurs when the lithium ions chemically bind to the components of the SEI or become trapped in the boundaries between SEI components, the layers become too electrically insulating, or if the thickness of the layer prevents lithium intercalation.

Overview of Experimental Techniques

Among the techniques that probe the average or long-range structure, powder X-ray diffraction (PXRD) and neutron diffraction (ND) will be briefly discussed. Techniques that provide local, atomistic information that will be mentioned are nuclear magnetic resonance (NMR) spectroscopy, X-ray photoelectron spectroscopy, X-ray absorption spectroscopy, and pair distribution function (PDF) analysis. A brief introduction to the underlying theory of each technique will be provided along with relevant examples to illustrate the type of information that can be extracted from each measurement. A particular emphasis will be placed in NMR spectroscopy and the various experiments that can be used to extract structural and dynamical information. The section will conclude with a few examples of the successes in first-principles calculations in determining the structure and properties of relevant materials.

Diffraction Methods

Two routine methods to determine the crystal structure, sample phase purity, estimations of atomic disorder, and grain size are through X-ray and neutron diffraction. Initially, these measurements could only be performed ex-situ for cycled battery samples. Currently, diffraction patterns of relevant materials can be taken in-situ, as the battery is being cycled. Fundamentally, the techniques rely on the measurement of the intensity and angles of scattered X-rays or neutrons after the beams come in contact with a solid. X-ray diffractometers are readily available, allowing for the technique to be routine in the study of condensed matter. In contrast, neutron diffraction measurements require a special facility – either a spallation source or nuclear reactor.

The X-rays and neutrons scatter elastically in these diffraction experiments, meaning the energy of the incident and diffracted radiation is comparable. Bragg reflections occur when the incident and scattered waves of radiation interfere constructively in accordance with Bragg's law. Bragg's law describes the condition whereby you have constructive interference between successive atomic planes in a solid separated by an interlayer spacing d and is given by

$$n\lambda = 2d\sin(\theta), \tag{8.4}$$

where n is an integer greater than or equal to 1, and θ is the scattering angle. Atoms arranged periodically in a 3-D structure will impose selection rules on the type of reflections that are observed based on the symmetry of the crystal lattice. After considering the various symmetry operations possible in a 3-D arrangement of atoms, it is possible to classify all crystals into 1 of 230 so-called space groups. Additionally, one can define a unit cell, which contains atoms in the appropriate geometry that represents the simplest repeat unit of the crystal structure. The unit cell is parameterized by the lengths of the cell edges as well as the angles between them. The positions of the atoms inside the unit cell are given by a set of coordinates (x_i, y_i, z_i) measured from a specific lattice point known as the origin. Displacement parameters accompany the atomic positions to indicate anisotropic or spatially dependent positional disorder from thermal motions in the crystal. Information about the space group, unit cell parameters, and atomic positions uniquely identifies a material. Figure 8.1 shows in-situ diffraction patterns of the olivine compound $LiFePO_4$/C composite material collected by Myung et al. as a function of Li content [20]. As Li continues to be deintercalated, reflections due to a secondary $FePO_4$

phase appear, while the intensity of the $LiFePO_4$ reflections diminishes. By the end of the charge, $FePO_4$ becomes the major phase. As Li is introduced into the $FePO_4$ phase, the process is reversed, and the $LiFePO_4$ phase becomes dominant again. The reflections remain at the same 2θ, indicating that the lattice parameters are essentially constant upon cycling, in sharp contrast to the lithium transition metal (Co, Ni, Mn) oxide materials.

One of the main differences between X-ray and neutron diffraction is that X-rays scatter from the electron density of atoms within the crystal, whereas neutrons scatter from atomic nuclei. This implies that heavier elements, those with a larger atomic number, scatter and absorb the X-rays more effectively, making it difficult to detect light elements like lithium. A second implication is that contrast between neighboring elements in the periodic table or two different elements whose oxidations states produce the same electronic configuration will be poor. Neutrons posses spin, which can interact with the magnetic moment of electrons leading to so-called magnetic scattering. In this manner, so-called magnetic reflections can appear in the diffraction pattern, which aid in elucidating the magnetic structure of materials. Unlike X-rays, the neutron-scattering length does not vary in a systematic manner. Consequently, neutrons may diffract relatively strongly with light elements. Additionally, adjacent elements, which would be difficult to discern through X-ray diffraction experiments, may possess nuclear scattering properties that allow for their identification. A relevant example of this contrast is between Fe and Co. An added benefit of neutrons over X-rays is that relatively few elements, or more precisely isotopes, have significant absorption cross-sections. Since neutrons scatter from the nucleus, there is a significant dependence of the diffracted intensity on the particular isotope of the element. The sensitivity of the technique to different isotopes offers the possibility of performing a null scattering experiment, where a specific ratio of two isotopes is chosen, having neutron-scattering lengths of opposite sign, that effectively cancels the contribution of the element to the diffracted intensity.

In X-ray diffraction, the intensity of the diffracted beam I_{hkl} is proportional to the so-called structure factor F_{hkl} given by

$$F_{hkl} = \sum_{n=1}^{N} f_n e^{2\pi i(hx+ky+lz)}, \tag{8.5}$$

where f_n is the scattering factor for atom n within the unit cell (repeat unit of the crystal), and h, k, l are the Miller indices, which describe the virtual planes of atoms in a crystal lattice. For neutron diffraction, a similar expression is obtained for the structure factor, with the exception that the neutron-scattering factor b replaces f. The nuclei of atoms act as pin-point scatterers, unlike the electron clouds which have a finite size relative to the wavelengths of the radiation used in the experiment. The parameter b is independent of the scattering angle, in sharp contrast to f in the analogous X-ray expression which decreases with increasing angle. Thus, d-spacings, atomic positions, and their corresponding thermal parameters are more

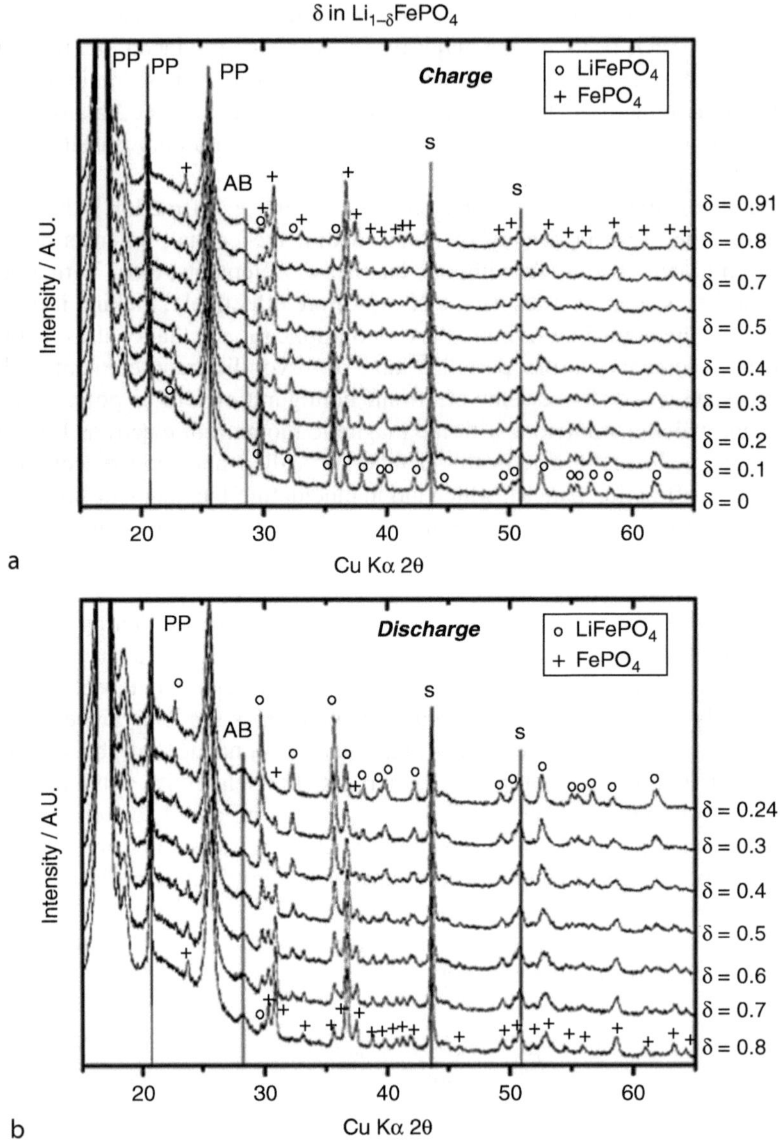

Fig. 8.1 In-situ X-ray diffraction patterns of $Li_{1-\delta}FePO_4$ as a function of lithium content δ (From [20])

precisely determined through powder neutron diffraction. A breakthrough in structure analysis using diffraction techniques was realized with the Rietveld method [21] which allowed the full profile of the diffraction pattern to be refined according to a specific structural model.

One method that is becoming increasingly useful in the study of solid, disordered materials is pair distribution function (PDF) analysis. The PDF can be understood as the distribution of bond lengths in a sample, weighted by the scattering power of the contributing atoms. It provides the probability of finding an atom at a distance r from another atom. The PDF is obtained from the sine Fourier transform of the total scattering pattern $S(Q)$, which includes both the Bragg and diffuse scattering, of a sample. Formally, this is given by

$$G(r) = 4\pi r[\rho(r) - \rho_0] = \frac{2}{\pi}\int_0^\infty Q[S(Q) - 1]\sin(Qr)dQ, \tag{8.6}$$

where $\rho(r)$ is the microscopic pair density, ρ_0 is the average number density, and Q is the magnitude of the scattering vector. For elastic scattering, $Q = 4\pi\sin(\theta)/\lambda$ with 2θ being the scattering angle and λ being the wavelength of the radiation used. A data plot includes $G(r)$ as a function of distance in angstroms. Since data cannot be collected to infinite Q, the real-space resolution of the PDF is decreased; Fourier transformation at a finite Q yields artifacts known as termination ripples. Additionally, the Q resolution results in an exponential dampening of peaks in the PDF with increasing r. As a result, it is desirable to perform experiments at high energy neutron or synchrotron X-ray sources to access high values of $Q_{\max}$ in order to enhance resolution. An added benefit of neutron PDF studies is that isotopic substitution can provide chemically specific pair distribution functions. For a sample containing two atoms of interest, say A and B, independent measurements can provide A–A, A–B, and B–B pairwise correlations.

Several analyses are available after obtaining an experimental PDF. Among these are bond length information and coordination number of atoms through a simple fit of the data. Peak widths can provide information regarding bond length distributions or correlated motions of atoms. PDFs are also used to refine or compare structural models of materials at the local- and medium-range length scales. Given a specific structural model, a calculated PDF, $G_c(r)$ is given by,

$$G_c(r) = \frac{1}{r}\sum_i\sum_j\left[\frac{b_i b_j}{\langle b\rangle^2}\delta(r - r_{ij})\right] - 4\pi r\rho_0, \tag{8.7}$$

where the sum extends over all ij atom pairs in the model crystal separated by an internuclear distance r_{ij}, b_i is the scattering power of atom i, and $\langle b\rangle$ is the average scattering power of the sample. For neutron scattering, b_i is the neutron-scattering length; for X-rays, it is the atomic form factor evaluated for a given Q. At $Q = 0$, the value of b_i in the X-ray scattering case is the number of electrons in the i atom. To model the finite Q in the data, a Q-dependent termination function of the form $\sin(Q_{\max}r/r)$ is convoluted with the calculated PDF $G_c(r)$. An exponential damping function can also be multiplied to mimic the loss of resolution in reciprocal space.

A few methods are available for a full profile refinement of the PDF obtained from a structural model. Structural parameters such as lattice constants, atomic positions, atomic displacement parameters, and site occupancies can be refined over a distance r using the program PDFFIT [22]. The program works well in instances where the model structure can be defined with a relatively small number of atoms. For more complex cases, in which many atoms are needed to match a structural model, the Reverse Monte Carlo (RMC) [23] technique can be employed. This algorithm involves the calculation of the PDF from a specific crystal configuration and a corresponding "goodness-of-fit" parameter χ^2 given by,

$$\chi^2 = \sum_{i=1}^{N} \frac{[G_e(r_i) - G_c(r_i)]^2}{\sigma^2} \tag{8.8}$$

where G_e and G_c are the experimental and calculated PDFs, respectively, and σ is a weighting factor corresponding to experimental uncertainty. The best calculation is that which produces a χ^2 minimum. The algorithm proceeds to a random selection of an atom in the model crystal. The occupancy or displacement of the atom is changed by a random amount. The new PDF and χ^2 are subsequently recalculated, along with the change in the agreement factor $\Delta\chi^2$. If the change made lowers χ^2, then the change is accepted. On the other hand, if the change increases χ^2 by $\Delta\chi^2$, it is not automatically rejected but accepted with a probability of $P = \exp(-\Delta\chi^2/2)$. This formalism ultimately ensures that the structural model does not get trapped in a local minimum, but converges to a global minimum. An RMC simulation serves to maximize the amount of disorder (entropy) in the atomic configuration while maintaining an agreement with experimental data. It is important to emphasize that this method can produce a range of atomic configurations, each with varying degrees of disorder that are consistent with experimentally obtained PDFs. Therefore, it is essential to have complementary information from other techniques which can place constraints on both local- and long-range structures to minimize the number of plausible configurations.

X-Ray Photoelectron Spectroscopy

X-ray photoelectron spectroscopy (XPS) is a quantitative technique which measures the number and kinetic energy of photoelectrons emitted from a sample that was irradiated with X-ray radiation. It is routinely used to determine the surface elemental composition or empirical formula of materials. XPS requires ultrahigh vacuum (UHV) conditions, on the order of $<10^{-9}$ torr, in order to detect the emitted electrons and minimize contamination or side reactions on the surface of the material. The spectrum involves a plot of electron count versus electron binding energy. The binding energy of electrons is specific to the element, so that peaks in the spectrum occur at characteristic energies corresponding to the electron configuration in a specific chemical bonding arrangement. Two peaks, with various integrated intensity ratios, are observed for electrons residing in *p*, *d*, or *f* orbitals. The separation of the two peaks, known as spin-orbit coupling, and their relative intensity assist in the identification of the elements in the sample. This is due to the fact that these values are characteristic and approximately constant for a core level of a given element, irrespective of the compound.

Variations in the binding energy of a core electron for an element, although small (on the order of a few electron volts eV), are still measureable and form the basis of the determination of the chemical speciation of a particular element, i.e., the bonding arrangement, type of nearest neighbors, and oxidation state. Briefly, the binding energy of an electron is determined primarily by the magnitude of the electrostatic interaction between itself and the nucleus. This interaction strength can be modulated by electrostatic shielding effects from other electrons, which in turn can depend on the removal or addition of charge brought about by chemical bonding. Removal of valence electron charge from an element (oxidation) results in an increase in binding energy; addition lowers the binding energy. This sensitivity to chemical speciation is demonstrated using the results from Dedryvère et al. (see Fig. 8.2), who studied various organic and inorganic decomposition products of the SEI in a CoO battery [24]. Peaks corresponding to C–C, C–O, O–C=O, and O_2C=O were clearly identified. Given quantitative information about the different carbon local environments, decomposition mechanisms that form the constituents of the SEI can be proposed.

The photoemission process that forms the basis of an XPS measurement can be deconstructed into three basic steps: (1) photon absorption and ionization, (2) atomic response with the generation of a photoelectron, and (3) transport of the ejected electron to the surface. These three steps are often labeled initial state, final state, and extrinsic state respectively. The intensity of an electron emitted at a depth d from the surface of material will be attenuated according to the Beer-Lambert Law. One can define the sampling depth as the minimum distance from the surface in which 95% of all the emitted photoelectrons are scattered. This parameter is intimately tied to the inelastic mean free path of an electron in a solid. The mean free path depends on the kinetic energy of the photoelectron and sample studied. For XPS, typical sampling depths are on the order of a few nanometers, rendering the technique most useful for thin films and surface studies. Composition-depth profiling can be performed by combining XPS with Ar^+ ion sputtering to study species away from the surface. In interpreting the results of XPS depth profiling, it is essential to consider two important factors that can affect the results: (1) the sputtering efficiency is dependent on the type of material being investigated – usually lower for stable inorganic solids and higher for organic matter, and (2) sputtering will induce surface chemical reactions and sample decomposition.

Quantitative measurements of an element ($\sim \pm 10\%$) can be made by relating the intensity of a specific peak "p" for an element "i" (I_i) using the relation:

$$I_i = CN_i\sigma_i\lambda_i \tag{8.9}$$

where C represents a constant which captures instrumental factors related to the signal acquisition, N_i is the atomic concentration of element "i," σ_i is the

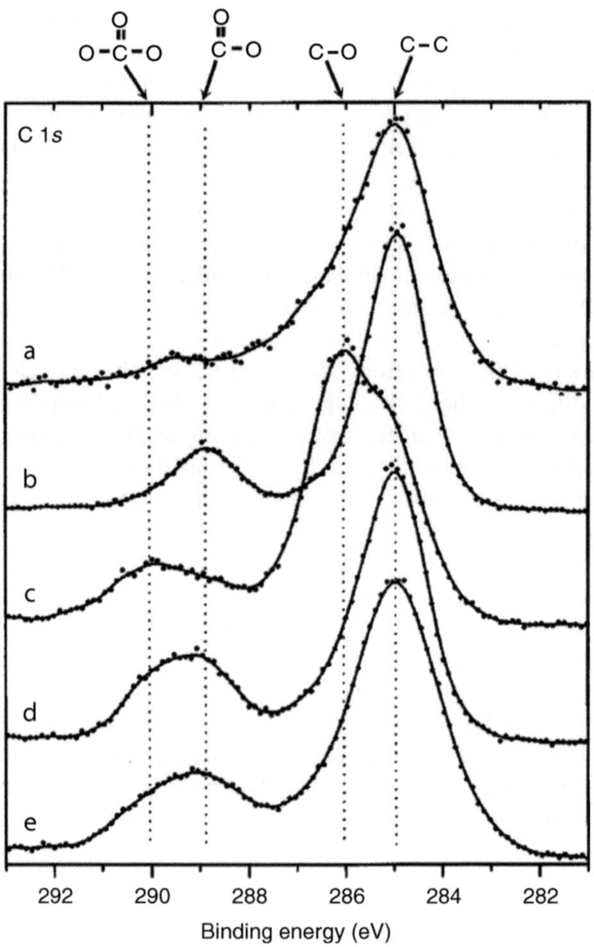

Fig. 8.2 Carbon 1*s* XPS spectra of the first charge/discharge cycle of CoO reacting with Li in a CoO-liquid electrolyte-Li metal electrochemical cell: (**a**) starting CoO powder, (**b**) after insertion of 1 Li per mole ($x = 1$), (**c**) after total discharge at 0.02 V, (**d**) after charge at 1.8 V, and (**e**) after charge at 3.0 V, from [24]. The assignments of the peaks are shown at the *top* of the figure

photoelectron cross-section for the element "i" expressed by peak "p" (commonly referred to as the Scofield factor), and λ_i is the inelastic mean free path of a photoelectron emitted by element "i" as expressed by peak "p."

X-Ray Absorption Spectroscopy

X-ray absorption spectroscopy (XAS) is performed using synchrotron radiation sources, where the X-ray energies are tunable and the intensity is large. XAS measurements of transition metals are most often performed at convenient energy ranges, i.e., without the use of ultrahigh vacuum conditions, although a notable exception includes the study of the oxygen K-edge in a lithium cobalt oxide-based cell [25]. The technique is involves tuning the incident X-ray radiation to excite core electrons of a specific element. The primary use of this absorption spectroscopy in battery science is to determine the oxidation state of a metal and bonding configuration between the metal of interest and the nearest neighboring ions or molecules, oftentimes referred to as ligands. Bonding between metals and ligands vary in degrees of covalency, a measure of how electron density is shared among atoms.

The energy profile of a sample depends on several factors: the element, the oxidation state of the element, the specific atomic orbital in which the electron resides, as well as the geometry and strength of the bonding to ligands. Depending on the type of core electron that is excited, the energy is given a specific label, called an "edge." The principal quantum numbers, $n = 1, 2, 3$, correspond to the K-, M-, and L-edges, respectively [26].

An XAS spectrum consists of normalized absorption plotted against the incident energy of the X-ray radiation. There are three main features to the absorption plot. The most prominent feature is called the "rising edge" and corresponds to a pronounced increase in absorption arising from $1s$ to p electronic transitions. A second region, known as the "pre-edge," occurs at slightly lower energies with respect to the rising edge. The pre-edge features arise from weak transitions between electrons in $1s$ and d atomic orbitals. Spectroscopy using energies in the pre-edge and rising edge regions is often called X-ray absorption near edge structure (XANES) or near-edge X-ray absorption fine structure (NEXAFS). The third region occurs at higher energies from the rising edge and is referred to as extended X-ray absorption fine structure (EXAFS). The EXAFS region is characterized by oscillations that arise from backscattering events that modulate the amplitude and wavelength of the ejected photoelectron from the source atom. Both the type of neighboring atoms and their distance to the source of the ejected photoelectron affect the degree of backscattering, so that bonding geometry and local coordination chemistry of the sample can be elucidated.

Given a homogeneous sample with a uniform thickness x, the absorption coefficient $\mu(E)$ can be related to the intensity of the incident (I_0) and transmitted (I) radiation using

$$\frac{I}{I_0} = e^{-\mu(E)x} \tag{8.10}$$

XAS can measure either transmission or fluorescence. The absorption coefficient is typically a smooth function of the energy of the radiation E, with the exception of the behavior at the absorption edge of a particular element. $\mu(E)$ can be approximated by

$$\mu(E) \approx \frac{\rho Z^4}{ME^3}, \tag{8.11}$$

where ρ is the density of the sample, Z is the atomic number of the element, and M is the corresponding atomic mass.

However, the true power and utility of the technique lies in the fact that the total absorption of the sample is not a simple summation of the atomic absorption coefficients, but is strongly affected by the neighboring atoms – a phenomenon particularly relevant in the EXAFS region. One can express the total absorption coefficient $\mu(E)$ as the absorption coefficient of an isolated atom $\mu_0(E)$ multiplied by a correction factor χ such that $\mu = \mu_0(1 + \chi)$. The parameter χ can be thought of as the fractional change in absorption coefficient induced by the neighboring atoms. In the single scattering approximation, Sayers et al. have introduced [27] an expression for χ – what is now known as the standard EXAFS equation:

$$\chi(k) = \sum_j \frac{N_j f_j(k) e^{-2k^2\sigma_j^2}}{kr_j^2} \sin\left[2kr_j + \delta_j(k)\right] \tag{8.12}$$

where k is the wave number of the photoelectron ($k = \sqrt{\frac{2m(E-E_0)}{\hbar^2}}$, E is the energy of the electron, E_0 is the absorption edge energy, m is the mass of the electron), N_j is the number of j atoms surrounding the element under investigation, r_j is the distance from the observed element to the nearest neighbors, σ_j^2 is a measure of the disorder in the distance to neighboring j atoms, $f(k)$ is the scattering amplitude of the wave and $\delta_j(k)$ is the corresponding phase shift. In essence, by simply measuring the scattered amplitude and phase shift, the number of neighboring atoms, and their distance, a measure of the atomic disorder can be estimated. Moreover, EXAFS will be sensitive to the type of neighboring element since the scattering amplitude depends on Z.

Figure 8.3a displays the normalized Ni K-edge absorbance spectrum of $Li_{1-x}Ni_{1/3}Co_{1/3}Mn_{1/3}O_2$ as a function of lithium content as performed by Tsai et al. [28]. Figure 8.3b shows the corresponding second derivative absorbance. Similar measurements can be made at the Co and Mn K-edges. The energy position of the pre-edge and absorption edge peaks of $LiNi_{1/3}Co_{1/3}Mn_{1/3}O_2$ are similar to those found in the reference compound NiO, consistent with the presence of Ni^{2+} in the pristine material. During delithiation, these peaks shift gradually to higher energies and eventually surpassing the energy position of the absorption edge of $LiNiO_2$ starting at the composition $Li_{0.15}Ni_{1/3}Co_{1/3}Mn_{1/3}O_2$, implying the existence of Ni^{4+} cations. During lithiation, these peaks shift to their original energy positions, indicating that during charging and discharging either a two-step (Ni^{2+}/Ni^{3+} followed by Ni^{3+}/Ni^{4+}) or single step (Ni^{2+}/Ni^{4+}) redox pair reaction with Ni occurs.

The authors then proceeded to analyze the corresponding EXAFS data. The k^2-weighted Fourier transformed spectra of the Ni K-edge of $Li_{1-x}Ni_{1/3}Co_{1/3}Mn_{1/3}O_2$ as a function of x is shown in Fig. 8.4. A total of four peaks are observed in each sample, which were ascribed to single and multiple scattering events. The first peak at ~1.5 Å is assigned to the shortest Ni-O distance in the octahedral site occupied by Ni. The second

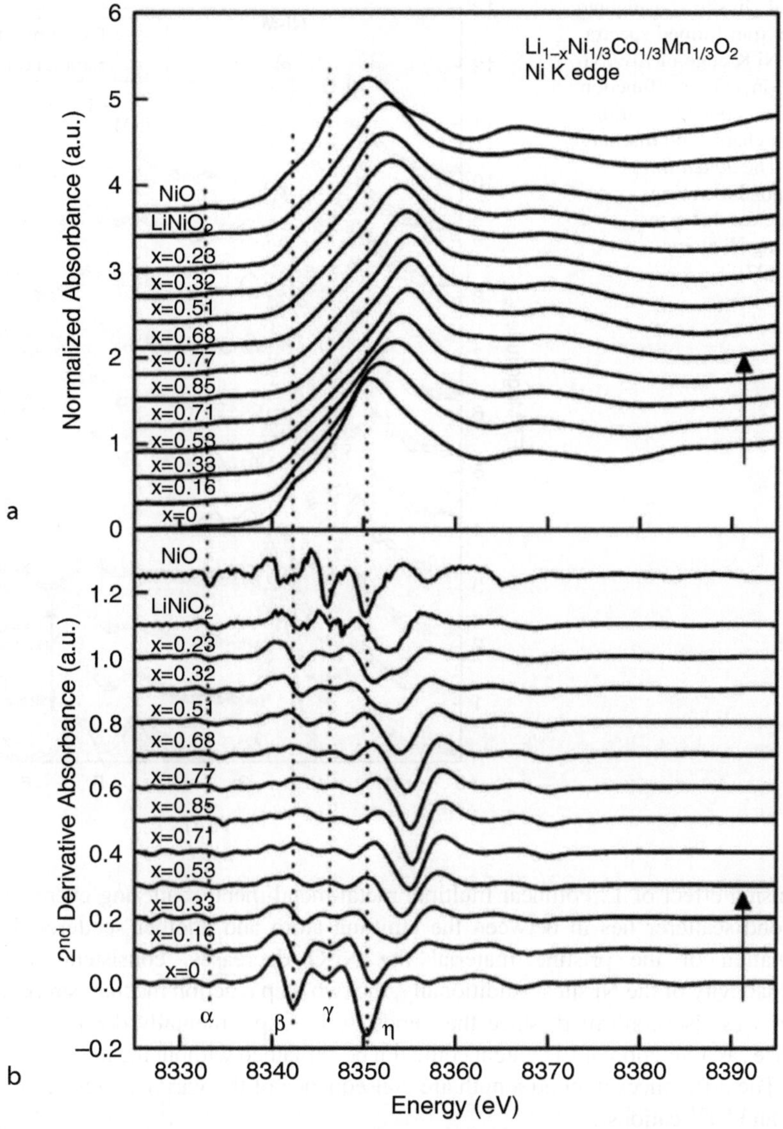

Fig. 8.3 (**a**) Normalized in situ absorbance and (**b**) second derivative absorbance of Ni K-edge for $Li_{1-x}Ni_{1/3}Co_{1/3}Mn_{1/3}O_2$ as a function of Li content x. The plots for $LiNiO_2$ and NiO are also presented to provide reference spectra for Ni^{3+} and Ni^{2+} local environments respectively (From [27])

peak at ~2.5 Å describes the distance from Ni to the nearest Ni/Co/Mn in the transition metal layer. The weak peak at 4.5 Å is dominated by combined effects of six metal-metal scattering within the transition metal layer, and six metal-metal scattering from two adjacent transition metal layers. The second weak peak at ~5.3 Å is attributed to

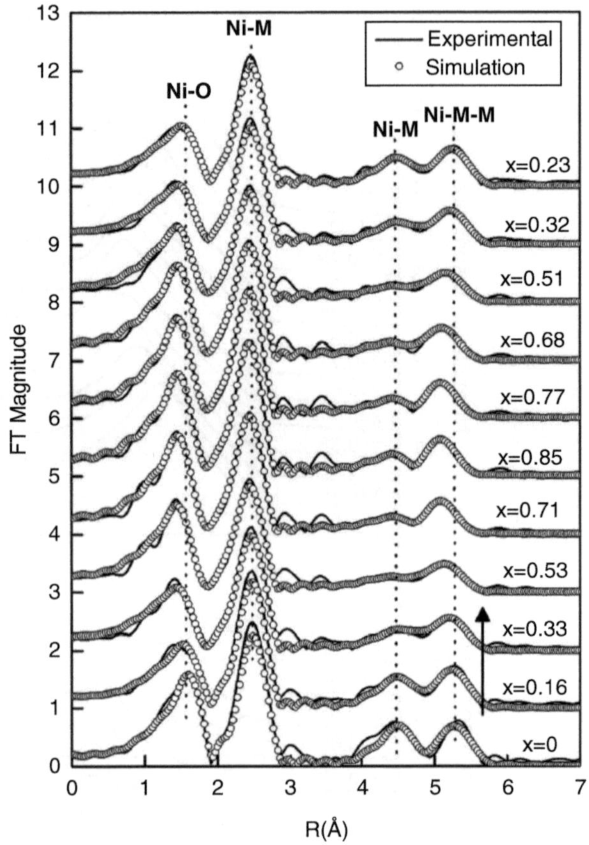

Fig. 8.4 In situ k^2-weighted Fourier transformed spectra of the Ni K-edge of $Li_{1-x}Ni_{1/3}Co_{1/3}Mn_{1/3}O_2$ as a function of lithium content x during the first charge and discharge cycle. The *dotted lines* correspond to various internuclear distances involving Ni atoms (From [27])

a focusing effect of 12 collinear multiple metal-metal-metal scattering events, where a second scatterer lies in between the emitting atom and another scatterer. During delithiation of the pristine material the Ni-O decreases, consistent with the electroactivity of the Ni sites. Additionally, the two-step reaction mechanism proposed earlier was also confirmed, since the trends in the experimentally determined Ni-O distances as a function of Li content could not be explained without the incorporation of Ni^{3+}. The differences in bond length are consequence of the various sizes of the Ni^{2+}, Ni^{3+}, and Ni^{4+} cations.

NMR Spectroscopy

Over the years, numerous materials have been investigated as potential electrodes in new generation Li-ion batteries using nuclear magnetic resonance (NMR) spectroscopy [29]. The primary nucleus of investigation for the electrode materials has been lithium. However, any nucleus possessing spin can potentially be studied, such as ^{31}P, ^{51}V, ^{13}C, ^{1}H, ^{19}F, etc. There are two spin-active isotopes of lithium which can be observed using NMR techniques – ^{6}Li and ^{7}Li. The natural abundance of ^{7}Li is approximately 92.41%, while that of ^{6}Li is 7.59%. The gyromagnetic ratio γ, in units 10^7radT^{-1}s^{-1}, of ^{7}Li and ^{6}Li is 10.40 and 3.94, respectively. Both nuclei are quadrupoles, with spin $I > ½$ ($I = 1$ and $I = 3/2$ for ^{6}Li and ^{7}Li, respectively), and are affected by the quadrupole interaction. The quadrupole moment of ^{7}Li is significantly larger than that of ^{6}Li. Consequently, the effects of the quadrupole interaction are more pronounced in ^{7}Li spectra. The interaction is anisotropic and produces characteristic broadening and line shapes in the static and magic angle–spinning (MAS) spectra, which reflect the local site symmetry of the lithium ion. If sample spinning speeds do not exceed the static line width of the spectra, a series of peaks, termed spinning sidebands, can be observed from the satellite transitions ($\pm\frac{3}{2} \leftrightarrow \pm\frac{1}{2}$) and to a lesser extent, the magnetic dipole broadened central transition ($+\frac{1}{2} \leftrightarrow -\frac{1}{2}$). The chemical shift range of lithium in diamagnetic compounds is relatively small, which may hinder the identification of chemically distinct sites. In contrast, the Li spectra of materials for Li-ion batteries, particularly the paramagnetic transition metal oxide cathode materials, are dominated by interactions involving the nucleus and the electronic spins.

The nucleus can interact with the unpaired electron spins of a nearby paramagnet through hyperfine interactions or with electrons in the conduction band of materials, the latter referred to as the Knight shift. These interactions can lead to significant peak broadening and shifts of the resonances. Nevertheless, these interactions contain information about the electronic structure and local environment of the Li atoms and can be theoretically used to track changes in each contribution upon cycling.

The observed NMR shift, expressed as $\Delta\omega/\omega_0$, induced by the Fermi contact hyperfine interaction in materials containing 3*d*-transition metals is proportional to the unpaired electron spin density at the nucleus site. The magnitude of the interaction is directly proportional to the Fermi constant A_c and the time-averaged electron spin $\langle S_z \rangle$ by

$$\frac{\Delta\omega}{\omega_0} = -\frac{A_c}{\omega_0 \hbar}\langle S_z \rangle \qquad (8.13)$$

The term $\langle S_Z \rangle$ can be expressed as

$$\langle S_Z \rangle = -\frac{B_0}{\mu_0 g N_A \mu_B}\chi_M \qquad (8.14)$$

where B_0 is the strength of the external magnetic field, χ_M is the molar susceptibility, μ_0 is the permeability of free space, g is the g-factor of the electron, N_A is Avogadro's number, and μ_B is the Bohr magneton.

The Fermi constant is a measure of the amount of spin density that is present at the site of interest, which in this case is the Li ion. The constant also determine the direction of the shift, i.e., whether it is positive or negative. The time average of the electron spin of the paramagnetic ion is of importance since at ambient temperatures, in systems that

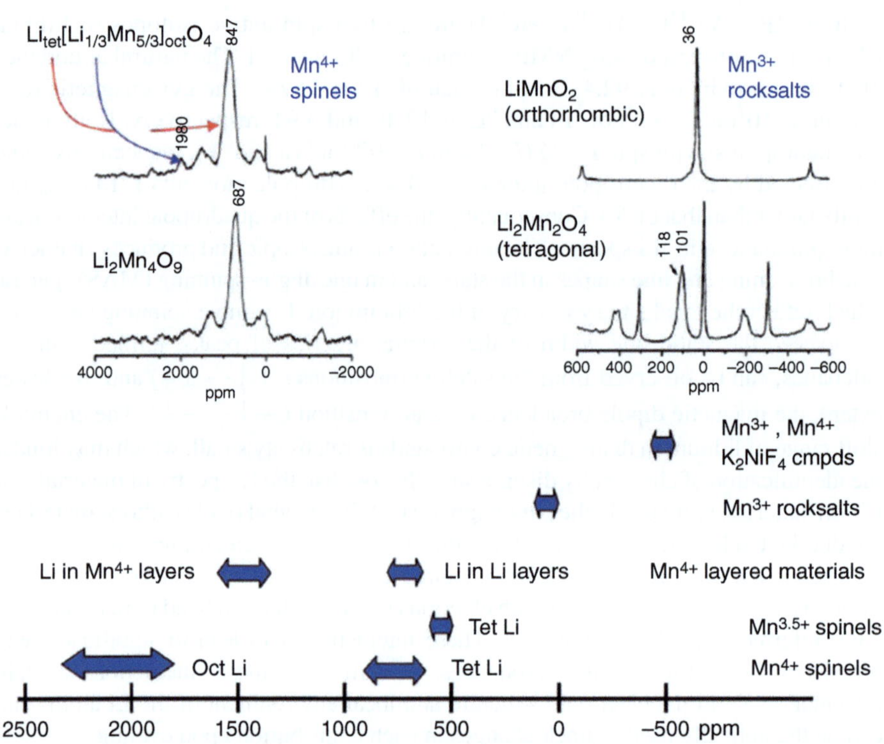

Fig. 8.5 Lithium hyperfine shifts observed in model Mn-containing compounds for four-coordinate and six-coordinate lithium environments (From [29])

can be studied by nuclear magnetic resonance spectroscopy, the relaxation of the electrons occurs many orders of magnitude faster than the coupling frequency A_c/h. The formalism presented is strictly true for systems where the orbital angular momentum is sufficiently suppressed so that the magnetic moments could be reasonably calculated using the "spin-only" contributions. For transition metal ions found in battery materials, this is usually a reasonable approximation. The Fermi contact shift in many materials has been found to be additive, so that the total observed shift arises from the summation of individual shift contributions induced by each paramagnetic ion. In this manner, shift scales can be created for the Li ion in different environments using the shifts of model compounds. This scale can then be used to make tentative assignments of Li spectra in battery materials. Figure 8.5 shows a Li shift scale for Mn-containing compounds compiled by Grey et al. [30].

The Li atoms in LiMO_2 (M = Co, Cr, Fe, Mn, Ni) phases, and their respective solid solutions, may interact with transition metal ions in the first and second cation coordination shells. These interactions are termed the 90° and 180° interactions, respectively, and are named according to the angle of the M–O–Li bond. In the discussion of the shift mechanisms, the labeling conventions used by Carlier et al. will be employed [31]. The 3d atomic orbitals of a metal in an ideal octahedral site

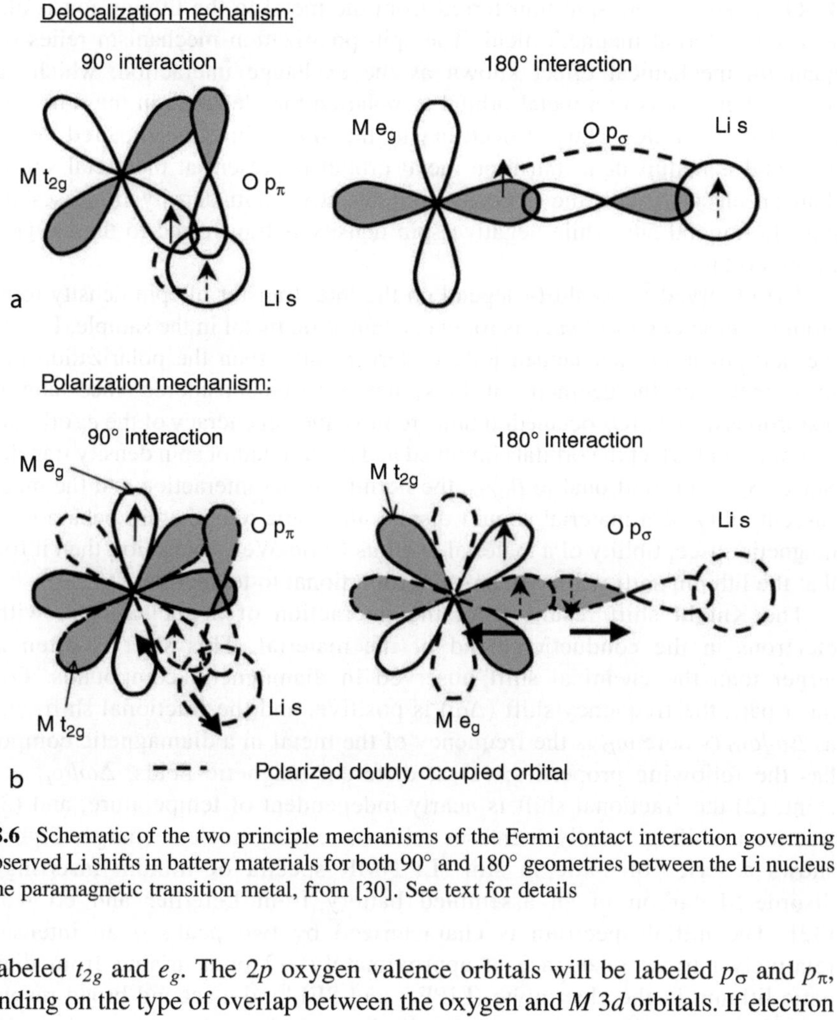

Fig. 8.6 Schematic of the two principle mechanisms of the Fermi contact interaction governing the observed Li shifts in battery materials for both 90° and 180° geometries between the Li nucleus and the paramagnetic transition metal, from [30]. See text for details

are labeled t_{2g} and e_g. The $2p$ oxygen valence orbitals will be labeled p_σ and p_π, depending on the type of overlap between the oxygen and M $3d$ orbitals. If electron density is present along the O-M internuclear axis, it is labeled p_σ. If electron density lies above and below the O-M internuclear axis, the O $2p$ valence orbital is labeled p_π. The Li valence atomic orbital is labeled s. Overlap of the M e_g, O $2p$ and Li s atomic orbitals leads to a bonding and antibonding molecular orbital, e_g-p_σ-s and $e_g{}^*$-p_σ -s respectively. The bonding orbital has a large contribution from the oxygen, while the antibonding orbital has a significant contribution from the metal. The M t_{2g} atomic orbitals are nonbonding to a first approximation, though Fermi contact shifts in some materials cannot be explained without mixing of these orbitals with Li s.

The spin transfer from the metal to the Li occurs via two concurrent mechanisms – spin delocalization and spin polarization (see Fig. 8.6). In the spin delocalization mechanism, the Li, O, and M atomic orbitals of appropriate symmetry overlap in the lattice. Consequently, a positive spin polarization is maintained along

M–O–Li, so that the spin transferred from the metal to the lithium ion is aligned with the external magnetic field. The spin polarization mechanism relies on the quantum mechanical effect known as the exchange interaction, which causes unpaired electrons in a metal orbital to polarize the electrons in the other doubly occupied 3*d* orbitals. Thus an electron with the *same* spin as the unpaired electron in a second nonequivalent transition metal orbital is present at the metal site rather than an electron with the opposite spin. Positive spin density increases on the transition metal site while negative spin density is transferred to the oxygen and lithium orbitals.

The observed $^{6,\,7}$Li shifts depend on the total transfer of spin density to the Li atom by these two mechanisms for every transition metal in the sample. In general, the delocalization mechanism induces larger shifts than the polarization mechanism. However, the geometry of the system must be considered since Jahn-Teller distortions of the MO_6 octahedral units remove the degeneracy of the e_g orbitals, for example, and affect the orbitals involved and the amount of spin density transferred. Since $\langle S_z \rangle$ is proportional to $B_0\chi_{\mathrm{M}}$, the Fermi contact interaction and the magnetic susceptibility of a material should display the same temperature behavior. If the magnetic susceptibility of a material exhibits Curie-Weiss behavior, then it follows that the lithium shift will be inversely proportional to temperature.

The Knight shift results from the interaction of a nuclear spin with the electrons in the conduction band of the material. This shift is often much larger than the chemical shift observed in diamagnetic compounds. For the most part, the frequency shift ($\Delta\omega$) is positive, and the fractional shift, defined as $\Delta\omega/\omega_D$ (where ω_D is the frequency of the metal in a diamagnetic compound) has the following properties: (1) at different magnetic fields, $\Delta\omega/\omega_D$ is constant, (2) the fractional shift is nearly independent of temperature, and (3) the fractional shift generally increases with increasing nuclear charge. Figure 8.7 shows a series of static in situ ^{7}Li NMR spectra of lithium inserting into disordered carbon in an assembled battery from Letellier and co workers [32]. The initial spectrum is characterized by two peaks – an intense and relatively narrow resonance at approximately −2 ppm arising from diamagnetic lithium in the electrolyte ($LiPF_6$) and SEI, and a second broad resonance at 263 ppm arising from metallic lithium used as the anode. A series of relatively weak and broad resonances appear during cycling of the battery, corresponding to various amounts of lithium inserting into the disordered carbon. These resonances vary smoothly during the charge and discharge cycles. The shifts of the Li metal and the various Li_xC_y compounds, each of which contains varying degrees of metallic character, from the diamagnetic 0 ppm region are due to the Knight shift effects.

NMR spectroscopy is very sensitive to a wide timescale – on the order of seconds to nanoseconds. An accurate mathematical treatment of motional averaging in NMR spectroscopy is highly dependent on the relative timescale of the motional process, with respect to the Larmor time scale (the reciprocal of the NMR frequency), since nonsecular Hamiltonian terms (i.e., those that do not commute with the Hamiltonian) are mainly responsible for relaxation. Motional

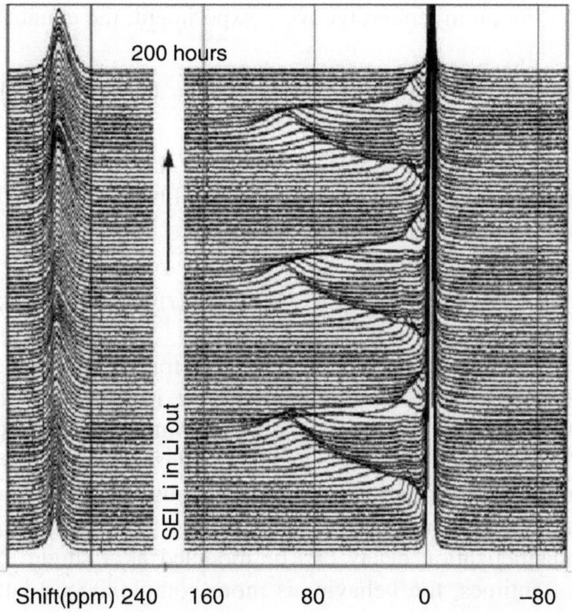

Fig. 8.7 In situ static ^{7}Li NMR of a disordered carbon-Li metal battery cell taken over three charge and discharge cycles. Peaks at ~0 ppm and 260 ppm correspond to lithium in the electrolyte/SEI and Li metal anode respectively. Resonances in the range of 0–140 ppm are assigned to Li environments inserted into the disordered carbon (From [31])

processes that are faster than the Larmor timescale, such as vibrations or librations, average the spin Hamiltonians *before* the secular approximation is made. Motion that is slower than the Larmor timescale, such as rotations and translations, average the spin Hamiltonian *after* the secular approximation has been made.

After the application of a radiofrequency pulse, which manipulates the nuclear magnetization, the spins begin to preferentially align with the external, static magnetic field, or to state it differently – the populations of spins in the different spin-states reach thermal equilibrium. Three relaxation times are typically measured: (1) the spin-lattice or longitudinal relaxation (T_1), (2) spin-spin or transverse relaxation (T_2), and (3) the relaxation in the rotating frame ($T_{1\rho}$). The rotating frame is a coordinate system which rotates about the z-axis (the direction of the static external magnetic field B_0) at the Larmor frequency of the nucleus under observation. This frame of reference is used to simplify the analysis of the trajectory of nuclear spin magnetization.

The T_1 relaxation refers to the time it takes for the z-component of the magnetization M_z to achieve an equilibrium value M_0. Two different equations could be used to extract T_1 from a series of measurements, depending on whether a saturation-recovery experiment or an inversion recovery experiment (explained in the subsequent section) is used. In the former case, the following equation is used:

$$M_z(t) = M_0\left(1 - e^{-\frac{t}{T_1}}\right) \tag{8.15}$$

For an inversion recovery experiment, the equation changes slightly to,

$$M_z(t) = M_0\left(1 - 2e^{-\frac{t}{T_1}}\right) \tag{8.16}$$

The T_2 relaxation time refers to the return of the observable, transverse (*xy*-plane) magnetization M_{xy} to an equilibrium value. This process could be modeled by a simple exponential

$$M_{xy}(t) = M_{xy0}e^{-\frac{t}{T_2}} \tag{8.17}$$

Two factors influence the loss of transverse magnetization – magnetic, molecular interactions and inhomogeneity in magnetic field. The observed decay in transverse magnetization is thus a combination of these two effects.

Both spin-lattice and spin-spin relaxation occur simultaneously, with $T_2 \leq T_1$ in the majority of cases. In heterogeneous samples, there are often multiple T_1 and T_2 relaxation times due, in part, to the presence of many local environments. $T_{1\rho}$ magnetization decay can be modeled as a single exponential much like T_2, but oftentimes, the behavior is more complex. Spin-lattice relaxation, evidenced by changes in the characteristic time T_1, occurs when motion is on the Larmor time scale. On the spectral timescale, typically on the order of μs to ms, line shapes of a spectrum are greatly affected, resulting in either line broadening or narrowing. Processes that are relatively very slow – millisecond to second timescales – do not result in line-shape changes. However, they can still be detected by observing their effect on the dynamics of longitudinal magnetization, specifically using a two-dimensional (2-D) exchange experiment [33].

To illustrate the line-shape changes occurring for motion on the spectrum timescale, a two-site symmetrical exchange will be considered. A nuclear spin is transported from site A to site B and back again at a rate *k*, i.e., A → B and B → A both occur at a rate *k*. The two sites occur with equal probability, and the exchange is fast enough so that no intermediate chemical states are considered. The two different chemical environments induce two chemical shifts. The angular chemical shift frequencies of the two sites are expressed as ω_A and ω_B for sites A and B, respectively. If the rate constant *k* is the same order of magnitude as the angular chemical shift frequency, i.e., $k = \omega_{\text{diff}} = \omega_A - \omega_B$, the line shapes will change drastically. Under these circumstances, the system is referred to as being in the intermediate exchange regime.

Within this intermediate exchange regime, one can define three regions – the crossover point, slow intermediate exchange, and the fast intermediate exchange regimes. The crossover point occurs when $k = |\omega_{\text{diff}}/2|$. For slow and fast intermediate exchange, $k < |\omega_{\text{diff}}/2|$ and $k > |\omega_{\text{diff}}/2|$, respectively. In the slow intermediate exchange regime, as one approaches the crossover point, motional broadening occurs. In the two-site symmetric exchange example, this would result in broadening of the two peaks along with coalescence as *k* increases to the crossover point. At the crossover point, a single broad resonance will be present at the average of the two chemical shifts. As k increases past the crossover point to the fast intermediate exchange regime, the line width of the single resonance is reduced – a phenomenon known as motional narrowing. Motional narrowing in this regime occurs because the spins jump so frequently between the two sites, that no significant phase difference in the transverse magnetization is accumulated as a result of the different local environments. The spin experiences a well-defined average precession frequency, so that the coalesced peak continues to narrow [33].

In ideal cases, the temperature dependent rate constants $k(T)$ display Arrhenius behavior such that the data could be fitted with,

$$k(T) = C \exp\left(-E_A / \left(N_A k_B T\right)\right) \tag{8.18}$$

where C is a constant, E_A is the activation energy of the jump, N_A is Avogadro's number, k_B is Boltzmann's constant and T is temperature. For an asymmetric two site exchange, where the rates of A → B and B → A are different, similar phenomena occur, with the exception that in the fast intermediate exchange limit, the coalesced peak is positioned at the average of the chemical shift of the original two peaks, weighted by the equilibrium concentrations of the two species.

Another advantage of NMR spectroscopy is the ability to measure self-diffusion coefficients. Diffusion measurements are overwhelmingly made in the study of liquid electrolytes. The basic principle of the experiment lies in the fact that nuclear spins placed in a nonuniform magnetic field will experience different local fields depending on their spatial location. Differences in the strength of local magnetic fields will induce slightly different procession frequencies of the nuclear spin under investigation.

When a simple Hahn-echo pulse sequence is applied to the sample in a non-uniform magnetic field, the presence of motion can be inferred by the echo signal intensity taken as a function of a delay. If a spin has not diffused, it will have a constant precession frequency so that at the end of the dephasing period 2τ, the spin magnetization is said to have been completely refocused. If a spin moves along a magnetic field gradient the precession frequency will change, resulting in an attenuated echo intensity at 2τ. By performing a series of experiments where τ is varied, the reduced intensity of the resonance could be used to calculate the diffusion coefficient of the nucleus using the equation [34],

$$M(2\tau) = M_0 e^{\frac{2\tau}{T_2}} e^{-\frac{2}{3}\left(\gamma \frac{\delta H}{\delta Z}\right)^2 D \tau^3} \tag{8.19}$$

where M_0 is some initial magnetization, T_2 is the spin-spin relaxation time, γ is the gyromagnetic ratio of the nucleus under investigation, $(\delta H/\delta Z)$ is the magnetic field gradient, and D is the diffusion coefficient. In the limit where $\tau << T_2$, the reduction of the echo signal intensity is dominated by the diffusion process. In samples with short T_2 values, it is difficult to separate the effects of spin-spin relaxation and diffusion in the attenuation of the echo signal.

These spin-echo experiments are most commonly performed using pulse field gradients, introduced by Stejskal and Tanner [35], or using the so-called "static fringe-field method" [36, 37]. In general, T_2 times greater than 2 ms for each species is required to perform the diffusion measurements and extract precise diffusion coefficients. Pulse-field gradient spin-echo methods require special amplifiers which control the strength, duration, and time interval between gradient pulses. The gradients that can be implemented typically range from 10 to 70 mT/cm. In this experiment, the delay between the Hahn-echo pulses is kept constant, so that the contribution of T_2 relaxation on the attenuation of the echo signal intensity is constant. Since the experiment is performed in the "sweet spot" of the magnet, where the magnetic field is the most homogenous, the spectral resolution is relatively high. The fringe-field method requires the probe to be positioned outside of the sweet spot of the magnet, where large static magnetic field gradients are present – typically greater than 100 mT/cm. The relatively large gradients allow relatively small diffusion coefficients to be measured, usually less than 10^{-12} m^2/s. Unfortunately, the line widths of the resonances are significantly large due to inhomogenous line broadening. Consequently, the sensitivity and resolution decreases. Additionally, the T_2 of the chemical species must be measured independently in a uniform magnetic field.

Conductivity and ionic diffusion could be related using the modified Nernst-Einstein equation

$$\sigma_{\mathrm{calc}} = \frac{Nq^2(D_{\mathrm{Li}} + D_{\mathrm{an}})\alpha}{kT} \tag{8.20}$$

where σ_{calc} is the calculated ionic conductivity, N is the number of ions per unit volume, q is the charge of the ions, D_{Li} and D_{an} are the diffusion coefficients of the lithium ion and anion respectively, a is the degree of dissociation, k is Boltzmann's constant, and T is temperature. In many instances, the calculated conductivities using the measured diffusion coefficients from pulse-field gradient NMR are much higher than values measured directly [38–43]. The discrepancy could arise from the differences in the migration distance probed using the two different techniques. Moreover, electrical conductivity detects the motion of charged ions, without discrimination of element or sign of the charge. Diffusion coefficients contain contributions from ions in both charge-neutral and dissociated species of a particular probe nucleus.

The measurement of diffusion coefficients is particularly important in polymer electrolyte research. One of the desired properties of a potential electrolyte material is the selective mobility of the Li^+ cation. Numerous strategies have been implemented to attain this property [1, 44, 45]. One method of quantifying the selective mobility is the calculation of transference numbers, which is the fraction of the total electric current that the anions and cations carry while passing through an electrolyte. Using the notation of Kalita et al. [46], the transference number for lithium, t^+, could be expressed as,

$$t^+ = \frac{D^+}{D^+ + D^-} \tag{8.21}$$

where D^+ and D^- are the cations and anion diffusion coefficients respectively. As implied above, determining the transference number for free Li^+ is not straightforward using NMR or electrical conductivity measurements.

Computational Methods

Computational methods have advanced significantly over the past few years, allowing for the accurate prediction and explanation of various properties of materials relevant to the battery field, including lithium and electron mobilities [47–49], voltage profiles [50–53], average insertion potentials [50, 54–58], and phase stabilities [59–61], using first-principles calculations. A few methods are available for first-principles calculations of electronic structure – Hartree-Fock, Møller-Plesset Perturbation theory, highly correlated methods, and density functional theory. Typical basis sets used for these calculations include Slater-type functions ($Y_l^m r^{n-1} \exp(-\xi r/a_0)$), Gaussian-type functions ($Y_l^m \exp(-(\alpha r)^2)$), or plane waves ($\exp(ig{\cdot}r)$), where Y_l^m denotes the spherical harmonics and r is the radial distance of the electron from the nucleus. The formalism within which many of these calculations are conducted in condensed matter is typically density functional theory (DFT), using plane waves and pseudopotentials. The intricate details of the theory are beyond the scope of this article. For the purposes of this discussion, it is sufficient to understand DFT as a reformulation of many-body quantum mechanics in terms of the probability density rather than the wave function. Specifically, it is used to investigate the electronic structure of condensed phases using functionals, i.e., functions of another function. The primary challenge in DFT is the expression of functionals to accurately describe the exchange and correlation energies of electrons. Approximations that have yielded accurate molecular geometries and ground-state energies include the generalized gradient approximation (GGA) method [62]. Pseudopotentials, which serve as approximate, effective potential functions to deal with the core electrons of an atom, that have been successful in the calculations of battery materials have been generated from the so-called projector-augmented-wave (PAW) method [63, 64] as utilized from the software package Vienna Ab initio Simulation Package (VASP) [65].

Survey of Experimental Results and Their Interpretation

The following section will summarize additional examples in recent literature which made use of at least one of the experimental techniques outlined above. Clearly, the body of work available in the field cannot be adequately summarized in a single section. Examples will be grouped by the major components of a Li-ion battery – the electrolyte, SEI, cathode, and anode materials. These particular experiments were selected not only to demonstrate how each technique can provide unique information about the structure of a material but also to emphasize the necessity of utilizing multiple techniques to refine a structural model or determine a chemical reaction pathway.

SEI

XPS has been used extensively to study the composition of the SEI and decomposition products of the electrolyte on electrode surfaces. In the overwhelming majority of cases, the binding energy of the C 1*s* photoelectron peak position is used to identify various carbon-containing decomposition products. Among the most prevalent species to be identified in the SEI are lithium carbonates – Li_2CO_3, $ROCO_2Li$, and carbonyl carbons (C=O) [66]. There have been a few studies using the O 1*s* and F 1*s* positions to identify LiF [67] and ether linkages [68]. The choice of electrolytes and lithium salts has a profound effect on the decomposition products [69]. Ethylene carbonate electrolytes are rich in $(CH_2OCO_2Li)_2$. XPS has been vital in determining the decomposition rate of electrolytes containing various additives [70] and as a function of temperature [71, 72]. Traces of Li_2O and LiOH were observed in a few instances, but could be the result of side reactions caused by Ar+ sputtering or trace amounts of water [66].

Components of the SEI were also investigated using NMR techniques [73]. ^{7}Li studies, while able to distinguish between lithium atoms in the SEI and those in paramagnetic transition metal compounds using the shift of the resonance, does not provide a lot of information regarding the different types of diamagnetic Li environments. This is because the shift range for Li^+ in salts and other compounds present in the solid electrolyte interphase is small and typically around 0 ppm. Nevertheless, Dupré et al. have used ^{7}Li MAS NMR to characterize interphase layers on electrodes as a function of aging [74–76] mostly using relaxation measurements and line-shape analyses. Leifer et al. have used ^{13}C MAS NMR to characterize electrolyte breakdown products using batteries composed of C-13-enriched carbonates stored at various temperatures [77]. Carbon NMR has a larger chemical shift range, relative to Li, rendering the nucleus more sensitive to changes in bonding or chemical speciation. The carbonates used in the study (a mixture of ethylene carbonate and diethyl carbonate) were selectively enriched in the carbonyl carbon position. By incorporating only one enriched carbonate at a time in the electrolyte mixture, the authors were able to identify the source of specific decompositions products. Strong signals emanating from this carbon site on SEI residing on the anode surface demonstrated that the formation of CO_2 is not a unique breakdown product for the carbonyl groups. The results suggested the formation of a new set of electrolyte breakdown products, including acetals, orthocarbonates, orthoesters, and fluorine-containing alkoxy compounds (from reaction with the breakdown products of the $LiPF_6$ salt used in the electrolyte).

Electrolytes

Carbonate-containing liquid electrolytes are primarily chosen for their ability to dissolve lithium salts and their relatively low viscosity (which facilitates Li-ion diffusion between electrodes). Their flammability has in part led to interest in the use of room-temperature ionic liquids (ILs) as replacements. ILs can potentially operate in a higher voltage window relative to carbonates and also have the added benefit of being more thermally stable and having low vapor pressure. The main drawback of this class of compounds is a high viscosity. Additionally, carbonates may have to be introduced at certain voltages to form a suitable SEI for operation.

Fernicola et al. have synthesized and characterized *N*-butyl-*N*-ethylpiperidinium *N*, *N*-bis(trifluoromethane)sulfonimide (PP_{24}TFSI) IL [78]. A lithium bis(trifluoromethane)sulfonimide (Li-TFSI) salt was dissolved in the ionic liquid and was introduced to a polymer matrix. The dynamics of ^{1}H, ^{19}F, and ^{7}Li, as well as the transport properties of the mobile species in neat PP_{24}TFSI, Li-TFSI mixed with PP_{24}TFSI, and Li-TFSI-PP_{24}TFSI mixed in a polymer membrane were studied using NMR. Spin-lattice relaxation measurements as a function of temperature probed short-range motion on small time scales, while diffusion measurements probed long-range motion on a larger time scale. Comparison of the Arrhenius plots of the ^{1}H spin-lattice relaxation time across the sample series indicates an increase in restrictive motion from the neat ionic liquid to the salt mixture and to the mixture in the membrane. This trend is due to the inherent high viscosity of the ionic liquid and IL-membrane interactions. The T_1 minimums shift toward higher temperatures for the Li-TFSI and membrane samples, reaffirming the dramatic change in the dynamics of the systems. Similar trends were observed for ^{19}F and ^{7}Li across the sample series (the ^{19}F relaxation times were assumed to be above the T_1 minimum). The dynamics of the membrane were well described with single relaxation processes, indicating that all the IL was subject to interactions with the membrane. Diffusion data were consistent with the previous spin-lattice relaxation measurements. The diffusivity of the PP_{24} and TFSI ions was comparable in all samples. The Li diffusivity in the salt mixture was lower than both the PP_{24} and TFSI ions, owing to a relatively stronger electrostatic interaction between Li^+ and TFSI. However, the TFSI/Li self-diffusion coefficient ratios decreased from 2 to approximately 1, going from the salt mixture to the membrane, resulting in a much higher Li transference number in the matrix. This increase in Li transference number is highly desirable in battery applications.

NMR spectroscopy has also been used extensively to study various electrolytes. Challenges and strategies for successful electrolytes have been previously described in literature [1, 17, 79, 80]. In a few instances, these materials consist of solid polymers [17], which are typically of high molecular weight or cross-linked, gel electrolytes, or composite materials containing additives [81–85]. In recent years, nanoscopic additives have gained recent attention [86–88]. A significant contribution of NMR spectroscopy to the polymer electrolyte community arose from line-width measurements of various resonances as a function of temperature [89–92]. The results showed a correlation between ionic conduction and the degree of mobility of the polymer chains. Moreover, the studies demonstrated that ion conduction predominately occurs in the amorphous phase of the electrolyte at temperatures above the glass transition temperature T_g. The proposition of this "segmental motion-assisted transport mechanism" in the amorphous phase has become broadly accepted among the community.

Lithium transport has also been studied in mechanically [93] or magnetically [94] oriented polymers. An analysis of angular dependent line widths in the static NMR spectra of these materials has shown that the induced reordering of the polymers along the stretching direction affects the ordering of the lithium cations. Additionally, lithium diffusion coefficients were shown to increase along the stretched axis [95]. As mentioned previously, lithium transport generally occurs more readily in amorphous polymer phases. This amorphous transport mechanism, mediated through the motion of the polymer chains, is in sharp contrast to the early model of ion transport in polyethylene oxide which asserted that the cations moved through ordered helical channels throughout the material [96]. The faster diffusion in the oriented samples is not unprecedented however. Gadjourova et al. have shown that for certain compositions of polyethylene electrolyte, the more crystalline phase has a higher conductivity above T_g [97]. Though amorphous phase transport is the dominant transport mechanism in randomly oriented polymer chains, the mechanism of conduction in well-ordered, aligned helices may depend on contributions from many other factors, warranting further basic research and a reexamination of previous models.

Cathode Materials

Olivines and their solid solutions have generated considerable interest as replacement cathode materials [98–100]. In particular, $LiFePO_4$ shows great promise as a low-cost, non-toxic alternative to ubiquitous $LiCoO_2$. Aside from facilitating lithium conduction by "opening up" the framework, substituting a XO_4^{n-} anion for an oxygen allows the redox couple to be tuned using the so-called "inductive effect." [101] Tucker et al. have used ^{7}Li MAS NMR at low field to study the nature of the hyperfine fields in pristine polycrystalline $LiMPO_4$ (M = Fe, Co, Ni, Mn) [102]. The temperature dependence of the inverse magnetic susceptibility was determined for all samples using a SQUID magnetometer. All lithium metal phosphates displayed Curie-Weiss behavior. The susceptibility data was fitted using a linear equation to extract Weiss constants and effective magnetic moments. Temperature-dependent NMR measurements revealed isotropic shifts closely following the Curie-Weiss trend of bulk susceptibility for all compositions, confirming the hyperfine coupling between the ^{7}Li nuclei and the paramagnetic transition metals in these materials. Due to the lowering of symmetry of the MO_6 octahedra from O_h to C_s in these materials [100], the d-orbitals of the transition metals split to two sets – one set of three orbitals at relatively higher energy and another for the remaining two. Assuming a high-spin electron configuration for the materials, the shift contribution for all unpaired electrons in both sets of orbitals was estimated. At 37°C, the unpaired electrons in the orbitals at a higher relative energy contribute approximately −24 to −28 ppm to the total shift. The unpaired electrons in the set containing the remaining two orbitals contribute 70–79 ppm to the shift. The hyperfine coupling constants for each sample were estimated and showed considerable variation with the type of transition metal. Additionally, the constants were of much smaller magnitude than those measured in lithium manganese oxide spinels [103]. These observations were explained on the basis of the longer *M*–O bonds in Li*M*PO_4 and the polarization of the oxygen electrons to the phosphate structural unit, both of which reduce the degree of covalency of the *M*–O bond. Wilcke et al. have reported variable temperature ^{31}P MAS NMR of the Li*M*PO_4 series along with ^{7}Li MAS NMR data of $LiFe_{1-x}Mn_xPO_4$, where $0 \leq x \leq 1$, using fast sample spinning rates and low static magnetic fields [104]. The ^{31}P shifts were much larger than the corresponding ^{7}Li shifts, demonstrating an increased sensitivity to the P-O-*M* bond covalency which may ultimately be correlated with the degree of Jahn-Teller distortions present in the material, electrochemical performance, and mechanisms that lead to failure.

Leifer et al. [105]. used ^{7}Li MAS NMR to study the structure of lithiated silver vanadium oxide, $Li_xAg_2V_4O_{11}$, where $x = 0.72$, 2.13, and 5.59. This compound is used in biomedical applications as a primary battery, particularly as the power source for implantable cardiac defibrillators (ICDs). Silver vanadium oxide is a vanadium bronze with semiconducting properties. It has been used successfully as a cathode material in the battery of ICDs due to its high rate capability and its high theoretical capacity (315 mAh/g) to 2 V. Electrochemical and structural studies of the average structure were performed by various authors who concluded that the systems undergoes a multistep reduction mechanism and forms silver metal in the early stage of the overall reaction [106–108].

At least three isotropic resonances were identified, two at approximately −1.5 ppm and −4.4 ppm for all samples studied and a third resonance present at −14 ppm in $Li_{0.72}Ag_2V_4O_{11}$ and $Li_{2.13}Ag_2V_4O_{11}$, which shifts to −54 ppm for $Li_{5.59}Ag_2V_4O_{11}$. Many of the spectral features observed were consistent with previously reported Li spectra of lithiated vanadium oxide materials [109–112]. The three plateaus in the electrochemical profile at 3.2, 2.7, and 2.4 V have been attributed to Ag reduction (3.2 V) and V reduction (2.7 V and 2.4 V) [107, 113]. The inserted Li ions reduce and displace the Ag^+ in the oxygen layers of $Li_xAg_2V_4O_{11}$ [114]. The peak at −1.5 ppm was assigned to Li in the solid electrolyte interphase and lithium salts LiF and Li_2CO_3, with the exception of Li_2O, which would have resonated at 2.8 ppm. The resonance at −4 ppm was assigned to Li in octahedral sites, some of which were occupied by Ag. The lack of shift upon lithiation of this resonance is consistent with the assignment, since the octahedral sites are further away from the vanadium ions and would therefore be least affected by changes in V oxidation state. The peak at −14 ppm was assigned to Li ions in tetrahedral sites near the V atoms as found in $\varepsilon - Li_xV_2O_5$ [115]. The broadening and concomitant shift of this resonance was attributed to the formation of V^{4+} or V^{3+} upon lithiation, whose presence is consistent with previous literature [109,

116, 117]. The profound broadening of the resonance is thus attributed to the paramagnetic vanadium ions [118] and shorter Li-V distances in the material that are a consequence of a collapse in the structure after the displacement of silver during discharge. The lithium shifts that occur at high lithiation levels are also consistent with the reduction of the silver prior to reduction of the vanadium. A line shape analysis of the vanadium spectrum from ^{51}V NMR showed that the crystallinity of the materials decreased steadily as they were lithiated. The $Li_{0.72}Ag_2V_4O_{11}$ had features that were close to the pristine material, implying a simple displacement of the Ag atoms with Li, while preserving the structural integrity and order. The spectral broadening and the loss of definition in the line-shape discontinuities were the most profound in $Li_{2.13}Ag_2V_4O_{11}$ and $Li_{5.59}Ag_2V_4O_{11}$.

XAS and XANES measurements at the Ag K-edge confirmed phase segregation of metallic silver from the SVO structure with an increase in lithium content. The XANES data, normalized to a per atom basis, of silver foil, SVO, and lithiated SVO displayed isosbestic points. This demonstrates that the SVO and Ag are related linearly by stoichiometry. In contrast, the normalized XANES data at the vanadium K-edge of SVO and $Li_{5.59}Ag_2V_4O_{11}$ did not display any isosbestic points, implying no linear stoichiometric relationship between vanadium and lithium between the two samples during lithiation. Shifts in the V pre-edge features of the XANES spectra of samples were consistent with the change in vanadium oxidation state from +5 to +3. Given the full-width half-maximum of the pre-edge peak in the samples, the authors postulated that vanadium could exist in multiple oxidation states, including V^{4+}.

Yamakawa et al. has recently implemented ^7Li MAS NMR to study conversion reactions of binary copper-II compounds (CuO, CuF_2, CuS) [119] and iron fluoride [120]. The results of the binary copper compounds will be summarized first. Prior to this work, two separate mechanisms were proposed for the conversion reaction of CuS by Chung et al. [121]. and Debart et al. [122]. The former reaction mechanism proceeded via a two step process. The first is an insertion reaction to form Li_xCuS, which reacts with Li to produce Li_2S and $Cu_{1.96}S$. The second process involves the subsequent conversion of $Cu_{1.96}S$ to Li_2S and copper metal. Debart et al. proposed that Cu_2S is produced as an intermediate phase, before reacting to produce Cu metal and Li_2S. Possible discharge reaction mechanisms for the CuO system have been proposed by various authors, primarily through the analysis of X-ray diffraction data [123, 124].

Based largely on the Li NMR results of Yamakawa et al., in combination with diffraction and electrochemical analysis, the following reaction mechanisms were proposed for the binary copper compounds. In the CuF_2 system, a simple single-step conversion reaction of the form

$$CuF_2 + 2Li^+ + 2e^- \rightarrow Cu + 2LiF, \tag{8.22}$$

takes place. In CuO, the following reaction could be written based on the preliminary Li NMR studies:

$$CuO + xLi \rightarrow Li_xCuO \tag{8.23}$$

The CuS system undergoes more complicated processes. In the first step, the reactions

$$xLi + \mathrm{CuS} \rightarrow \mathrm{Li}_x\mathrm{CuS} \qquad (0 < x \leq 0.16) \tag{8.24}$$

$$\mathrm{Li}_x\mathrm{CuS} + (y - x)\mathrm{Li} \rightarrow \mathrm{Li}_y\mathrm{CuS} \qquad (y \sim 1) \tag{8.25}$$

The second process involves the decomposition of the Li_yCuS phase via

$$(1 - z)\left(\mathrm{Li}_y\mathrm{CuS} + \mathrm{Li}^+ + e^- \rightarrow \frac{1+y}{2}\mathrm{Li}_2\mathrm{S} + \mathrm{Cu}\right)(y \sim 1) \tag{8.26}$$

and the concurrent disproportionation reaction

$$z\left(\mathrm{Li}_y\mathrm{CuS} \rightarrow 0.5\mathrm{Cu}_{1.96}\mathrm{S} + 0.02\mathrm{Cu} + \frac{y}{2}\mathrm{Li}_2\mathrm{S}\right) \tag{8.27}$$

Upon charging, assuming $z \sim 3$ and the Li content extracted from the electrochemical data, the reaction could be written as follows:

$$0.85\mathrm{Li}_2\mathrm{S} + 0.76\mathrm{Cu} + 0.15\mathrm{Cu}_{1.96}\mathrm{S} \rightarrow \mathrm{Li}_{y}, \mathrm{CuS} \rightarrow \mathrm{Li}_{0.35}\mathrm{CuS} + 1.35\mathrm{Li} \tag{8.28}$$

Electrochemical activity of iron-III fluoride (FeF_3) was first reported by Arai et al. [125]. The compound has theoretical capacity of approximately 200 mAh/g in the 4.5–2.5 V region and 400 mAh/g in the 2.5–1.5 V region. Li et al. [12] and Amatucci and co-workers [13, 14, 126] have since proposed reaction mechanisms for the higher and lower voltage windows:

$$\mathrm{FeF}_3 + \mathrm{Li} \rightarrow \mathrm{LiFeF}_3, \tag{8.29}$$

for the 4.5–2.5 V region and

$$\mathrm{LiFeF}_3 + 2\mathrm{Li} \rightarrow \mathrm{Fe} + 3\mathrm{LiF}, \tag{8.30}$$

for the 2.5–1.5 V region. In the higher voltage region lithium is inserted into FeF_3 in a two-phase reaction to form $Li_{0.5}FeF_3$ and then through a single-phase reaction to form $LiFeF_3$. Badway et al. have proposed that during the conversion reaction the iron metal and lithium fluoride react to form a, "rutile-like FeF_2" structure [14]. Doe et al. have performed first-principles calculations suggesting that intermediate phases could form during the electrochemical process [127].

The ^{7}Li MAS NMR spectra of lithiated FeF_3 samples, taken at cutoff voltages between 3.3 and 1.7 V, display two resonances at approximately 0 ppm and another between 104 and 172 ppm. The resonance at higher frequencies is relatively broader than the peak at 0 ppm. The larger shifts were attributed primarily to the Fermi contact interaction between the Li nuclear spins and the unpaired Fe electrons, which occur via the Li-F-$Fe^{3+/2+}$ bonds. The larger line width of these peaks is attributed to the through-bond and through-space interactions between the ^{7}Li nuclei and rapidly fluctuating electronic dipole moments of the iron, which efficiently relaxes the nuclei and consequently reduces T_1 and T_2. Previous $^{6,7}Li$ NMR spectroscopy performed by Grey and co-workers on lithium iron oxides showed that typical shifts for lithium were between 280 and 520 ppm [128, 129]. The smaller shifts observed in the iron fluoride systems was attributed to the presence of Fe^{3+} and Fe^{2+} coordinated to Li and the decreased covalency of the Fe-F bond relative to the Fe-O bond.

The samples arrested at 3.3 and 3.0 V both display a peak at 172 ppm, whose intensity increases with lithium content. The constant shift is indicative of an average Fe oxidation state around the Li coordination environment that remains unchanged. These results are consistent with the proposed two-phase region involving a Li_yFeF_3 ($y \sim 0.5$) phase and an unlithiated FeF_3 phase. The peak at 172 ppm gradually shifts to lower frequency between Li = 0.5 and 0.75, corresponding to voltages of 3.0 and 2.5 V respectively. This observation is in agreement with a solid solution reaction in which Li is inserted to the $Li_{0.5}FeF_3$ phase and a reduction from Fe^{3+} to Fe^{2+} occurs. The resonance does not increase in intensity but noticeably broadens into the baseline at Li = 1.0 (1.7 V), which is ascribed to very fast nuclear relaxation due to the presence of high-spin Fe^{2+} ions.

The ^{7}Li MAS NMR spectra of samples with Li = 1.0–3.0 display a single resonance at 0 ppm with spinning sidebands spanning over 5,000 ppm. Upon discharge, the width of the signal manifold increases as does the broadening of each individual peak. These observations are consistent with the formation of iron nanoparticles. Due to the relative ease of the measurement, it was rationalized that the nano-sized iron metal was not ferromagnetic but superparamagnetic. Liao et al. have shown that particles less than or equal to 6.6 nm display superparamagnetic properties [130].

^{6}Li MAS NMR was performed on the same sample set and revealed a second resonance for the Li = 1.7 sample, at approximately 86 ppm. This peak was assigned to Li in an insertion phase with a higher Fe^{2+}/Fe^{3+} ratio. An additional resonance at 104 ppm was observed in the ^{6}Li spectrum of the Li = 1.0 sample. The relative intensity of the 86 ppm peak was smaller than the 104 ppm peak of the other samples presumably due to the onset of the conversion reaction. Upon charging, the ^{6}Li MAS NMR spectra of Li = 0.5–3.0 revealed resonances in similar positions as the discharged samples, i.e., peaks at 0 ppm and between 80 and 168 ppm, with a gradual shift from 80 to 168 ppm for the Li = 2.0–0.5 samples. The similarity in the peak positions of the charged and discharged samples imply that the Li ions return to very similar local environments and that Fe^{2+} and Fe^{3+} are both present at the end of the charge, having an average iron valence state of $Fe^{2.5+}$. Moreover, the total width of the sideband manifold for the 0 ppm resonance decreases, suggesting that the concentration of superparamagnetic Fe decreases upon charging.

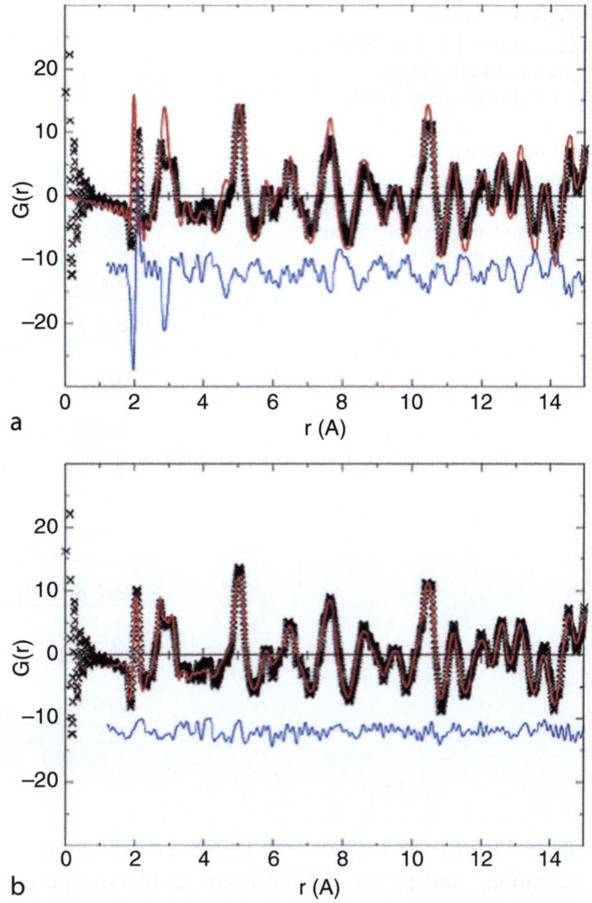

Fig. 8.8 Neutron PDF data of $^{7}LiNi_{0.5}Mn_{0.5}O_2$ that was fitted (**a**) using an ideal $LiCoO_2$ structural model where the Ni, Mn, and Li atoms are randomly distributed in the transition metal layers and (**b**) after a series of Reverse Monte Carlo (RMC) calculations which accounted for cation ordering. The raw data is shown in *black*, the fit in *red*, and the difference between the two in *blue* (From [131])

A study by Bréger et al. on $LiNi_{0.5}Mn_{0.5}O_2$, employing X-ray and neutron PDF analysis in conjunction with Li NMR serves as an ideal illustration of employing complementary techniques to obtain clear structural constraints and generate a detailed model of the material [131]. In order to determine the local environment of each transition metal independently, three samples were prepared for neutron diffraction measurements: $^{6}LiNi_{0.5}Mn_{0.5}O_2$, $^{7}LiNi_{0.5}Mn_{0.5}O_2$ and $^{7}Li^{ZERO}Ni_{0.5}Mn_{0.5}O_2$, corresponding to a Li-6-enriched, Li-7-enriched, and a Li-7-enriched sample containing appropriate amounts of ^{62}Ni to natural abundance nickel in order to produce null scattering from the element, referred to as $^{7}Li^{ZERO}$. The PDF data revealed distributions in the metal-oxygen bond lengths arising from local distortions which were not captured from the average $LiCoO_2$ structure. The PDF data also took into account the superstructure reflections, arising from the non-random distribution of transition metals in the layers, in the scattering data that could not be indexed (labeled) using the $LiCoO_2$ structural model in Rietveld refinements. NMR results and RMC calculations provided further evidence that the cations were non-randomly distributed in the transition metal layers (see Fig. 8.8). On average, Ni atoms are surrounded by

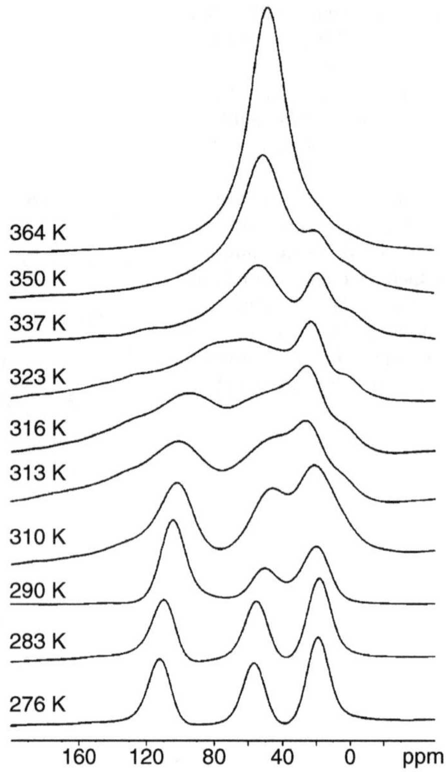

Fig. 8.9 Variable temperature ^{7}Li MAS NMR spectra of $Li_3V_2(PO_4)_3$ displaying the temperature dependence of the shift and the coalescence of the various lithium environments due to exchange, from [136]. The resonance around 0 ppm is due to a diamagnetic impurity

more Mn atoms in the first cation coordination shell (nearest transition metal neighbors) and by Li and Ni atoms in the second coordination shell (next nearest transition metal neighbors). Manganese atoms tend to be surrounded by Li and Ni in the first coordination shell and by other Mn atoms in the second coordination shell. The raw data can be compared to different transition metal–ordering schemes proposed from first-principles calculations, transmission electron microscopy, and previous NMR results [132–135]. The two-ordering schemes proposed were the so-called honeycomb and flower arrangements. While the numbers of metal-metal contacts extracted from the experimental data show that a completely random distribution of the transition metals is a poor structural model, it was difficult to distinguish between the flower and honeycomb ordering schemes.

Cahill et al. have used variable temperature ^{7}Li MAS NMR spectroscopy to study the lithium dynamics of monoclinic $Li_3V_2(PO_4)_3$ [136]. The three lithium sites at 103, 52, and 17 ppm (labeled Li3, Li2, and Li1 respectively) are clearly resolved and unambiguously assigned at a sample spinning speed of 25 kHz and a magnetic field strength of 7 T. The relative intensity of the peaks, including contributions from the respective side-band intensities was 1:1:1, as predicted from the crystal structure. One-dimensional ^{7}Li MAS NMR spectra taken from 276 to 364 K (see Fig. 8.9) display a linear shift dependence with respect to temperature indicative of Curie-Weiss behavior. The electronic insulating nature of the materials as well as the aforementioned Curie-Weiss behavior as a function of temperature confirmed that the origin of the lithium shifts is due to hyperfine interactions. Additionally, the resonances begin to coalesce as the temperature is increased. This implies the hopping of Li atoms between different sites in the crystal lattice. According to the data, Li1 and Li2 begin exchanging first at relatively lower temperatures. Based on the separation of the peaks before coalescence at 316 K, an estimation of approximately 6.5 kHz (corresponding to a correlation time τ_c of 154 μs) was determined as the lithium hopping rate between the two crystallographic sites. By 364 K, all three Li resonances have coalesced, implying that all lithium atoms are exchanging between the three sites.

Through an analysis of 2D Exchange spectra of the sample as a function of temperature, the authors were able to extract activation energies for lithium conduction between Li1-Li2, Li2-Li3, and Li1-Li3 sites. The activation energies extracted from the NMR measurements can be correlated with structural factors which could affect Li-ion mobility, such as the internuclear distance between the Li sites and bottlenecks in the conduction pathway. The activation energy of the hopping between two Li sites increased when the internuclear distance between Li sites increased and when the distance of the shortest Li-O contact decreased. The lithium ions in $Li_3V_2(PO_4)_3$ can be visualized as existing on sheets along the crystallographic ac plane. The implication of these NMR results is that conduction primarily occurs in one dimension, in a straight line across different Li sites using a series of oxygen atoms along the channel to facilitate the transport. These quantitative measurements are invaluable for rationalizing conduction mechanisms in these materials and understanding transport properties along at small length scales.

Anode Materials

Lithiated graphite has been one of the most widely investigated anode materials [137–139]. The first reported study of lithium intercalated into graphite was by Conard and Estrade [140]. The insertion results in changes in the electronic properties which can be monitored by the Knight shift of the intercalated ^{7}Li using NMR techniques. Increasing the performance of graphitic anodes in Li-ion batteries has been a long-standing research initiative. Partially oxidized or pyrolytic graphite, along with hard carbon, have been investigated to improve performance [141–145]. The chemical and structural modification offered some degree of control over the reversible and irreversible capacity changes in the cell or improving the cycling degradation by facilitating SEI formation.

Most Li NMR studies are performed ex-situ, whereby a material has been investigated at discrete points in the charge-discharge profile. This requires multiple batteries to be made and cycled at those specific points, before being sacrificed for the NMR study. Studies employing such a protocol are time-consuming and can be complicated by sample variability. Moreover, the products of short-lived processes cannot be easily identified or quantified. Recently, Li NMR spectroscopy has been applied in-situ under routine operating conditions. To date, these *in-situ* experiments have been under static conditions, i.e., no magic angle spinning. The first *in-situ* NMR experiment was performed by Gerald et al. [146]. An NMR toroid was used for both spectroscopy and imaging. These initial experiments struggled from poor signal-to-noise ratios, in part due to the incompatibility of a standard lithium ion battery with the toroidal design. Chevallier et al. and Letellier et al. circumvented some of the initial issues by implementing a so-called, "plastic lithium-ion battery" [147, 148] which introduced a flexibility to the battery and eliminated the need for external pressure to maintain contact between the various components of the cell. These initial *in-situ* studies were performed on modified carbonaceous materials designed to replace graphitic carbon as the anode.

Another replacement of graphitic carbon in lithium ion batteries is silicon, due to its large volumetric and gravimetric energy density – 8,322 mAh cm^{-3} (calculated from the original volume of Si) and 3,572 mAh g^{-1} respectively. The high capacitance induces large volumetric changes in the electrode (~300%) which results in issues of capacitance retention, particle fracture and battery design [149]. The Li-Si binary phase diagram contains four reported crystalline lithium silicides – $Li_{12}Si_7$, Li_7Si_3, $Li_{13}Si_4$, and $Li_{21}Si_5$ [150]. These complex materials lie on the border between intermetallic compounds and Zintl phases and contain orbitals delocalized over many lithium ions. Key et al. [149]. have collected ^{7}Li of these model compounds in order to correlate shifts with local environments in silicon based electrodes. The spectrum of $Li_{21}Si_5$ has very broad resonances at approximately 60–100 ppm, owing to the Knight shift. In contrast to $Li_{21}Si_5$, the other lithium silicide phases have single, broad resonances between 6 and 19 ppm, indicative of vastly different electronic properties of each compound. The authors hypothesized that the various lithium sites in the model compounds were not well-resolved due to a possible combination of factors – slow Li motion, residual ^{7}Li-^{7}Li dipolar coupling, and structural disorder. Nevertheless, a trend toward smaller shifts, implying a greater electron density around the Li atom and more shielded environments, is observed as the Li/Si ratio increases. The resonances with larger shifts in the semi-conductors were found in compounds with Si-Si bonds, whereas smaller shifts were observed in the samples with isolated Si atoms.

The same system was investigated by in-situ ^{7}Li NMR using a plastic battery to monitor real-time changes during a complete charge and discharge at a C/75 rate. The data set largely agrees with the ex situ measurements with slight variations in the shifts of the peaks which were ascribed to the relatively poor resolution of static NMR and small susceptibility effects from metal components in the battery. One striking difference between the in situ and ex situ data was the presence of a new resonance at approximately −10 ppm, at the bottom of the discharge profile (<30 mV), where diffraction measurements displayed evidence for crystalline $Li_{15}Si_4$. The intensity of the −10 ppm resonance decreased rapidly upon charging. A second battery that was fully discharged ex-situ was then monitored using *in-situ* ^{7}Li NMR. The −10 ppm resonance disappeared as the battery relaxed over a period of approximately 10 h.

The phenomenon was attributed to a reaction between the metastable $Li_{15}Si_4$ phase with the electrolyte through a "self-charge" mechanism. The reactivity of the amorphous lithium silicides and the self-discharge mechanism may lead to capacity loss or safety concerns. However, the authors noted that the use of carboxymethylcellulose (CMC) as a binder inhibited the discharge process significantly, implicating the reduction of side reactions with the electrolyte and, hence, capacity losses.

Future Directions

In order to meet the ever-increasing demand for energy and portable electronic devices, innovation in the battery community must be a priority. A prerequisite for innovation is a fundamental understanding of chemical reaction pathways and structure determination on a broad length scale. As a natural response to this challenge, there has been a surge of multidisciplinary collaboration. These efforts have been exceedingly successful in elucidating the atomistic details of a plethora of materials and providing a clear direction for future investigations. Breakthroughs are contingent upon a continued and open dialogue between scientists spanning a wide array of expertise. Electrochemical behavior is inextricably related to structure. Consequently, there will be an increasing need to accurately probe and quantify changes in chemical speciation, bond angles, bond lengths, and oxidation state. A particular challenge is to model disordered and nano-sized materials. Implementation and advancement of techniques sensitive to the local environment will be of utmost importance to place structural constraints in these classes of compounds. Such constraints will be invaluable to the computation community in their efforts to predict the phases and electrochemical properties of new materials. Investigation of various room-temperature ionic liquids, for replacement of commonly used carbonate electrolytes, as well as a long-term investment into novel material design and characterization will dominate the battery research community for many years to come. It is only through a concerted effort among materials engineers, analytical, and computational scientists that we are able address the issues of today and those of tomorrow.

Acknowledgments The authors thank past and present members of the solid-state NMR group at Hunter College for their contributions and gratefully acknowledge support from the US Department of Energy and the U.S. Office of Naval Research.

Bibliography

1. Goodenough J, Kim Y (2010) Challenges for rechargeable Li batteries. Chem Mater 22:587–603
2. Whittingham MS (1976) Electrical energy storage and intercalation chemistry. Science 192(4244):1126–1127

3. Dahn JR, von Sacken U, Juzkow MW, Al-Janaby H (1991) Rechargeable $LiNiO_2$/carbon cells. J Electrochem Soc 138:2207–2211
4. Delmas C, Peres J, Rougier A, Demourgues A, Weill F, Chadwick A, Broussely M, Perton F, Biensan P, Willmann P (1997) On the behavior of the Li_xNiO_2 system: an electrochemical and structural overview. J Power Sources 68(1):120–125
5. Saadoune I, Delmas C (1996) $LiNi_{1-y}Co_yO_2$ positive electrode materials: relationships between the structure, physical properties and electrochemical behaviour. J Mater Chem 6(2):193–199
6. Capitaine F, Gravereau P, Delmas C (1996) A new variety of $LiMnO_2$ with a layered structure. Solid State Ion 89(3–4):197–202
7. Rossen E, Jones C, Dahn J (1992) Structure and electrochemistry of $Li_xMn_yNi_{1-y}O_2$. Solid State Ion 57(3–4):311–318
8. Liu Z, Yu A, Lee J (1999) Synthesis and characterization of $LiNi_{1-x-y}Co_xMn_yO_2$ as the cathode materials of secondary lithium batteries. J Power Sources 82:416–419
9. Yoshio M, Noguchi H, Itoh J, Okada M, Mouri T (2000) Preparation and properties of $LiCo_yMn_xNi_{1-x-y}O_2$ as a cathode for lithium ion batteries. J Power Sources 90(2):176–181
10. Poizot P, Laruelle S, Grugeon S, Dupont L, Tarascon JM (2000) Nano-sized transition-metaloxides as negative-electrode materials for lithium-ion batteries. Nature 407(6803):496–499
11. Li H, Richter G, Maier J (2003) Reversible formation and decomposition of LiF clusters using transition metal fluorides as precursors and their application in rechargeable Li batteries. Adv Mater 15(9):736–739 (Weinheim, Germany)
12. Li H, Balaya P, Maier J (2004) Li-storage via heterogeneous reaction in selected binary metal fluorides and oxides. J Electrochem Soc 151(11):A1878–A1885
13. Badway F, Cosandey F, Pereira N, Amatucci G (2003) Carbon metal fluoride nanocomposites – high-capacity reversible metal fluoride conversion materials as rechargeable positive electrodes for Li batteries. J Electrochem Soc 150(10):A1318–A1327
14. Badway F, Pereira N, Cosandey F, Amatucci G (2003) Carbon-metal fluoride nanocomposites – structure and electrochemistry of FeF_3: C. J Electrochem Soc 150(9): A1209–A1218
15. Badway F, Mansour A, Pereira N, Al-Sharab J, Cosandey F, Plitz I, Amatucci G (2007) Structure and electrochemistry of copper fluoride nanocomposites utilizing mixed conducting matrices. Chem Mater 19(17):4129–4141
16. Armand M (1994) The history of polymer electrolytes. Solid State Ion 69:309–319
17. Armand M (1986) Polymer electrolytes. Annu Rev Mater Sci 16:245–261
18. Fenton DE, Parker JM, Wright PV (1973) Complexes of alkali-metal ions with poly(ethylene oxide). Polymer 14(11):589
19. Tarascon J-M, Armand M (2001) Issues and challenges facing rechargeable lithium batteries. Nature 414:359–367
20. Myung ST, Komaba S, Hirosaki N, Yashiro H, Kumagai N (2004) Emulsion drying synthesis of olivine $LiFePO_4$/C composite and its electrochemical properties as lithium intercalation material. Electrochim Acta 49(24):4213–4222
21. Rietveld HM (1969) A profile refinement method for nuclear and magnetic structures. J Appl Crystallogr 2:65–71
22. Proffen T, Billinge SJL (1999) PDFFIT, a program for full profile structural refinement of the atomic pair distribution function. J Appl Crystallogr 32:572–575
23. McGreevy RL, Pusztai L (1988) Reverse monte carlo simulation: a new technique for the determination of disordered structures. Mol Simul 1:359–367
24. Dedryvere R, Laruelle S, Grugeon S, Poizot P, Gonbeau D, Tarascon JM (2004) Contribution of X-ray photoelectron spectroscopy to the study of the electrochemical reactivity of CoO toward lithium. Chem Mater 16(6):1056–1061
25. Alamgir FM, Petersburg CF, Daniel RC, Jaye C, Fischer DA (2009) Soft X-ray characterization technique for Li batteries under operating conditions. J Synchrotron Radiat 16:610–615

26. Stern EA (1974) Theory of extended X-Ray-absorption fine-structure. Phys Rev B 10(8):3027–3037
27. Sayers DE, Stern EA, Lytle FW (1971) New technique for investigating noncrystalline structures – Fourier analysis of extended X-ray – absorption fine structure. Phys Rev Lett 27(18):1204–1207
28. Tsai YW, Hwang BJ, Ceder G, Sheu HS, Liu DG, Lee JF (2005) In-situ X-ray absorption spectroscopic study on variation of electronic transitions and local structure of $LiNi_{1/3}Co_{1/3}Mn_{1/3}O_2$ cathode material during electrochemical cycling. Chem Mater 17(12):3191–3199
29. Yoon W, Kim N, Yang X, McBreen J, Grey C (2003) Li-6 MAS NMR and in situ X-ray studies of lithium nickel manganese oxides. J Power Sources 119:649–653
30. Grey CP, Lee YJ (2003) Lithium MAS NMR studies of cathode materials for lithium-ion batteries. Solid State Sci 5(6):883–894
31. Carlier D, Menetrier M, Grey C, Delmas C, Ceder G (2003) Understanding the NMR shifts in paramagnetic transition metal oxides using density functional theory calculations. Phys Rev B 67 174103:1–14
32. Letellier M, Chevallier F, Beguin F, Frackowiak E, Rouzaud J (2004) The first in situ Li-7 NMR study of the reversible lithium insertion mechanism in disorganised carbons. J Phys Chem Solids 65(2–3):245–251
33. Levitt MH (2001) Spin dynamics: basics of nuclear magnetic resonance. Wiley, New York
34. Slichter CP (1990) Principles of nuclear magnetic resonance. Springer, Berlin
35. Stejskal E, Tanner J (1965) Spin diffusion measurements: spin echoes in the presence of a time-dependent field gradient. J Chem Phys 42:288–292
36. Kimmich R, Unrath W, Schnur G, Rommel E (1991) NMR measurement of small self-diffusion coefficients in the fringe-field of superconducting magnets. J Magn Reson 91(1):136–140
37. Demco D, Johansson A, Tegenfeldt J (1994) Constant-relaxation methods for diffusion measurements in the fringe-field of superconducting magnets. J Magn Reson Ser A 110(2):183–193
38. Gorecki W, Jeannin M, Belorizky E, Roux C, Armand M (1995) Physical properties of solid polymer electrolyte PEO(LiTFSI) complexes. J Phys Cond Matter 7(34):6823–6832
39. Johansson A, Gogoll A, Tegenfeldt J (1996) Diffusion and ionic conductivity in $Li(CF_3SO_3)$ PEG(10) and $LiN(CF_3SO_2)(2)$PEG(10). Polymer 37(8):1387–1393
40. Hayamizu K, Aihara Y, Price W (2000) Correlating the NMR self-diffusion and relaxation measurements with ionic conductivity in polymer electrolytes composed of cross-linked poly (ethylene oxide-propylene oxide) doped with $LiN(SO_2CF_3)(2)$. J Chem Phys 113(11):4785–4793
41. Hayamizu K, Aihara Y, Price W (2001) NMR and ion conductivity studies on cross-linked poly(ethyleneoxide-propyleneoxide) and branched polyether doped with $LiN(SO_2CF_3)(2)$. Electrochim Acta 46(10–11):1475–1485
42. Adebahr J, Forsyth M, Gavelin P, Jacobsson P, Oradd G (2002) Ion and solvent dynamics in gel electrolytes based on ethylene oxide grafted acrylate polymers. J Phys Chem B 106(47):12119–12123
43. Gorecki W, Roux C, Clemancey M, Armand M, Belorizky E (2002) NMR and conductivity study of polymer electrolytes in the imide family: P(EO)/Li[N(SO_2CnF2n + 1)(SO_2CmF2m + 1)]. Chemphyschem 3(7):620–625
44. Croce F, Appetecchi G, Persi L, Scrosati B (1998) Nanocomposite polymer electrolytes for lithium batteries. Nature 394:456–458
45. Zhou F, MacFarlane D, Forsyth M (2003) Boroxine ring compounds as dissociation enhancers in gel polyelectrolytes. Electrochim Acta 48(12):1749–1758
46. Kalita M, Bukat M, Ciosek M, Siekierski M, Chung S, Rodriguez T, Greenbaum S, Kovarsky R, Golodnitsky D, Peled E, Zane D, Scrosati B, Wieczorek W (2005) Effect of calixpyrrole in PEO-$LiBF_4$ polymer electrolytes. Electrochimia Acta 50(19):3942–3948

47. Koudriachova MV, Harrison NM, de Leeuw SW (2002) Density-functional simulations of lithium intercalation in rutile. Phys Rev B 65 235423:1–12
48. Kang K, Ceder G (2006) Factors that affect Li mobility in layered lithium transition metal oxides. Phys Rev B 74 094105:1–7
49. Tibbetts K, Miranda CR, Meng YS, Ceder G (2007) An ab initio study of lithium diffusion in titanium disulfide nanotubes. Chem Mater 19(22):5302–5308
50. Aydinol MK, Kohan AF, Ceder G, Cho K, Joannopoulos J (1997) Ab initio study of lithium intercalation in metal oxides and metal dichalcogenides. Phys Rev B 56(3): 1354–1365
51. Courtney IA, Tse JS, Mao O, Hafner J, Dahn JR (1998) Ab initio calculation of the lithium-tin voltage profile. Phys Rev B 58(23):15583–15588
52. Meng YS, Wu YW, Hwang BJ, Li Y, Ceder G (2004) Combining ab initio computation with experiments for designing new electrode materials for advanced lithium batteries: $LiNi_{1/3}Fe_{1/6}Co_{1/6}Mn_{1/3}O_2$. J Electrochem Soc 151(8):A1134–A1140
53. Arroyo-DeDompablo ME, Van der Ven A, Ceder G (2002) First-principles calculations of lithium ordering and phase stability on Li_xNiO_2. Phys Rev B 66 064112:1–9
54. Wolverton C, Zunger A (1998) Prediction of Li intercalation and battery voltages in layered versus cubic Li_xCoO_2. J Electrochem Soc 145(7):2424–2431
55. Hwang B, Tsai Y, Carlier D, Ceder G (2003) A combined computational/experimental study on $LiNi_{1/3}Co_{1/3}Mn_{1/3}O_2$. Chem Mater 15(19):3676–3682
56. Kganyago KR, Ngoepe PE, Catlow CRA (2003) Ab initio calculation of the voltage profile for LiC_6. Solid State Ion 159(1–2):21–23
57. Launay M, Boucher F, Gressier P, Ouvrard G (2003) A DFT study of lithium battery materials: application to the β-VOXO4 systems (X = P, As, S). J Solid State Chem 176(2):556–566
58. Zhou F, Cococcioni M, Marianetti CA, Morgan D, Ceder G (2004) First-principles prediction of redox potentials in transition-metal compounds with LDA + U. Phys Rev B 70 235121:1–8
59. Reed J, Ceder G (2002) Charge, potential, and phase stability of layered Li(Ni0.5Mn0.5)O-2. Electrochem State Lett 5(7):A145–A148
60. Arroyo-DeDompablo ME, Ceder G (2003) First-principles calculations on Li_xNiO_2: phase stability and monoclinic distortion. J Power Sources 119121:654–657
61. Carlier D, Van der Ven A, Delmas C, Ceder G (2003) First-principles investigation of phase stability in the O-2-$LiCoO_2$ system. Chem Mater 15(13):2651–2660
62. Perdew JP, Burke K, Ernzerhof M (1996) Generalized gradient approximation made simple. Phys Rev Lett 77(18):3865–3868
63. Blochl PE (1994) Projector augmented-wave method. Phys Rev B 50(24):17953–17979
64. Kresse G, Joubert J (1999) From ultrasoft pseudopotentials to the projector augmented-wave method. Phys Rev B 59(3):1758–1775
65. Kresse G, Furthmuller J (1996) Efficient iterative schemes for ab initio total-energy calculations using a plane-wave basis set. Phys Rev B 54(16):11169–11186
66. Bar-Tow D, Peled E, Burstein L (1999) A study of highly oriented pyrolytic graphite as a model for the graphite anode in Li-ion batteries. J Electrochem Soc 146(3):824–832
67. Peled E, Tow DB, Merson A, Gladkich A, Burstein L, Golodnitsky D (2001) Composition, depth profiles and lateral distribution of materials in the SEI built on HOPG-TOF SIMS and XPS studies. J Power Sources 97–98:52–57
68. Lu M, Cheng H, Yang Y (2008) A comparison of solid electrolyte interphase (SEI) on the artificial graphite anode of the aged and cycled commercial lithium ion cells. Electrochim Acta 53(9):3539–3546
69. Peled E, Golodnitsky D, Ulus A, Yufit V (2004) Effect of carbon substrate on SEI composition and morphology. Electrochim Acta 50(2–3):391–395
70. Ota H, Kominato A, Chun WJ, Yasukawa E, Kasuya S (2003) Effect of cyclic phosphate additive in non-flammable electrolyte. J Power Sources 119:393–398

71. Andersson AM, Edstrom K (2001) Chemical composition and morphology of the elevated temperature SEI on graphite. J Electrochem Soc 148(10):A1100–A1109
72. Andersson AM, Edstrom K, Rao N, Wendsjo A (1999) Temperature dependence of the passivation layer on graphite. J Power Sources 82:286–290
73. Meyer BM, Leifer N, Sakamoto S, Greenbaum SG, Grey CP (2005) High field multinuclear NMR investigation of the SEI layer in lithium rechargeable batteries. Electrochem Solid State Lett 8(3):A145–A148
74. Dupre N, Martin JF, Degryse J, Fernandez V, Soudan P, Guyomard D (2010) Aging of the $LiFePO_4$ positive electrode interface in electrolyte. J Power Sources 195(21):7415–7425
75. Dupre N, Martin J, Oliveri J, Soudan P, Guyomard D, Yamada A, Kanno R (2009) Aging of the $LiNi_{1/2}Mn_{1/2}O_2$ positive electrode interface in electrolyte. J Electrochem Soc 156(5):C180–C185
76. Dupre N, Martin JF, Guyomard D, Yamada A, Kanno R (2008) Detection of surface layers using Li-7 MAS NMR. J Mater Chem 18(36):4266–4273
77. Leifer N, Smart MC, Prakash GKS, Gonzalez L, Sanchez L, Smith KA, Bhalla P, Grey CP, Greenbaum SG (2011) 13C solid state NMR suggests unusual breakdown products in SEI formation on lithium ion electrodes. J Electrochem Soc 158(5):A471–A480
78. Fernicola A, Weise F, Greenbaum S, Kagimoto J, Scrosati B, Soleto A (2009) Lithium-ion-conducting electrolytes: from an ionic liquid to the polymer membrane. J Electrochem Soc 156(7):A514–A520
79. Scrosati B (1995) Battery technology – challenge of portable power. Nature 373:557–558
80. Gray FM (1991) Solid polymer electrolytes: fundamentals and electrochemical applications. VCH, New York
81. Kalita M, Bukat M, Ciosek M, Siekierski M, Chung SH, Rodriguez T, Greenbaum SG, Kovarsky R, Golodnitsky D, Peled E, Zane D, Scrosati B, Wieczorek W (2005) Effect of calixpyrrole in PEO-$LiBF_4$ polymer electrolytes. Electrochim Acta 50(19):3942–3948
82. Periasamy P, Tatsumi K, Shikano M, Fujieda T, Saito Y, Sakai T, Mizuhata M, Kajinami A, Deki S (2000) Studies on PVdF-based gel polymer electrolytes. J Power Sources 88(2):269–273
83. Pawlowska A, Zukowska G, Kalita M, Solgala A, Parzuchowski P, Siekierski M (2007) The effect of receptor-polymer matrix compatibility on properties of PEO-based polymer electrolytes containing a supramolecular additive. Part 1. Studies on phenomenon of compatibility. J Power Sources 173(2):755–764
84. Bloise A, Donoso J, Magon C, Rosario A, Pereira E (2003) NMR and conductivity study of PEO-based composite polymer electrolytes. Electrochim Acta 48(14–16):2239–2246
85. Masuda Y, Seki M, Nakayama M, Wakihara M, Mita H (2006) Study on ionic conductivity of polymer electrolyte plasticized with PEG-aluminate ester for rechargeable lithium ion battery. Solid State Ion 117(9–10):843–846
86. Dai Y, Wang Y, Greenbaum S, Bajue S, Golodnitsky D, Ardel G, Strauss E, Peled E (1998) Electrical, thermal and NMR investigation of composite solid electrolytes based on PEO, LiI and high surface area inorganic oxides. Electrochim Acta 43(10–11):1557–1561
87. Best A, Adebahr J, Jacobsson P, MacFarlane D, Forsyth M (2001) Microscopic interactions in nanocomposite electrolytes. Macromolecules 34(13):4549–4555
88. Adebahr J, Best A, Byrne N, Jacobsson P, MacFarlane D, Forsyth M (2003) Ion transport in polymer electrolytes containing nanoparticulate TiO_2: The influence of polymer morphology. Phys Chem Chem Phys 5:720–725
89. Berthier C, Gorecki W, Minier M, Armand M, Chabagno J, Rigaud P (1983) Microscopic investigation of ionic conductivity in alkali metal salts-poly(ethylene oxide) adducts. Solid State Ion 11(1):91–95
90. Chung S, Jeffrey K, Stevens J (1991) A Li-7 nuclear magnetic resonance study of $LiCF_3SO_3$ complexed in poly(propylene-glycol). J Chem Phys 94(3):1803–1811

91. Donoso J, Bonagamba T, Panepucci H, Oliveira L, Gorecki W, Berthier C, Armand M (1993) Nuclear magnetic relaxation study of poly(ethylene oxide)-lithium salt based electrolytes. J Chem Phys 98(12):10026–10036
92. Johansson A, Tegenfeldt J (1996) Segmental mobility in complexes of $Pb(CF_3SO_3)(2)$ and poly(ethylene oxide) studied by NMR spectroscopy. J Chem Phys 104(13): 5317–5325
93. Chung S, Wang Y, Greenbaum S, Golodnitsky D, Peled E (1999) Uniaxial stress effects in poly(ethylene oxide)-LiI polymer electrolyte film – a Li-7 nuclear magnetic resonance study. Electrochem Solid State Lett 2(11):553–555
94. Golodnitsky D, Livshits E, Kovarsky R, Peled E, Chung S, Suarez S, Greenbaum S (2004) New generation of ordered polymer electrolytes for lithium batteries. Electrochem Solid State Lett 7(11):A412–A415
95. Golodnitsky D, Livshits E, Ulus A, Barkay Z, Lapides I, Peled E, Chung S, Greenbaum S (2001) Fast ion transport phenomena in oriented semicrystalline LiI-P(EO)n-based polymer electrolytes. J Phys Chem A 105(44):10098–10106
96. Armand MB, Chabagno JM, Duclot MJ (1979) Fast ion transport in solids. Elsevier, New York
97. Gadjourova Z, Andreev Y, Tunstall D, Bruce P (2001) Ionic conductivity in crystalline polymer electrolytes. Nature 412:520–523
98. Padhi A, Nanjundaswamy K, Goodenough J (1997) Phospho-olivines as positive-electrode materials for rechargeable lithium batteries. J Electrochem Soc 144(4):1188–1194
99. Morgan D, Van der Ven A, Ceder G (2004) Li conductivity in Li_xMPO_4 (M=Mn, Fe, Co, Ni) olivine materials. Electrochem Solid State Lett 7(2):A30–A32
100. Yamada A, Chung S, Hinokuma K (2001) Optimized $LiFePO_4$ for lithium battery cathodes. J Electrochem Soc 148(3):A224–A229
101. Padhi A, Nanjundaswamy K, Masquelier C, Okada S, Goodenough J (1997) Effect of structure on the Fe^{3+}/Fe^{2+} redox couple in iron phosphates. J Electrochem Soc 144(5):1609–1613
102. Tucker M, Doeff M, Richardson T, Finones R, Cairns E, Reimer J (2002) Hyperfine fields at the Li site in LiFePO4-type olivine materials for lithium rechargeable batteries: a Li-7 MAS NMR and SQUID study. J Am Chem Soc 124(15):3832–3833
103. Gee B, Horne CR, Cairns EJ, Reimer JA (1998) Supertransferred hyperfine fields at Li-7: Variable temperature Li-7 NMR studies of $LiMn_2O_4$-based spinels. J Phys Chem B 102(50): 10142–10149
104. Wilcke SL, Lee YJ, Cairns EJ, Reimer JA (2007) Covalency measurements via NMR in lithium metal phosphates. Appl Magn Reson 32(4):547–563
105. Leifer N, Colon A, Martocci K, Greenbaum S, Alamgir F, Reddy T, Gleason N, Leising R, Takeuchi E (2007) Nuclear magnetic resonance and X-ray absorption spectroscopic studies of lithium insertion in silver vanadium oxide cathodes. J Electrochem Soc 154(6): A500–A506
106. Crespi A, Skarstad P, Zandbergen H (1995) Characterization of silver vanadium oxide cathode material by high-resolution electron microscopy. J Power Sources 54(1):68–71
107. Takeuchi K, Marschilok A, Davis S, Leising R, Takeuchi E (2001) Silver vanadium oxides and related battery applications. Coord Chem Rev 219–221:283–310
108. Ramasamy R, Feger C, Strange T, Popov B (2006) Discharge characteristics of silver vanadium oxide cathodes. J Appl Electrochem 36(4):487–497
109. Vijayakumar M, Selvasekarapandian S, Nakamura K, Kanashiro T, Kesavamoorthy R (2004) Li-7 MAS-NMR and vibrational spectroscopic investigations of $Li_xV_2O_5$ (x = 10, 12 and 14). Solid State Ion 167(1–2):41–47
110. Holland G, Buttry D, Yarger J (2002) Li-7 NMR studies of electrochemically lithiated V_2O_5 xerogels. Chem Mater 14(9):3875–3881

111. Holland G, Yarger J, Buttry D, Huguenin F, Torresi R (2003) Solid-state NMR study of ion-exchange processes in V_2O_5 xerogel, polyaniline/V_2O_5, and sulfonated polyaniline/V_2O_5 nanocomposites. J Electrochem Soc 150(12):A1718–A1722
112. Nakamura K, Nishioka D, Michihiro Y, Vijayakumar M, Selvasekarapandian S, Kanashiro T (2006) Li-7 and V-51 NMR study on Li+ ionic diffusion in lithium intercalated $Li_xV_2O_5$. Solid State Ion 177(1–2):129–135
113. West K, Crespi A (1995) Lithium insertion into silver vanadium-oxide, $Ag_2V_4O_{11}$. J Power Sources 54(2):334–337
114. Rozier P, Galy J (1997) $Ag_{1.2}V_3O_8$ crystal structure: Relationship with $Ag_2V_4O_{11-y}$ interpretation of physical properties. J Solid State Chem 134(2):294–301
115. Rozier P, Savariault J, Galy J, Marichal C, Hirschinger J, Granger P (1996) Epsilon-$Li_xV_2O_5$ bronzes ($0.33 \leq x \leq 0.64$) a joint study by X-ray powder diffraction and 6Li, 7Li MAS NMR. Eur J Solid State Inorg Chem 33(1):1–13
116. Kuwabara K, Itoh M, Sugiyama K (1986) Ionic-electronic mixed conduction in $Li_xV_2O_5$. Solid State Ion 20(2):135–139
117. Garcia-Alvarado F, Tarascon J (1994) Lithium intercalation in $Ag_2V_4O_{11}$. Solid State Ion 73(3–4):247–254
118. Stallworth P, Kostov S, denBoer M, Greenbaum S, Lampe-Onnerud C (1998) X-ray absorption and magnetic resonance spectroscopic studies of $Li_xV_6O_{13}$. J Appl Phys 83(3):1247–1255
119. Yamakawa N, Jiang M, Grey CP (2009) Investigation of the conversion reaction mechanisms for binary copper(II) compounds by solid-state NMR spectroscopy and X-ray diffraction. Chem Mater 21(14):3162–3176
120. Yamakawa N, Jiang M, Key B, Grey CP (2009) Identifying the local structures formed during lithiation of the conversion material, iron fluoride, in a Li ion battery: a solid-state NMR, X-ray diffraction, and pair distribution function analysis study. J Am Chem Soc 131(30):10525–10536
121. Chung JS, Sohn HJ (2002) Electrochemical behaviors of CuS as a cathode material for lithium secondary batteries. J Power Sources 108(1–2):226–231
122. Debart A, Dupont L, Patrice R, Tarascon JM (2006) Reactivity of transition metal (Co, Ni, Cu) sulphides versus lithium: the intriguing case of the copper sulphide. Solid State Sci 8(6):640–651
123. Ikeda H, Narukawa S (1983) Behavior of various cathode materials for non-aqueous lithium cells. J Power Sources 9(3–4):329–334
124. Grugeon S, Laruelle S, Herrera-Urbina R, Dupont L, Poizot P, Tarascon J (2001) Particle size effects on the electrochemical performance of copper oxides toward lithium. J Electrochem Soc 148(4):A285–A292
125. Arai H, Okada S, Sakurai Y, Yamaki J (1997) Cathode performance and voltage estimation of metal trihalides. J Power Sources 68(2):716–719
126. Cosandey F, Al-Sharab J, Badway F, Amatucci G, Stadelmann P (2007) EELS spectroscopy of iron fluorides and FeFx/C nanocomposite electrodes used in Li-ion batteries. Microsc Microanal 13(2):87–95
127. Doe R, Persson K, Meng Y, Ceder G (2008) First-principles investigation of the Li-Fe-F phase diagram and equilibrium and nonequilibrium conversion reactions of iron fluorides with lithium. Chem Mater 20(16):5274–5283
128. Nielsen U, Paik Y, Julmis K, Schoonen M, Reeder R, Grey C (2005) Investigating sorption on iron-oxyhydroxide soil minerals by solid-state NMR spectroscopy: A Li-6 MAS NMR study of adsorption and absorption on goethite. J Phys Chem B 109(39):18310–18315
129. Kim J, Nielsen U, Grey C (2008) Local environments and lithium adsorption on the iron oxyhydroxides lepidocrocite (gamma-FeOOH) and goethite (alpha-FeOOH): A H-2 and Li-7 solid-state MAS NMR study. J Am Chem Soc 130(4):1285–1295
130. Liao P, MacDonald B, Dunlap R, Dahn J (2008) Combinatorially prepared [LiF](1-x)Fe-x nanocomposites for positive electrode materials in Li-ion batteries. Chem Mater 20(2):454–461

131. Breger J, Dupre N, Chupas P, Lee P, Proffen T, Parise J, Grey C (2005) Short- and long-range order in the positive electrode material, Li(NiMn)(0.5)O-2: a joint X-ray and neutron diffraction, pair distribution function analysis and NMR study. J Am Chem Soc 127(20):7529–7537
132. Yoon W, Paik Y, Yang X, Balasubramanian M, McBreen J, Grey C (2002) Investigation of the local structure of the $LiNi_{0.5}Mn_{0.5}O_2$ cathode material during electrochemical cycling by X-ray absorption and NMR spectroscopy. Electrochem Solid State Lett 5(11):A263–A266
133. Yoon W, Iannopollo S, Grey C, Carlier D, Gorman J, Reed J, Ceder G (2004) Local structure and cation ordering in O_3 lithium nickel manganese oxides with stoichiometry $Li[Ni_xMn_{(2-x)/3}Li_{(1-2x)/3}]O_{-2}$ – NMR studies and first principles calculations. Electrochem Solid State Lett 7(7):A167–A171
134. Van der Ven A, Ceder G (2004) Ordering in $Li_{-x}(Ni_{0.5}Mn_{0.5})O_{-2}$ and its relation to charge capacity and electrochemical behavior in rechargeable lithium batteries. Electrochem Commun 6(10):1045–1050
135. Meng YS, Ceder G, Grey CP, Yoon W-S, Shao-Horn Y (2004) Understanding the crystal structure of layered $LiNi_{0.5}Mn_{0.5}O_2$ by electron diffraction and powder diffraction simulation. Electrochem Solid State Lett 7(6):A155–A158
136. Cahill LS, Chapman RP, Britten JF, Goward GR (2006) Li-7 NMR and two-dimensional exchange study of lithium dynamics in monoclinic $Li_3V_2(PO_4)(3)$. J Phys Chem B 110(14):7171–7177
137. Matsumura Y, Wang S, Mondori J (1995) Interactions between disordered carbon and lithium in lithium ion rechargeable batteries. Carbon 33(10):1457–1462
138. Wang S, Kakumoto T, Matsui H, Matsumura Y (1999) Mechanism of lithium insertion into disordered carbon. Synth Met 103(1–3):2523–2524
139. Nakagawa Y, Wang S, Matsumura Y, Yamaguchi C (1997) Li-7-NMR study of lithium charged in carbon electrode. Synth Met 85(1–3):1363–1364
140. Conard J, Estrade H (1977) Résonance magnétique nucléaire du lithium interstitiel dans le graphite. Mater Sci Eng 31:173–176
141. Peled E, Menachem C, BarTow D, Melman A (1996) Improved graphite anode for lithium-ion batteries – chemically bonded solid electrolyte interface and nanochannel formation. J Electrochem Soc 143(1):L4–L7
142. Menachem C, Wang Y, Flowers J, Peled E, Greenbaum S (1998) Characterization of lithiated natural graphite before and after mild oxidation. J Power Sources 76(2):180–185
143. Wang Y, Yufit V, Guo X, Peled E, Greenbaum S (2001) Li-7 nuclear magnetic resonance study of lithium insertion in pristine and partially oxidized graphite. J Power Sources 94(2):230–237
144. Gotoh K, Maeda M, Nagai A, Goto A, Tansho M, Hashi K, Shimizu T, Ishida H (2006) Properties of a novel hard-carbon optimized to large size Li ion secondary battery studied by Li-7 NMR. J Power Sources 162(2):1322–1328
145. Persson K, Sethuraman VA, Hardwick LJ, Hinuma Y, Meng YS, van der Ven A, Srinivasan V, Kostecki R, Ceder G (2010) Lithium diffusion in graphitic carbon. J Phys Chem Lett 1(8):1176–1180
146. Gerald RE, Klingler RJ, Sandi G, Johnson CS, Scanlon LG, Rathke JW (2000) Li-7 NMR study of intercalated lithium in curved carbon lattices. J Power Sources 89(2):237–243
147. Chevallier F, Letellier M, Morcrette M, Tarascon JM, Frackowiak E, Rouzaud JN, Beguin F (2003) In situ Li-7-nuclear magnetic resonance observation of reversible lithium insertion into disordered carbons. Electrochem Solid State Lett 6(11):A225–A228
148. Letellier M, Chevallier F, Clinard C, Frackowiak E, Rouzaud JN, Beguin F, Morcrette M, Tarascon JM (2003) The first in situ Li-7 nuclear magnetic resonance study of lithium insertion in hard-carbon anode materials for Li-ion batteries. J Chem Phys 118(13):6038–6045

149. Key B, Bhattacharyya R, Morcrette M, Seznec V, Tarascon JM, Grey CP (2009) Real-time NMR investigations of structural changes in silicon electrodes for lithium-ion batteries. J Am Chem Soc 131(26):9239–9249
150. Nesper R (1990) Structure and chemical bonding in zintl-phases containing Lithium. Prog Solid State Chem 20(1):1–45

Further Readings

Balbuena PB, Wang Y (2004) Lithium-ion batteries: solid electrolyte interphase. Imperial College Press, London
Linden D, Reddy TB (2002) Handbook of batteries, 3rd edn. McGraw-Hill, New York

Chapter 9
Lithium-Ion Batteries, Safety

Brian Barnett, David Ofer, Suresh Sriramulu, and Richard Stringfellow

Glossary

Abuse	External stress applied to a battery or cell that is not anticipated to occur under normal operating conditions and may cause the cell to experience thermal runaway.
Cascading	Process by which one cell in a battery releases sufficient heat that thermal runaway of one or more neighboring cells in a pack ensues.
Field-failure	*Infrequent* safety incident that occurs in lithium-ion cells/batteries in the field under "normal" operating conditions and does not appear to have an obvious external trigger, but is often caused by an internal short circuit. Such internal short circuits are often caused by foreign metal particles.
Safe Zone	For a given cell construction and materials, and for a given heat transfer environment, conditions of energy and power associated with an internal short such that a thermal runaway is not possible.
Thermal runaway	A process of uncontrolled heat release and rapid temperature rise.
Threshold energy	For a given cell design and heat transfer environment, an energy value dissipated in an internal short below which no thermal runaway is possible.
Threshold power	

This chapter was originally published as part of the Encyclopedia of Sustainability Science and Technology edited by Robert A. Meyers. DOI:10.1007/978-1-4419-0851-3

B. Barnett (✉) • D. Ofer • S. Sriramulu • R. Stringfellow
TIAX LLC, 35 Hartwell Avenue, Lexington, MA 02421-3102, USA
e-mail: barnett.b@tiaxllc.com

R.J. Brodd (ed.), *Batteries for Sustainability: Selected Entries from the Encyclopedia of Sustainability Science and Technology*, DOI 10.1007/978-1-4614-5791-6_9,
© Springer Science+Business Media New York 2013

	For a given cell design and heat transfer environment, an internal short power below which no thermal runaway is possible.
Trigger for thermal runaway	A stimulus that initiates thermal runaway in a Li-ion cell. Examples of potential triggers include a variety of abuses as well as internal short circuits created by foreign metal particles.

Definition of the Subject and Its Importance

Safety of lithium-ion batteries is a critical topic that has not received adequate attention in the past, largely due to the fact that data regarding safety failures have been severely restricted. As a result, there are numerous misunderstandings in a field that has not received the same degree of scientific and technical rigor as other areas of lithium-ion battery technology development. However, safety of lithium-ion batteries will become even more important as lithium-ion technology enters transportation markets. Under suitable triggers, Li-ion cells can experience thermal runaway, i.e., the rapid increase in cell temperature accompanied by venting, vent-with-flame, ejection of cell parts, fire, and explosion. Safety failures of lithium-ion cells can result from a variety of triggers including overcharging, overheating, crushing, mechanical impact, and external short circuits. Safety tests have been devised for all these abuses, with varying degrees of fidelity. However, most safety incidents that have taken place with lithium-ion batteries occur due to the slow and rare development in cells of internal short circuits that mature to the point that they result in thermal runaway. Most safety tests carried out in the laboratory or factory do not replicate the conditions by which safety incidents actually occur in the field. These issues are characterized in detail, and an improved overall framework for considering lithium-ion battery safety is suggested.

Introduction

Since their introduction in 1991, lithium-ion batteries have become the dominant rechargeable battery technology for portable products and are beginning to make inroads in transportation and stationary storage applications. Lithium-ion battery technology provides the highest available energy density of any long-cycling rechargeable battery, which has been the most significant factor leading to its widespread adoption. As of this writing, over 4.5 billion lithium-ion cells (almost 40,000 MWh) [1] will be expected to be produced in 2011, the bulk of which are used in consumer electronics applications such as laptop computers and cell phones. While the demand for such cells is increasing at the rate of 15% on a year-on-year basis, emerging applications can significantly increase the Li-ion market. It is anticipated that emerging transportation and stationary storage markets will lead to dramatic increases in the volume of Li-ion production. One recent estimate places automotive Li-ion demand in 2020 at 150,400 MWh, in contrast to 72,500 MWh projected for portable applications in that year. The transportation and stationary applications are expected to use larger capacity cells than those employed for consumer electronics applications [1].

Despite the obvious success of Li-ion technology, safety concerns remain [2]. Under suitable triggers, Li-ion cells can undergo venting, vent-with-flame, ejection of cell parts, fire, and explosion as the consequence of a process of uncontrolled heat release, termed "thermal runaway." This risk is not surprising given the possible sources of heat release within a cell. Essentially, Li-ion batteries store a very large amount of electrochemical energy, which can be released as heat, for example, through an internal short. The electrochemical energy content of state-of-the-art Li-ion cells is by itself sufficient to raise cell temperature to 700°C if completely dissipated within the cell under adiabatic conditions. But furthermore, Li-ion cells contain energetic materials that can undergo exothermic decomposition reactions within the cell, as well as containing combustible organic solvents (with heat of combustion that is about a third of the heat of combustion of gasoline) and combustible carbonaceous anode materials, both of which can burn in atmospheric oxygen if exposed under suitable ignition conditions. The heat release from such reactions can be as much as ten times the stored electrochemical energy.

Safety failures of lithium-ion cells can result from a variety of triggers. Overcharging, overheating, crushing, mechanical impact, and external short circuit all represent forms of external triggers (often termed "abuse" conditions) that can lead to safety incidents. These external abuses have been extensively studied. Battery engineers have developed several cell-level and pack-level tests to evaluate how a cell/battery responds to the abuse, and they have also developed technologies and approaches to manage the cell and battery response to such external abuses [3–11]. Standardized tests also have been developed to assess cell and battery "tolerance" to external abuses [12, 13]. In general, this chapter shall try to draw distinctions between tests used to screen cells/batteries for their tolerance to a given trigger and the actual tolerance of a cell/battery to the given trigger (i.e., the effectiveness of the test).

Safety concerns have been heightened by highly publicized safety incidents and ensuing widespread recalls of lithium-ion batteries used in laptop computers and cell phones [14, 15]. When these rare safety incidents occur, lithium-ion batteries operating under otherwise normal conditions undergo what appear to be spontaneous thermal runaway events, often with violent flaming and extremely high temperatures. Moreover, these failures usually involve cells and cell designs that have passed extensive abuse testing, including the standardized abuse safety tests. *Most such Li-ion safety incidents in the field are not preceded by any obvious external abuse*. We refer to these spontaneous safety incidents as "field-failures".

The fact that cells involved in field-failures generally have passed all standardized safety tests is the first clue that these incidents should be given special consideration. Internal shorts – not external abuses – cause field-failures far more frequently than is generally appreciated. Such shorts are a major focus of this chapter.

Given the near ubiquitous use of Li-ion batteries, field-failures may result in significant property damage and may also prove dangerous to users, not to mention the enormous costs of the recalls they have prompted. Furthermore, the advent of new transportation and stationary applications employing cells larger than those used in portable applications, and often with many cells in series, raises the threat of even more severe outcomes.

Although field-failures and the ensuing recalls have resulted in significant financial losses for the battery industry, relatively little is known or published about them, or the underlying causes of such incidents. Outside of the battery companies and their customers, very few organizations have had an opportunity to carry out careful postmortems of cells and batteries involved in these incidents. As a result, a very large fraction of the lithium-ion battery community, including materials companies, national laboratories, and the worldwide academic community, are not able to approach the lithium-ion battery safety issue with the same scientific and technical rigor that they apply to other aspects of lithium-ion technology. We believe that if better background information were made available to engineers and scientists across the battery community, there would be a profound impact on widely held perspectives regarding safety, how to test for safety, and what actually makes a lithium-ion cell/battery safe. Importantly, most safety tests carried out in the laboratory or factory do not replicate the conditions by which safety incidents actually occur in the field. We will describe this conundrum in greater detail below.

We (at TIAX) are one of the few organizations that have had an opportunity to carry out a significant number of detailed investigations of actual lithium-ion safety incidents [16–21]. Although confidentiality requirements preclude us from showing specific examples of field-failure cell postmortems, we will describe general observations and conclusions in this chapter. We are hopeful that the lessons learned from such work can enhance the overall perspective of the lithium-ion battery community and support some of the insights developed in this chapter. Finally, an improved overall framework for considering lithium-ion battery safety is suggested.

Basis for Li-Ion Safety Concerns

Li-ion cells are widely understood to have higher energy content than other rechargeable batteries, but this perception is generally focused only on their higher *electrochemical* energy density. While that perception is correct, the heat release associated with the exothermic reactions of all of a cell's contents must also be

Table 9.1 Summary of contributions to heat release in a lithium-ion cell

Process	Temperature range (°C)	Energy release	Energy relaese in a 10 Wh 18650 cell
Anode decomposition and reaction with electrolyte	80–120	~400 J/g-anode[a, b]	~11 KJ
	150–300	~1,300 J/g-anode[a, b]	
Cathode decomposition and reaction with electrolyte	150–300	~1,600 J/g-cathode[a]	~23 KJ
Self-reaction of salt with solvent	250–400	~900 J/g electrolyte[c]	~4 KJ
Complete combustion of solvent[d]	Auto-ignition temperature ~450	~18 KJ/g Solvent	~110 KJ

[a]Values estimated from DSC and ARC measurements of cell components at TIAX, for charged active (graphite and $LiCoO_2$) in contact with standard carbonate electrolyte ($LiPF_6$ in carbonate solvents)
[b]There are two parts to the anode decomposition reactions, a low temperature part typically attributed to the decomposition of SEI film, and a higher temperature reaction involving the reaction of the anode with the electrolyte
[c]Values estimated from DSC and ARC measurements of standard Li-ion electrolyte
[d]There is insufficient oxygen available inside a typical 18650 cell, even with a metal oxide cathode, to effect complete combustion of the solvent that would be present in the cell. However, if vented at high temperatures or vented in the presence of an ignition source, the solvent can burn outside the cell

considered when evaluating safety. In the case of Li-ion technology, this heat release can exceed ten times the stored electrochemical energy content if in the field-failure process, all the cell's combustible contents are reacted with air. For example, Table 9.1 summarizes the typical contributions to the heat release in a generic Li-ion cell. The table lists the key chemical reactions and provides estimates of the following:

- Temperature range over which the rates of these reactions are significant
- Total heat release from these exothermic reactions
- Heat release that can be attributed to the particular reaction in a 10-Wh 18650 cell (electrochemical energy of 36 kJ)

As can be seen from the table, the anode and cathode reactions with electrolyte and the self-reaction of electrolyte contribute significantly to the heat release relative to the electrochemical energy content of the cell. The mechanism and kinetics of these reactions have been studied extensively in the literature [22–29]. However, complete combustion of the solvent results in much higher heat release. Note that even in Li-ion cells with metal oxide cathodes there is insufficient oxygen available within the cell to completely combust the solvent, a factor not always recognized by many safety investigators. Thus, a major fraction of the solvent combustion can only occur outside the cell. An even greater release of heat occurs if the graphitic anode content of the cell is

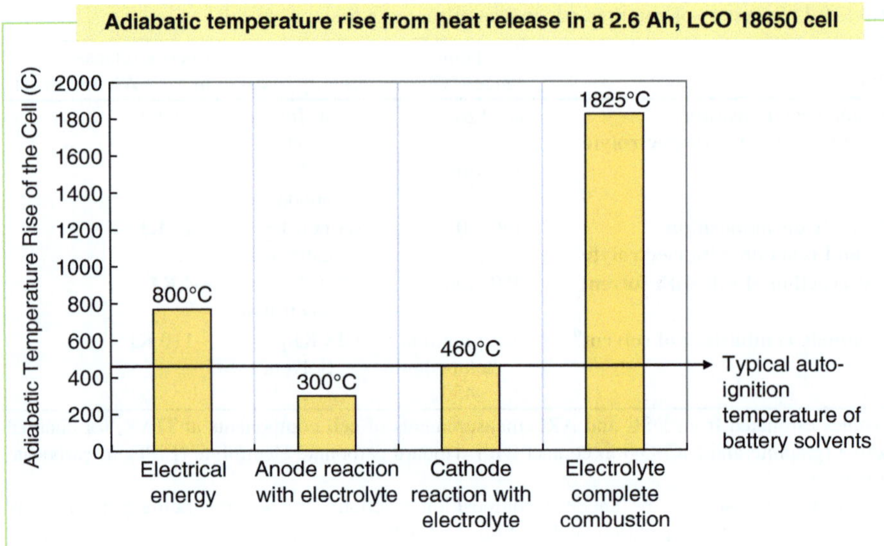

Fig. 9.1 Estimate of the adiabatic temperature rise from heat release in a 10-Wh Li-ion 18650 cell

combusted upon exposure to atmosphere; postmortem audits show that this does in fact occur in many field-failures. Graphite's heat of combustion is nearly 33 kJ/g, meaning that complete combustion of a 2.6-Ah cell's graphite content will generate about 260 kJ.

Adding graphite's combustion heat to the energies listed in Table 9.1, the total energy release by an 18650 cell can exceed 400 kJ. The heat that can potentially be released within the cell absent of atmospheric combustion, shown in Table 9.1, can itself be significant and can drive severe cell temperature increases. The highest temperatures to which cells might be driven by these processes can be estimated by considering the adiabatic temperature rise associated with the different sources of energy release discussed above, as summarized in Fig. 9.1. Release of electrochemical energy alone, as by full discharge of the cell through an internal short, can raise the cell temperature by more than 700°C under adiabatic conditions. Furthermore, the anode and cathode exothermal reactions together can also raise the cell temperature by 700°C. These temperatures are significantly higher than the autoignition temperature of the solvent, and hence, if the solvent is ejected from the cell at these temperatures, it is likely to combust outside the cell, further stimulating the development of very high temperatures, and igniting the graphitic anode with explosive combustion.

Table 9.1 and Fig. 9.1 illustrate the powerful energetics underlying thermal runaway events. An increase in temperature can stimulate significant heat release from the exothermic anode, cathode, and electrolyte decomposition reactions. If the rate of heat removal is slower than the rate of heat generation, thermal runaway will eventually occur, accompanied by enormous heat release and very high temperatures.

Triggers for Li-Ion Thermal Runaway

In general, some form of trigger, or initiating process, is required to produce a thermal runaway event. The previous section amply illustrates that triggers producing increased temperature of either the whole cell or a significant region of the cell can result in thermal runaway. Table 9.2 summarizes some common triggers. Ideally, battery developers and implementers can utilize safety tests that specifically screen for a cell's or battery's tolerance to a given trigger.

The triggers that have received the greatest attention include abuse conditions such as external heating, overcharging, external shorting, impact, penetration, etc. Some of these triggers are relatively easy to anticipate as risks, and to study and to develop screening tests for, often by selecting a set of abuse conditions that are reasonably plausible. In fact, numerous standardized (or semi-standardized) abuse tests have been developed to assess abuse tolerance (a broader all-encompassing term might be "trigger tolerance"). Examples of such tests include the hot box test, the forced overcharge test, the external short circuit test, etc. [12, 13] These tests are not reviewed in detail, except to point out that they do not generally replicate the conditions in which actual safety events occur in the field. Certain other triggers present more difficult challenges with respect to development of truly effective safety screening tests. In addition, some safety tests can be "gamed," i.e., manipulated to favorably alter the outcome by carrying out the tests under certain conditions, or by modifying cells so that they can pass the tests rather than demonstrate true tolerance to the particular abuse. Situations where definition of plausible screening safety tests may *not* be possible are highlighted. Table 9.2 identifies (rows 4 and 5) two such triggers.

Technologies and components such as PTCs, CIDs, vents, and safety circuits have been incorporated in today's lithium-ion cells and packs in order to manage safety with respect to many external abuse triggers.

Table 9.2 Summary of triggers that can lead to thermal runaway of Li-ion batteries

Trigger	Why can this occur?	Is this managed?
Overcharge	Defective connections, failure of charging circuit	Yes, battery management system Yes, cell-level safety devices
Overheating from external sources	Battery back placed too close to a heat source	Yes, cell-level safety devices open the cell at suitable internal pressure
Cell crushing creating massive internal shorts	Physical abuse of battery pack	Yes, design enclosures are built more tolerant to specific abuses
Internal short-circuits (a.k.a., field failures)	Internal-short caused by manufacturing defects	No, new technologies needed
Cascading of thermal energy release	Affected cell can raise the temperature of surrounding cells	No, new technologies needed

The internal short is yet another trigger that can lead to thermal runaway. Essentially, cell discharge through the internal short causes i^2R heating and high temperatures in the vicinity of the short. High temperatures in the vicinity of the short circuit can stimulate additional heat release from the exothermic decomposition reactions described above, which can then result in thermal runaway. Thermal runaway following internal shorting is discussed in greater detail in a subsequent section describing simulation tools and results. Evidence from investigations of field-failures suggests that they are predominantly triggered by internal short circuits that are very clearly not present or detectable in fresh cells.

Tests (e.g., see Refs. [30–36]) are being developed to simulate an internal short circuit by deliberately placing particles in cells or by controlled pinch tests. There are also some advocates for the use of the nail penetration test or the round bar crush test as means of screening cells for tolerance to internal shorts. However, these tests are not standardized at the moment, and it can be argued that there are fundamental challenges in mimicking the behavior of an internal short circuit that develops only during normal charge/discharge cycling in the field. Later sections will discuss in greater detail the testing issues with respect to internal shorts, and reasons are suggested as to why cells cannot be screened for safety with respect to this trigger. Furthermore, currently there are no strategies to manage internal short-related thermal runaway of Li-ion cells. We will suggest some strategies later in this chapter.

Thermal runaway of a single cell in a lithium-ion pack will frequently stimulate a "cascading" event. In this phenomenon, some time after one of the cells in a pack undergoes thermal runaway consequent to an external or internal trigger, other cells in the battery pack also go into thermal runaway, sometimes as a series of discrete events. The enormous amount of heat release from a cell undergoing thermal runaway can cause the temperature of one or more nearby cells to rise to the point that they too are stimulated to thermal runaway. Although cascading has been recognized as a risk for Li-ion batteries, adequate cascading screening safety tests have not been developed, and anticascading technologies are not routinely incorporated into battery packs.

The remainder of this manuscript focuses on the internal short and its critical role in triggering thermal runaway. The review of key observations summarized from our audits of field-failure incidents leads to the deduction that the majority of such safety events are due to internal short circuits caused by foreign metal particles in lithium-ion cells, and accordingly, most of this chapter focuses on them. This summary is followed by discussions of the likely mechanism of internal short formation as probed by experimental measurements. Then, simulations are described that elucidate the factors controlling thermal runaway following an internal short circuit, and that provide the basis for a novel construct allowing the first "quantification" of Li-ion battery safety associated with any given cell and its operating environment. The key differences between abuse tolerance and field-failures are then described, followed by discussion of an overall framework to improve the safety of Li-ion batteries.

Key Observations from Analysis of Field-Failures

This section summarizes key observations from our direct investigations of lithium-ion battery field-failures, including work carried out with some of the largest battery and portable device manufacturers in the world. While confidentiality restrictions prevent us from incorporating specific details of these safety incidents, we are able to highlight major observations and insights gained from this work, and from other sources. We have learned that:

- Field-failures occur at a frequency of one failure in 5–10 million cells for the most accomplished manufacturers. Such failures are overwhelmingly attributable to internal short circuits occurring during otherwise "normal" operation, with no outward signs of impending failure.
- These internal shorts are not discernible at the point of manufacture but appear to develop over time. Typically, the failure occurs only after a significant service time in the field (e.g., more than 3 months) and may not occur for 1–3 years. Thus, "fresh" cells do not necessarily exhibit the conditions that lead to a safety event.
- A high proportion of these internal short circuits are caused by extraneous (foreign) metal particles introduced into the cell during the manufacturing process. Evidence for such particles can be found in cell postmortems, except that a high level of destruction often hinders or prevents satisfactory investigation. While metal particles may enter the cell-making activity with incoming materials, they also can be produced by a wide variety of the material processing and metal cutting and shaping operations used in cell manufacturing. The exact composition of the metal causing a given field-failure is often not known with certainty, but iron, nickel, and zinc have all been known to cause field-failures; in typical $LiCoO_2$/graphite cylindrical cells, these elements are not normally present in the jelly roll, and thus have been implicated from postmortem analysis of cells.
- Various processes are employed by battery manufacturers to reduce the frequency of metal particle generation and/or incorporation in cells. These processes include extensive shrouding of key operations, strategic placement of magnets, use of higher grade metals in various operations, and aggressive monitoring of replacement schedules for all cutting and forming tools. Additional measures, such as high-pot and formation aging tests, are employed to catch and reject cells exhibiting internal shorts during cell production and qualification [37, 38].
- A surprisingly high percentage of lithium-ion cells have foreign metal particles in them. Given the rarity of field-failures, it is clear that not all particles in the cells lead to internal shorts or to thermal runaway. In fact, it is likely that the vast majority of internal short circuit defects occurring after some time in the field are manifested as a "dead" cell/pack, rather than as a violent field-failure.

- In field-failures, an induction period lasting on the order of minutes is often reported, during which self-heating of the battery is discernible by users/observers, prior to violent flaming and explosion.
- Postmortem examinations of actual field-failures indicate that all or portions of a cell have experienced very high temperatures. For example, cells from incidents routinely exhibit multiple localized silver-colored metallic beads. These beads are recondensed aluminum, which melts at 660°C. In some instances, there are areas in which copper current collectors have also experienced melting. Copper melts at 1,085°C.
- Neither the actual mechanism of short circuit initiation nor the mechanism by which an internal short leads to a thermal runaway is fully understood yet. After the fact, postmortems show a level of destruction that renders it impossible to ascertain the exact morphology that existed at the point of thermal runaway. We use experimental and simulation results to discuss the likely mechanism of short formation, and the factors that control thermal runaway following an internal short.

Implications for Managing Field-Failures

Most safety incidents that have been observed and evaluated involved portable computer battery packs. Although some incidents have taken place in mobile phone batteries, computers use many more cells, with cells in series and in parallel. Our discussion of the mechanism of short initiation will show that such failures are also possible in HEV, PHEV, and BEV cells. Without suitable safety technologies in place, safety incidents in HEV, PHEV, and BEV packs are likely to be more severe than those observed in laptops owing to the larger cell sizes employed (approximately 2–30 times on an energy basis) and the lower impedance of the cells. Furthermore, with roughly 75–80 cells per pack (for cells in a single string) and the total energy per pack approximately 20–200 times larger than that of computer packs, the consequences of thermal event "cascading" is a major concern. Field-failure in a vehicle pack could inflict far greater damage than that caused by notebook computer batteries.

Because the consequences of these rare safety events can be so serious, especially with increased use of lithium-ion in transportation applications, it is necessary to recognize that no manufacturing environment can always operate with zero defects. Therefore, given the recognition that foreign particles in cells can never be completely eliminated, there is a need to develop safety technologies that are "on guard" to catch their deleterious effects as early as possible. A useful perspective is to recognize that achieving Li-ion systems which are totally free from the threat of spontaneous safety events is *not* a quality problem. For example, the rate of field-failures for cells produced by the most experienced manufacturers is already at a very low level – one failure in 5–10 million cells produced, which already far exceeds six sigma manufacturing quality guidelines. These manufacturers typically benefit from the most powerful elements of a learning curve – they have produced large numbers of cells (over a billion in several cases) over a very long period of time (more than a decade). It is reasonable to assume that less-experienced manufacturers will eventually approach similar levels, but no manufacturer will achieve zero defects in perpetuity.

Manufacturing improvements should, of course, continue to be implemented whenever possible, but it should be recognized that improvements in the cleanliness or quality control of manufacturing lines are not the fail-safe solution. Manufacturing "quality" is already maintained at a very high level. The defect rate is already very low and not readily correlated with the level of small extraneous metal particles in cells. Greatly reducing the level of particles in cells could probably only be accomplished with significant (and unacceptable) additional processing costs. (For example, consider the manufacturing quality levels that must be achieved in the semiconductor industry, and the associated costs.) Clearly, a new approach is needed to make significant progress toward improving the safety of Li-ion batteries and, in particular, to deal with the insidious role of the internal short circuit in causing lithium-ion safety events.

We have been developing tools and techniques to enable the development of new technologies to manage field-failures. In the following sections, some insights are described that have been gained from exercising these tools; namely, the likely mechanism by which a foreign metal particle forms an internal short, and the mechanism by which an internal short then stimulates thermal runaway.

Likely Mechanism of Internal Short Formation in a Field-Failure

The exact mechanism of field-failure due to foreign metal particles in cells may never be known with certainty, given the rarity of events and the extensive destruction that takes place in affected cells. The realization that field-failures are rare has a number of practical consequences for studies of internal short formation, for attempts to identify electrical signals warning of developing internal shorts, and for development of any safety technology. In particular, a reliable method(s) to generate internal shorts "similar" to those that lead to thermal runaways in the field is essential. Although not discussed in detail in this chapter, we have developed several methodologies to create internal shorts in cells. This section describes the most likely mechanism for initiation of the internal short circuit from metal particle contamination and how we have used internal short implantation methods to investigate that mechanism.

The most likely mechanism for initiation of the internal short involves the formation of a metal dendrite between the anode and cathode. In this process, a metal particle present in or on the cathode (positive) will dissolve by an electrochemical oxidation process, and the resulting metal ions will diffuse to the anode

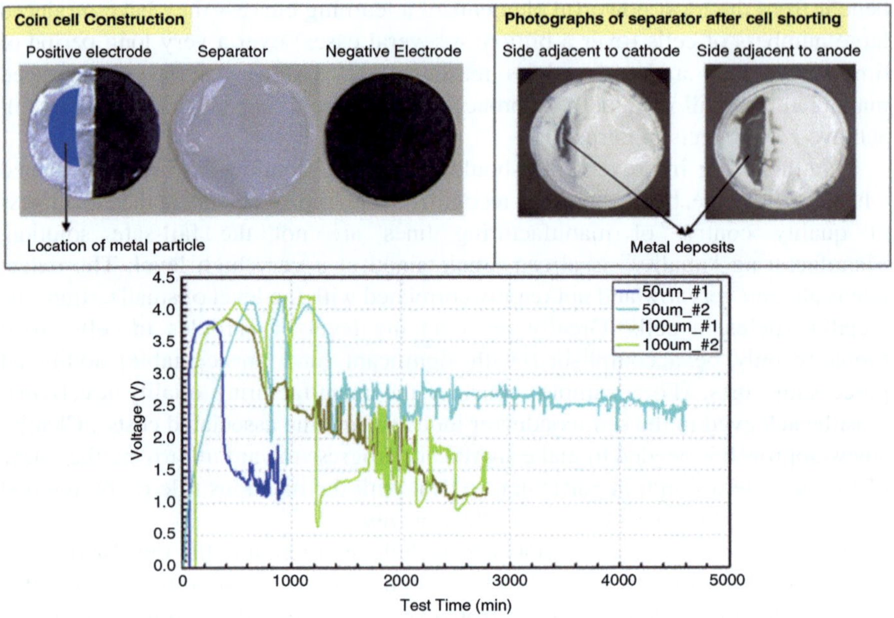

Fig. 9.2 *Internal short formation in a coin cell.* Photographs in the *top left panel* show the coin cell construction for these tests. A crescent-shaped Ni particle was placed on a scraped-off portion of the positive electrode. The cycling data on the bottom panel show the formation of an internal short circuit in the cell during normal charge/discharge cycling, for two different thicknesses of the Ni particle (two different coin cells for each particle thickness). The photographs in the *top right panel* are of the separator from one of the coin cell tests. The metal deposits can be clearly seen. The cathode active material was $LiCoO_2$, the separator was Celgard 2325, and the anode material was MCMB2528

(negative) where they will plate, often over an extended period of time and many charge/discharge cycles, resulting in the formation of a metallic dendrite that eventually makes a durable electrical contact back to the cathode and shorts the cell. In the field, the exact location of the short in a cell is also a critical factor with respect to occurrence of thermal events. We have confirmed this mechanism by implanting particles of iron and nickel in Li-ion cells.

Figure 9.2 provides experimental evidence supporting this mechanism. The top left panel of Fig. 9.2 illustrates a cell constructed with a large (oversized relative to particles implicated in field-failures) nickel particle with thickness of either 50 or 100 μm placed on the cathode current collector. The bottom pane of Fig. 9.2 shows the subsequent cycling of this cell to the point of development of an internal short. Postmortem examination of the cell was very instructive. The top right panel of Fig. 9.2 shows deposits of plated metal adhered to the separator. Inspection of both sides of the separator shows that the deposit's area is larger on the anode side than on the cathode side, clearly indicating the direction of its growth from the anode to the cathode. Furthermore, EDAX of the anode itself shows deposition of nickel on the anode,

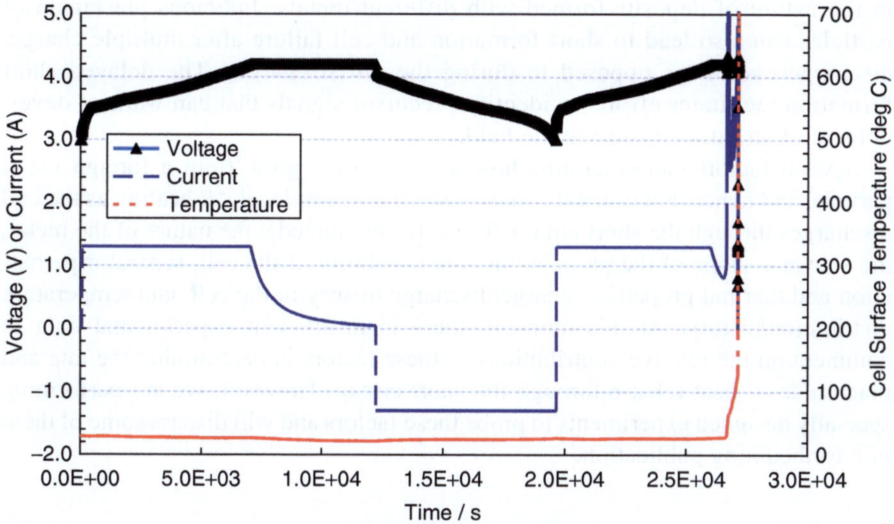

Fig. 9.3 *Response of an 18650 cell following deliberate implantation of a metal particle on the cathode of the cell.* In this experiment, a fully discharged commercial 18650 cell was removed from its can. The jelly roll was partially unwound, and a metal particle was placed on the cathode side. The jelly roll was then placed in a suitably sized can and normally cycled. The cell experienced thermal runaway during normal charge/discharge cycling

which of course was not present prior to cell assembly. Taken together, these results clearly show the operative mechanism for formation of internal shorts. Metal particles on (and possibly in) the cathode dissolve and plate out on the anode, growing back through the separator, leading to an internal short.

It is not surprising that a nickel metal particle present in the cathode would result in shorting of the Li-ion cell. The standard potential for Ni/Ni^{2+} is 2.9 V (vs Li/Li^{+}), implying that Ni will dissolve at typical cathode potentials >3.5 V (vs Li/Li^{+}) and plate at typical anode potentials <1.5 V (vs Li/Li^{+}). Similar behavior is also expected of Fe; the standard potential for Fe/Fe^{2+} is 2.6 V (vs Li/Li^{+}). Separate cyclic voltammetry experiments on Ni and Fe foils in Li-ion battery electrolytes also confirmed this mechanism. This mechanism also explains why internal shorts can develop even on open circuit and explains how the shorts can "grow" over time and not be present for some time after manufacture of the cells. Plating and dendrite growth is a stochastic process, and induction of the overall cell shorting process is also likely to be so. Thus, several factors can limit the rate of dendrite growth and probably even its occurrence.

It has also been experimentally confirmed that metal particles placed on the cathode in 18650 cells can cause internal short circuits capable of inducing thermal runaway (sample result shown in Fig. 9.3). These experiments will be detailed in a future publication. Using this sort of test platform, the resistance of shorts and the evolution of that resistance during tests can be estimated. It has also been possible to evaluate variations in behavior of various metals and variations in the nature of deposits formed with different metals. Judicious placement of particles can also lead to short formation and cell failure after multiple charge/discharge cycles (as opposed to during the initial charge). The delayed short formation facilitates efforts to identify precursor signals that can warn of development of an internal short in the field.

Several factors can determine how a short might grow from a foreign metal particle, and hence determine the magnitude and rate of local i^2R heating as the cell discharges through the short circuit. These factors include: the nature of the metal, the size and shape of the particle, the rate capability of the cell, separator permeation and thermal properties, charge/discharge history of the cell, and temperature distribution/history. At this moment, there is insufficient experimental data to comment on the relative contributions of these factors in determining the rate and magnitude of heat release through the short circuit. However, we are performing specially designed experiments to probe these factors and will discuss some of them in a forthcoming publication.

Other Possible Mechanisms of Internal Shorting

There are various ways in which an apparently normal cell might develop an internal short in the field [39, 40]. However, the nature of the internal shorts formed by these different routes, and the extent to which they may or may not be likely to induce thermal runaway, might be very different. Mechanisms cited in the literature or generally discussed by battery engineers that implicate foreign metal particles are summarized below:

- *Lithium deposition/dendrites* form during cell operation for a number of possible reasons, including charge processes that drive the anode to the lithium potential or as a result of lithium imbalance in the cell. Unintentional overcharge due to a faulty BMU can also result in lithium deposition on the anode, with repeated overcharging leading to lithium dendrite growth. There is also evidence reported for foreign metal particles on the anode surface serving as nucleation sites for lithium deposition. There is no doubt that lithium deposition occurs in lithium-ion cells under a variety of circumstances. We suspect that lithium dendrites can lead to internal shorts, resulting in premature cell failure. However, lithium dendrites are less likely to dissipate sufficient power to enable thermal runaway than are dendrites of higher melting point metals. Our simulations show that the temperature of the dendrite itself can significantly exceed the melting temperature of lithium metal (180°C). Hence, it is less likely that lithium deposits/dendrites could support powers similar to those possible with metal contaminants such as Ni or Fe (melting temperatures >1,400°C).
- *Particle migration and separator puncture.* A metal particle in the cell (presumably in/on the anode) can migrate somewhat as a result of the volume changes that occur during charge and discharge and can translate sufficiently to pierce the

separator and create a short. In this mechanism, the nature of the contact between the metal and the cathode could potentially be quite different from that in the mechanism of metal particle dissolution at the cathode followed by plating back to the cathode from the anode. A complex question raised by the formation of internal shorts by metal particles (and a probable driver of the process's highly stochastic nature) is how a durable contact is eventually established between the high voltage cathode and the relatively easily oxidized and dissolved metal. In the dissolution/plating mechanism, the growing dendrite might contact some of the original metal particle remaining on and contacting the cathode, and furthermore, the ongoing dissolution/plating process could result in multiple and repeated contacts being made between the growing dendrite and the cathode, providing many opportunities for a durable short to be established. The mechanism of particle migration and separator puncture might be less likely to yield a durable cathode contact that is capable of supporting a thermal runaway-inducing power dissipation level.

- *Lower power particle short induces separator melting, creating a higher power short*. It has been postulated that a small particle-based internal short, rather than itself dissipating sufficient power to stimulate thermal runaway, generates only enough heat to cause localized separator melting and a larger area face-to-face interelectrode short. This larger short then stimulates thermal runaway.

A number of technologies have been developed to address some of the above scenarios. High puncture strength and high melt strength separators have been developed. Numerous organizations across the battery industry have elected to deploy ceramic layers, implemented as porous coatings on electrodes or separators in cells, as a means of enhancing lithium-ion battery safety. Although such efforts may be effective in preventing the separator puncture or melting scenarios, there is a danger in viewing them as a panacea. These separators and ceramic layers are all necessarily highly porous (otherwise they will compromise a cell's performance), and so the mechanism of particle dissolution and growth of dendrites to the point of internal shorting will still be operative. In our limited testing of cells containing ceramic layers, we have indeed observed formation of internal shorts, including those that mature to thermal runaway, in cells that have been seeded with a particle on the cathode. Therefore, technologies such as ceramic layers may lessen the frequency of internal short occurrence but should not be viewed as eliminating the problem.

Factors Influencing Thermal Runaway from Internal Shorts

As we described earlier, an internal short circuit can cause the local temperature to rise "in the vicinity" of the short, which then stimulates additional heat release due to exothermic anode reaction with electrolyte and cathode reaction with electrolyte.

Whether or not thermal runaway follows from such an internal short is a particularly complex problem because:

1. An internal short initiates with a highly nonhomogeneous temperature distribution in the cell due to heat generation in a very small area (a "hot spot")
2. Local temperatures get high enough, fast enough to stimulate exothermic decompositions on a local basis
3. Whether thermal runaway follows from this nonhomogeneous temperature profile is determined by a complex interaction among instantaneous values of temperature, concentrations, and rates of reactions, as well as applicable heat transfer both within the cell and out of the cell through its surface

Simulations allow us to understand this complicated relationship and to identify the critical parameters that can alter this relationship to prevent thermal runaway.

Model Description

We have developed a model that is capable of fully three-dimensional, transient simulations of the thermal response of the cell following an internal short circuit. The model employs a commercial finite element analysis (FEA) code, Abaqus, augmented with our own subroutines. Key features of the model are highlighted in Fig. 9.4. Abaqus was augmented with subroutines for calculating local heat generation from the short circuit and from anode and cathode decomposition reactions. An effective model must consider not just the total heat release which was discussed earlier but also the kinetics (how fast can the heat be released?). Kinetics submodels for heat generation can be derived from differential scanning calorimetry (DSC) or accelerating rate calorimetry (ARC) measurements of charged anode and cathode materials in contact with standard electrolytes. The measurement techniques and conversion to useful kinetic rate expressions will be detailed in a future publication. The model also incorporates representative thermal properties (specific heat, thermal conductivity, density) for the various components of the Li-ion cell. These were either measured in our labs or taken from the published literature (including Refs. [41, 42]). Dissection of state-of-the-art 18650 cells provided estimates of cell properties such as weights of anode and cathode materials, current collector dimensions, can dimensions, etc. The model also incorporates a submodel for separator shutdown kinetics and is being augmented to incorporate a submodel for electrochemical kinetics and pressure, including pressure release from the local temperature rise and decomposition reactions.

The model allows the internal short to be placed anywhere in the cell and typically represents the short as a constant power, localized heat source that acts for a specified period of time. (The ability to terminate the short after an arbitrary period of time is a unique and valuable benefit of a model, for reasons revealed

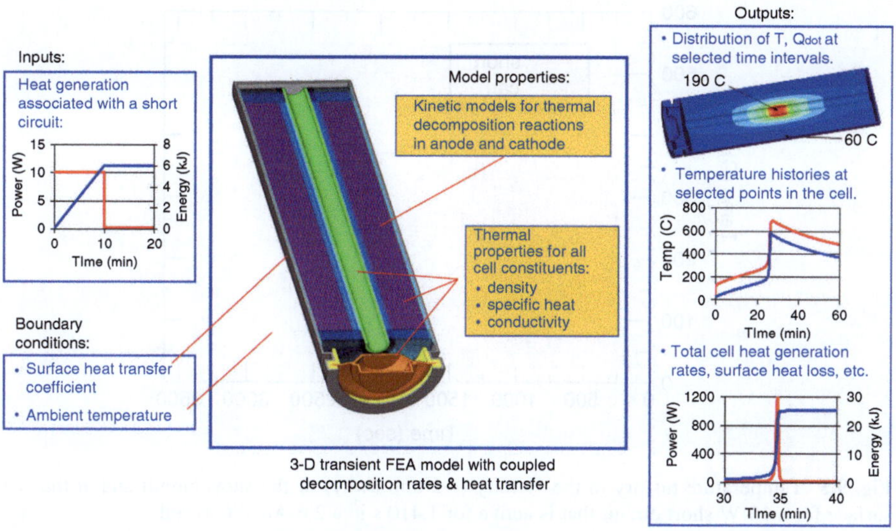

Fig. 9.4 Key features of our FEA model to simulate thermal runaway of Li-ion cells

below, that is not available in actual short circuits). Surface heat transfer coefficients and ambient temperatures are specified as boundary conditions. The initial cell temperature is also specified. The transient heat transfer model calculates and keeps track of the local values of temperature, species concentrations, extents of reaction, and heat generation rates at every time step. The model can be two-dimensional or fully three-dimensional, depending on the degree of symmetry afforded by the geometry of the cell and the conditions that are being evaluated, such as the location of the short. For example, for an 18650 cell, if the short is near the center of the cell, the results of a two-dimensional, axisymmetric model are essentially the same as those of a fully three-dimensional analysis of the same conditions. This opportunity for simplification is not available for a short that is located near the outer surface of the jelly roll.

Thermal Runaway Induced by an Internal Short

The following sample simulation illustrates the progression of thermal runaway following an internal short circuit. Consider an 18650 cell which has a 10-W internal short circuit starting at time t = 0 near the center of the cell, and lasting for 1,410 s in this simulation. For this two-dimensional, axisymmetric simulation, Fig. 9.5 shows the time dependence of temperature in the vicinity of the short and at the cell surface (midway between the top and bottom of the cell). Figure 9.6 shows the temperature distribution in the cell at select times, and Fig. 9.7 shows the

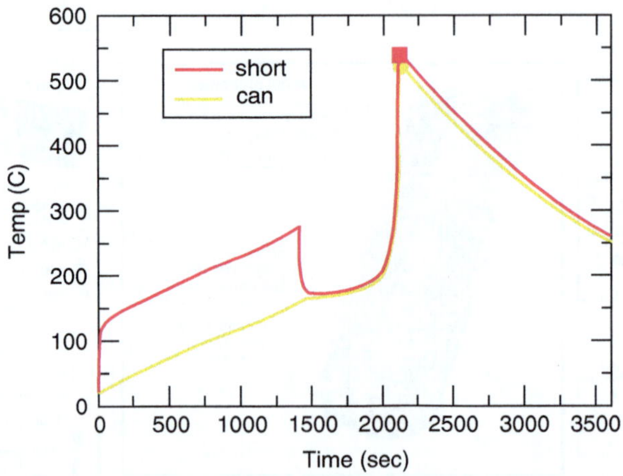

Fig. 9.5 Temperature history in the vicinity (~2 mm away) of the short circuit and at the can surface for a 10-W short circuit that is active for 1,410 s in a 2.6-Ah 18650 cell

cumulative heat release and extent of reaction for the anode and cathode decomposition reactions. A surface heat transfer film coefficient of 10 W/m^2-K was employed, corresponding to nearly quiescent ambient conditions. The initial temperature and ambient temperature were both set to 20°C.

The results shown in Figs. 9.5–9.7 can be considered in terms of three distinct time periods. In the first time period, the temperature in the vicinity of the short circuit is higher than that at the outer extremities of the cell, owing to heat generation from the short circuit (which has been placed near the center of the cell). The internal short is a localized heat source, or hot spot. In practice, this heat source can be driven from other cells in the battery pack or just from the shorted cell itself in the case of an isolated cell. In addition to electrical heating, heat release from the anode and cathode decomposition reactions, occurring wherever critical onset temperatures in the cell have been exceeded, also contributes to the temperature rise. As shown in Fig. 9.6, the iso-temperature contours are oblong in this time frame due to significantly higher thermal conductivity in the axial direction of the jelly roll compared to the radial direction.

Once the short power is reduced to zero at 1,410 s, the cell temperature quickly becomes uniform, as heat transfer internal to the cell equalizes the temperature within the cell.

In the second time period (from about 1,410–2,000 s), there is no longer a contribution from the local heat source because the internal short has been turned off. The temperature continues to increase, albeit at a much slower rate. In this time frame, the rate of heat release from the decomposition reactions is only slightly higher than the rate of heat loss at the cell surface, resulting in a gradual rise of the cell temperature. As can be seen from the results in Fig. 9.7, it is primarily the exothermal anode decomposition reactions that contribute to this heat release; the

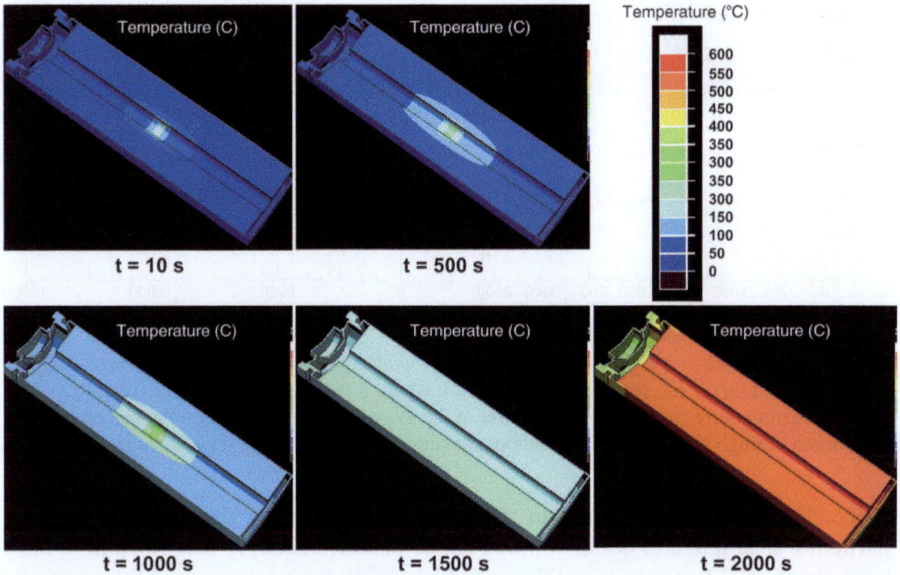

Fig. 9.6 Temperature distributions corresponding to the simulation in Fig. 9.5 showing the progressive change in temperature in a cell following the formation of an internal short circuit that leads to eventual thermal runaway

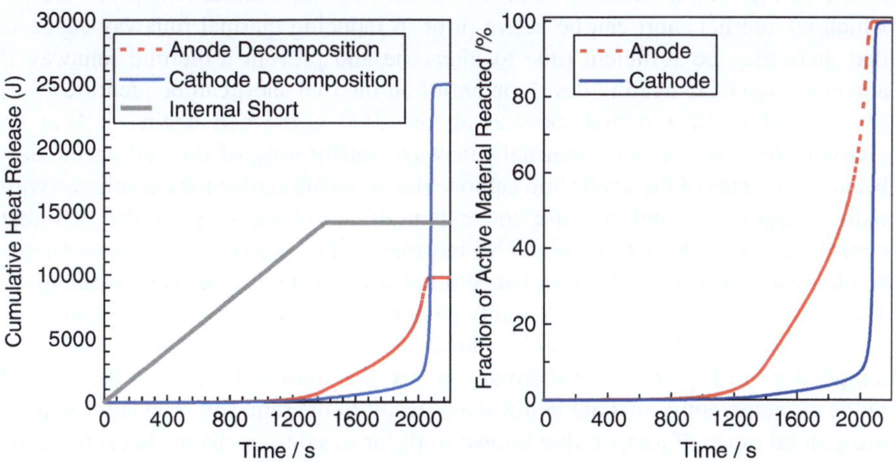

Fig. 9.7 Cumulative heat release (plot on the *left*) and the extent of reaction (plot on the *right*) for the anode and cathode decomposition reactions, corresponding to the simulation results in Fig. 9.5

energy contributed from the cathode decompositions is insignificant in this time frame. The anode decomposition reactions assume a triggering role because their rates (and hence the associated heat release) become significant at a lower temperature than the cathode decomposition reactions, and hence, the anode reactions

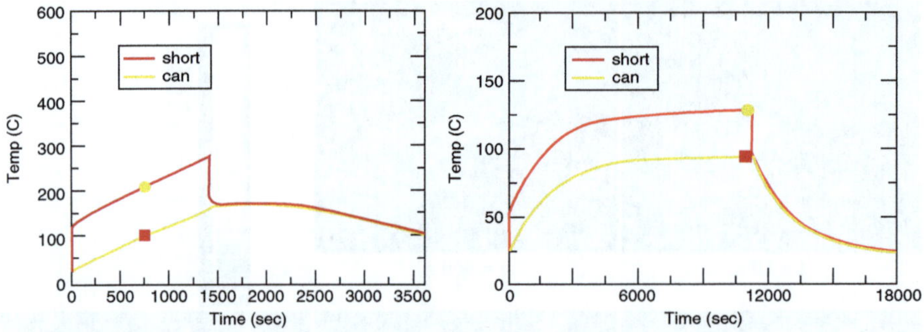

Fig. 9.8 *Simulated temperature profiles for two conditions where thermal runaway does not occur*. The plot on the left corresponds to the case where a 10-W short was active for 1,400 s. The plot on the right corresponds to a short power of 3.5 W, which was active until the cell completely discharged through the short circuit without causing thermal runaway

proceed to a greater extent than the cathode reactions in this time frame. The particular example and corresponding set of conditions being discussed here have been chosen to illustrate a condition in which a sufficient percentage of the material in the cell has been elevated to a temperature that is just above the "point of no return" on the path toward thermal runaway. In an equivalent simulation in which the internal short is turned off just 10 s sooner at 1,400 s, discussed below and shown in Fig. 9.8, no thermal runaway ensues. The substantial length of time for which an internal short can be active prior to inducing thermal runaway suggests that there may be sufficient time to intervene and prevent a thermal runaway if adequate signals indicating development of an internal short can be identified.

In the third time period, beyond about 2,000 s, the cell begins to heat up dramatically and goes into thermal runaway. Self-heating of the cell accelerates because the rates of the anode and cathode decomposition reactions accelerate with rising temperature and because the extent of active material in the cell that contributes to this heat generation also increases. The combination of these factors results in an unstable increase in temperature which is known as thermal runaway. The rapid rise in temperature is roughly what witnesses report when they have been in proximity to lithium-ion field-failures. Note not only the very rapid rise in temperature in Fig. 9.5 once thermal runaway has initiated, but also the general range of temperatures that are realized as a result of the runaway. The latter reflects the general range of temperature known to occur in safety incidents because of the evidence (previously cited) that is obtained from cell postmortems.

The above discussion illustrates the complexity of thermal runaway induced by an internal short. Furthermore, the rapid rise in temperature at ~2,000 s illustrates one of the reasons that material-level heat release from DSC or ARC measurements by themselves are not likely to be relevant predictors of relative safety of various cathode materials. Note how in Figs. 9.6 and 9.7 the progression to a thermal runaway is inevitable beyond about 1,410 s, at which point the short has been turned off in this simulation, and note how the cell quickly traverses a wide range of temperatures (in this case, from about 150°C to about 550°C) as runaway proceeds.

As mentioned above, this model has also been used to explore the conditions under which thermal runaway will not occur. Figure 9.8 shows results for two conditions where thermal runaway does not occur. The figure on the left corresponds to a 10-W short, otherwise identical to the one depicted in Fig. 9.5, which is active for 1,400 s, whereas the figure on the right corresponds to a 3.5-W short that is active until the cell completely discharges through the short. All other conditions are identical to the simulations corresponding to Figs. 9.5–9.7.

Comparing the results in Fig. 9.5 with the left hand plot of Fig. 9.8 shows that the energy dissipated in the short is a key factor in determining thermal runaway. Essentially, when the 10-W short is active for 1,400 s (Fig. 9.8), thermal runaway does not occur. But when the 10-W short is active for only 10 s longer (Fig. 9.5), thermal runaway occurs. Thus, there is a minimum energy (or "threshold energy") that must be dissipated in the short before thermal runaway will occur. The specific value of that threshold energy is related to cell design, materials employed, and operative heat transfer conditions. Of course, slight variations in any of these parameters might result in a different threshold energy.

The right-hand plot in Fig. 9.8 shows a case in which short power has been reduced in successive simulations until thermal runaway does not occur no matter how long the short is maintained in an active state (a "threshold power"). Note the parallel lines in Fig. 9.8 beyond about 5,000 s, corresponding to the state in which the rate of heat generation in the cell is exactly equal to the rate of heat loss from the outer surface of the cell. Since the experiment is bounded by the total energy of the cell, the short essentially terminates when there is no capacity remaining in the cell. As can be seen in the figure, the cell temperature then decays to room temperature.

The Concept of a "Safe Zone"

The results presented in Figs. 9.5, 9.7, and 9.8 have been chosen from a larger number of simulations because, in combination, they illustrate how a new figure can be constructed from simulations to demarcate a "Safe Zone" for a given cell construction and cell materials and for the associated heat transfer conditions. Figure 9.9 illustrates the construct for a generic 18650 cell. In this figure, various short powers are represented on the x-axis. The y-axis corresponds to the relative energy dissipated in the short as a percentage of the total electrochemical energy in the cell.

Figures 9.5, 9.7, and 9.8 represent conditions that can be placed on this chart. The 10-W short begins at t = 0 corresponding to zero energy release on the y-axis and moves vertically with increasing time. At 1,400 s, there has been insufficient energy input to stimulate a thermal runaway, but at 1,410 s, thermal runaway ensues. A threshold energy has been exceeded. Running a series of such simulations, a family of such situations can be constructed, demarcating a threshold energy curve on the plot.

Similarly, Fig. 9.8 (right) has already demonstrated existence of a threshold power below which no thermal runaway is possible, no matter how long the short is

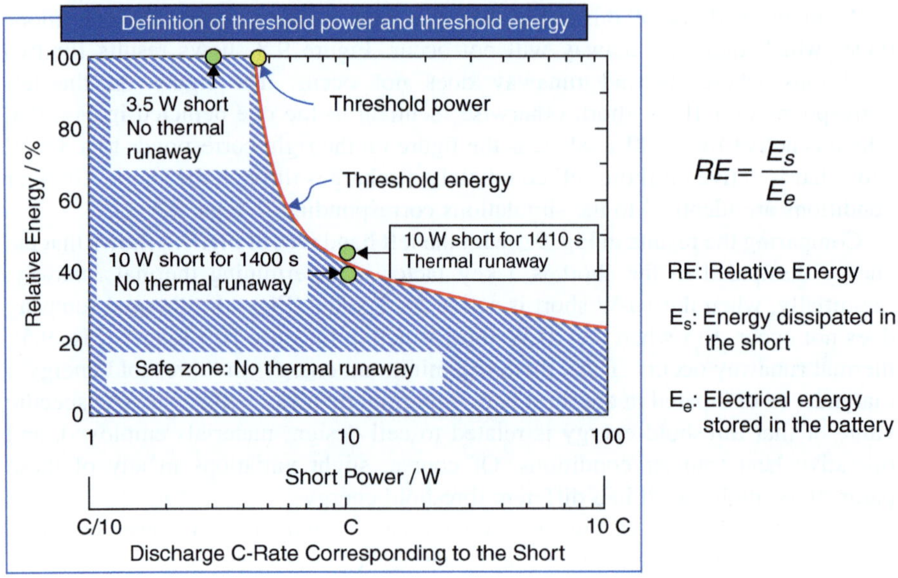

Fig. 9.9 The Safe Zone construct

in place. The existence of a threshold power implies that there will be a finite range of short resistance capable of dissipating sufficient localized heat to stimulate thermal runaway. If short resistance is above this critical range, insufficient current is supplied to the short. The critical short resistance range will be defined by the characteristic of the cell and pack.

Note in Fig. 9.9 how the threshold energy curve rises sharply as short power is reduced to values approaching the threshold power. The resulting cross-hatched area is a "Safe Zone." Internal shorts with power or energy characteristics within this region will not induce thermal runaway. In practice, the threshold power is a more critical parameter, since in the field there are no practical means to interrupt an internal short once it has been initiated. Implications of the threshold energy curve's shape will be discussed in a future publication.

Note that a "Safe Zone" plot can be constructed for any lithium-ion cell, representing the particular cell size and design (i.e., form factor, capacity, electrode dimensions, etc.) and materials employed, as well as the particular heat transfer conditions that apply. One can use Safe Zone plots to compare the relative safety of various cells and/or materials, as well as to evaluate the relative safety of cells under various heat transfer conditions.

Illustrative Applications of the Safe Zone Concept

The Safe Zone concept allows a quantitative assessment of safety in a way that has not previously been possible. There are numerous ways that the Safe Zone concept

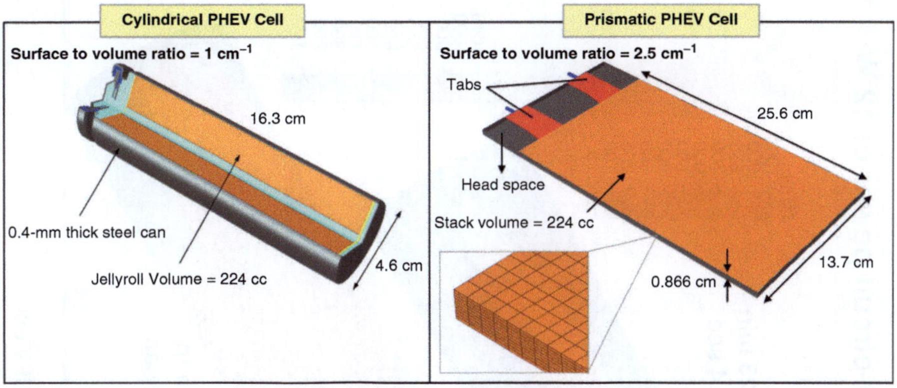

Fig. 9.10 Cylindrical and prismatic 33-Ah PHEV cell designs employed for simulating thermal runaway following an internal short circuit

may be employed, including as a tool to design safer cells, to assess the safety impact of various system design changes, to assess the impacts of various materials changes, and to consider the relative risk of cascading in a battery pack. The following describes an example of one such use of the Safe Zone concept.

In the last few years, there have been multiple discussions regarding the relative safety of cylindrical cells versus prismatic cells. The subject is attracting a great deal of attention with respect to transportation applications. It has generally been accepted that prismatic cells are safer and will allow superior cooling. But how can this advantage be quantified, and how can it be linked to cooling conditions as well as cooling requirements? The Safe Zone concept, and the associated concept of a threshold power, provides the necessary quantitative basis to allow these problems to begin to be assessed as engineering issues, rather than as a matter of intuition.

Cylindrical and prismatic cell designs with identical capacity (33 Ah, as might be used in PHEV applications) were modeled with identical internal short circuits to assess the threshold power parameters. Figure 9.10 illustrates the cell designs.

Initial cell temperature was set at 20°C, the surface heat transfer coefficient was set for both cells at 10 W/m^2-K, and the short was assumed to be located near or at the middle of each cell. Simulations were carried out for various short power levels in order to understand and quantify the power levels associated with safety events in these two cells. Figure 9.11 illustrates the results for an internal short of 12 W, in particular showing the rapid transition of temperatures throughout the cylindrical cell over the critical few seconds when the cell goes into thermal runaway. Similarly, Fig. 9.12 illustrates the temperature transitions in the prismatic cell with a 53 W internal short around the time at which thermal runaway occurs. Figures 9.11 and 9.12 illustrate the rapid temperature increase that accompanies thermal runaway for both cell

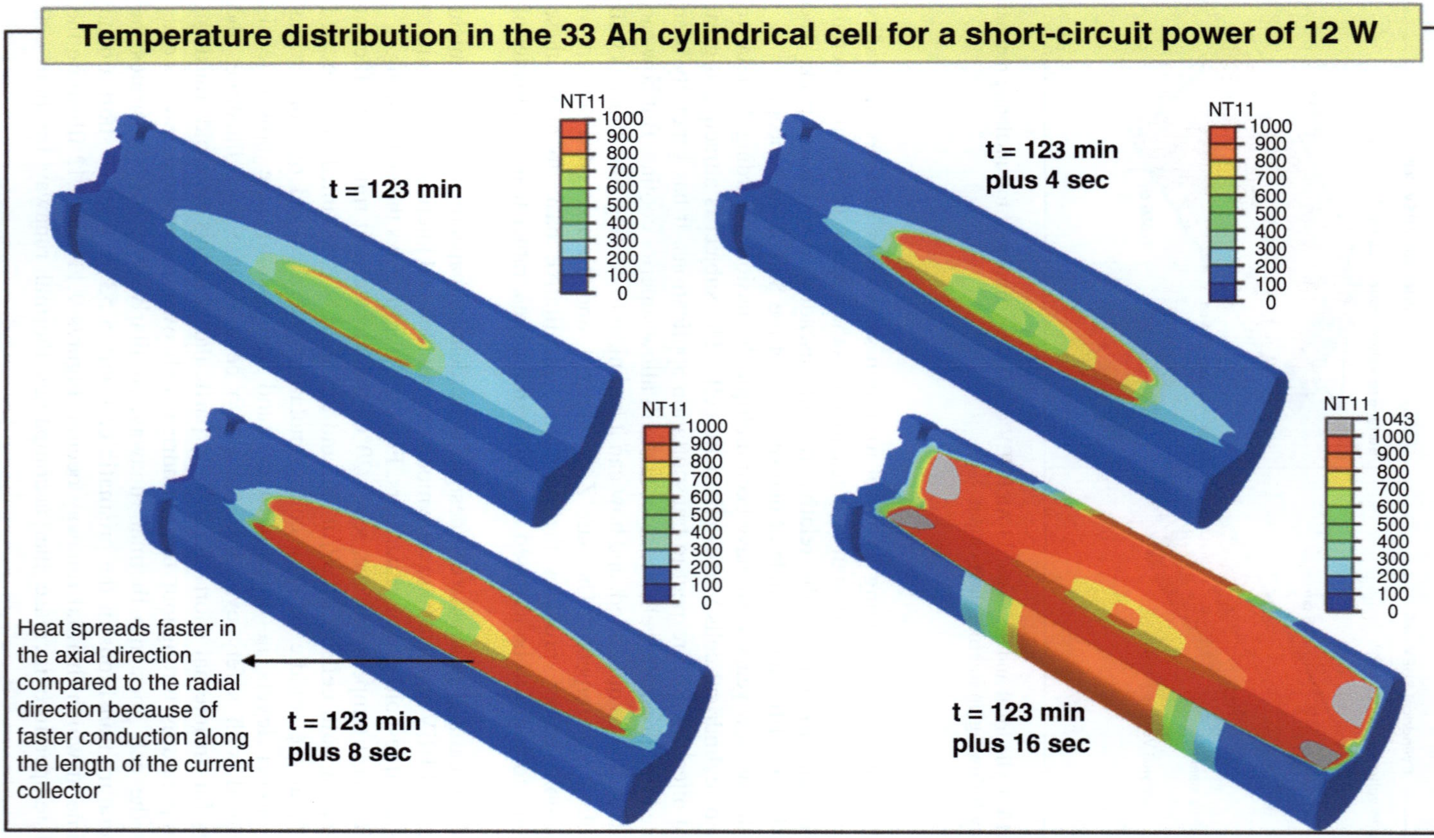

Fig. 9.11 Temperature distributions at selected times following a 12-W internal short circuit in the cylindrical PHEV cell

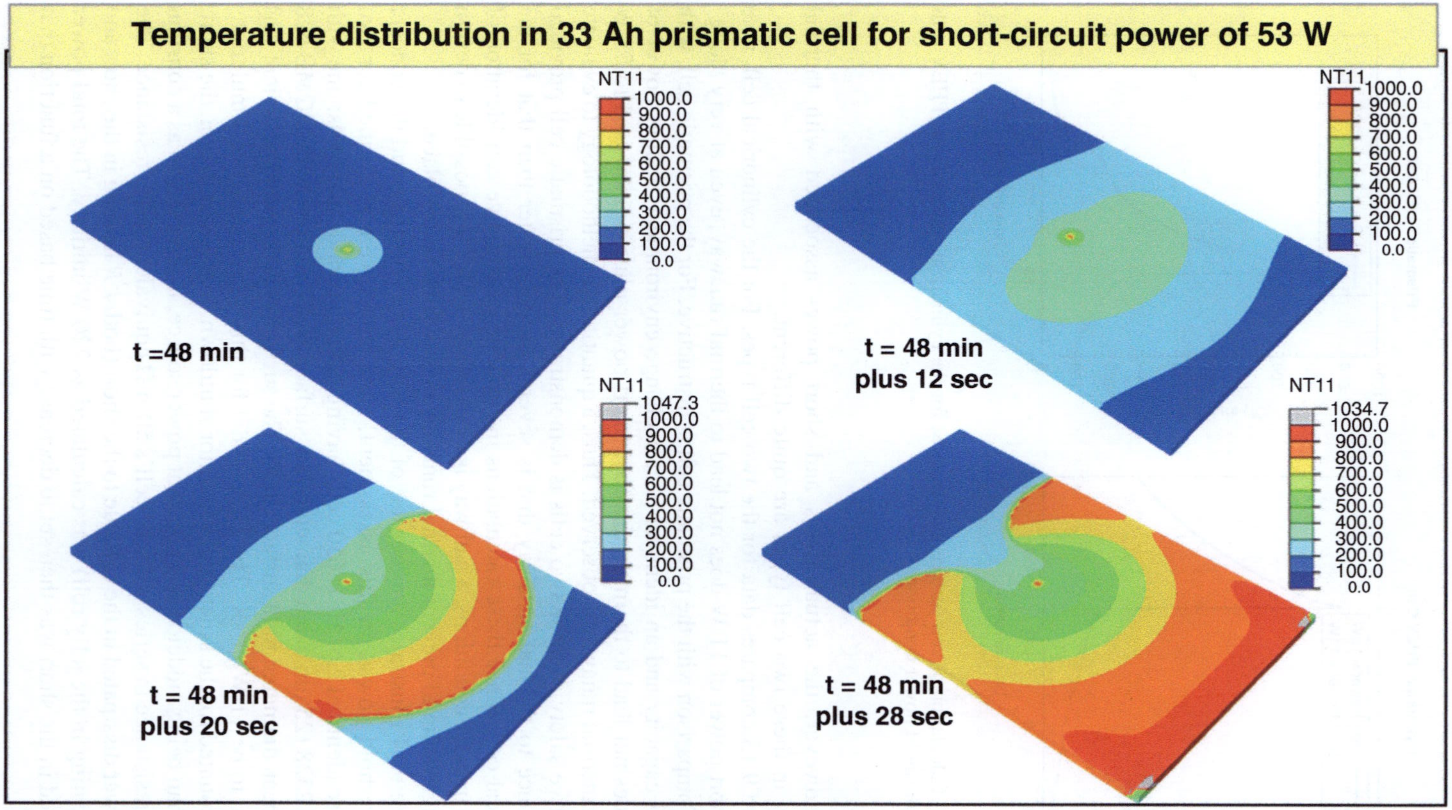

Fig. 9.12 Temperature distributions at selected times following a 53-W internal short circuit in the prismatic PHEV cell

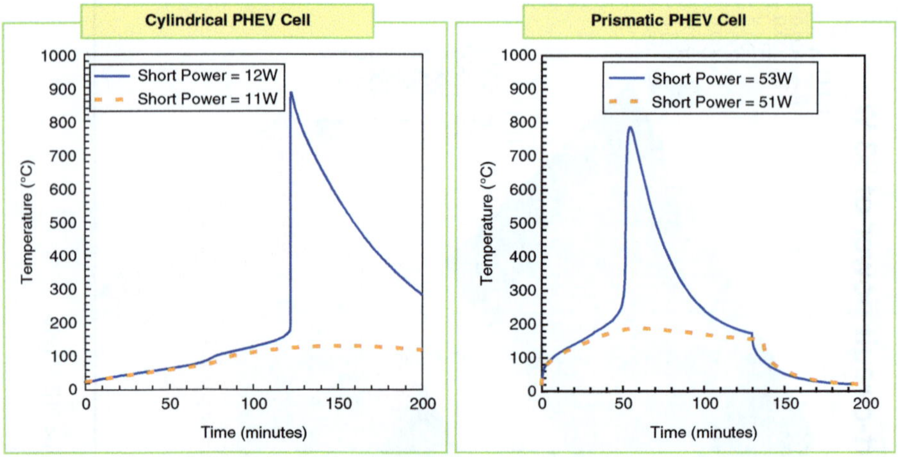

Fig. 9.13 Calculation of cell surface temperatures for the cylindrical and prismatic PHEV cells near the threshold power conditions

types; however, the actual times and short power associated with thermal runaway in these two cell types are quite different.

Figure 9.13 compares data for the two cell types. For the cylindrical cell, note that a short power of 11 W does not lead to thermal runaway, even at very "long" times. Comparison with the prismatic cell is instructive. For the prismatic cell, with identical capacity and an identical heat exchange environment, a short power of 51 W does not lead to thermal runaway. Short power must be increased to 53 W before a thermal runaway is observed. Thus, a quantitative methodology to evaluate the relative safety of these two cells is demonstrated. The prismatic cell provided a resistance to thermal runaway that is several factors greater than that for the cylindrical cell. Since these calculations incorporate quantitative consideration of the heat transfer conditions, a pathway is available by which it is possible to design a cooling system to prevent thermal runaway events from taking place.

Figure 9.14 illustrates another use of the Safe Zone construct and the concept of a threshold power. In this assessment, simulations were performed for two otherwise identical 2.6-Ah 18650 cells having different anode materials: one with MCMB 2528 (25 μm mean diameter, low surface area), and one with MCMB 628 (6 μm mean diameter, high surface area). The surface heat transfer coefficient was assumed to be 11 W/m^2-K. The FEA model for this particular set of simulations also accounted for the kinetics of separator shutdown. In this simulation, the short, rather than being modeled as a constant power source, is represented as a constant 25-mΩ resistance in series with the cell's 50-mΩ equivalent series resistance. The total power dissipated in the cell due to the short (both i^2R heating in the short and Joule heating in the jelly roll) was calculated as 220 W initially. The total power dissipated in the short was then set to decrease with time based on a function that

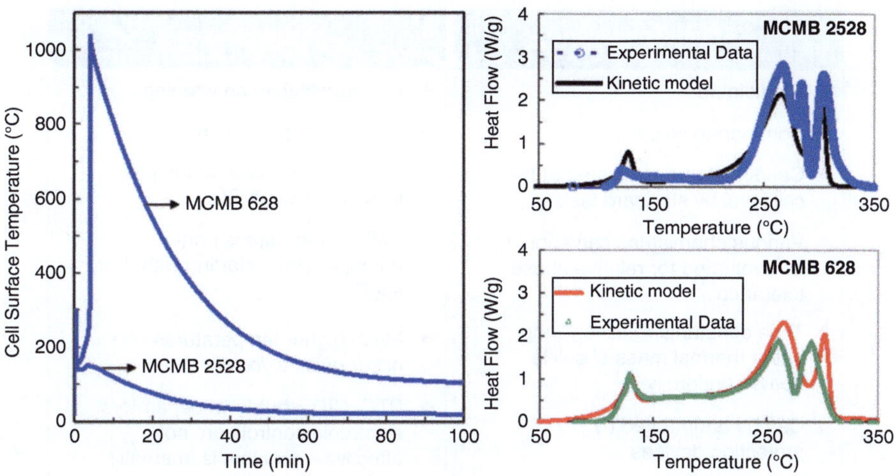

Fig. 9.14 Simulation results showing the effect of the choice of anode material on thermal runaway. The plot on the *left* shows the cell surface temperature time dependence for two different anode submodels

includes the relative volume of the jelly roll in which the temperature had risen past the point at which the separator would shut down (130°C).

The right hand charts in Fig. 9.14 show DSC data for these two anode materials and the associated anode models that were employed in the simulations. The results of the simulations are shown on the left hand side of the figure. The results are striking – a simple change in anode material has a dramatic impact on the safety of this cell, with the cell employing MCMB 628 going into thermal runaway while the cell employing MCMB 2528 does not. DSC data on the right of Fig. 9.14 suggest the reasons for this result. The MCMB 628 shows significantly higher heat release between 100°C and 200°C than the MCMB 2528. This example shows the extent to which the anode material acts as the trigger for thermal runaway when internal shorts and field-failures occur. Furthermore, the cell surface temperature profiles on the left side of Fig. 9.14 illustrate another critical factor that is not well appreciated: the cell transitions rapidly into thermal runaway well below 200°C – well before any significant exothermic decomposition of cathode material has taken place.

Differences Between Abuse Tolerance and Field-Failures

The distinction between abuse tolerance and field-failure (internal shorts) is important for a number of reasons. It helps bring into focus the extent to which historic testing for safety is focused on abuse testing, even though the actual safety events that occur in the field are quite different. From a mechanistic perspective, the triggers are fundamentally different and the resulting responses

Abuse Tolerance	Field Failures
◆ Predictable ◆ Common to all cells ◆ Can/should be evaluated at the cell level by standard tests ◆ Various chemistries can/should be evaluated for relative abuse tolerance ◆ Time constants relatively long, entire thermal mass of cell is heated uniformly ◆ Can be augmented by protection devices ◆ Externally-driven	◆ Not predictable, no warning ◆ One-in-a-million (or less) ◆ Difficult to evaluate at the cell level, or through QC ◆ Cell temperature is non-homogeneous, starting with "hot spot" ◆ Much higher temperatures can occur *quickly, locally* ◆ PTC, CID, shutdown separators, electronic controls are not effective – problem is internally driven.

Fig. 9.15 Summary of the key differences between abuse tolerance and field-failures

are different. The general distinctions between abuse tolerance and field-failure are summarized in Fig. 9.15 and discussed below.

Field-failures are overwhelmingly attributable to internal short circuits that occur during otherwise "normal" operation, i.e., there is no external abuse or trigger. In contrast, heat exposure or forced overcharge are examples of an external abuse driving a potentially unsafe situation. The response to these abuse conditions should be identical for all cells of the same cell design employing the same materials.

In response to most abuses (note that penetration- and impact-based abuses are exceptions), a cell's temperature throughout an abuse is homogeneous, until the cell fails or the abuse is terminated. All cells of the same cell design and same materials should respond to an abuse in a roughly identical manner. In contrast, field-failure involves a nonhomogeneous temperature distribution in a cell's response to an internally driven trigger – the internal short, or hot spot.

The internal short represents a particularly insidious failure mechanism not only because it takes place during otherwise normal operation but also because it is self-driven and cannot be protected against by typical safety protection devices such as PTCs, CIDs, etc. It is also how most safety incidents are initiated in the field, and yet the battery industry does not have adequate tests to "predict" the risk for such failures.

The lack of availability of suitable tests for the internal short should not be a surprise – any event that takes place rarely, is stochastic *and* exhibits some form of incubation process, arguably does not exhibit a detectable precondition at the point of manufacture and certainly does not exhibit a predisposal to the problem *in every cell*. Thus, the practice of removing a few cells from the manufacturing line to test for safety is fine for abuse triggers (for abuse tolerance) but is clearly inappropriate as a measure of susceptibility to development of an internal short in the field.

Perhaps most challenging is the tension between the desire of those in the battery community to have a test that they can use to screen cells for field-failure (the internal short) and the statistical impossibility of picking cells to test that have the necessary preconditions for possible internal short-induced thermal runaway, putting aside the issue of the extent to which actual charge/discharge, voltage/time, and temperature effects may impact the probability of development of an internal short. That tension has led many across the battery and application/user communities to advocate that tests such as nail penetration, round bar crush impact tests, or other custom tests mentioned earlier, be employed to screen cells for tolerance to internal shorts. There are many reasons why this approach is an inappropriate surrogate for field-failure. One reason is that the condition that leads to internal shorts leading to thermal runaway is not present in fresh cells. In addition, it is statistically impossible to identify cells that actually have some form of precondition *and* for which that precondition has matured to the point at which an internal short exists. Most importantly, the nature of the internal shorts produced in such tests is nothing like that of the shorts involved in field-failures. These abuse tests often involve multiple shorts and hard shorts; conditions that are not representative of how internal shorts occur in the field. Both modeling and actual observations have shown that a number of different variable parameters can be critical to whether a given short will cause runaway in a given cell, and thus, the possibility of thermal runaway is a complex, multidimensional response surface. The single protocol, "one size fits all" approach of internal short circuit abuse tests means that such tests fix a number of these parameters and can therefore be gamed. Thus, relative safety under internal short circuit abuse test conditions is not likely to be relevant to the tendency for field-failure.

Nevertheless, the nail penetration test, the round bar crush test, and the drop test do reflect potential triggers (in this case, abuses) that must be considered, especially for transportation applications. In such applications, it is easy to imagine impact conditions that might lead to multiple internal shorts in cells. It is clearly desirable to understand the tolerance of cells to such abuses. The challenge, however, is to understand how representative such tests may be for the wide variety of impacts (speeds, angles, penetrations, etc.) that might occur in the field (putting aside the issue of potential gaming of the tests). Similar problems have been addressed with regard to transport of large quantities of flammable liquid in vehicles' gasoline tanks.

This abuse tolerance versus field-failure framework is also a useful way to characterize testing work carried out by various researchers, or tests that purport to address safety. For example, various tests can be assessed with respect to whether they reflect or measure a property of all cells of the same design and materials (an abuse), or whether they reflect a rare condition that develops after time in the field.

An Overall Framework for Consideration of Li-Ion Battery Safety

The fact that lithium-ion field-failures (1) are so rare, (2) cannot be completely eliminated, and (3) can cause severe damage has led us to develop an integrated framework for development of lithium-ion battery safety technologies. The framework is especially important in transportation applications. It is clear that field-failures continue to occur in portable computer applications (multiple CPSC and newspaper references can be found) and seem to attract only modest attention. However, a single field-failure in a PHEV or EV is likely to be more violent and represents greater threat to human safety. Thus, it is highly desirable to develop means to assure that such field-failures do not occur and that their impact is minimized.

The overall framework is depicted in Fig. 9.16 and advocates that several categories be employed to consider lithium-ion safety technologies. These categories help organize specific approaches. First are technologies that preempt thermal runaway, examples of which might include specific cell designs or specific separator types. Next, technologies to detect the earliest stages of development of internal shorts might, for example, involve identification of specific electrical signals and the use of specific sensors. Detection can be of value only if accompanied by access to effective means for intervening and preventing development of a full-blown thermal runaway. When all else fails, it is still very desirable to confine thermal runaway to just a single cell, namely to prevent cascading phenomena.

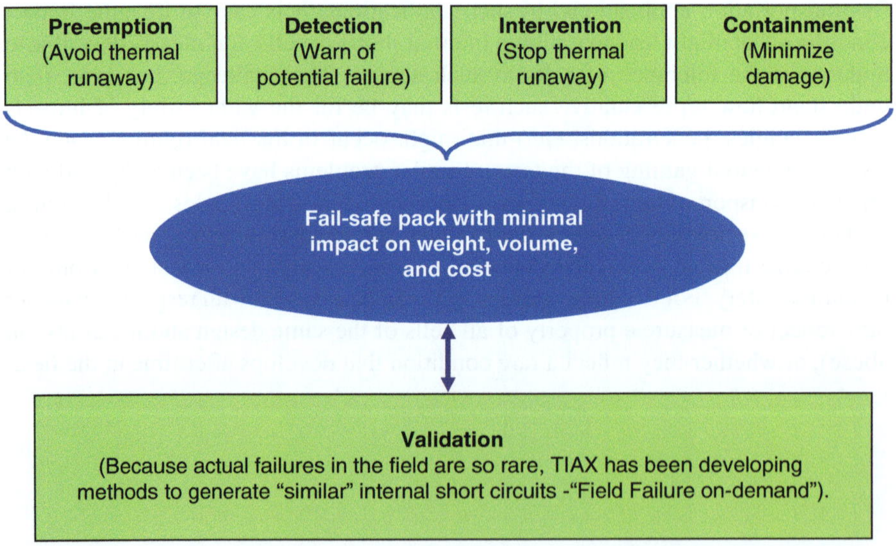

Fig. 9.16 Framework for consideration of Li-ion battery safety technologies

Figure 9.16 also captures the difficult challenge of validating any new safety technology that is developed. How can one validate a safety technology intended to address events that occur on the order of once for every 5–10 million cells without producing and testing many millions of cells? The challenge is fundamental, and thus, it is important to have available means of producing cell/battery failures that are arguably "similar" to those that occur in the field. For these reasons, verification of the failure mechanism is also important.

Future Directions

Abuse tolerance in lithium-ion cells is critical to public safety, and a number of screening tests for lithium-ion cells and batteries have been devised. Battery developers and users then test cells and batteries to screen for tolerance with respect to the given abuse, and then design cells and batteries with improved abuse tolerance.

In contrast, safety in lithium-ion cells/batteries requires deliberate new strategies to deal with the possible occurrence of internal shorts that can progress to thermal runaway. These strategies need to recognize the extent to which the safety tests most typically utilized in the battery industry do not reflect the most common and most dangerous mechanism by which safety incidents actually occur in the field – field-failures.

The battery community does not utilize an overall strategy to intervene, for safety's sake, as temperature rises in a lithium-ion cell or battery in advance of a thermal runaway. Although literature publications may consider the relative temperatures at which particular battery materials decompose, there is no consideration given to how these temperatures may be prevented.

The work presented in this chapter clearly shows that when thermal runaway occurs, it proceeds so fast that cell temperature progresses from less than about 150–180°C to over 600°C almost instantly. This result also indicates the fallacy in imputing safety advantages to one battery material that exhibits an onset temperature for decomposition (as captured, e.g., in DSCs) that is somewhat higher than that of another battery material (e.g., a cathode material with an onset temperature of 220°C is suggested to result in a safer cell than one with an onset temperature of 190°C). For example, the assumption is often made that cathode materials with higher onset temperatures will automatically make a cell safer or eliminate safety incidents. This work illustrates why that view has limited validity; first because anode decomposition is generally the lower temperature trigger for initiating thermal runaway subsequent to internal heating generated by an internal short, and second, because higher cathode decomposition temperature can at best increase the threshold energy for inducing thermal runaway (not the threshold power), but cannot eliminate the threat altogether.

Looking to the future, we suggest that abuse tolerance and field-failures be managed as distinct concerns. With both categories, a reasonable strategy involves identifying and pursuing all efforts to prevent the final runaway process from maturing.

In that sense, a new approach to safety should also recognize one additional factor; namely that lithium-ion cells are no longer useful when their temperature exceeds about 100–110°C, and thereafter, they should/must be removed from service. Abuse tolerance can be further enhanced by focusing all efforts on identifying and shutting down all offending safety-related triggers by, for example, discharging the cell as soon as a cell temperature exceeds about 110–120°C. If an abuse/trigger is removed reasonably quickly, the cell simply cools down, and no safety event occurs.

In the case of field-failures (the internal short), this approach does not work because the cell skin temperatures may not reach these levels until thermal runaway is inevitable and/or in progress. Because it can be self-driven, thermal runaway from internal shorts must be detected during charge/discharge cycles prior to the final charge cycle that results in an internal short "mature" enough to lead to a thermal runaway. Therefore, it is imperative to identify early warning signals and to develop sensors that can warn of developing internal shorts well before a thermal runaway initiates and that are capable of integration at low cost into battery management systems.

We hope this chapter encourages the battery community to pursue more fundamental work into safety testing as well as understanding of safety-related processes in lithium-ion batteries. Work is needed to develop more sophisticated models, to measure material properties that contribute to enhanced safety, to develop improved methods of calorimetry, and to develop new safety technologies that will help assure safety of lithium-ion batteries. With greater visibility of the issues, perhaps this field will be taken up in graduate schools as an important topic that should yield rich problems and productive theses.

Acknowledgments The authors gratefully acknowledge support from the US Department of Energy for portions of this work. In addition, countless discussions with present and past colleagues at TIAX over several years have resulted in many insights regarding battery safety captured in this work. Specifically, the authors gratefully acknowledge Drs. Christopher H. McCoy, Mehmet Rona, Jane Rempel, Ms. Rosalind Takata, Drs. Bookeun Oh, S.K. Singh, Karen Thomas-Alyea, and Per Onnerud. Investigation of various safety incidents also led to very productive discussions with major battery companies and portable product companies which, though anonymous, are gratefully acknowledged.

Bibliography

1. Takeshita H (2011) Worldwide market update on secondary batteries for portable devices, automotive and ESS. Tutorial Presented at the 28th international battery seminar and exhibit, Ft. Lauderdale, FL
2. Zhang J (2011) Li-ion in EDV and safety perspectives. Presentation at the 28th international battery seminar and exhibit, Ft. Lauderdale, FL

3. Roth EP (2008) Abuse testing of high power batteries. Presented at the DOE vehicle technologies peer review, Gaithersburg, MD
4. Nishiyama Y, Tanaka T, Nakajima K (2006) Numerical simulations of thermal behavior with safety improvements of Li-ion battery. Presentation at the 210th ECS meeting, Cancun, Mexico
5. Spotnitz RM, Weaver J, Yeduvaka G, Doughty DH, Roth EP (2007) Simulation of abuse tolerance of lithium-ion battery packs. J Power Sources 163(2):1080–1086
6. Botte GG, Johnson BA, White RE (1999) Influence of some design variables on the thermal behavior of a lithium-ion cell. J Electrochem Soc 146(3):914–923
7. Spotnitz R, Franklin J (2003) Abuse behavior of high-power, lithium-ion cells. J Power Sources 113:81–100
8. Hatchard TD, MacNeil DD, Basu A, Dahn JR (2001) Thermal model of cylindrical and prismatic lithium-ion cells. J Electrochem Soc 148(7):A755–A761
9. Ichimura M (2007) The safety characteristics of lithium-ion batteries for mobile phones and the nail penetration test. In: Proceedings of the 29th international telecommunications energy conference, Rome, IEEE
10. Horn Q, White K, Singh S (2010) Assessing thermal stability of commercial lithium-ion cells. IMLB, Montreal
11. Kim G-H, Pesaran A, Spotnitz R (2007) A three-dimensional thermal abuse model for lithium-ion cells. J Power Sources 170(2):476–489
12. US Automotive Battery Consortium (2005) Freedomcar electrical energy storage system abuse test manual for electric and hybrid electric vehicle applications
13. UL 1642, published by the Underwriters Laboratory
14. Jacoby M (2007) Chem Eng News 85(51):26
15. Hockenberry J (2007) Building a better battery. Wired Magazine, November 2007
16. Stringfellow R, Ofer D, Sriramulu S, Barnett B (2010) New framework for lithium-ion battery safety. IMLB, Montreal, Canada, invited talk (TIAX LLC)
17. Stringfellow R, Ofer D, Sriramulu S, Barnett B (2010) 218th meeting of the Electrochemical Society, Las Vegas
18. Barnett B, Sriramulu S (2010) A perspective on Li-ion safety and opportunities for portable and electric vehicle applications. Presentation at the 27th international battery seminar and exhibit, Ft. Lauderdale, FL
19. Barnett B, Ofer D, Oh B, Stringfellow R, Singh SK, Sriramulu S (2008) On the role of the active materials in thermal runaway from internal short circuits. IMLB 2008 international meeting on lithium batteries, Tianjin
20. Barnett B, Sriramulu S, Singh SK (2007) Influence of active material heat release kinetics on thermal runaway following an internal short circuit. 212th ECS meeting, Washington DC, 7–12 Oct 2007
21. Barnett B, Doughty D, Thomas-Alyea K, Roth P (2006) Safety for lithium-ion: abuse tolerance versus field failure. IMLB 2006 international meeting on lithium batteries, 18–23 June 2006
22. Harris SJ, Timmons A, Pitz WJ (2009) A combustion chemistry analysis of carbonate solvents used in Li-ion batteries. J Power Sources V193:855–858
23. Haik O, Ganin S, Gershinsky G, Zinigrad E, Markovsky B, Aurbach B, Halalay I (2011) On the thermal behavior of lithium intercalated graphites. J Electrochem Soc 158(8):A913–A923
24. MacNeil DD, Dahn JR (2001) Test of reaction kinetics using both differential scanning and accelerating rate calorimetries as applied to the reaction of Li*x*CoO2 in non-aqueous electrolyte. J Phys Chem A 105:4430–4439
25. Maleki H, AlHallaj S, Selman JR, Dinwiddie RB, Wang H (1999) Thermal properties of lithium-ion battery and components. J Electrochem Soc 146(3):947–954
26. Lu W, Belharouak I, Vissers D, Amine K (2006) In situ thermal study of Li1 + x[Ni_1/3Co_1/3Mn_1/3](1 − x)O − 2 using isothermal micro-clorimetric techniques. J Electrochem Soc 153(11):A2147–A2151

27. Yang H, Bang H, Amine K, Prakash J (2005) Investigations of the exothermic reactions of natural graphite anode for Li-ion batteries during thermal runaway. J Electrochem Soc 152(1): A73–A79
28. Arai H, Tsuda M, Saito K, Hayashi M, Sakurai Y (2002) Thermal reactions between delithiated lithium nickelate and electrolyte solutions. J Electrochem Soc 149(4):A401–A406
29. MacNeil DD, Dahn JR (2001) Test of reaction kinetics using both differential scanning and accelerating rate calorimetries as applied to the reaction of Li_xCoO_2 in non-aqueous electrolyte. J Phys Chem A 105:4430–4439
30. Maleki H, Howard JN (2009) Internal short circuit in Li-ion cells. J Power Sources 191(2):568–574
31. Battery Association of Japan (2008) Activities for safety of Li-ion batteries. Presentation at UN informal working group meeting, Washington, DC, 11–13 Nov 2008
32. Cai W, Wang H, Maleki H, Howard J, Lara-Curzio E (2011) Experimental simulation of internal short circuit in Li-ion and Li-ion-polymer cells. J Power Sources 196(18):7779–7783
33. Orendorff CJ, Roth EP, Nagasubramanian G (2011) Experimental triggers for internal short circuits in lithium-ion cells. J Power Sources 196(15):6554–6558
34. Keyser M (2011) Development of a novel test method for on-demand internal short circuit in a Li-ion cell. Presentation at the 11th international automotive battery conference, Pasadena, CA
35. Zhao F (2007) LIB safety study and improvement. Presentation at the IEEE symposium on product compliance engineering. Denver, CO, USA
36. Mikolajczak C, Harmon J, White K, Horn Q, Wu M, Shah K (2010) Detecting lithium-ion cell internal faults in real time. Power Electronics Technology, March 2010
37. Hayes TA, Mikolajczak C, Horn Q (2010) Key manufacturing practices and techniques to achieve high quality Li-ion cells. Presented at the 27th international battery seminar and exhibit, Ft. Lauderdale, FL
38. Hayes T, Mikolajczak C, Megerle M, Wu M, Gupta S, Halleck P. Use of CT scanning for defect detection in lithium-ion batteries. In: Proceedings, 26th international battery seminar and exhibit for primary and secondary batteries, small fuel cells, and other technologies, Fort Lauderdale, FL, March 16–19, 2009
39. Mikolajczak CJ, Harmon J, Priya G, Godithi R, Hayes T, Wu M (2010) From lithium plating to cell thermal runaway: a combustion perspective. Presentation at the 27th International Battery Seminar and Exhibit, Fort Lauderdale, Florida, March 2010
40. Darcy E (2007) Screening Li-ion batteries for internal shorts. J Power Sources 174:575–578
41. Chen SC, Wan CC, Wang YY (2005) Thermal analysis of lithium-ion batteries. J Power Sources 140:111–124
42. Kawaji H, Oka T, Tojo T, Atake T, Hirano A, Kanno R (2002) Low-temperature heat capacity of layer structure lithium nickel oxide. Solid State Ionics 152–153:195–198

Chapter 10
Lithium-Ion Battery Systems and Technology

Zhengming (John) Zhang and Premanand Ramadass

Glossary

Active material	The material in the electrodes of a cell or battery that takes part in the electrochemical reactions of charge or discharge.
Battery/battery pack	An assembly of any number of Li-ion cells, associated electronics, battery packaging, and connector(s).
Capacity	The total number of Ampere-hours (Ah) that can be withdrawn from a fully charged cell or battery under specified conditions of discharge.
Cell	Basic lithium ion unit providing a source of electrical energy by direct conversion of chemical energy that consists of electrodes, separator, electrolyte, container, and terminals, and that is designed to be charged electrically.
Current collector	An inert member of high electrical conductivity used to conduct current from or to an electrode during discharge or charge.
Cycle	The discharge and subsequent or preceding charge of a Li-ion cell or battery such that it is restored to its original conditions.
Cycle life	The number of cycles under specified conditions which are available from a Li-ion cell or battery before it fails to meet specified criteria as to performance.
Electrolyte	The medium which provides the ion transport mechanism between the positive and negative electrodes of a cell.

This chapter was originally published as part of the Encyclopedia of Sustainability Science and Technology edited by Robert A. Meyers. DOI:10.1007/978-1-4419-0851-3

Z. (John) Zhang (✉) • P. Ramadass
Celgard, LLC, 13800 South Lakes Dr., Charlotte, NC 28273, USA
e-mail: johnzhang@celgard.com; premramadass@celgard.com

R.J. Brodd (ed.), *Batteries for Sustainability: Selected Entries from the Encyclopedia of Sustainability Science and Technology*, DOI 10.1007/978-1-4614-5791-6_10,
© Springer Science+Business Media New York 2013

Negative electrode	The electrode acting as an anode when a cell or battery is discharging.
Overcharge	The forcing of a current through a Li-ion cell or battery pack after it has been fully charged.
Positive electrode	The electrode acting as a cathode when a cell or battery is discharging.
Separator	An ionically conductive, electronically insulative material which prevents electronic contact between electrodes of opposite polarity in the same cell.
Thermal runaway	The cell condition where the internal cell reactions generate more thermal heat than the cell can dissipate. The condition causes cell venting and premature failure.

Definition of the Subject

Lithium-ion battery (LIB) is one of rechargeable battery types in which lithium ions move from the negative electrode (anode) to the positive electrode (cathode) during discharge, and back when charging. It is the most popular choice for consumer electronics applications mainly due to high-energy density, longer cycle and shelf life, and no memory effect. With Li-ion batteries currently gaining much attraction in electric drive vehicle, the concern for global warming and a clean environment may be well served with advances in such systems.

Introduction

During the last 15 years, lithium-ion batteries have dominated the advanced energy sources by powering the modern portable electronics and replaced many other commercial battery systems in the market. The prime reasons for its rapid success and proliferation in consumer electronic market are its superior characteristics over other battery systems, namely, high voltage, high-energy density, and longer cycle and shelf life with no memory effect. The significant progress of Li-ion batteries is mainly due to numerous innovations and advancements in materials, designs, and safety. This entry is intended to provide an overview of Li-ion batteries on several aspects. Starting with a brief discussion on its history, commercial success, and working mechanism, all the critical components inside the cell have been discussed in adequate detail. The later part of this chapter will primarily focus on manufacturing process, recent advances, and modern trends in Li-ion systems and detailed discussion on Li-ion cell safety.

History of Li-Ion Batteries

The working mechanism of Li-ion cells is often related to that of a rocking chair since within the cell, the lithium ions swing between the negative electrode and the positive electrode through an organic liquid electrolyte, like a rocking chair from side to side. This "rocking-chair" concept was first proposed by M.S. Whittingham at Exxon in 1970s where he discovered the concept of intercalation electrodes using titanium(II) sulfide as the cathode and lithium metal as the anode [1]. The principal concept is based on the intercalation reaction and is rather different from other conventional secondary batteries which are based on chemical reactions.

Goodenough was the first to patent $LiCoO_2$ as a lithium intercalation cathode material in 1980 [2] and H. Ikeda of Sanyo was the first to patent an intercalation material in an organic solvent such as graphite in 1981 [3]. S. Basu of Bell Laboratories filed an US patent in 1982, based on his finding of lithium intercalation in graphite at room temperature [4]. I. Kuribayashi and A. Yoshino developed a new cell design using an intercalation carbon anode and a $LiCoO_2$ cathode and filed patents worldwide [5]. In 1991, Sony Energytec Inc. began to produce commercial lithium-ion cells under the leadership of Yoshio Nishi. Bellcore team under the leadership of J. M. Tarascon pioneered the modern concept of "polymer" (soft pack) Li-ion cells [6–9].

Soon after commercialization, the Li-ion battery (LIB) system became a popular choice because of its high-energy density, good performance, and no memory effect as occurred with nickel-cadmium (Ni-Cd) or nickel-metal hydride (Ni-MH) batteries. LIBs have been primarily used for portable electronics, especially cellular phones and notebook computers. Recently, the application area has been extended to power tools, electric bikes, and energy storage systems. Several companies are now working toward adapting the lithium-ion system for use in electric drive vehicle (EDV) applications.

Li-ion cells are being built in many different shapes and configurations—button, flat, rectangular (also referred as prismatic), and cylindrical. The cell components (electrodes, electrolyte, and separator) are designed to accommodate a particular cell shape and design. The separators are either stacked between the electrodes or wound together with electrodes to form jellyrolls, as shown in Fig. 10.1. Typical safety features included in cells with metal enclosure are the PTC and CID devices that are usually incorporated into the cell top plate and seal. The PTC is a positive temperature coefficient device that changes resistance at a set temperature or current flow and stops a thermal runaway condition from developing. The CID is a current interrupt device that is incorporated into the cell cap, and it interrupts the electrical connection between the cathode tab and the positive terminal when the internal pressure in the cell reaches a given pressure. Another safety feature present in Li-ion cell is the cell vent which activates at pressures just higher than the CID to release the volatile solvent and stops cell action.

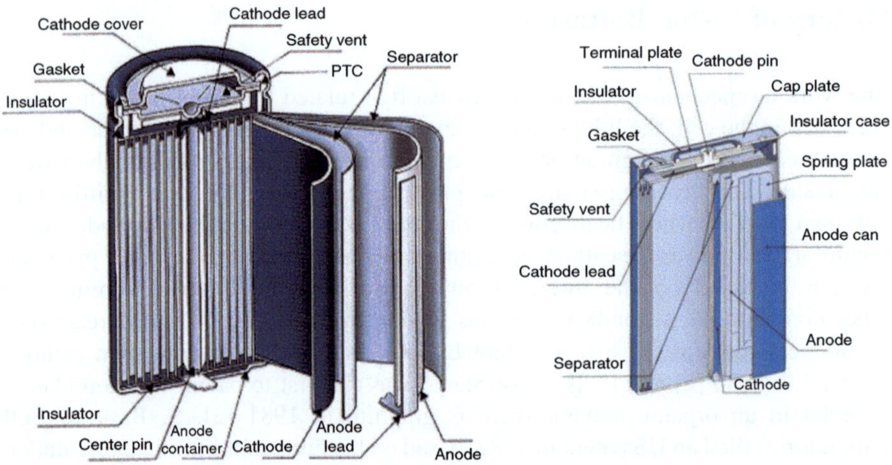

Fig. 10.1 Typical Li-ion battery configuration: (**a**) spirally wound cylindrical cell, (**b**) wound type prismatic cell

Commercial Success

The battery industry has seen enormous growth over the past 15 years in portable, rechargeable battery packs. The majority of this surge can be attributed to the widespread use of cell phones, laptop computers, tablet PCs, video games, and other wireless electronics. At present, this technology is gaining worldwide attention as a battery option for transportation applications, including electric drive vehicles namely HEVs, PHEVs, and BEVs.

The higher volumetric and gravimetric energy storage capability are key characteristics of the Li-ion battery system compared to the conventional sealed nickel-cadmium (Ni-Cd), nickel-metal hydride (Ni-MH), and valve-regulated lead acid (VRLA) battery systems. Figure 10.2 presents the energy density and specific energy comparisons of small sealed rechargeable battery systems. Some of the distinct advantages of Li-ion system over other commercial rechargeable systems are the choice of chemistry with highest energy and lightest weight, good cycle life, no memory effect, higher energy efficiency, and better high rate capability. Of course there are also certain issues for Li-ion system similar to any other high-energy storage devices that include higher charging times, thermal runaway concerns, relatively more expensive, and requiring advanced protection circuitry for safety and to prevent overcharge and over discharge.

Based on a recent report on rechargeable battery markets that Global Industry Analysts (GIA) announced, the global market for rechargeable batteries is forecast to reach US$16.4 billion by the year 2015 [10]. Factors driving the market growth include growing consumer acceptance of rechargeable battery technologies in various parts of the world, rapid growth in the electronics market, and increasing role of rechargeable batteries in the automotive sector. It should be noted that in 2006, the

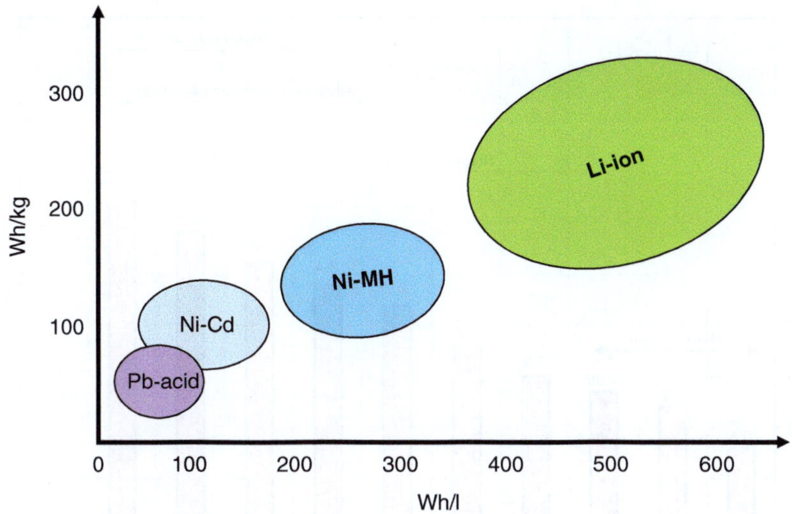

Fig. 10.2 Volumetric energy density (Wh/l) and gravimetric energy density (Wh/kg) for major rechargeable battery systems

market size of rechargeable battery systems for portable power applications was around US$6 billion out of which Li-ion comprises 75% of the market.

The market survey reports show that the lithium-ion battery is becoming an almighty rechargeable system and in 2010, a whopping sum of around 3.9 billion cells has been supplied all together by cell manufacturers around the world which is about twofold increase in supply compared to 2006 [11] and the supply is expected to grow by at least 15% during 2011. There is a steady increase in the demand for lithium-ion batteries for all portable electronic devices (almost 100% of cell phones and notebook PCs), and in addition, the Li-ion system also started penetrating more and more in other arenas like power tools, energy storage systems (ESS), and so on. Li-ion system now starts competing with Pb-acid systems in the energy storage systems arena primarily due to its superior cycle life and higher columbic efficiency. For the automotive applications, lithium-ion cells have become a hot topic and are now on the verge of being brought on board vehicles on a massive scale. Although the Li-ion battery for HEV application still has little experience in the real market, research and development activity for it is very dynamic at present worldwide. This fact must surely reflect the common recognition that the technology is one of the most promising for the HEV battery of the coming generation. It is predicted that the HEV will be significantly widespread by 2015 and is expected to take a certain portion of the application market at that time.

The tremendous progress in Li-ion cells is clearly visible with as much as a twofold increase in the volumetric and gravimetric energy density combined with a 50% reduction in price was achieved between 1995 and 2006. Figure 10.3 depicts the increase in capacity of the cylindrical 18650 cell (18 mm in diameter and 65.0 mm long) from 1992 to 2006. In 1995, an 18650 cell with a capacity of

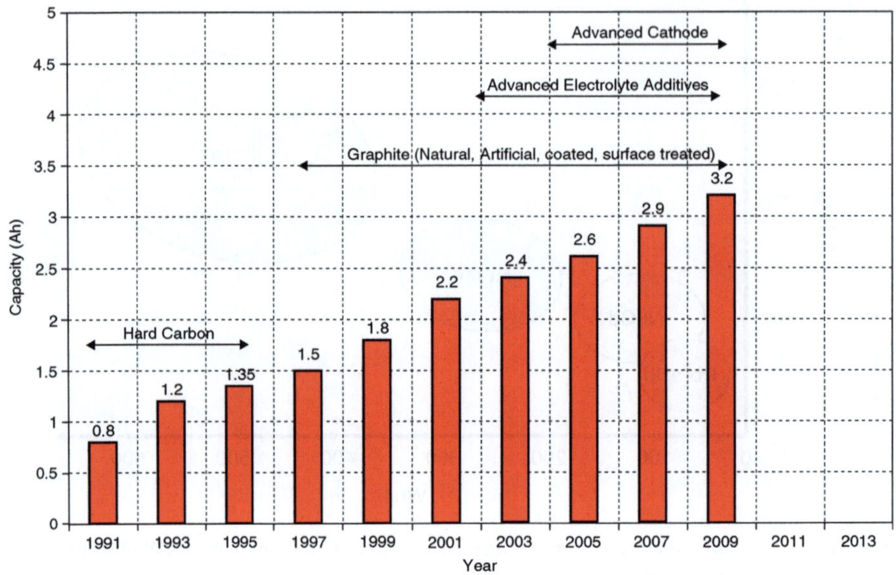

Fig. 10.3 Capacity (Ah) increase for 18650 lithium-ion cell resulting from advances in active materials and functional additives to electrolyte

around 1.5 Ah was sold for $8, while in 2006, the same size cell with 2.6 Ah sold for about $4. The early LIBs had a capacity of 800 mAh and an end-of-charge voltage of 4.1 V. The initial cells used hard-carbon anode materials which had a capacity of about 200 mAh/g, and the $LiCoO_2$ had a capacity of nearly 110 mAh/g due to 4.1 V charging voltage. In 2003, the capacity of the 18650 cell reached 2.4 Ah. This corresponds to an energy density of over 200 Wh/kg or 500 Wh/L, respectively. These values were reached in part by increasing the cell-operating voltage higher than 4.2 V due to the availability of improved graphite anode materials, electrolyte additives, and a stabilized $LiCoO_2$.

The cell producers accomplished the performance improvements through engineering improvements in cell design, new electrode materials, and automated high-speed production to reduce the cost. The capacity of the 18650 cell had reached 2.9 Ah in 2007 based on treated graphite anode and planar-nickel-based cathode and with several kinds of electrolyte additives [12]. With further continuous improvements in all the cell components that includes silicon alloy–type anode materials, lithium-nickel-cobalt-aluminum and nickel-manganese-cobalt cathode materials, novel electrolyte and/or additives, some cell manufacturers are currently able to achieve a maximum capacity of up to 3.4 Ah for the same 18650 cell design.

The major Li-ion cell manufacturers are from Japan, China, and S. Korea. Figure 10.4 presents the global market share for bare cell shipment volumes from the three countries cell manufacturers for the consumer portable power applications in 2010. Some of the major cell manufacturers include Sanyo, Sony, and Panasonic from Japan; Samsung SDI and LGChem from Korea; and BYD, ATL, and Lishen from China.

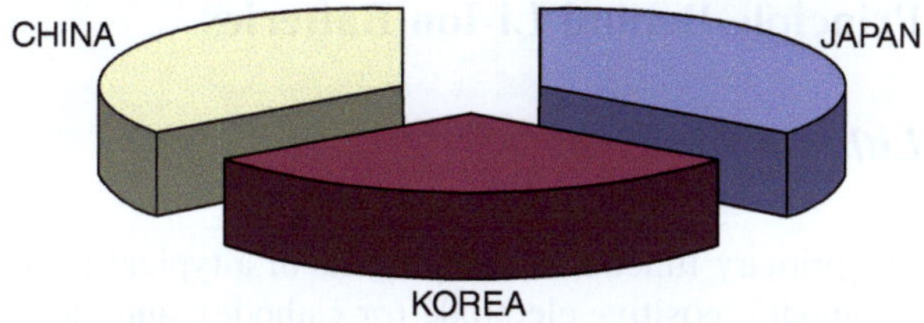

Fig. 10.4 Major cell manufacturers and their lithium-ion battery movement globally in 2010

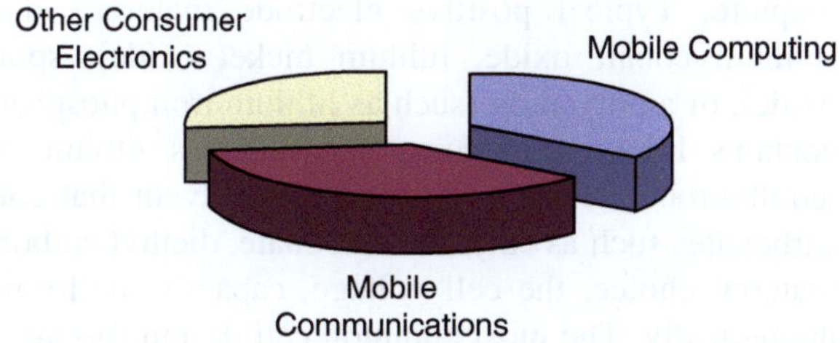

Fig. 10.5 Movements in cell shipment volumes for each lithium-ion battery application globally in 2010

Figure 10.5 shows the split-up of small format LIBs supplied for portable power applications namely mobile computing, mobile communications, and other consumer electronics including power tools during 2010. The thin, rectangular polymer/laminate cell construction has been found increasingly favorable in the modern portable consumer electronics market (e.g., Iphone, Ipad smart phones, and so on). Overall, for the consumer electronics market sector, Japan and Korea dominates the production of 18650 type cells and China dominates in prismatic type cell production. There are no major Li-ion manufacturers in the United States or in Europe, even though they constitute large markets for devices powered by Li-ion batteries [13]. Production activity in the United States so far is limited to companies that supply the niche medical and military markets.

There are many emerging market opportunities for Li-ion cells in transportation. Li-ion batteries can provide very high power and are therefore a good candidate for EDV applications. The automotive power market is viewed as the largest future growth opportunity for Li-ion batteries. Their high voltage, low weight, small volume, and potentially long life make them very attractive. Different cell designs and geometries are currently being adapted by cell manufacturers for automotive applications. A few common designs include rectangular flat pouch cell and larger cylindrical cell design.

The cell's capacities for EDV applications are in the range of 25–100 Ah. The typical energy requirement for plug-in hybrid electric vehicle (PHEV) is between 5 and 20 kWh and for all electric vehicle (EV), the energy requirement is more than 20 kWh with a voltage of 300 V or more. A number of modules with energy in the range of 1–2 kWh are necessary. The advantage of Li-ion over Ni-MH for hybrid electric vehicle (HEV) applications that has an energy requirement in the range of 3–5 kWh is still quite not clear, but with decreasing cost as production volume increases, Li-ion cells are considered to be economically competitive. Once Li-ion system finds its place, the transportation market could dominate over the present portable device market.

Principle Behind Li-Ion Batteries

Lithium-Ion Cell

The primary functional components of a typical Li-ion cell are negative electrode (or anode), positive electrode (or cathode), and electrolyte. Most commonly used negative electrode materials include hard carbon, graphitic carbon, and treated graphite. Typical positive electrode materials may include a layered oxide (lithium cobalt oxide, lithium nickel oxide), spinel (like lithium manganese oxide), or a polyanion (such as lithium iron phosphate). The electrolyte in general contains lithium-containing salt such as lithium hexafluorophosphate, lithium tetrafluoroborate, etc. dissolved in a solvent that comprises a mixture of organic carbonates such as ethylene carbonate, diethyl carbonate, etc. Depending upon the material choice, the cell voltage, capacity, performance, and safety can change dramatically. The most common cell design formats of Li-ion cells include cylindrical, prismatic, and pouch. In this section, the working mechanism of Li-ion cells has been discussed along with an overview of each cell component (negative electrode, positive electrode, electrolyte, and separator). Later part of this chapter would add more information on each of them including the modern trend.

Li-ion Cell Mechanism

A typical Li-ion cell consists of a positive electrode composed of a thin layer of powdered metal oxide (e.g., $LiCoO_2$) coated on aluminum foil and a negative electrode formed from a thin layer of powdered graphite, mounted on a copper foil. The two electrodes are separated by a porous polyolefin film called as separator soaked typically in a electrolyte made of LiPF6 salt dissolved in a mixture of organic solvents such as ethylene carbonate (EC), ethyl methyl carbonate (EMC), or diethyl carbonate (DEC).

During cell operation, all the three functional components participate in the electrochemical reaction. Both the negative and positive electrodes are materials into which, and from which, lithium can migrate. During insertion (or intercalation), lithium moves into the electrode and during extraction (or deintercalation), lithium moves back out. During discharge, the lithium ions are deintercalated from the negative electrode moved through the electrolyte and intercalated into the positive electrode. The reverse happens during charging in which lithium ions deintercalate from positive electrode and gets intercalated into the negative electrode. Figure 10.6 shows the schematic representation of the operation of a lithium-ion cell during discharge process. Useful work can only be extracted if electrons flow through a closed external circuit.

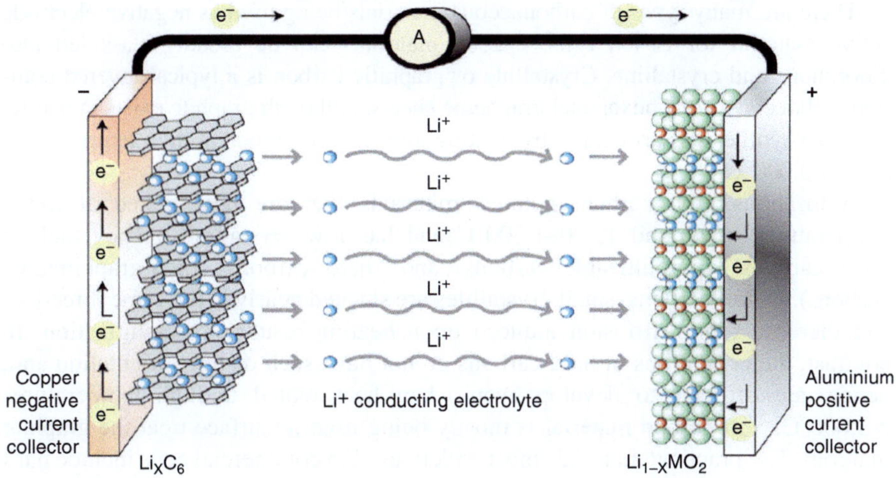

Fig. 10.6 Schematic representation of Li-ion cell operating mechanism

The positive electrode half-cell reaction (with charging being forward) can be represented as:

$$\mathrm{LiCoO_2} \Leftrightarrow \mathrm{Li}_{1-x}\mathrm{CoO_2} + x\mathrm{Li}^+ + xe^-$$

The negative electrode half-reaction (with charging being forward) is:

$$x\mathrm{Li}^+ + xe^- + 6\mathrm{C} \Leftrightarrow \mathrm{Li}_x\mathrm{C}_6$$

where x refer to moles of lithium. The overall cell reaction can thus be represented as:

$$\mathrm{LiCoO_2} + 6\mathrm{C} \Leftrightarrow \mathrm{Li}_{1-x}\mathrm{CoO_2} + \mathrm{Li}_x\mathrm{C}_6$$

where the forward reaction corresponds to charging and the reverse reaction occurs during discharge.

Negative Electrode Material (Anode)

The basic requirements for a negative electrode material to be used in lithium-ion cell are that the material should have minimal volume expansion and stress associated with it during charge/discharge process, higher electronic conductivity, lower irreversible capacity loss during first charging or intercalation process, stable under wide operating temperature window in a highly reducing environment, and lower specific surface area (typically 2 m^2/g) for optimal performance and safety.

There are many types of carbonaceous materials being used as negative electrode active material for Li-ion cells. Carbon materials can be broadly classified into amorphous and crystalline. Crystalline or graphitic carbon is a typical layered compound that consists of hexagonal grapheme sheets. Although graphite exists in nature, it can be synthesized artificially by treating a pyrolyzed carbon at high temperatures around 3,000°C.

Amorphous kind carbon includes materials that are heat-treated at lower temperatures of around 1,100–1,300°C and has low crystallinity. They include "soft carbons" (graphitizable carbons) and "hard carbons" (non-graphitizable carbons). In soft carbons, small crystallites are stacked nearly in the same direction, and thereby, subtle diffusion induced upon heating results in graphitization. In contrast, the crystallites of hard carbons do not have such distinct orientation and, hence, are difficult to develop even when heat-treated at high temperatures >3,000°C. Soft carbon material is mostly being used to surface treat the graphite material. The practical materials most widely used in commercial cells include hard carbon, natural graphite, synthetic graphite, and amorphous carbon coated graphite.

Solid Electrolyte Interface (SEI)

During the first charging after cell assembly, a protective film has been formed on the surface of the negative electrode through reductive decomposition of the electrolytic solution [14, 15]. The passivation film often called as solid electrolyte interface (SEI) acts as an interface between the electrode surface and the electrolyte solution and has the properties of a solid electrolyte with high electronic resistivity. Other favorable properties of the SEI include good mechanical stability/flexibility and good adhesion to the anode. In general, SEI film consists of insoluble and partially soluble reduction products of the electrolyte components. Formation of SEI film suppresses further solvent decomposition and plays a beneficial role in improving the safety and cyclability of lithium ion cells. The SEI formation on negative electrodes thus is a prerequisite for their stable charging and discharging; however, it is the primary cause for the irreversible capacity loss during the first charge through consumption of a considerable amount of lithium. The electrolyte often has additives to enhance SEI formation.

Figure 10.7 shows a typical charge/discharge characteristic of graphitic carbon material. The solid line represents the intercalation or charging, and the dotted line shows the deintercalation or discharging. During first intercalation (charging), the potential drops rapidly after subtle retardation at ~0.8 V (vs. Li^+/Li). Most of the Li intercalation and deintercalation takes place at potentials 0.25 V (vs. Li^+/Li), accompanied by three potential plateaus. Charge consumed upon first charging is not fully recovered upon discharging and the capacity that cannot be recovered is called irreversible capacity which is observed for any carbonaceous material due to the formation of a passive film over the surface of electrode referred as solid electrolyte interface (SEI). At the second and subsequent cycles, graphite shows good

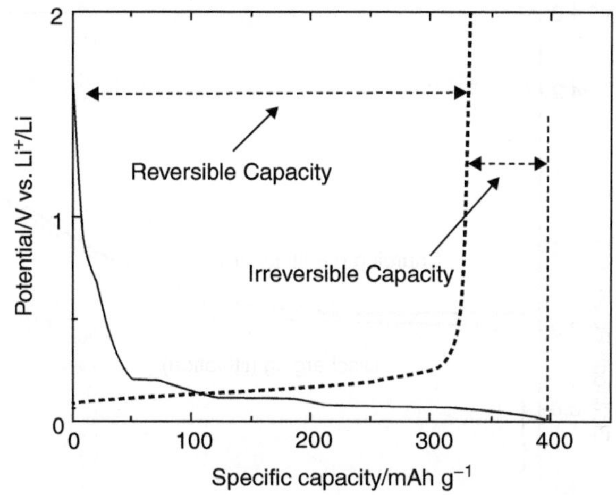

Fig. 10.7 Typical charge/discharge characteristics of graphitic carbon material

Table 10.1 Properties of few commonly used negative electrode materials for lithium-ion cell

Anode type	First charge capacity (mAh/g)	Irreversible capacity loss	Practical capacity (mAh/g)
Hard carbon	250	25–30%	180–190
Natural graphite	370	16–18%	300–320
Surface treated natural graphite	370	7–9%	340
MCMB graphite	350	10–12%	310–320
Massive artificial graphite (MAG)	360	5–8%	330–340
Mesophase carbon fiber (MCF)	350	10–12%	310

reversibility and the reversible specific capacity ranges from 300 to 370 mAh/g. Table 10.1 shows the properties of a few commonly used negative electrode materials of Li-ion cells. Graphitic carbons can insert lithium up to a stoichiometry of LiC_6 corresponding to 372 mAh/g.

Positive Electrode Material (Cathode)

In general, positive electrodes in commercially available Li-ion cells utilize a lithiated metal oxide as the active material. Most of the research and commercialization efforts has centered on two classes of materials. The first group includes materials with layered structure that enables two-dimensional diffusion of lithium ions or a spinel structure which enables three-dimensional diffusion as the transition-metal ions are ordered in all layers. Examples include $LiCoO_2$, $LiNiO_2$, $LiNi_{1-y}Co_yO_2$, etc. for layered compounds and $LiMn_2O_4$ for spinel type compound. The second group includes materials with more open structures like vanadium oxide, transition-metal phosphates namely olivine ($LiFePO_4$). The first group,

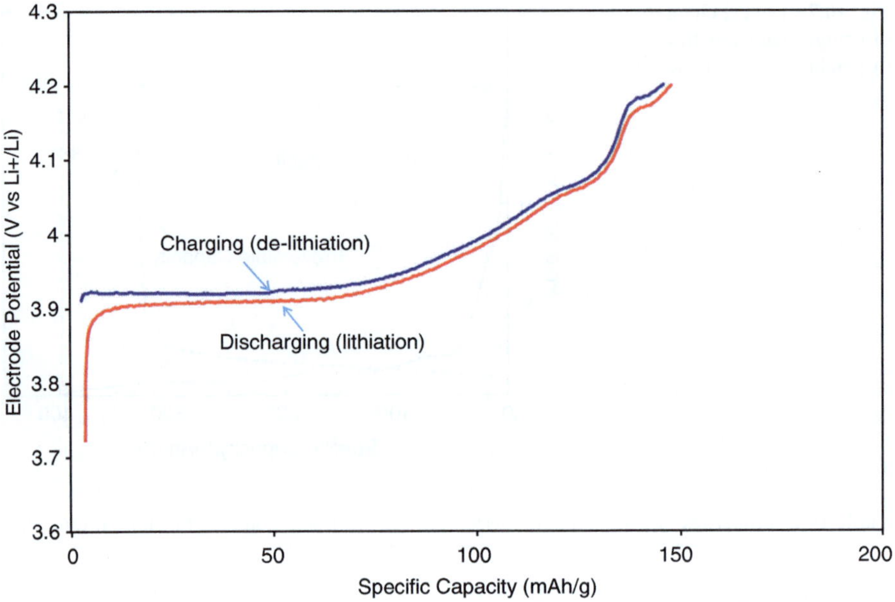

Fig. 10.8 Typical charge/discharge characteristics of lithium cobalt oxide ($LiCoO_2$) positive electrode material. Charging corresponds to lithium deintercalation, and discharging refers to lithium intercalation

Table 10.2 Properties of few commonly used positive electrode materials for lithium-ion cell

Positive electrode material	Practical capacity for EOCV 4.2 V (mAh/g)	Density (g/cc)
$LiCoO_2$	140	5.05
$LiNiO_2$	160	4.8
$LiNi_{0.8}Co_{0.2}O_2$	140	4.85
$LiNi_{0.8}Co_{0.15}Al_{0.05}O_2$	160	4.8
$LiNi_{1/3}Co_{1/3}Mn_{1/3}O_2$	165	4.7
$LiMn_2O_4$	100	4.2
$LiFePO_4$	120[a]	3.7

[a]Charge cutoff voltage is ~3.65 V

because of their more compact lattices, will have an inherent advantage in energy stored per unit of volume, but the second group compound like $LiFePO_4$ is cheaper [16]. Figure 10.8 shows the charge/discharge characteristics of $LiCoO_2$ material. The theoretical specific capacity of the $LiCoO_2$ cell is relatively low at around 130 mAh/g because only around 0.5 Li/Co can be reversibly cycled without causing cell capacity loss due to changes in the $LiCoO_2$ structure associated with the phase changes that cause low reaction rates and also poor stability of the electrode at low lithium contents. Table 10.2 shows the properties of a few commonly used positive electrode materials of Li-ion cells.

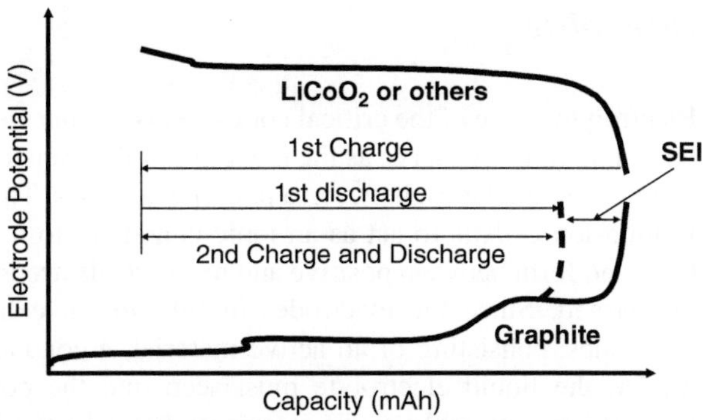

Fig. 10.9 Schematic showing charge/discharge characteristic of Li-ion cell electrode and state of charge operating window after SEI formation

The ideal requirements for a positive electrode active material used in lithium-ion cell are that the material needs to contain a readily reducible/oxidizable atom, like a transition metal; should react with lithium in a reversible manner without changing the host structure; be a good electronic conductor; be able to react with lithium very rapidly to provide higher power density; should react with lithium with a high free energy of reaction to obtain high voltage, high capacity, and hence high-energy storage; and be stable under wide voltage and temperature window. There are several physical properties that determine the quality of positive electrode active material and also influence the battery characteristics. They include particle size, shape, distribution of particle size, water content, tap density, specific surface area, crystallinity, impurity level, and so on.

Cell Balance

Lithium-ion cells operate by cycling lithium ions between two insertion electrode hosts having different insertion energies. Figure 10.9 shows the schematic of charge discharge characteristics of Li-ion cell electrodes (e.g., $LiCoO_2$ as positive and graphite as negative electrode) [17]. After assembling and filling with electrolyte, the cell will be in a completely discharged state where all the lithium ions are present in the positive electrode. During initial charging, some portion of the lithium ions goes into the irreversible SEI film formation reaction and rest of the lithium ions gets intercalated into the negative electrode. The negative to positive electrode capacity ratio is very critical in determining the cell performance and safety with continuous cycling. Smaller capacity ratios would present safety hazard as the negative electrode can get overcharged as more lithium ions are available for intercalation than is desirable. Higher capacity ratios may prevent optimal utilization of the negative electrode. The typical negative to positive electrode capacity ratio used by most cell manufacturers is around 1.15. The first charge capacity of negative electrode (Li-ion intercalation) needs to be at least 15% greater than the first charge capacity of the positive electrode (Li-ion deintercalation) to accommodate the SEI formation.

Electrolyte

Electrolyte is one of the critical components for any electrochemical device, and the basic function is to serve as the medium for the transfer of charges, which are in the form of solvated ions, between a pair of electrodes. The role of liquid electrolytes in lithium-ion cells is to act as an ionic conductor to transport solvated lithium ions back and forth between positive and negative electrodes as the cells are charged and discharged. Since the electrodes in lithium-ion cells are the porous composite electrodes, consisting of an active material, a conductive material and a polymer binder, the liquid electrolyte must seep into the porous electrodes and transfer lithium ions smoothly at the interfaces between the liquid and solid phases. The minimal requirements of the electrolyte used in Li-ion cells include good ionic conductivity and electronic insulator property, stability over a wide electrochemical voltage window and operating temperature. Most commonly used salt materials for the electrolyte include $LiPF_6$, $LiBF_4$, and the most common solvents are carbonate based, namely, EC, DMC, DEC, EMC, and PC.

An ideal electrolyte solvent for Li-ion cells shall meet the following minimal criteria, namely, high dielectric constant, to be able to dissolve salts of sufficient concentration, lower viscosity for facile ion transport, inert to all cell components, especially the charged surfaces of the electrodes, lower melting point, and higher boiling point to remain liquid in a wide temperature range.

An ideal electrolyte solute in lithium-ion cells completely dissolves and dissociate, in the nonaqueous media, and the solvated ions should be able to move in the media with high mobility, should be stable against oxidative decomposition at the positive electrode, should be inert to electrolyte solvents and other cell components, and should be nontoxic and remain stable against thermally induced reactions with electrolyte solvents and other cell components. LiPF6 is one of the most commonly used salts on commercial Li-ion cells. The success of LiPF6 was not achieved by any single outstanding property but, rather, by the combination of well-balanced properties, namely, conductivity, ionic mobility, dissociation constant, thermal stability, and electrochemical/chemical stability.

Electrolyte Additives

Electrolyte additives are compounds that add extra functions to the liquid electrolytes in addition to the fundamental function as an ionic conductor. Examples of role-assigned electrolytes include anode passivation film-forming agents, cathode protection agents, overcharge protection agents, wetting agents, flame retardant agents, and so on. It should be noted that small amount of the electrolyte additives dramatically improves the performances of lithium-ion cells. Since the influence of electrolyte additives depends not only on the electrode materials but also on cell designs, the research and development of the additives is quite challenging.

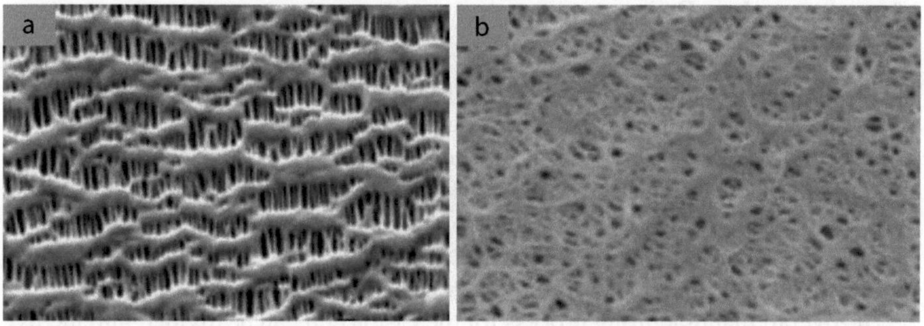

Fig. 10.10 SEM images of separator membrane produced through (**a**) dry and (**b**) wet process methods

Separators

A separator is a porous membrane physically placed between electrodes of opposite polarity, permeable to ionic flow but preventing electric contact between the electrodes. As batteries have become much sophisticated, separator function also has become more demanding and complex. The main function of separator in Li-ion cell is to keep the positive and negative electrodes apart to prevent electrical short circuits and, at the same time, allow rapid transport of ionic charge carriers that are needed to complete the circuit during the passage of current in an electrochemical cell. They should be very good electronic insulators and have the capability of conducting ions by soaking electrolyte. Li-ion battery separators can be produced through dry and wet processes. Figure 10.10 presents the micrographs of separator membrane produced through dry and wet process methods.

A wide variety of properties are required for separators used in Li-ion cells [18]. They include: electronic insulator, minimal electrolyte (ionic) resistance, sufficient mechanical and dimensional stability and physical strength to allow easy handling, chemical stability to degradation by electrolyte, impurities, and electrode reactants and products, readily wetted by electrolyte and effective in preventing migration of particles or soluble species between the two electrodes. In order to achieve good performance of Li-ion cells, the separators should have higher ionic conductivity with electrolyte, low shrinkage, and uniform pore structure. Cells with high-resistance separators will perform poorly during high rate discharge and also will increase the cell-charging time.

Microporous polyolefin membranes in current use are thin (30 um) and are made of polyethylene (PE), polypropylene (PP), or laminates of polyethylene and polypropylene. They are made up of polyolefin materials because they provide excellent mechanical properties, chemical stability, and acceptable cost. They have been found to be compatible with the cell chemistry and can be cycled for several hundred cycles without significant degradation in chemical or physical properties.

Manufacturing Process

The basic principles and processes of cell design and fabrication are well known and quite similar among cell manufacturers. The International Electrotechnical Commission (IEC) has established a common nomenclature for describing the various cell sizes and chemistry. For example, the most common cylindrical Li-ion cell ICR18650 translates into: I is for Li-ion technology, C for cobalt-based cathode, R is for a round cell, 18 is for the cell diameter in millimeters, and 650 is for the cell height in tenths of a millimeter.

The overall manufacturing process for the Li-ion cell can be divided into the five major processes:

1. Mixing, kneading, coating, pressing, and slitting processes of the positive electrode and negative electrode materials
2. Winding process of the positive electrode, negative electrode, and separator
3. Insertion of the wound cell core and electrolyte injection into the battery case
4. Cell closing or sealing process
5. Formation, aging, and cell selection

Electrode Fabrication and Winding Process

Figure 10.11 shows an outline of the initial process of Li-ion cell manufacturing that includes electrode preparation to winding [19]. In this process, the active materials are coated onto metal foils and calendered for thickness. The positive electrode material consists of active material, such as $LiCoO_2$, $LiNiO_2$, or $LiMn_2O_4$; a carbon-conductive additive like acetylene black; and a binder material namely polyvinylidene difluoride (PVdF). The active material and conductive agent are first dried and are then fed into a ball mill or other mixer along with the previously prepared binder solution (e.g., PVDF dissolved in *N* -methyl pyrrolidone), and then thoroughly stirred. The ball mill contains ceramic balls, e.g., glass, zirconia, about 2–3 mm in diameter, to assist in mixing and the mixing condition can strongly influence the performance of the cell.

The process for the negative electrode follows essentially similar to that of the positive electrode but with different materials. Carbon or graphite is used for the negative electrode-active material. PVDF, carboxymethylcellulose (CMC), or styrene butadiene rubber (SBR) is generally used for the binder that is usually dissolved in solutions like NMP depending on the type of binder.

A better result in mixing may be obtained by first kneading the materials into a stiff paste and then adjusting the viscosity by adding the solvent so that it can be applied easily in the coating process. The electrode slurries (viscosity from 10,000 to 20,000 cps) are uniformly coated onto both sides of the current collector (aluminum foil for the positive and copper foil for the negative) of

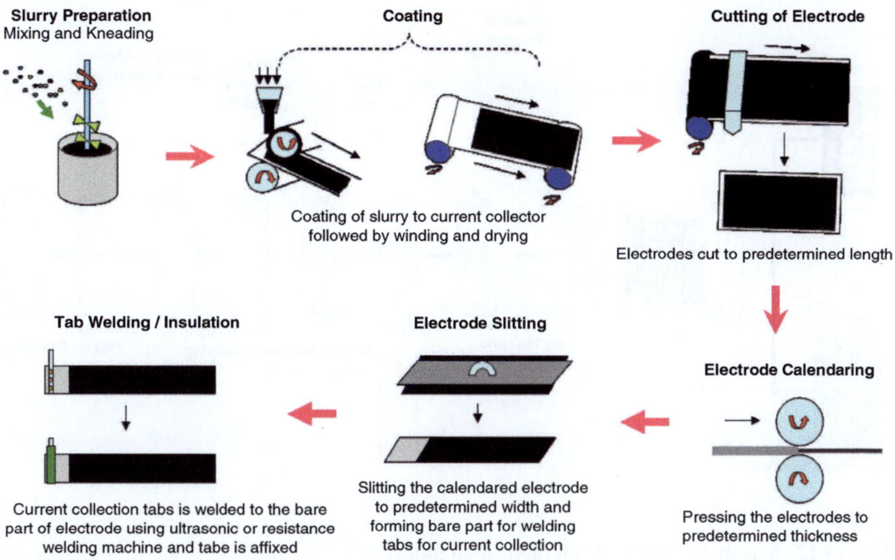

Fig. 10.11 Schematic of battery electrode fabrication process

10–20 μm thickness. Proper mixing techniques are very critical to obtain uniform distribution of the components of the active mass during the coating operation.

Coating operations can use a slot die, reverse roll coating, or doctor blade coating equipment. Close control of coating thickness is essential to ensure that all the parts fit into the can during final assembly. The coating thickness for different cell designs can vary from 50 to 300 μm, depending on the cell design. The coated material is dried in a continuous process. The coating solvent is collected to prevent emission of toxic vapors. Next, the dried electrodes undergo calendaring process where it is compressed with a roller press machine to provide accurate thickness control of the electrode thickness and to increase the density of the electrode mass. After calendaring, the master roll is slit to the width specified for cell construction and wound onto a roll for the winding operations.

18650 Cell Fabrication Process

Figure 10.12 presents the schematic of the assembly of cylindrical cells [19]. The reels of the slit anode, cathode, and separator stock are mounted on the winding machine. The strips have been sized for length, width, and thickness that match the cell design. Once the foils are threaded into the machine, the machine starts and continues to operate automatically until the reels are used up. An aluminum tab is

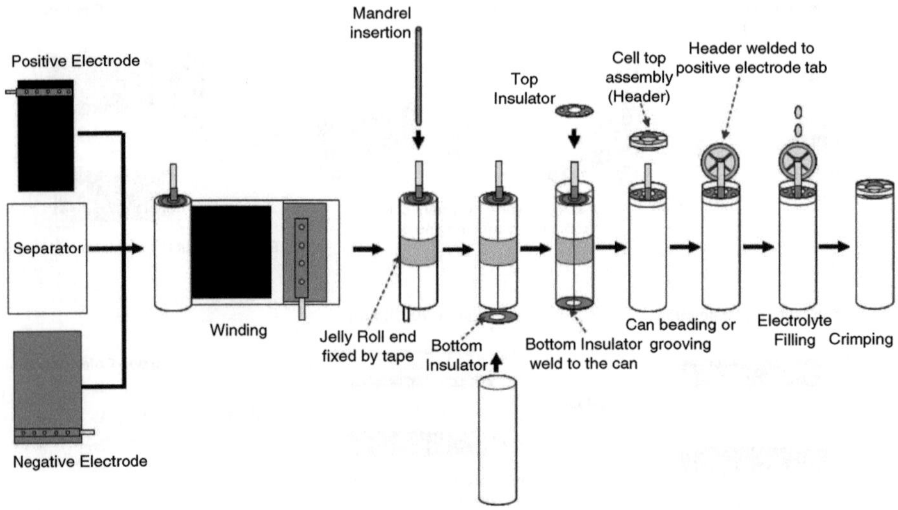

Fig. 10.12 Schematic of 18650 cell fabrication process

affixed to the bare part of the aluminum foil positive electrode by ultrasonic welding before winding, and similarly, a nickel or copper tab is affixed to the copper negative electrode collector by ultrasonic welding.

The winding machine then combines the two electrodes and separator strip on a mandrel (also referred as pin) and winds the combination in jellyroll fashion into a tight bobbin or cell core. Cylindrical cells are wound on a round mandrel while the prismatic cells use a flat paddle. Winding requires constant tension on the coil as it is formed and grows to the final size. Any irregularity leads to a gap between the separator and electrode, resulting in a nonuniform current distribution that can lead to a malfunction or shortened cycle life. The wound coil is checked for internal shorts before being inserted into the can with a hi-pot or impedance tester. In the last step, the cell core is taped to keep it tightly wound before inserting it into the cell case. It is important that the active material neither peels or sloughs off of the active material nor the separator twists during the winding process. Early rejection of potential cell faults is economical and prevents investing more work on bad/faulty cells.

Following insertion into the cell case, a tubular mandrel with a serration may be inserted in the winding core. The function of this mandrel is to improve the integrity of the cell core and for safety of the battery. When the internal pressure rises, gas has free passage through the hollow center of the mandrel and is released through the vent. When the cell is crushed, the mandrel causes both electrodes to short-circuit and instantaneously discharge. Any moisture contamination of the cell has a deleterious effect on its operation. Therefore, all cell assembly operations are usually carried out in a dry room or dry box. Alternatively, the cell then may be put in a heated vacuum oven for certain durations in order to extract residual water from the cell core before electrolyte filling. In the next process, the cell is filled with the electrolyte using a vacuum injection apparatus.

After filling, the cell cap, also called the header, is welded to the aluminum tab and the cell is finally sealed by crimping the header to the can. The header contains the polymer spacer, vent, PTC, and CID safety devices designed to activate and prevent dangerous temperatures and pressures from developing internal to the cell. After sealing, the cells are completely washed with isopropyl alcohol or acetone in order to remove any adhering electrolyte. An electrolyte leakage test is done using a smell sensor apparatus in order to confirm a good seal. Finally, the cell is subjected to an aging process to select out cells with microshorts and to categorize cells by capacity for assembly into battery packs. The temperature, length of storage, and cell checking methodology vary. This process is designed to identify cells with internal cell faults and microshorts that are not found during cell fabrication.

Large Format Li-Ion Cell Manufacturing Process: Flat Plate Pouch Configuration

Li-ion polymer and flat plate cells are produced in small sizes for cellular phones (about 0.5 Ah and higher) and large sizes (up to 200 Ah) for energy storage and motive power applications. A common characteristic is that the polymer separator holds the electrolyte and forms a physical barrier between the anode and cathode to prevent internal shorting. Most polymer cells are flat plate (prismatic) construction.

There is no one general cell fabrication process for polymer, flat or pouch, cells as there is for the liquid electrolyte cylindrical and prismatic cells. Each manufacturer has a slightly different process flow. Cell designs are essentially laminated constructions produced by several different fabrication processes. A typical example was shown in Fig. 10.13 [19]. Cell assembly follows the general outline as that for cylindrical cells. The cell assembly could be either a wound design or a stack design. Wound design follows closely to that of cylindrical cell except that the mandrel would be a thin flat plate depending upon the width requirement. In stack design, positive and negative electrodes along with separator are stacked together to give the desired cell capacity. Alternatively, a Z-fold construction may also be used with the electrodes bonded to the separator [20]. The stacks then are bonded together using ultrasonic or heat to ensure uniform bonding and dimensional control.

The cell case is formed from an aluminum-polymer laminated film. The film is fusion bonded, the electrode stack is inserted, and the exterior case is fused by a heat or an ultrasonic sealing apparatus. The assembly then is checked for internal shorts, the moisture is removed by heating in a vacuum dryer, the electrolyte (polymer or conventional electrolyte depending upon the cell kind) is then injected, and the cell is vacuum-sealed in a two-chamber configuration. After formation, any gas formed is removed by a vacuum, the cell resealed, and the gas chamber discarded. Finally, the cell is subjected to

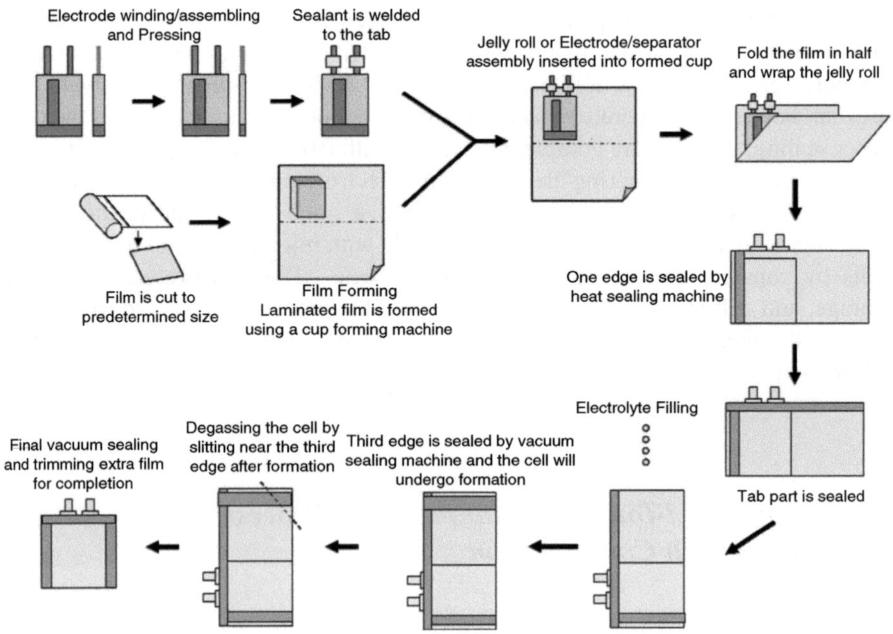

Fig. 10.13 Schematic of pouch cell fabrication process

top sizing, both side bending, leakage tested with an odor sensor, cleaning and drying, inspection of any coil winding misalignment using an X-ray apparatus, lot number printing, storage, charging and discharging, and then final inspection.

Formation and Aging

Once the cell assembly process is complete, the final step in the overall production process shifts to the formation and aging of the cells. This applies to cylindrical, prismatic, flat plate, and polymer cell constructions. Li-ion cells are assembled in the discharged condition and must be activated by charging. The first charge is called "formation," which activates the active materials in the cells and establishes their ability to function. The first charge typically starts at a lower current to properly form the protective solid electrolyte interface (SEI) layer on the graphite/carbon anode and then increases to the normal current at about 30% into the charge period.

The cell voltage is measured after the first charge and stored (aged) for a set time period. The storage time and temperature vary from one manufacturer to another. The voltage and capacity measurements are stored for use in the cell-selection/-matching process. These will be used later to sort cells out with internal microshorts or other manufacturing defects. Differences in voltage at the start and end of the storage period can be used to identify problem cells with low voltage and low capacity. Cells with low voltage result from "soft" or "micro" internal shorts and are discarded. The exact first charge regimen and cell-selection procedure may vary with manufacturer. Some manufacturers give the cell one or two cycles after formation to check the capacity for cell matching in assembly of battery packs. Others use the voltage measurement after first charge.

In order to keep up with the market demands, tough quality, performance, and safety requirements, in the cell fabrication methods and process control techniques, are followed. Optical and X-ray systems are currently being used to continuously monitor critical processes like winding, and ensure precise placement of components. The electrolyte-filling process is completely automated by using precision pump combined with advanced multistep vacuum/pressure cycles to ensure that the electrolyte permeates and completely fills the porosity in the separator and electrode structures. The internal construction of every cell produced is inspected by X-ray for proper top cap installation, misalignment of the winding from processing, the improper bending of the tabs, and so forth, which potentially could cause an internal short circuit or cell malfunction.

Safety

Lithium-ion batteries have been commercially available for over 2 decades and currently represent state-of-the-art power source for all modern consumer electronic devices. Due to its advanced chemistry, Li-ion cells exhibit superior performance characteristics over most other rechargeable battery systems. The lithium-ion technology offers a high energy and power density, long life, and reliability that makes it attractive for electric drive vehicle (EDV), military, and aerospace fields, and large format Li-ion cells and battery packs are currently under development for such applications.

Like any other energy storage devices, one of the concerns about the Li-ion technology is the safety associated with the system and is the primary focus for cell/pack manufacturers apart from cost and durability. Li-ion cells can undergo thermal runaway if subjected to unreasonable conditions for which they are not designed as it involves a combination of highly energetic materials with flammable electrolyte solutions.

With several billion Li-ion cells being used to power cell phones, laptops, and other consumer electronics, there are bound to be concerns over field incidents [21–23]. Battery failures that vent with fire are taken very seriously and manufacturers choose a conservative approach by adopting safety recalls. Considering the number of lithium-ion batteries used in the market, this energy storage system has caused little harm in terms of damage and personal injury.

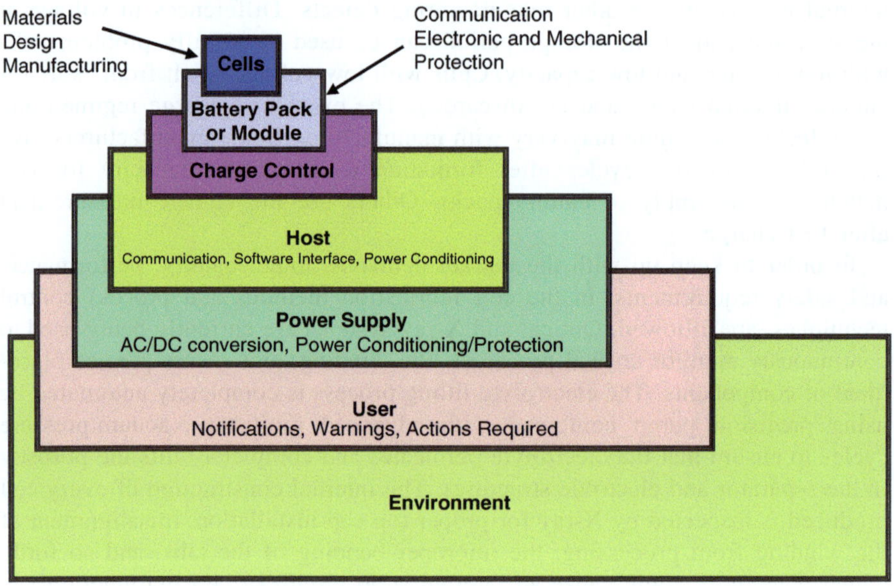

Fig. 10.14 System approach to Li-ion design and use (IEEE)

Li-ion cells have had an excellent safety record in the field with very few documented safety events in over one billion cells in the market place. Safety is continuously improving since production rates are increasing. This excellent safety record has come about due to care on the part of manufacturers and due to regulatory standards like IEEE, IEC, and UL governing the safety of cells. Cell manufacturers are not only trying packing more energy into the pack but also attempting to make it safer. However, there is still room for further improvement in safety as Li-ion system moves into several new arenas like electric cars and trucks, stationary energy storage, military and satellite applications that mostly utilize larger format batteries.

System Level Safety

In general, ensuring a safe product begins at cell level and ends with the user. The IEEE 1625 standard for laptops and the IEEE 1725 standard for mobile phones have recently focused on conveying the concept that Li-ion battery pack safety is a function of the entirety of the cell, pack, system design, and manufacture [24, 25]. A system level approach thus becomes very essential in addressing the safety of Li-ion batteries. System level safety includes the combination of cell, battery pack, host device, power supply, or adapter, end user/environment, and each of these aspects has a role to play in ensuring pack safety. Figure 10.14 presents the schematic of the system approach to Li-ion design

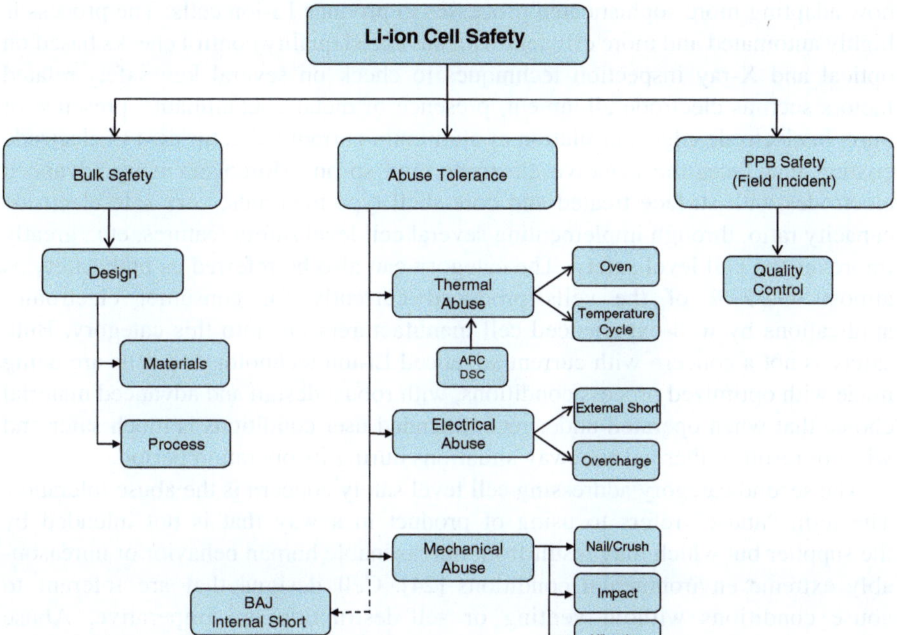

Fig. 10.15 A schematic flowchart that represents the cell level safety in Li-ion cells for consumer electronics applications

and use. The level of assurance for protection and safety on system level depends on the implementation by the manufacturer/supplier and also on the actions of the end user. This part of system level safety is about understanding the intended use of a product, and mitigating reasonable and foreseeable misuse that may occur. IEEE P1625, the Standard for Rechargeable Batteries for Portable Computing, was the first standard to encompass all levels of the battery manufacturing process and include the customer experience.

Cell Level Safety

The safety of lithium-ion batteries is an essential requirement for commercial use. Cell manufacturers continue to work toward designing safer cells, through modifications of cell components and through improvements in manufacturing practices that ultimately resulted in cells exhibiting excellent safety record in the field. On cell level, thermal runaway failures can occur for a number of reasons, including poor cell design (electrochemical or mechanical), cell manufacturing flaws, external abuse of cells, etc.

The cell level safety can be broadly classified into three categories shown in Fig. 10.15. The first one represents safety that depends upon cell design aspects, namely, materials and process conditions. Cell manufacturers around the world are now adapting more sophisticated processes to produce Li-ion cells. The process is highly automated and more efficient with advanced quality control checks based on optical and X-ray inspection techniques to check on several key safety-related factors such as electrode alignment, presence of metal contaminants, presence of burrs in electrode edges, insulation of aluminum current collector next to electrode coating that faces the negative electrode, and so on. Moreover, using advanced electrodes with surface treated and core-shell type materials, very safe electrode capacity ratio, through implementing several cell level safety features, etc., greatly improves the cell level safety. The category can also be referred as bulk safety as almost 99.999% of the cells produced currently for consumer electronics applications by well-experienced cell manufacturers fall into this category. Bulk safety is not a concern with current advanced Li-ion technology as cells are being made with optimized process conditions, with robust design and advanced material choice that when operated under recommended user conditions is much safer and will not result in thermal runaway situations during its operation period.

The second category addressing cell level safety concern is the abuse tolerance. The term "abuse" refers to using of product in a way that is not intended by the supplier but which may result from unreasonable human behavior or unreasonably extreme environmental conditions [24]. Cell designs that are tolerant to abuse conditions without venting or self-destruction are imperative. Abuse tolerance tests for consumer electronics cells are defined by International Electrotechnical Committee (IEC), Underwriters Laboratories Inc. (UL), and Japan Storage Battery Association (JSBA). The tests are typically classified as thermal, electrical, and mechanical abuse tests. Thermal abuse tests include high-temperature ramp also referred as hot-oven test and temperature cycling test. Characterization test methods, namely, ARC and DSC, are widely used to study the thermal properties of the cell components such as self-heat rate, onset runaway temperature, and so on that are directly linked to the thermal stability of the materials, and these test methods provide insight on thermal abuse tolerance response of Li-ion cells. Overcharge and external short circuit tests are categorized as electrical abuse tests and the most common mechanical abuse tests include crush, nail penetration, crush, impact and vibration tests. For more details regarding each test condition, UL document 1642 could be referred [26]. Cell manufacturers in general adopt a scaling method in quantifying the abuse response of Li-ion cells. The response may be ranging from least severe ones, namely, no change, vent with electrolyte leakage, to the most severe ones like smoke with fire and explosion. For abuse tolerance tests, the criterion for the cell to pass is no fire and no explosion.

The third category is the PPB or parts per billion level safety that refers to the field incidents of Li-ion cells that were subjected to normal recommended operating conditions. Field failures typically represent unprovoked battery explosions, and according to industry experts, such events are rare and it is estimated that one in ten million lithium-ion batteries fail that way [27]. Stringent quality control measures and new safety evaluation methods are being recommended that would help in controlling PPB level. Battery

Association of Japan has developed a test methodology to analyze internal short occurring in a Li-ion cell and to characterize the failure associated with it [24]. With this test method, the field incident caused by internal short circuit can be simulated practically [28]. With the increase in the energy of lithium-ion batteries, more advances in safety technology are also necessary to reduce the field incidents still further and gain more consideration from EDV, satellite, and other specialty applications.

More Details on Cell Components and Modern Trend

Negative Electrode Material

Hard Carbon

One of the typical examples for hard-carbon materials is the thermal decomposition product of polyfurfuryl alcohol resin (PFA). Hard-carbon materials can offer higher capacity beyond the theoretical limit of LiC_6 at lower potentials but have not achieved broad acceptance because of their higher irreversible capacity and sloping discharge voltages [29]. They are highly sensitive to moisture and are difficult to process; however, the material is practically usable and able to offer around 200 mAh/g capacity. Significant research work has been carried out on hard carbon with the purpose of overcoming the drawbacks through surface modifications and pretreatments. But graphitic carbon is still considered the best choice in practical batteries because of its higher performances.

Graphite Intercalation Compounds (GIC)

Theoretically, graphitic carbons can insert lithium up to a stoichiometry of LiC_6 corresponding to 372 mAh/g. The process is called intercalation and Li-ions occupy sites between graphene planes. Morphology of graphitic materials has a strong impact on electrochemical behavior and strongly influences critical stages in the precipitation of the surface films and their passivation properties. In general, graphite particles with some degree of disorder show high reversibility and stability in Li insertion process and are less dependent on solution composition when compared to highly ordered materials.

Graphite intercalation compounds (GIC) have a significant feature called the staging phenomenon, which is characterized by a periodic sequence of intercalant layers (lithium ions) between graphite layers. The nth-stage compound consists of intercalant layers arranged between every n graphite layers. The first-stage lithium graphite intercalation compound has the stoichiometry of LiC_6 with the specific capacity of 372 mAh/g, a theoretical saturated value of lithium storage for graphite under normal pressure. The staging phenomenon can be easily monitored and controlled by the electrochemical reactions of carbons in Li^+-containing electrolytes, for instance, conducting galvanostatic (constant current) charge/discharge [30] and slow cyclic voltammetric(CV) [31, 32] tests, and analyzing the differential capacity (dQ/dV vs V, where Q is capacity and V is voltage) has proven to be a particularly useful electrochemical method. In conjunction with the electrochemical techniques, some physical methods have been applied to shed light on the stage occurrence and transitions during lithium intercalation into and deintercalation from graphite host. These methods include in situ XRD [30, 33], ex situ XRD [34], in situ laser Raman spectra [35], and STM [36].

Natural Graphite

Natural graphite is chosen for lithium-ion battery anode materials mainly because of its low cost, low and flat potential profile, high Coulombic efficiency in proper electrolytes, and relatively high reversible capacity (330–350 mAh/g). On the other hand, it has two main shortcomings, namely, low rate capacity and incompatibility with PC-based electrolytes.

The low rate capacity is primarily due to its high anisotropy causing unfavorable orientation of particles that leads to sluggish of Li^+ intercalation and inadequate electronic contact between graphite particles and copper substrate [37, 38]. To overcome this issue, mechanical millings have been applied to pulverize the natural graphite flakes into small pieces [39], through which the preferred orientation of crystallites within each natural graphite flake particle can be distorted to some extent by separated graphite fragments. One other limitation of natural graphite is its incompatibility with PC-based electrolytes. Studies have been carried out to overcome this through modifying the electrolytes with additives and also modifying the graphite by coating amorphous carbon layers through thermal vapor decomposition technique [40–42].

Synthetic Graphite

Synthetic graphite has many properties that are similar to those of natural graphite. Besides, it has many unique merits such as high purity, variety of structures suitable for smooth Li^+ intercalation and diffusion, and so forth. Nevertheless, it is more expensive because of the high-temperature treatment (>2,800°C) on soft carbon precursors, and its reversible capacity is a little smaller than natural graphite.

Graphitized MCMB, mesophase-pitch-based carbon fiber (MCF), and vapor-grown carbon fiber (VGCF) are the representatives of benchmark synthetic graphite anode materials for Li-ion batteries in the market today. Graphitized MCMB has many advantages, namely, high packing density that guarantees high-energy density, and small surface area that decreases the irreversible capacity corresponding to electrolyte decomposition; most surfaces of MCMB spheres are composed of edge-plane surfaces, thus facilitating Li^+ intercalation and higher rate capability, ease of spreading onto copper foil.

Although amorphous carbon and spherical mesophase base graphite have been used mainly as anode materials in Li-ion cells, artificial graphitic carbon also called as massive artificial graphite (MAG) with lower voltage and the higher energy density is being used mainly to fulfill the difficult requirement for mobile equipment in recent years. MAG has been developed for the mass production of unique graphite for the purpose of battery application [43].

The MAG particles consist of spherical aggregation of fine, flat crystals. It has an average diameter of 20–30 μm and an aspect ratio of about one. MAG particles have a pseudo-isotropic texture where primary flat particles are randomly aggregated that also aids in tolerating expansion and shrinkage during charge/discharge operation. The discharge capacity of MAG exhibits a value of 362 mAh/g, which is very close to the theoretical one (372 mAh/g) of graphite. Graphite with a high crystallinity, which has a smaller interlayer spacing in the crystal, has relatively higher discharge capacity. This could be attributed to the fact that the graphite with the higher crystallinity is mainly composed of regular stacking structure, which is suitable for the formation of Li-GIC. Studies also have shown that the movement of Li^+ ions in MAG occurs in the vertical direction to the collector during the charge/discharge and Li^+ insertion into graphite occurring at the edge of graphite crystal, thereby having an easier lithium transfer and hence better high rate performance compared to flaky natural graphite [43].

Lithium Titanate Anode

Lithium titanate ($Li_4Ti_5O_{12}$, referred to as LTO) is another promising anode material for certain niche applications that require high C rates, fast charging capability. Some of the challenges for the LTO-based cells include lower voltage (2.5 V vs. LCO and 1.9 V vs. LFP), lower capacity, and excellent high-temperature stability. It has superior high rate performance with very long cycle life. It is being developed for automotive and energy storage applications.

Silicon-Based Anode

Existing lithium-ion batteries using graphite anodes have inherent limited Ah capacity. Silicon-based anodes theoretically offer as much as a tenfold capacity improvement over graphite. Silicon, when converted to the compound $Li_{21}Si_5$ by reaction with lithium in an electrochemical cell, has a maximum capacity of 4,200 mAh/g [44]. Due to its relatively low cost, low toxicity, and higher theoretical capacity, silicon is one of the promising candidates to replace graphite anode in Li-ion systems.

However, implementation of Si as an anode material is greatly hampered by several factors such as large volume change, low intrinsic electrical conductivity, and poor cyclability. The loss of capacity on repeated cycling is attributed to mechanical fragmentation and electrical isolation of silicon species due to the volumetric expansion and contraction associated with lithium insertion and extraction [45]. Many experimental techniques were investigated to analyze the cycling behavior and investigate mechanical degradation upon cycling [46, 47]. The reaction of silicon with lithium produces various alloy-type compounds (LixSiy) and causes enormous volume changes of up to 300% during the charge/discharge process, and this, together with the low electrical conductivity of silicon, obstructs the commercial use of silicon as a negative electrode material [48].

There are several approaches used to date for investigating silicon-based anodes. The thin-film approach involves deposition of an active silicon layer as a continuous media on a specially roughened metal or carbon current collector using ion sputtering [49, 50] or vapor deposition techniques [51, 52]. The film formation method, which results in strong adhesion between the current collector and very thin silicon layer, leads to the formation of a stable, self-organized columnar structure during the first volumetric expansion. This structure remains stable over cycles and aids in providing cycling stability of the films. Despite such attractive features of thin silicon film electrodes, they are yet to be commercialized. The thin silicon film electrodes are an order of magnitude thinner than cathodes, which might cause technological problems [53] and does not seem to be easily scalable. Moreover, solid electrolyte interface (SEI) film formation mechanism has yet to be studied in detail [54] and might highlight potential complications for real LIB employing such electrodes, especially during long-term cycling.

The approach is the composite electrode approach that relies on the use of silicon in the form of tiny silicon "cores," uniformly distributed within an electrochemically active or inactive buffering media, which prevents silicon-silicon interaction and the formation of large clusters. The core-shell structured anode materials comprising Si/SiO_2 blended with graphite material are good examples for the composite concept and have been gaining much attention primarily due to its ability to deliver higher capacity with continuous cycling, longer and stable cycling performance with much lower volume change than other silicon-based systems under study. Mixtures of silicon and different blends of graphite were studied and it was found that graphite is important in improving cycling performance [55, 56]. To suppress the volume effects of silicon upon cycling, techniques such as pyrolyzed PVC reactions and high-energy mechanical milling (HEMM) were adopted to disperse ultrafine silicon into a feasible carbonaceous matrix that shows promising results on improving the coulombic efficiency after the first cycle [57]. Further optimization of Si-composite based anodes could lead to practical lithium-ion batteries with high-energy density.

Positive Electrode Material

$LiCoO_2$

Lithium cobalt oxide (LCO) is the primarily used positive electrode–active material of Li-ion rechargeable cells. The simplicity involved in manufacturing and the balanced cell characteristics, namely, cycle performance, rate capability, high and low-temperature performance, safety, and so on, are some of the reasons why LCO is a leading player among positive electrode–active materials. However, as the Li-ion market is driven by cost reduction both in consumer electronics and in the vehicle and stationary storage segment combined with the present higher market price of cobalt, the application of LCO material is decreasing and focus now shifts more toward nickel, manganese, and iron type materials and cobalt-doped layered mixed oxide materials.

$LiCoO_2$ is obtained using a cobalt compound and a lithium compound as the raw materials and heating. Today cobalt oxide (Co_3O_4) is used as the cobalt compound and lithium carbonate (Li_2CO_3) as the lithium compound due to the stability of their quality and supply, the easiness of handling. The reaction scheme for the $LiCoO_2$ can be indicated as:

$$2Co_3O_4 + 3Li_2CO_3 + 1/2O_2 \rightarrow 6LiCoO_2 + 3CO_2$$

Goodenough recognized that $LiCoO_2$ had a structure similar to the layered structures of the dichalcogenides and showed that the lithium could be reversibly removed electrochemically, thus making it a viable cathode material [58].

$LiNiO_2$

Lithium nickel oxide is isostructural with lithium cobalt oxide and has higher specific capacity. However, it is much more susceptible to thermal runaway and has not been used in commercially lithium-ion cells.

$LiMn_2O_4$ (Spinel)

James Hunter of Eveready Battery Co. was the first to patent spinel cathode material. The application of material to Li-ion system has been developed by J. M. Tarascon [59] and extensively studied by M. Thackeray [60]. Generally, lithium spinel oxides suitable for the cathode are limited to those with a normal spinel in which the lithium ions occupy the tetrahedral (8a) sites and the transition-metal ions reside at the octahedral (16d) sites. Currently, spinel is the center of much interest as the cathode material for large format lithium-ion cell for hybrid electric vehicle applications where high power, safety, and low cost are the strongly required features.

Olivine Compound ($LiFePO_4$)

A number of researchers have pursued the oxides of iron as potential cathode materials for Li-ion cells. The first cathode material produced in this category is $LiFePO_4$ [61]. Its low cost and environmentally benign characteristics could have a major impact in electrochemical energy storage. Orthorhombic $LiFePO_4$ of the olivine structure forms $FePO_4$ during charging/discharging, and two crystal phases exist during charging/discharging; thus, it exhibits a flat discharge curve with a discharge potential of 3.4 V [62]. The specific capacity of $LiFePO_4$ is lower compared to nickel and cobalt-based materials. Substitution of iron by manganese transforms the shape of the charge/discharge curves, and this proved effective in increasing the energy density [63, 64]. The capacities of the lower voltage (3.5 V) and higher voltage region (4.0 V) correspond to the content of iron and manganese. Modified $LiFePO_4$ material too makes it more stable and compatible with a wide range of electrolyte solvents that greatly enhances performance and rate capability.

Mixed Oxide Cathodes

Nickel Cobalt Dioxide, $LiNi_{1-y}Co_yO_2$, and Nickel Cobalt Aluminum oxide, $LiNi_{1-y-z}Co_yAl_zO_2$

In general, the introduction of cobalt into nickel oxide structures leads to more stability and nickel oxide structures are less likely to lose oxygen than the pure nickel oxide, and studies on the structural details and physical properties of the mixed oxide system showed that there is an increased ordering with increase in the cobalt concentration. An issue with all these layered oxides is their electronic conductivity, which is not uniformly high across the lithium composition range or nickel substitution. For example, cobalt substitution in $LiNiO_2$, as in $LiNi_{0.8}Co_{0.2}O_2$, reduces the electronic conductivity at 4.2 V significantly. These dramatic changes demand that a conductive diluent be added to the cathode-active material, which reduces both the energy storage and the power capabilities.

Substituted nickel oxides, such as $LiNi_{1-y-z}Co_yAl_zO_2$ (NCA), are prime candidates for the cathode of advanced lithium batteries for use in large-scale systems as required for hybrid electric vehicles. They currently are the battery of choice for satellite applications. On charging these mixed oxides, the nickel is oxidized first to Ni^{4+} then the cobalt to Co^{4+} [65]. The significant feature of NCA type material is its increased thermal stability by doping with cobalt and aluminum. The temperature of the exothermic peak with oxygen evolution increased from 200°C of $LiNiO_2$ to 310°C of $LiNi_{0.8}Co_{0.15}Al_{0.05}O_2$ due to the decrease in the nickel content and the aluminum doping [66]. The thermal stability has been considerably improved and there is no oxygen evolution observed below 300°C. Moreover NCA type material has a higher capacity, better cycle life and also rate capability than that of the conventional cobalt-based cathode.

Nickel Cobalt Manganese Oxide, $Ni_xCo_yMn_zO_2$

One other commonly used mixed oxide material as positive electrode is cobalt- and manganese-doped lithium nickel oxide. In general, addition of nickel suppresses manganese dissolution to a great extent, thereby enhancing the cycling performance especially at higher temperature. The most common stoichiometry is the 333 compound $LiNi_{1/3}Mn_{1/3}Co_{1/3}O_2$ synthesized using coprecipitation method by Ozhuku et al. [67] and Jeff Dahn in Canada (3 M). There have been many studies on the 333 compound, which have been made by a number of synthetic techniques over a wide range of temperatures. Most of these studies show similar behavior with the capacity of around 165 mAh/g increasing with increasing charging potential [68–70]. The material attracts much attention because of the large battery capacity of about 200 mAh/g, and is a promising material in future. However, the material has certain disadvantages associated with the physical property, namely, lower tap density, smaller particle size that leads to lower energy density and thus mostly preferred to be used in the mixed cathode material. Similar to NCA type material, for the NMC material also, the DSC peak temperature related to the oxygen evolution shifts to the higher temperature as the increase in manganese doping amount and the amount of heat generation decreases extremely. The thermal stability of the cathode material thus could be improved by manganese doping [71].

High-Energy Structured Type and Layered Cathode Materials

There are several approaches currently being pursued to improve the cathode energy density, high rate performance, and cycling performance. One of the means to achieve the trade-off properties, namely, higher capacity and superior thermal stability, is through core-shell structured type electrodes. Higher capacity would be delivered by the core material, and structural/thermal stability would be achieved by the shell material [72–74]. A typical example is NCM type material in core to provide higher capacity surrounded by Ni-/Mn-based layered oxide as shell to offer good thermal stability.

One other approach is a composite layered oxide synthesized through continuous, low-cost highly homogenous coprecipitation process that provides higher capacity and cell voltage. One typical example is $xLi_2MnO_3.(1–x)LiNiO_2$ that claims to deliver around 250 mAh/g with longer cycle life and is able to operate at 4.6 V with surface modification [75, 76]. Other approaches include: surface treated cathode like AlF_3-coated cobalt-doped material to improve cycle life and power when operated at elevated temperatures and pseudo co-crystal-like structure with particle surface doping (e.g., $LiCoO_2$ type core material with the surface being treated with dopants like magnesium) to improve stability of the structure. Coating the positive electrode with materials such as Al_2O_3, $AlPO_4$, and ZrO_2 is also being actively pursued and used in commercial Li-ion cells [77–79]. Such surface coatings provide superior charge discharge characteristics, increased rate capability, long cycle life, and enhancement in structural stability, and this also elevates thermal stability.

Electrolyte and Additives

For the battery electrolyte, more focus is on developing additives to carry out additional functions to improve the performance and safety also referred as role-assigned electrolytes. Sanyo [80] and Saft [81] companies were the pioneers in proposing Vinylene carbonate (VC) as a solvent additive for graphite based electrochemical systems. VC is one of a popular additive for battery electrolytes typically being used as an anode passivation film-forming agent offering excellent stability and improvement in cycle life. The addition of VC has shifted the decomposition temperature of the SEI to a higher temperature, and this enhanced thermal stability can explain the suppression of the reaction between the electrolyte and the anode when the SEI is broken and repaired. It is considered that the relatively thinner homogeneous SEI formed by VC is an origin of the better cell performances, although the quality of its SEI is strongly dependent on the kind of graphitic carbons and charging conditions.

Sulfite compounds, namely, ethylene sulfite (ES), propylene sulfite (PS), dimethyl sulfite (DMS), and diethyl sulfite (DES), were also being used as passivation film-forming agents especially for PC-based electrolyte, and these compounds were reduced at about 2 V versus Li^+/Li to form the passivation layers, which hindered the PC co-intercalation into the graphene layers [82–84]. Cyclic sulfonate compounds, namely 1,3-propane sulfone (1,3-PS), were reported to give a good SEI film, which resulted in improving cycle and storage performances of cylindrical hard-carbon/ $LiMn_2O_4$ cells [85].

Some key additive materials used as overcharge protection agents include biphenyl (BP), cyclohexyl benzene (CHB), and hydrogenated *m*-terphenyl (H-mTP). BP polymerizes [86, 87] on the positive electrode during overcharge and the liberated protons migrate to the anode, generating hydrogen gas. This mechanism is used for the overcharge protection: In case of cylindrical cells with pressure-activated current intermitted devices (CID), the internal pressure rise by H_2 evolution helps to operate them and prismatic cells utilize the current shutdown mechanism by the increase in the internal resistance by the polymerized film and the meltdown of polyethylene separators [88].

For protecting the cells from overcharge, one other method is including redox shuttle-type compounds as additives in electrolyte. A compound "R" with a reversible redox potential is added to the electrolyte solution. The R is oxidized to compound "O" on the positive electrode, and then O migrates to the negative electrode and is reduced to the original form R. The requirements for such a compound are that the redox potential should be slightly positive (0.1–0.2 V) to the formal potential of the positive electrode at the end-of-charge and the redox reactions during overcharge should be kinetically reversible on both the electrodes. Some examples include aromatic compounds like 2,4-difluoroanisole (DFA) [87], 4-bromo-1,2-dimethoxybenzene [89], and 2,7-dibromothianthrene [90].

Wetting characteristics are very critical for cell design to obtain optimal performance. In general, one might expect that the solvents and liquid electrolytes will be attractive to hydrophilic surfaces such as cathode materials, but repulsive to hydrophobic ones such as polyolefin separators and carbonaceous anode materials [91]. The separator wettability can limit the performance of batteries by increasing the separator and cell resistance. The contact angle is a good measure of wettability. The uptake of electrolyte can be enhanced by wetting agents such as methyl phenyl carbonate (MPC) [92] and di-octyl carbonate (DOC) [93].

The additives used in electrolyte as a flame-retarding agents include trimethyl phosphate (TMP), phosphazene derivatives, namely, hexamethoxycyclotriphosphazene (HMPN), tris(2,2,2-trifluoroethyl)phosphate (TFP), and hexamethylphosphoramide (HMPA). Due to the passivation layer formed by these flame-retarding additives [94–96], the exothermic peaks were depressed and also the self-heat rate occurring at elevated temperature thereby provided safety during abuse situations. In addition fluorinated ester group materials (e.g., FEC) are also being used as additive to electrolyte to improve upon high-temperature stability [97–99]. There are also several ongoing research activities on molten salt electrolytes to gain the benefits of the nonflammable characteristics, but overcoming the conductivity issues is still challenging.

Separators

The process for making Li-ion battery separators can be broadly divided into dry and wet processes. Both processes usually employ one or more orientation steps to impart porosity and/or increase tensile strength. Dry process involves melting a polyolefin resin, extruding it into a film, thermal annealing to increase the size and amount of lamella crystallites, and precisely stretching to form tightly ordered micropores. In this process, a row lamellar crystal structure is generated in the polymer in the initial extrusion step. This nonporous structure is highly oriented as a result of extrusion and annealing conditions. In the next step, film is stretched to form micropores. This microporous structure is continuous throughout the bulk interior of the membrane. No solvents are required to produce separator membranes through dry process method.

Wet process also called as phase inversion process involves mixing of hydrocarbon liquid or some other low-molecular-weight substance generally with a polyolefin resin, heating and melting the mixture, extruding the melt into a sheet, orienting the sheet either in the machine direction or biaxially, and then extracting the liquid with a volatile solvent. The structure and properties of the membranes can be changed by controlling the composition of the solutions and the evaporation or subtractions of solvents in the gelation and solidification processes. The separators made by wet process use ultrahigh-molecular-weight polyethylene (UHMWPE). The wet process method is not applicable for polymers such as polypropylene, which do not dissolve in conventional solvents.

Shutdown separators are one of the safety devices inside the cell and act as a last line of defense. The separator shutdown is irreversible, which is fine for polyethylene-based separators, which melt at around 130°C. The impedance of the separator increases by two to three orders of magnitude due to an increase in cell temperature, which results from cell abuse (e.g., short circuit, overcharge). The separator not only should shut down around 130°C, but it also should maintain its mechanical integrity at higher temperatures. The high-temperature melt integrity of separators also is a very important property to keep the cell safe during extended overcharge or during extended exposure to higher temperatures.

Many technology advancements and improvements are achieved for the separators that also enable the Li-ion cell to operate more reliably under wide ranges of voltage and temperature for prolonged cycles. Separators existing in current market exhibit superior uniformity in thickness and pore distribution, higher strength, and dimensionally more stability at elevated temperatures. The separator membrane shows higher thermal stability through blending with high melting polymer resins. Ceramic material coating on top of polyolefin separator is currently becoming a popular choice for large format Li-ion cells for EDV applications to improve safety. Ceramic layer controls the shrinkage of the separator significantly especially at temperatures close to the melting points of polyolefin, thereby enabling the separator to isolate the electrodes for much higher temperatures.

Battery Management Systems

Owing to the fact that no two cells in a battery are exactly identical, their parameters such as capacity, voltage may vary by a fraction even when manufactured under similar conditions with advanced quality control protocols. In the case of a new battery, these factors may not be very noticeable, but play a major role in determining the performance of the pack with continuous charge/discharge cycles. This problem is easier to detect in a battery of up to 6–12 cells as the case in NBPC battery packs. A faulty cell can be easily disguised in a large battery pack consisting of tens or hundreds of series-connected cells as will be used for EDV applications. Thus, it becomes very essential to monitor the batteries individually and detect faults early.

A battery management unit (BMU) monitors and controls a battery pack. To increase the capacity and voltage, cells are connected in series-parallel arrangements. A typical laptop battery pack would contain a total of six 18650 cells where three cells are connected in series and two of the three cell series combination are connected in parallel. This setup is commonly denoted as 3S2P configuration. In the case of electric vehicle applications, it is common to have several hundred cells in series-parallel arrangements to get the desired voltage of 300–400 V and capacity.

The basic functionalities in a BMU consist of safety functions, voltage and current measuring, state of charge (SOC), and temperature monitoring. For Li-ion battery packs, battery management is the most essential as cells are needed to be controlled individually. When all cells are in parallel, the voltages are forced to be equal according to Kirchhoff's circuit laws, and hence, no balancing is needed. When the cells are connected in series, the voltages will differ and hence need balancing.

Cells in a battery pack may have slightly different voltages and capacity after complete charge due to minor differences in capacity, impedance, and rate of self-discharge. Cell balancing has been used to reduce the voltage differences among cells, thereby enhancing capacity and cycle life. Overvoltage safety detection functions will abort charging of the battery when any cell reaches its overvoltage threshold. Therefore, other cells may stay undercharged. One consequence is that the battery's full capacity will not be available for discharge. The same occurs when the weakest cell reaches the lower threshold, resulting in capacity not being used. Without balancing, the different cell capacities will start to drift with every charge/discharge cycle, leading to a decrease in the battery pack's total performance.

Cell balancing can be carried out using current bypass by burning up energy through a resistor as heat. This is often referred as passive balancing where to even out differences; the lowest cell determines the level of capacity after balancing. Passive balancing shall not be applied during discharge as it reduces the available capacity of the battery. One other means of balancing is active balancing where energy would be passed from a cell with higher voltage to a cell with lower voltage.

In electric vehicle applications, battery management is essential and much more demanding than for portable batteries. It has to interface with a number of other onboard systems, has to work in real time in rapidly changing charge/discharge conditions as the vehicle accelerates and brakes, and it has to work in harsh and uncontrolled environments. Determining the battery state of charge is particularly critical in electric vehicle applications since these batteries require both high-power charge capability (regenerative braking) and high-power discharge capability for start or acceleration. Therefore, they must be maintained at a SOC that allows delivering the required power, while still having enough "headroom" to accept regenerative charge without overcharging the cells.

In addition to current and voltage monitoring, the BMU also monitors the temperature of the cells in all modes of operation (drive, charge, etc.) and controls the pump, fans, and heater to manage the temperature of the battery. Battery thermal management is critical for the battery life and the vehicle performance, and it is done through the battery electronic control unit. The goal is to make a battery pack work at an optimum average temperature (with life and performance trade-off) with even temperature variations in the modules and within the pack.

Bibliography

1. Whittingham MS (1976) Electrical energy storage and intercalation chemistry. Science 192(4244):1126
2. Goodenough JB (1980) U.S. Patent 4,302,518 (Issued 31 Mar 1980)

3. Ikeda H, Narukawa K, Nakashim H (1981) Japanese Patent 1769661 (Issued 18 June 1981)
4. Basu S (1982) U.S. Patent 4,423,125 (Issued 13 Sept 1982)
5. Yoshino A, Jitsuchika K, Nakashima T (1985) Japanese Patent 1989293 (Issued 10 May 1985)
6. Gozdz AS, Schmutz CN, Tarascon JM (1994) U.S. Patent 5,296,318 (Issued 22 Mar 1994)
7. Gozdz AS, Schmutz CN, Tarascon JM, Warren PC (1995) U.S. Patent 5,418,091 (Issued 23 May 1995)
8. Gozdz AS, Schmutz CN, Tarascon JM, Warren PC (1995) U.S. Patent 5,456,000 (Issued 10 Oct 1995)
9. Gozdz AS, Tarascon JM, Warren PC (1995) U.S. Patent 5,460,904 (Issued 24 Oct 1995)
10. Global Industrial Analysts Report. http://www.prweb.com/releases/2011/1/prweb8040051.htm. Accessed Feb 2011
11. Institute of Information Technology, Ltd. (IIT) report, Year 2010, Q4, Chapter 7
12. Brodd RJ (2009) In: Yoshio M, Brodd RJ, Kozawa A (eds) Li-ion batteries: science and technologies. Springer, New York, pp 1–7
13. Takeshita H (2006) 23 rd international seminar on primary and secondary batteries, Ft. Lauderdale, Mar 2006
14. Peled E (1979) The electrochemical behavior of alkali and alkaline earth metals in nonaqueous battery systems—the solid electrolyte interphase model. J Electrochem Soc 126:2047
15. Peled E (1999) In: Besenhard JO (ed) Handbook of battery materials. Wiley-VCH, Weinheim, pp 419–458
16. Whittingham MS (2004) Lithium batteries and cathode materials. Chem Rev 104:4271–4301
17. Zhang ZJ (2011) Li-ion application in EDV and its safety perspectives, Pacific power sources symposium 2011, Hawaii
18. Zhengming (John) Z, Ramadass P (2009) In: Yoshio M, Brodd RJ, Kozawa A (eds) Li-ion batteries: science and technologies. Springer, New York, pp 367–412
19. Tagawa K, Brodd RJ (2009) In: Yoshio M, Brodd RJ, Kozawa A (eds) Li-ion batteries: science and technologies. Springer, New York, pp 181–193
20. Kim J, Hong JJ, Koh S (2006) Proceedings of the 42nd power sources conference, Philadelphia, June 2006
21. Tullo A (2006) Dell recalls lithium batteries. In: Chemical and engineering news: American Chemical Society, 21 Aug 2006
22. Hales P (2006) Dell laptop explodes at Japanese conference. In: The inquirer, June 2006 Staff (27 July 2007). Nokia – Retrieved 15 June 2010
23. N91 cell phone explodes Mukamo. In: Filipino news (blog), July 2007
24. IEEE 1625 (2008) IEEE standard for rechargeable batteries for multi-cell mobile computing devices
25. IEEE 1725 (2011) IEEE standard for rechargeable batteries for cellular telephones
26. UL 1642 (1995) UL standard for safety for lithium batteries, 3rd edn., Dated April 26, 1995
27. Jacoby M (2007) Burning batteries. Chem Eng News 85:26–28
28. Santhanagopalan S, Ramadass P, Zhang Z (2009) Analysis of internal short-circuit in a lithium ion cell. J Power Sources 194:550–557
29. Nishi H (1998) Lithium ion batteries. In: Wakihara M, Yamamoto O (eds) Fundamentals and performances. Codensha/Wiley, New York, pp 181–198
30. Dahn JR (1991) The phase diagram of Li_xC_6. Phys Rev B 44:9170
31. Tatsumi K, Zaghib K, Sawada Y, Abe H, Ohsaki T (1995) Anode performance of vapor-grown carbon fibers in secondary lithium-ion batteries. J Electrochem Soc 142:1090
32. Aurbach D, Levi M, Levi E (1997) The mechanism of lithium intercalation in graphite film electrodes in aprotic media. Part 1. High resolution slow scan rate cyclic voltammetric studies and modeling. J Electroanal Chem 421:79
33. Aurbach D, Ein-Eli E (1995) The study of Li-Graphite intercalation processes in several electrolyte systems using *In Situ* X-ray diffraction. J Electrochem Soc 142:1746

34. Ohzuku T, Iwakoshi Y, Sawai K (1993) Formation of Lithium-Graphite intercalation compounds in nonaqueous electrolytes and their application as a negative electrode for a lithium ion (Shuttlecock) cell. J Electrochem Soc 140:2490
35. Inaba M, Yoshida H, Ogumi Z, Abe T, Mizutani Y, Asano M (1995) *In Situ* raman study on electrochemical Li intercalation into graphite. J Electrochem Soc 142:20
36. Funabiki A, Inaba M, Abe T, Ogumi Z (1999) Nucleation and phase-boundary movement upon stage transformation in lithium–graphite intercalation compounds. Electrochim Acta 45:865
37. Xu K, Zhang S, Poese BA, Jow TR (2002) Lithium bis(oxalato)borate stabilizes graphite anode in propylene carbonate. Electrochem Solid-State Lett 5:A259
38. Kinoshita K (1998) Carbo: electrochemical and physico-chemical properties. Wiley, New York, p 70
39. Wang H, Ikeda T, Fukuda K, Yoshio M (1999) Effect of milling on the electrochemical performance of natural graphite as an anode material for lithium-ion battery. J Power Sources 83:141
40. Yoshio M, Wang H, Fukuda K, Hara Y, Adachi Y (2000) Effect of carbon coating on electrochemical performance of treated natural graphite as lithium-ion battery anode material. J Electrochem Soc 147:1245
41. Wang H, Yoshio M (2001) Carbon-coated natural graphite prepared by thermal vapor decomposition process, a candidate anode material for lithium-ion battery. J Power Sources 93:123–129
42. Wang H, Yoshio M, Abe T, Ogumi Z (2002) Characterization of carbon-coated natural graphite as a lithium-ion battery anode material. J Electrochem Soc 149:A499
43. Nishida T (2009) In: Yoshio M, Brodd RJ, Kozawa A (eds) Li-ion batteries: science and technologies. Springer, New York, pp 329–341
44. Wakihara M, Yamamoto O (1998) Lithium batteries-fundamentals and performance. Wiley-VCH, Germany
45. Besenhard JO, Yang J, Winter M (1997) Will advanced lithium-alloy anodes have a chance in lithium-ion batteries?. J Power Sources 68:87
46. Ozhuku T, Tomura H, Sawai K (1997) Monitoring of particle fracture by acoustic emission during charge and discharge of Li/MnO2 cells. J Electrochem Soc 144:3496
47. Ozhuku T, Matoba N, Sawai K (2001) Direct evidence on anomalous expansion of graphite-negative electrodes on first charge by dilatometry. J Power Sources 97–98:73
48. Yoshio M, Wang H, Fukuda K, Umeno T (2002) Carbon-coated Si as a lithium-ion battery anode material. J Electrochem Soc 149:A1598
49. Neudecker BJ, Zuhr RA, Bates JB (1999) Lithium silicon tin oxynitride (Li_ySiTON): high-performance anode in thin-film lithium-ion batteries for microelectronics. J Power Sources 81:27
50. Bordearu S, Brousse T, Schleich DM (1999) Amorphous silicon as a possible anode material for Li-ion batteries. J Power Sources 81:233
51. Green M, Fielder E, Scrosati B, Watchler M, Moreno JS (2003) Structured silicon anodes for lithium battery applications. Electrochem Solid State Lett 6:A75
52. Takamura T, Ohara S, Suzuki J, Sekine K (2002) The 11th international meeting on Lithium batteries, Monterey, 23–28 June 2002, Abs#257
53. Zaghib K, Kinoshita K (2004) 12th international meeting on Lithium batteries, Nara, 27 June–2 July 2004, Abs#7
54. Yonezu I, Tarui H, Yoshimura S, Fujitani S, Nohma T (2004) 12th international meeting on Lithium batteries, Nara, 27 June–2 July 2004, Abs#58
55. Yoshio M, Wang H, Fukuda K, Umeno T, Dimov N, Ogumi Z (2002) Carbon-coated Si as a lithium-ion battery anode material. J Electrochem Soc 149:A1598
56. Dimov N, Fukuda K, Umeno T, Kugino S, Yoshio M (2003) Characterization of carbon-coated silicon: structural evolution and possible limitations. J Power Sources 114:88
57. Liu Y, Hanai K, Yang J, Imanishi N, Hirano A, Takeda Y (2004) Morphology-stable silicon-based composite for Li-intercalation. Solid State Ionics 168:61
58. Mitzushima K, Jones PC, Wiseman PJ, Goodenough JB (1980) Li_xCoO_2 ($0<x<-1$): a new cathode material for batteries of high energy density. Mater Res Bull 15:783

59. Tarascon JM, McKinnon WR, Coowar F, Bowmer TN, Amatucci G, Guyomard D (1994) Synthesis conditions and oxygen stoichiometry effects on Li insertion into the spinel $LiMn_2O_4$. J Electrochem Soc 141:1421
60. Thackeray MM, David WIF, Bruce PG, Goodenough JB (1983) Lithium insertion into manganese spinels. Mater Res Bull 18:461
61. Padhi AK, Nanjundaswamy KS, Goodenough JB (1997) Phospho-olivines as positive-electrode materials for rechargeable lithium batteries. J Electrochem Soc 144:1188
62. Mukerjee S, Yang XQ, Sun X, Lee SJ, McBreen J, Ein-Eli Y (2004) In situ synchrotron X-ray studies on copper–nickel 5 V Mn oxide spinel cathodes for Li-ion batteries. Electrochim Acta 49:3373
63. Yamada A, Chung S-C (2001) Crystal chemistry of the olivine-type $Li(Mn_yFe_{1-y})PO_4$ and $(Mn_yFe_{1-y})PO_4$ as possible 4 V cathode materials for lithium batteries. J Electrochem Soc 148: A960
64. Li G, Azuma H, Tohda M (2002) $LiMnPO_4$ as the cathode for lithium batteries. Electrochem Solid State Lett 5:A135
65. Nakai I, Nakagome T (1998) In Situ transmission X-ray absorption fine structure analysis of the Li deintercalation process in $Li(Ni_{0.5}Co_{0.5})O_2$. Electrochem Solid State Lett 1:259
66. Yoshio M et al (2009) In: Yoshio M, Brodd RJ, Kozawa A (eds) Li-ion batteries: science and technologies. Springer, New York, pp 9–48
67. Ohzuku T, Ariyoshi K, Yamamoto S, Makimura Y (2001) A 3-volt lithium-ion cell with Li $[NiMn_{3/2}]O_4$ and $Li[Li_{1/3}Ti_{5/3}]O_4$: a method to prepare stable positive-electrode material of highly crystallized $Li[Ni_{1/2}Mn_{3/2}]O_4$. Chem Lett 30:1270
68. Ohzuku T, Makimura Y (2001) Layered lithium insertion material of $LiCo_{1/3}Ni_{1/3}Mn_{1/3}O_2$ for lithium-ion batteries. Chem Lett 30:642
69. Yabuuchi N, Ohzuku T (2003) Novel lithium insertion material of $LiCo_{1/3}Ni_{1/3}Mn_{1/3}O_2$ for advanced lithium-ion batteries. J Power Sources 119–121:171
70. Park SH, Yoon CS, Kang SG, Kim H-S, Moon S-I, Sun Y-K (2004) Synthesis and structural characterization of layered $Li[Ni_{1/3}Co_{1/3}Mn_{1/3}]O_2$ cathode materials by ultrasonic spray pyrolysis method. Electrochim Acta 49:557
71. Ohzuku T, Ariyoshi K, Yamamoto S, Makimura Y (2001) A 3-volt lithium-ion cell with Li $[Ni_{1/2}Mn_{3/2}]O_4$ and $Li[Li_{1/3}Ti_{5/3}]O_4$: a method to prepare stable positive-electrode material of highly crystallized $Li[Ni_{1/2}Mn_{3/2}]O_4$. Chem Lett 30:1270
72. Sun YK et al (2005) Synthesis and characterization of $Li[(Ni_{0.8}Co_{0.1}Mn_{0.1})_{0.8}(Ni_{0.5}Mn_{0.5})_{0.2}]O_2$ with the microscale core−shell structure as the positive electrode material for lithium batteries. J Am Chem Soc 127:13411
73. Sun YK et al (2006) Novel core−shell-structured $Li[(Ni_{0.8}Co_{0.2})_{0.8}(Ni_{0.5}Mn_{0.5})_{0.2}]O_2$ via coprecipitation as positive electrode material for lithium secondary batteries. J Phys Chem B 110:6810
74. Sun YK et al (2009) High-energy cathode material for long-life and safe lithium batteries. Nat Mater 8:320
75. Thackeray M (2009) US DOE-VT Annual merit review, Proj ID ES020
76. Amine K (2010) US DOE merit review, Proj. ID ES015
77. Ho-Jin Kweon, Jun-Won Suh, Won Il Jung (2004) U.S. Patent 6, 753, 111 (Issued 22 June 2004)
78. Cho J et al (2001) High-performance ZrO_2-coated $LiNiO_2$ cathode material. Electrochem Solid State Lett 4(10):A159
79. Cho J et al (2000) Novel $LiCoO_2$ cathode material with Al_2O_3 coating for a Li ion cell. Chem Mater 12(12):3788
80. Fujimoto M, Takahashi M, Nishio A (1992) Japan patent 3059832 (Issued 27 July 1992)
81. Simon A, Boeuve J-P (1997) U.S. patent 5,626,981 (Issued 6 May 1997)
82. Wrodnigg GH, Besenhard JO, Winter M (1999) Ethylene sulfite as electrolyte additive for lithium-ion cells with graphitic anodes. J Electrochem Soc 146:470

83. Wrodnigg GH, Wrodnigg TM, Besenhard JO, Winter M (1999) Propylene sulfite as film-forming electrolyte additive in lithium ion batteries. Electrochem Commun 1:148
84. Wrodnigg GH, Besenhard JO, Winter M (2001) Cyclic and acyclic sulfites: new solvents and electrolyte additives for lithium ion batteries with graphitic anodes?. J Power Sources 97–98:592
85. Kusachi Y, Utsugi K (2003) Extended abstracts of the 44th battery symposium in Japan ,Sakai, p 526 4–6 Nov (2003)
86. Xiao L, Ai X, Cao Y, Yang H (2004) Electrochemical behavior of biphenyl as polymerizable additive for overcharge protection of lithium ion batteries. Electrochim Acta 49:4189
87. Kim H-J, Yoo S-I, Cho J-J (2002) Extended abstracts of the 43 rd battery symposium in Japan, Fukuoka, 12–14 Oct 2002, p 78
88. Tobishima S, Ogino Y, Watanabe Y (2002) Effect of electrolyte additives to provide safety and discharge characteristics of lithium batteries. Electrochemistry 70:875
89. Adachi M, Tanaka K, Sekai K (1999) Aromatic compounds as redox shuttle additives for 4 V class secondary lithium batteries. J Electrochem Soc 146:1256
90. Lee D-Y, Lee H-S, Kim H-S, Sun H-Y, Seung D-Y (2002) Redox shuttle additives for chemical overcharge protection in lithium ion batteries. Korean J Chem Eng 19:645
91. Blomgren GE (1999) Electrolytes for advanced batteries. J Power Sources 81–82:112
92. Suzuki H, Shima N, Hasegawa K, Yoshida Y (1996) JP1996–306387A; JP3893627B
93. Noda D, Kotato M, Fuji T, Suzuki H (2002) JP2002–319433A
94. Lee CW, Venkatachalapathy R, Prakash J (2000) A novel flame-retardant additive for lithium batteries. Electrochem Solid-State Lett 3:63
95. Zhang S, Xu K, Jow TR (2003) Tris(2,2,2-trifluoroethyl) phosphite as a co-solvent for nonflammable electrolytes in Li-ion batteries. J Power Sources 113:166
96. Gozales SI, Li W, Lucht BL (2004) Hexamethylphosphoramide as a flame retarding additive for lithium-ion battery electrolytes. J Power Sources 135:291
97. Okuno H et al (1996) U.S. patent 5,525, 443 (Issued 11 June 1996)
98. Yamaguchi et al (2009) U.S. patent 7,491, 471 B2, 17 Feb 2009
99. Abe K et al (2011) U.S. patent application 20110045361 (Pub. Date 24 Feb 2011)

Chapter 11
Medical Device Batteries

Michael J. Root

Glossary

Anode	The negative electrode of a discharging cell or battery.
Cathode	The positive electrode of a discharging cell or battery.
Electrolyte	A solution or material that completes the electrical circuit in a cell by way of ionic conduction.
Hermetic seal	A way to seal implantable medical device batteries that is impermeable to fluids and usually includes a terminal feed through that is sealed in glass.
Implantable defibrillator	An implanted medical device that functions as a pacemaker and is also capable of delivering high energy shocks to the heart to treat ventricular tachycardia (abnormally fast heart rate) and fibrillation.
Neurostimulation or neuromodulation	Electrical stimulation of nerves to modify nerve activity.

This chapter was originally published as part of the Encyclopedia of Sustainability Science and Technology edited by Robert A. Meyers. DOI:10.1007/978-1-4419-0851-3

M.J. Root (✉)
Cardiology, Rhythm and Vascular Research and Development, Boston Scientific Corp., 4100 Hamline Ave. North, St. Paul, MN 55112, USA
e-mail: michael.root@bsci.com

R.J. Brodd (ed.), *Batteries for Sustainability: Selected Entries from the Encyclopedia of Sustainability Science and Technology*, DOI 10.1007/978-1-4614-5791-6_11,
© Springer Science+Business Media New York 2013

Pacemaker	An implanted medical device that delivers low level electrical stimulation to the heart for treatment of bradycardia (abnormally slow heart rate).
Primary cell	A cell that is intended to be discharged only.
Secondary or rechargeable cell	A cell that can be charged following depletion using an external electrical energy source.

Definition of the Subject

Medical device batteries serve an important role in modern health care. They power the devices that allow patients to function more normally by managing and improving their health or even survive life threatening disease conditions.

There are several ways to classify medical devices. Some provide therapeutic functions while others are used for diagnostic purposes. Some provide both functions. For therapeutic devices, devices may be further described as life sustaining or life enhancing (improves the quality of life).

Life sustaining devices provide therapy that keeps a patient alive. An implantable defibrillator is an example of a life sustaining device. Certain cardiac patients experience tachycardia (a very rapid heart beat) or ventricular fibrillation. Either condition limits the ability of the heart to effectively pump blood. Death can result if this type of arrhythmia is sustained and a normal heart rhythm is not restored by a defibrillator.

Life enhancing devices might treat conditions, like severe, chronic lower back pain, that do not threaten a patient's life, but prevent normal function and reduce their quality of life.

Improving the well-being of the patient is the most important use of wearable or implantable medical devices, of course. Medical devices can provide effective treatment or diagnostic information. Although it is not a necessary consideration, treating or monitoring a patient's condition remotely – that is, outside of the hospital or clinic – with such devices can also impact energy sustainability. Patients do not need to be transported to the hospital or clinic as frequently for treatment or monitoring.

In this entry are discussed a few of the specialized batteries for medical devices that are portable or wearable (carried with the patient, like hearing aids), or implantable (surgically placed inside the body as with neurostimulation pain management devices). There is a focus on the batteries designed for a few of the more common applications – implantable cardiac rhythm management (cardiac pacemakers and defibrillators), pain management, and hearing loss devices.

Introduction

Medical Devices that Use Batteries

There are a substantial number of wearable and implantable medical devices powered by batteries. These include devices for cardiac rhythm management (pacemakers, defibrillators, and heart failure devices), hearing loss, bone growth and fusion, drug delivery for therapy or pain relief, nerve stimulation for pain management, urinary incompetence and nervous system disorders, vision, diagnostic measurements and monitoring, and mechanical heart pumps.

This entry will be limited to three major categories of medical devices – cardiac rhythm, neurostimulation, and hearing devices. The first two are devices that apply electrical stimuli to muscle tissues or nerves and the last involves sound amplification. The battery systems used by these devices are used in other devices, as well. Some of these will be noted in the sections describing the batteries.

General Design Considerations for Medical Device Battery Performance

There are several important features that battery developers must consider when designing batteries for medical devices. Many of these are also important for most other battery types, as well.

Medical device batteries are fundamentally the same as any other battery designed for consumer electronics, military, or aerospace applications. All require the same three components to be able to function as an electrochemical power source – a negative electrode (or anode) material to supply electrons, a positive electrode (or cathode) material that takes up electrons, and an electrolyte that completes the electrical circuit through ionic conduction. The other components in a cell are necessary to make the cell perform efficiently, minimize its size, and make it safe and reliable. These components include one or more separators that are electrically insulating to prevent direct contact between the anode and cathode but allow ions to pass through, current collectors to convey electrons to or from the electrodes and various insulators to prevent short circuits.

Medical devices, especially implantable devices, often use batteries that are custom designed and built specifically for that device. Their sizes and shapes may not be standard. The electrode assemblies may consist of pellet or slug electrodes, spirally wound electrode foils, or stacked electrode plates (Fig. 11.1).

Batteries for implantable medical devices are hermetically sealed. Hermetic seals have long been used for certain cell types, like lithium-sulfur dioxide and lithium-thionyl chloride, where long shelf life is important, or exposure to corrosive and toxic materials could result if the cell leaks.

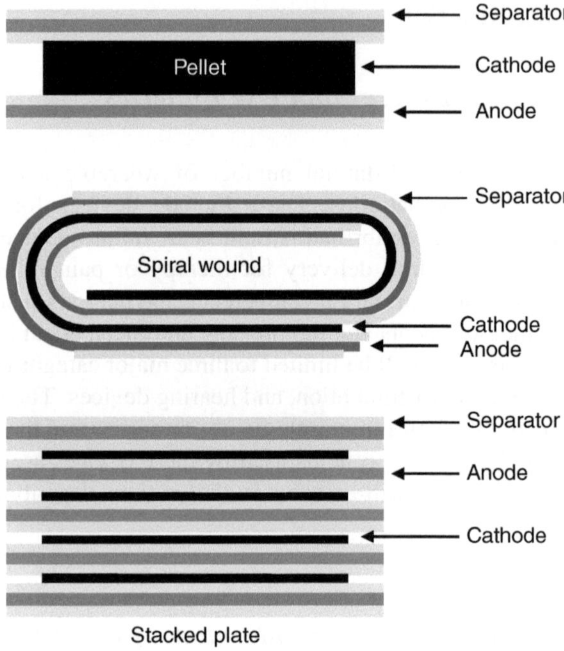

Fig. 11.1 Electrode assembly designs for prismatic medical device batteries

A feedthrough terminal is used in hermetically sealed cells as a way to connect to one of the battery electrodes to the external device circuit. A glass to metal seal insulates the terminal from the cell case. The glass is specially formulated to create a tight seal against the battery case and the terminal and resist attack by the cell electrolyte. Glass materials like TA-23 and Cabal-12 are common.

A number of different chemistries have been used for medical device batteries. For example, over 20 years ago there were at least 14 battery chemistries developed and used in implantable cardiac rhythm devices for treating cardiac arrhythmias (Fig. 11.2). Only a few of these survived and are in use today.

Electricity as Medicine

Physiological effects of electricity were known by the time Italian scientist Alessandro Volta first invented the battery in 1800. The introduction of the battery meant that a sustained source of electricity was available for the first time. Prior to that, electrostatic sources of electricity and storage devices, like the Leyden jar, were used. Such devices provided only momentary electrical current.

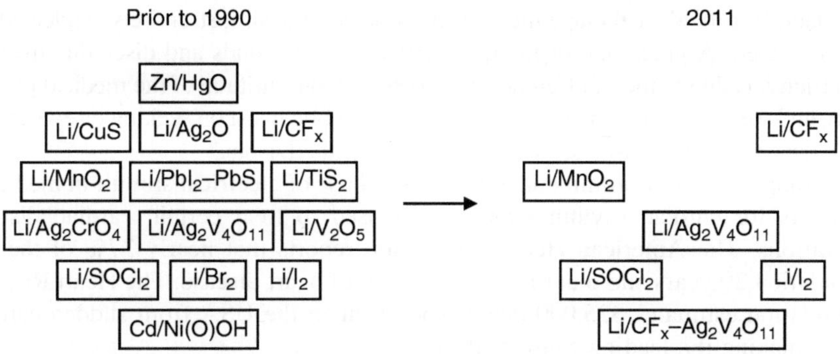

Fig. 11.2 Types of battery chemistries used for cardiac rhythm devices prior to 1990 and today

The electrochemical cell, or battery, enabled advances in many areas of research and development involving electricity during the nineteenth century including electrochemistry, electrophysiology, and the use of electricity as medical therapy.

The work of Guillaume-Benjamin-Amand Duchenne (de Boulogne), a French neurologist, was foundational to the advancement of electrophysiology. German neurologist Robert Remak was an early proponent of electrotherapy to treat diseases of the nervous system.

J. Althaus wrote in his 1860 book "A Treatise on Medical Electricity, Theoretical and Practical," [1] "We know that, whatever may be the properties of the nerves, they can be called into action by galvanism. But the effects are widely different according to the form of electricity that is used; again, the quality and intensity of electricity are both of great importance; not less so the mode in which it is transmitted to the human body, and the length of time during which its action is kept up. In fact, we are able, by merely varying modes of applying electricity, to arouse or to kill the vital power of the nerves, and to diminish or to increase their properties. Hence electricity can only be expected to be of service in the treatment of disease, if we are guided in its use by an exact knowledge of the physiological effects which it will invariably produce."

In 1873, Herbert Tibbits wrote, "There will be found here no new ground opened out, but only an earnest endeavor to sift the wheat of our existing knowledge from the chaff, and to make the reader as much at home with his electrical as with his other medical instruments; and further to lead him to estimate electricity at its fair and proved value in therapeutics, as an agent, not to be indiscriminately advocated as a panacea, nor, on the other hand, neglected by the inexperienced, but in appropriate cases to be regarded as one of the most powerful and serviceable weapons with which we can combat disease"[2].

Many claims of treatments using electricity were false, though. H. Lewis Jones in 1892 cautioned the medical community, [3] "One thing is certain, that without a thorough grounding in the physical part of the subject, no satisfactory advances can be made in a field of therapeutics which is at present almost entirely neglected by medical men. A great deal of the quackery which surrounds and discredits medical electricity, is due to the indifference and contemptuous attitude of the medical profession, and we have only ourselves to blame if the public insist on seeking elsewhere for treatment which is refused to them by their medical advisors."

Perhaps, the most prominent and successful use of electrical stimuli in medicine today is for cardiac rhythm management used to treat certain cardiac disease conditions. The American Heart Association reports that nearly 37% of the US population 20 years and older have some form of heart disease. The Heart Rhythm Foundation estimates 325,000 deaths each year in the USA from sudden cardiac arrest making it a leading cause of death.

There are three implantable devices used today to treat cardiac arrhythmias – the pacemaker, the implantable cardioverter defibrillator (ICD), and the cardiac resynchronization therapy (CRT) devices for heart failure patients.

Implantable Cardiac Pacemakers

Blood flow throughout the body occurs because the pumping action of the heart maintains the arterial blood pressure at a higher level than the venous blood pressure. Each heart beat involves the coordinated contraction of cardiac muscle cells as the result of the initiation and propagation of an electrical impulse [4].

Rather than contracting from nerve stimulation like skeletal muscles, cardiac muscle or myocardial cells respond to an electrical signal created by ionic (K^+, Na^+, and Ca^{2+}) concentration differences across their cell membranes. At rest, this potential, called an action potential, is about −90 mV. When the cell is activated, it rapidly depolarizes by redistributing the ionic concentrations across the cell membrane and contraction follows. Specialized muscle cells in the sinoatrial (SA) node of the heart start this process by spontaneously depolarizing once it reaches the critical firing level (CFL) of about −55 mV. The SA node is the primary natural pacemaker in normally functioning hearts. The signal is then quickly conveyed throughout the rest of the heart by way of several conduction pathways [4].

There are a number of conditions that cause the heart to beat abnormally. Electrocardiograms measure changes in potential that occur in the heart. An electrocardiogram of a normal heart rhythm is shown in Fig. 11.3. Electrocardiograms showing a normal rhythm, bradycardia, tachycardia, and ventricular fibrillation. The beats are regular and the time intervals between them are typical.

Bradycardia is a condition wherein the heart beats too slowly or skips beats altogether (for example, see Fig. 11.3). This can occur if the electrical impulses through the heart are slowed, too few impulses are transmitted, or they are blocked altogether [4].

Implantable cardiac pacemakers are electronic devices that treat bradycardia. External pacemakers were developed in the early 1950s. These were large devices that were not portable. Introduction of the transistor in the mid-1950s meant that

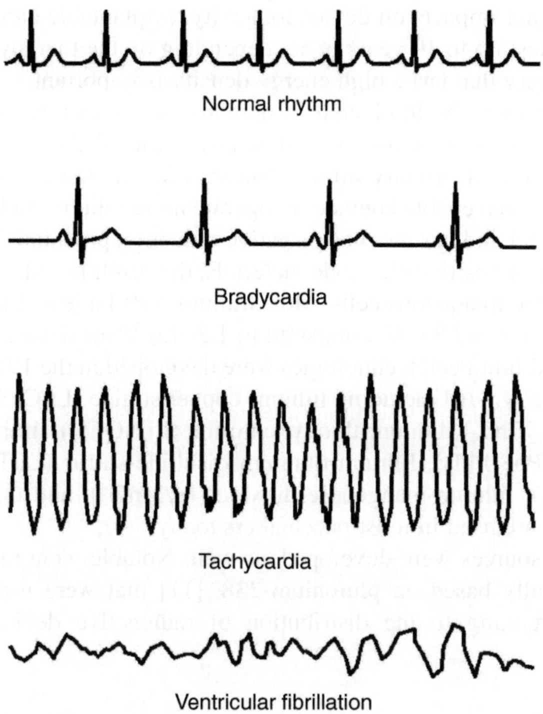

Fig. 11.3 Electrocardiograms showing a normal rhythm, bradycardia, tachycardia, and ventricular fibrillation

smaller devices could be built. Earl Bakken and coworkers later developed a smaller external pacemaker that could be carried by the patient [5]. The first cardiac pacemakers to be implanted in humans lasted hours to days [6]. In 1960, Wilson Greatbatch developed the first implantable cardiac pacemaker that lasted for longer than a few days [7].

Pacemakers today can pace in one or both chambers (right atrium and right ventricle) on the right side of the heart. The power demand on a pacemaker battery of any individual device varies based on device programming, the number of chambers paced, and the patient's cardiac condition. Cardiac pacing therapy is a relatively low power operation – generally between 20 and 100 μW on average. The amount of electrical energy used to stimulate the heart can be adjusted to meet the needs of the patient. The first pacemakers where asynchronous – meaning they paced at a single rate. Devices today adapt to the patient's activity level to pace more rapidly or more slowly as needed. Additional low power demands on these devices include sensing to detect a natural heart beat and other monitoring features.

Although requiring only low levels of power, it is the frequent application of pacing therapy, as well as the continuous background device operations, that have the most significant impacts on device longevity. Implantable cardiac pacemakers are expected to last up to 10 years or so, depending on the therapy required by the patient; so a battery that has a high energy density is important.

The first devices to be implanted in humans used either rechargeable nickel-cadmium (NiCd) cells or alkaline zinc-mercuric oxide (Zn/HgO) cells [8].

The development of primary lithium batteries for implantable medical devices was a big advance that enabled devices to operate more reliably and longer. Lithium is the lightest metal and has the most negative reduction potential. When combined with any number of positive electrode materials, the result is cells with high energy densities compared to aqueous cells. Most lithium cells have an initial open circuit voltage between 1.8 and 3.9 V, compared to 1.2–1.6 V for most aqueous cells.

A number of lithium cell technologies were developed in the 1970s [9] and some were used clinically, [10] including lithium-copper sulfide (Li/CuS), lithium-silver chromate (Li/Ag_2CrO_4), lithium-thionyl chloride ($Li/SOCl_2$), lithium-lead sulfide-lead iodide ($Li/PbS\text{-}PbI_2$), lithium-titanium disulfide-sulfur ($Li/TiS_2\text{-}S$), lithium-bromine (Li/Br_2), lithium-manganese dioxide (Li/MnO_2), and the lithium-iodine (Li/I_2) cells that are used in most pacemakers today.

Other power sources were developed, as well. Notable were radioactive power sources, principally based on plutonium-238, [11] that were used in the 1970s. Regulations pertaining to the distribution of radioactive devices limited their acceptance.

Implantable Cardiac Pacemaker Batteries

Lithium-Iodine

Lithium-iodine (Li/I_2) was proposed as a power source for implantable cardiac pacemakers in 1971 [12]. The first pacemaker run by a Li/I_2 cell was implanted in 1972 [6]. These cells were originally developed as more reliable and longer lived alternative to the zinc-mercuric oxide cells (see below) used in implantable cardiac pacemakers since they were introduced in 1960 and on into the mid-1970s. However, Li/I_2 cells have been the dominant power source for implantable cardiac pacemakers for more than 30 years.

The negative electrode reaction is:

$$Li \rightarrow Li^+ + e^-$$

and the positive electrode reaction is:

$$I_2 + 2e^- \rightarrow 2I^-$$

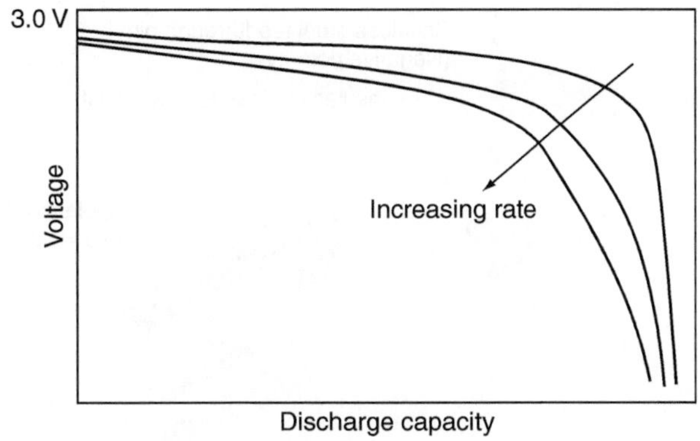

Fig. 11.4 Discharge voltage of Li/I_2 cells. The voltage is lower for higher discharge loads

Iodine is stabilized by mixing it with a polymeric pyridine – poly-2-vinylpyridine (abbreviated P2VP or PVP) – to form an I_2-P2VP charge transfer complex.

The overall cell reaction is:

$$2Li + I_2 \rightarrow 2LiI$$

When a Li/I_2 cell is built and the I_2-P2VP mixture is added, it reacts directly with Li to form in situ a layer of solid LiI between negative and positive electrode materials. This layer serves to protect the Li from further reaction with the I_2. LiI also functions both as an electrolyte and a separator.

The movement of Li^+ is rather slow in solid LiI with conductivities less than $10^{-6}\ \Omega^{-1}\ cm^{-1}$ [13]. This limits Li/I_2 cells to low rate applications, like implantable cardiac pacemakers.

The open circuit potential of the Li/I_2 cell is about 2.8 V. As the cell discharges, the thickness of the LiI electrolyte layer increases. This increases the internal resistance of the cell, so the discharge voltage decreases somewhat as the cell becomes depleted. The faster the discharge rate, the more sloped the discharge voltage becomes (Fig. 11.4).

The gradual voltage and resistance changes during discharge present the means to predict the extent of cell depletion, thereby allowing an accurate prediction of pacemaker longevity. This advanced warning allows sufficient time for the physician and the patient to schedule device replacement well ahead of complete battery depletion and the subsequent inability of the pacemaker to deliver therapy.

A cross section of a Li/I_2 prismatic cardiac pacemaker cell is shown in Fig. 11.5. In this design, a lithium sheet is placed between two I_2-P2VP electrodes. Energy densities for Li/I_2 cells are 210–280 Wh kg^{-1} and 810–1,030 Wh dm^{-3} [14].

Other devices that use Li/I_2 cells are implantable bone growth stimulators [15] and an implantable electromechanical hearing system [16].

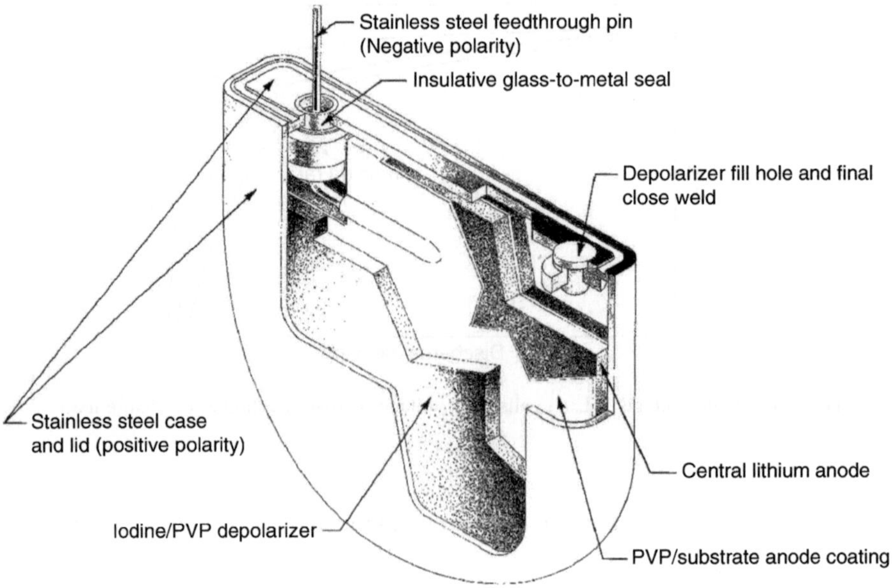

Fig. 11.5 Cross section of a Li/I_2 prismatic cardiac pacemaker cell. © Greatbatch, Inc (Reprinted with permission)

Lithium-Carbon Monofluoride

The most common positive electrode materials that have been used in all types of batteries, including those designed for medical devices (I_2 and O_2 notwithstanding), are usually metal oxides, such as silver vanadium oxide ($Ag_2V_4O_{11}$), manganese dioxide (MnO_2), mercuric oxide (HgO), and silver oxide (Ag_2O), or nonmetal oxides, like thionyl chloride ($SOCl_2$). Oxide compounds have a high energy density and can be chemically stable.

A cell pairing lithium (Li) with elemental fluorine (F_2) would have a theoretical open circuit voltage of 5.9 V. It is not practical to build a cell using elemental fluorine, which is a gas at room temperature and highly reactive, but fluoride compounds that retain some of the high energy density could be more realistic positive electrode materials [17].

Carbon monofluoride (CF_x) is one such fluoride compound that today is used as a positive electrode material in Li batteries for a number of different applications. For example, they are used in certain types of heart failure devices – implantable cardiac resynchronization therapy pacemakers (CRT-P). CRT-P devices can pace the right atrium and right ventricle, but they are also capable of pacing the left ventricle. Pacing three chambers requires more power than a Li/I_2 cell can deliver, so a different battery type is needed. Li/CF_x cells were developed in response to the increased power required by CRT-P devices. Vagal nerve stimulator devices also use a Li/CF_x cells.

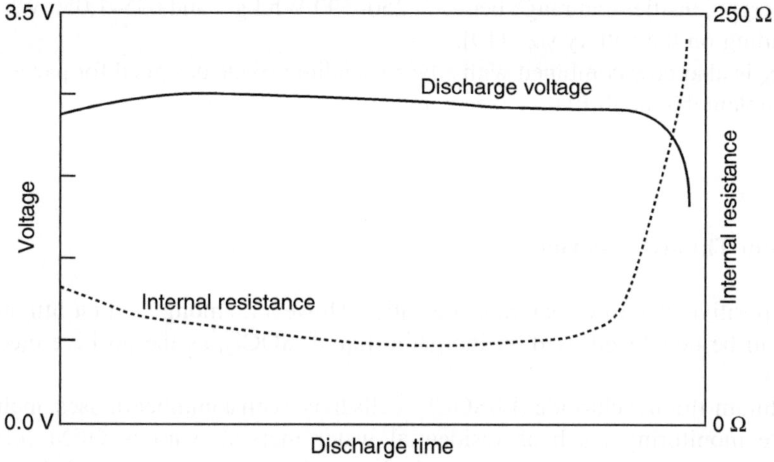

Fig. 11.6 Discharge voltage and internal resistance for a Li/CF_x coin cell

Li/CF_x cells were first launched/introduced commercially in the mid-1970s and are available as coin cells and spiral wound cylindrical cells in a variety of shapes and sizes.

CF_x used in batteries is synthesized by the direct reaction of fluorine gas with a carbon starting material, such as petroleum coke, at high temperatures – usually between 350°C and 600°C depending on the type of carbon starting material and the fluorination level. The value of x in CF_x used in batteries, where x is the average number of fluorine atoms per carbon atom throughout the CF_x material, is usually between 0.9 and 1.2.

Fluorine is a highly corrosive and hazardous gas and so is difficult to handle safely. As a result there are only a few companies that have the ability or desire to manufacture CF_x in large quantities.

The CF_x material is mixed with a conductive carbon, like acetylene black or graphite, and a binder to maintain the mechanical integrity of the electrode. Typical electrolyte solutions are comprised of lithium tetrafluoroborate ($LiBF_4$) dissolved in one or more organic solvents, generally propylene carbonate (PC) or gamma-butyrolactone (GBL) with 1, 2-dimethoxyethane (DME).

The overall cell reaction is:

$$x\text{Li} + \text{CF}_x \rightarrow \text{C} + x\text{LiF}$$

Lithium fluoride (LiF) precipitates as a reaction product and it can limit the utilization of the CF_x.

The nominal open circuit voltage for Li/CF_x is 3.0 V, but the discharge voltage is closer to 2.8 V. The internal resistance of the Li/CF_x cell decreases during the early stages of discharge and remains low throughout discharge (Fig. 11.6), [18] attributed to the formation of conductive carbon as a reduction product of CF_x.

Energy densities can range between 250–590 Wh kg^{-1} and 635–1,050 Wh dm^{-3} depending on the battery size [19].

CF_x is also now combined with silver vanadium oxide designed for pacemakers and implantable defibrillators (see below).

Lithium Thionyl Chloride

Most positive electrode materials are solids. However, among the first lithium cell types to be developed used an inorganic liquid, $SOCl_2$, as the positive electrode material.

Lithium-thionyl chloride ($Li/SOCl_2$) cells have seen a number of uses, including remote monitoring (such as residential water meters), various OEM (original equipment manufacturer) electronic devices, military, aerospace, and down-hole oil well monitoring applications. Medical device uses have included implantable heart monitors, drug infusion pumps, and some of the earliest implantable cardiac pacemakers [9, 10].

The positive electrode reaction is:

$$2SOCl_2 + 4e^- \rightarrow S + SO_2 + 4Cl^-$$

with the overall cell reaction:

$$4Li + 2SOCl_2 \rightarrow S + SO_2 + 4LiCl$$

Thionyl chloride serves as both positive electrode material and the electrolyte solvent. The electrolyte salt typically is lithium aluminum chloride ($LiAlCl_4$) or, sometimes, lithium gallium chloride ($LiGaCl_4$).

The design of the $Li/SOCl_2$ cell puts $SOCl_2$ in direct contact with the Li anode material. If the direct reaction between Li and $SOCl_2$ continues, the active electrode materials would produce no useful electrical energy. The reaction between Li and $SOCl_2$ is self-limiting, though. Similar to other lithium cell systems, reaction products form a passive layer on the Li that inhibits further reactions from occurring. The passive layer must be both electrically insulating yet ionically conductive for the battery to discharge efficiently.

In systems with solid cathodes, the active positive electrode material is usually mixed with a conductive carbon that conveys electrons from the current collector to the electrode active material. In this case (and similar to the zinc-air cell discussed below), liquid $SOCl_2$ is the active material and needs to come into contact with an electrode site (cathode) where it can take up electrons as the cell discharges. The cathode in $Li/SOCl_2$ cells is a porous carbon, such as acetylene black, and a PTFE binder. An aluminum screen can be used as a current collector to electrically connect the cathode and the positive cell terminal.

Fig. 11.7 Discharge voltage for a $Li/SOCl_2$ cell

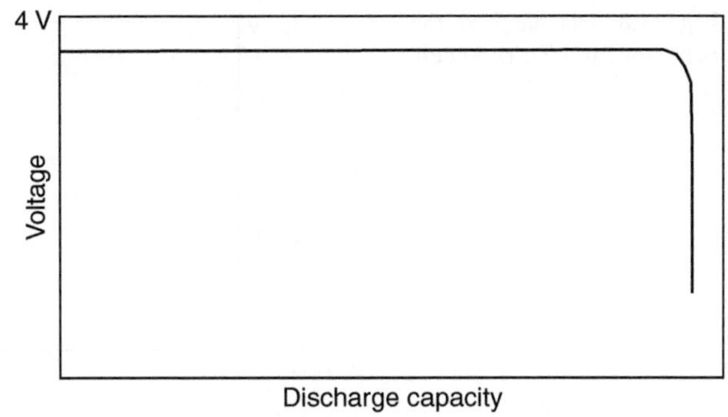

The lithium chloride (LiCl) and sulfur (S) that form precipitate and build up at the cathode. The cell capacity can be limited by these products if the cathode becomes blocked. Additionally, sulfur dioxide (SO_2) gas forms as a reaction product.

The $Li/SOCl_2$ cell has one of the higher open circuit potentials, 3.65 V, of any primary lithium cell. The voltage is quite flat throughout discharge (Fig. 11.7). The energy density is about 380 Wh kg^{-1} and 715 Wh dm^{-3}.[14] $Li/SOCl_2$ cells are produced in a wide variety of shapes and sizes for both high and low power applications, including prismatic and spiral-wound cylindrical cells used in pacemakers implanted during the 1970s.

Zinc-Mercuric Oxide

Although no longer used, zinc-mercuric oxide (Zn/HgO) cells were the power source of choice for the first commercially viable implantable cardiac pacemakers. More than 3 million Zn/HgO cells were implanted in the 16 years from when the first successful cardiac pacemaker was implanted in 1960 and 1976 [20]. They helped many bradycardia patients until they were eventually supplanted by lithium cells, particularly Li/I_2, so a brief description of Zn/HgO technology is included here. Zn/HgO cells were also once used in wearable hearing aids.

The use of mercuric oxide as a positive electrode material for Zn/HgO cells was patented by British inventor Charles L. Clarke in 1884 [21], and also by others like Danish scientist Johannes N. Brønsted [22] in the early 1900s. A practical Zn/HgO cell was developed in the 1940s by Samuel Ruben with manufacturing support from the Mallory Battery Co [23]. The Zn/HgO cell was designed to replace the Leclanché zinc-manganese dioxide (Zn/MnO_2) cells used in military communications equipment during World War II. The Zn/HgO cell had better storage life and performance compared to Leclanché cells, particularly in the high temperature and high humidity conditions of the South Pacific [24].

Assorted Zn/HgO button and cylindrical cell sizes were produced after the war for military and space applications, as well as a number of consumer applications, including calculators, watches, and cameras. Some of the first implantable cardiac

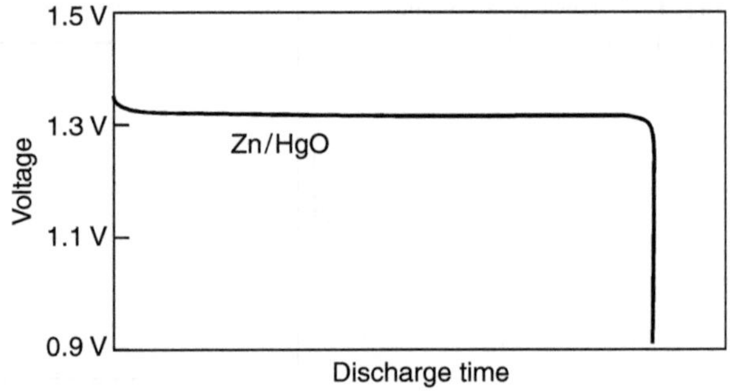

Fig. 11.8 Discharge voltage behavior for a Zn/HgO cell

pacemakers that were available commercially were powered by Zn/HgO batteries and they became a popular cell for use in wearable hearing aids.

The Zn negative electrode material, or anode, and electrolyte solution are similar to other primary alkaline battery types, like zinc-air and zinc-silver oxide (Zn/Ag_2O). Zinc powder is mixed with a gelling agent like polyacrylic acid and a KOH-ZnO-H_2O electrolyte.

The positive electrode reaction is:

$$HgO + H_2O + 2e^- \rightarrow Hg + 2OH^-$$

The HgO is usually mixed with a conductive carbon powder, like graphite, to improve the electrical conductivity of the cathode. Mercuric oxide forms Hg as it is reduced. Mercury metal is highly conductive, and so it does lower the internal resistance of the cell. It is also a liquid at normal operating conditions, and it tends to coalesce into droplets and pool in a way that can reduce performance. Additives to the cathode like manganese oxides or silver can minimize Hg pooling.

The overall cell reaction may be written:

$$Zn + HgO + 2KOH + H_2O \rightarrow K_2Zn(OH)_4 + Hg$$

or

$$Zn + HgO \rightarrow ZnO + Hg$$

A cross section of a Zn/HgO button cell is similar to that of a Zn/Ag_2O cell and is shown in Fig. 11.16.

The initial open circuit potential of Zn/HgO cells is about 1.35 V, but can be between 1.40 V and 1.55 V if MnO_2 is added to the cathode. The voltage remains rather constant throughout discharge (Fig. 11.8).

Energy densities for Zn/HgO button cells are about 100 Wh kg^{-1} and 470 Wh dm^{-3} [14].

Mercuric oxide cells are not readily available any longer. Government regulatory agencies throughout the world forced these cells from the market because of environmental concerns.

For wearable hearing aids, other cells are available, including Zn/Ag_2O and zinc-air discussed below. Lithium-iodine cells replaced Zn/HgO batteries for implantable cardiac pacemakers in the starting, in the early 1970s. The Zn/HgO cells tended to generate gas which is difficult to manage in an implantable device. Further, actual device longevity – generally in the range of 18–36 months – was lower than the expected 5 or more years [20]. Early battery depletion was usually the cited cause; however, some claimed device malfunction was actually responsible [20].

Another concern was the voltage characteristics of Zn/HgO cells during use. The end of battery life is difficult to predict because the discharge voltage is flat through battery life then quite rapidly decreases when the battery becomes completely depleted (Fig. 11.8). The lithium cells developed to take the place of Zn/HgO cells were more reliable, longer lived, and the end of battery life is more easily anticipated.

Implantable Cardioverter Defibrillators and Heart Failure Devices

The implantable cardioverter defibrillator (ICD) is a cardiac pacemaker. It can be used to pace one or both chambers on the right side of the heart. It has an additional feature, though. ICDs can also impart powerful shocks to the heart if it is beating too fast (tachycardia) or goes into ventricular fibrillation. Either condition means that blood cannot be pumped very efficiently, if at all. A number of major clinical studies were done that identified various categories of heart patients who could benefit from the therapies delivered by devices like the ICD.

The cardiac resynchronization therapy defibrillator device (CRT-D) provides the same pacing and defibrillation functions as an ICD, but can also pace the left ventricle for heart failure patients.

The need to deliver energetic shocks to the heart within seconds presents a challenge for battery designers. The battery must provide years of operation for the constant or frequent low power demands of pacing, sensing, and other device functions in the tens of mW range and also the infrequent, but high power, pulses to shock the heart. To deliver a shock, the battery must charge high voltage electrolytic capacitors generally somewhere between 600 and 800 V. The shock is delivered by discharging the capacitors into the heart tissue. Rapidly charging the capacitors requires power on the order of watts. Sometimes multiple shocks are required to get the heart back to a normal rhythm. The balance between energy density for longevity and rate capability for rapidly charging the high voltage capacitors requires careful selection of the chemistry as well as the mechanical design of the battery.

The first clinical studies of ICDs implanted in humans started in 1980 [6] following years of development by Mirowski and coworkers [25]. The first commercially available device was released in 1985. Since then, large clinical studies identified new patient populations that could benefit from ICDs and CRT-Ds.

The first commercial devices were up to 160 cm^3 in volume and needed to be implanted in the abdomen because of their size. Device features were limited and nonprogrammable. Devices typically lasted 2 years before replacement was necessary.

Devices today are about one-fifth the size of the first ICDs. The device industry continued to make improvements by developing smaller devices and while adding more features, such as multichamber pacing, monitoring patient cardiac performance with recorded electrocardiograms, and remote monitoring using radio frequency (RF) telemetry, while providing greater longevity. Devices now are small enough to be implanted in the pectoral area of the chest and can last for 5 or more years.

Improvements in device electronics partially helped to improve device longevities, but new battery systems were also needed. The earliest ICDs were powered by lithium-vanadium pentoxide (Li/V_2O_5) cells [6]. Longevities were insufficient using these cells, so a different battery chemistry, lithium-silver vanadium oxide ($Li/Ag_2V_4O_{11}$ or Li/SVO) was developed [26]. $Li/Ag_2V_4O_{11}$ cell was the cell of choice for ICDs and CRT-D devices since 1985 until recently when other cell chemistries have become more prevalent.

Implantable Cardioverter Defibrillator Batteries

Lithium-Silver Vanadium Oxide

The earliest implantable defibrillators used lithium-vanadium oxide (Li/V_2O_5) cells. The chemical stability of this type of cell was unsatisfactory, and they were soon replaced with lithium-silver vanadium oxide ($Li/Ag_2V_4O_{11}$ or Li/SVO) cells. Until only the last few years, $Li/Ag_2V_4O_{11}$ cells were by far the most common cell system used in implantable defibrillators.

Silver vanadium oxide is prepared by the high temperature reaction of V_2O_5 with a silver salt like $AgNO_3$ by a decomposition pathway [27]:

$$2V_2O_5 + 2AgNO_3 \rightarrow Ag_2V_4O_{11} + \text{nitrogen oxides}$$

or with Ag_2O through a combination mechanism [28]:

$$2V_2O_5 + Ag_2O \rightarrow Ag_2V_4O_{11}$$

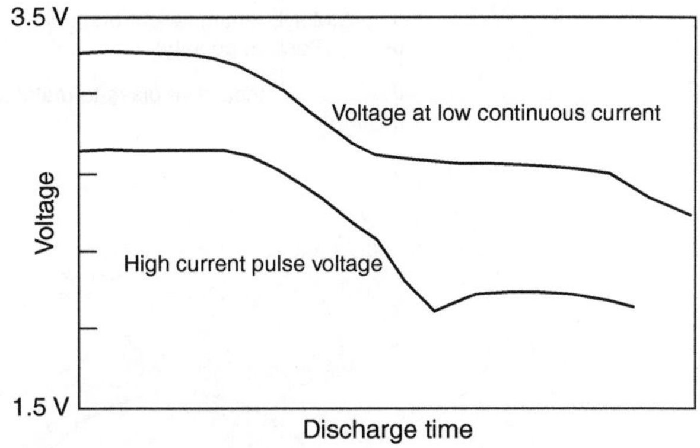

Fig. 11.9 Low continuous current and high current pulse discharge voltages for a lithium-silver vanadium oxide ($Li/Ag_2V_4O_{11}$) cell

Reduction to about 1.5 V vs. Li involves up to 7 electrons per mole of $Ag_2V_4O_{11}$:

$$Ag_2V_4O_{11} + 7e^- \rightarrow Ag_2V_4O_{11}{}^{7-}$$

though utilization is less under realistic use conditions.

The $Li/Ag_2V_4O_{11}$ cell discharge reaction occurs in several steps involving sequential reduction of Ag(I) to Ag(O) and V(V) to V(IV) [29]:

$$2Li + Ag^{I}{}_2V^{V}{}_4O_{11} \rightarrow Li_2Ag^{0}{}_2V^{V}{}_4O_{11}$$

$$4Li + Ag^{I}{}_2V^{V}{}_4O_{11} \rightarrow Li_6Ag^{0}{}_2V^{IV}{}_4O_{11}$$

Some V(III) may also form, especially if the discharge proceeds beyond $Li_6Ag_2V_4O_{11}$.

Silver(I) is reduced to silver(0) in the first discharge step. Silver metal has a high conductivity, so the resistance of the $Ag_2V_4O_{11}$ positive electrode material decreases rapidly during the early stages of discharge.

The different discharge steps occur at different cell potentials (Fig. 11.9) that result in a series of voltage steps. The distinct voltage levels offer a means of determining the state of charge for the battery, or at least a set range of states of charge. This, in turn, helps predict battery end of life.

However, a challenge of the $Li/Ag_2V_4O_{11}$ system is the increase in internal resistance that occurs in the middle of discharge. This becomes an issue during the high current demands while charging the high voltage capacitors when it is necessary to deliver high energy shocks to the heart. As the internal resistance increases, the high power pulse voltage decreases (Fig. 11.9). If the voltage drop is severe enough, it can delay delivery of therapy shocks.

The cause for much of the internal resistance increase is attributed to vanadium that becomes slightly soluble in the middle portion of discharge. The dissolved vanadium diffuses to and deposits on the Li surface. If allowed to build up, the

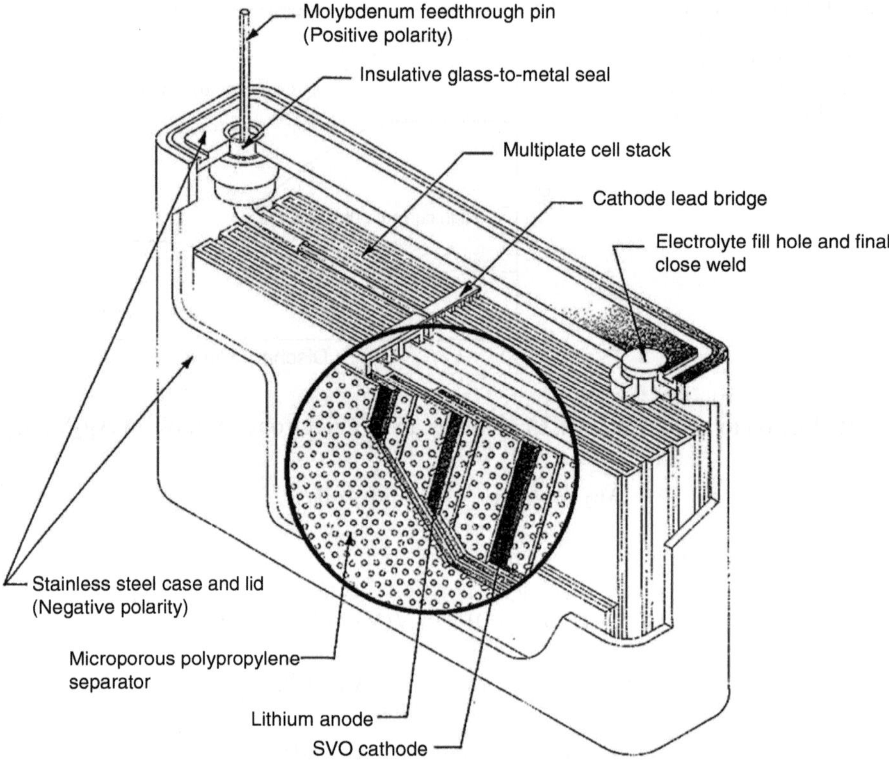

Fig. 11.10 Cross section of a lithium-silver vanadium oxide (Li/$Ag_2V_4O_{11}$) cell for implantable defibrillators. © Greatbatch, Inc (Reprinted with permission)

internal resistance increases [30]. The resistance can be lowered by pulling multiple high current pulses from the cell that presumably remove the vanadium as the Li dissolves during the pulse discharge [30].

Silver vanadium oxide is combined with a conductive carbon and a binder like PTFE to make the cathode. The usual electrolyte solution used is lithium hexafluoroarsenate ($LiAsF_6$) in mixed organic solvents, like propylene carbonate and 1,2-dimethoxyethane.

A cross section of a Li/$Ag_2V_4O_{11}$ defibrillator cell is shown in Fig. 11.10. Energy densities are on the order of about 270 Wh kg^{-1} and 780 Wh dm^{-3} [19].

Lithium-Manganese Dioxide

The lithium-manganese dioxide (Li/MnO_2) cell was originally developed and commercialized in the mid-1970s [31] for low power applications such as wrist watches, calculators, and computer memory backup. Later, cells were developed for

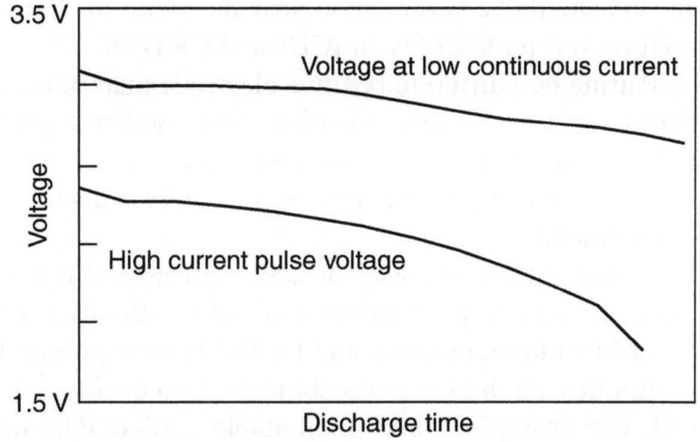

Fig. 11.11 Low continuous current and high current pulse voltages for a Li/MnO_2 cell

high power uses like cameras. Today, Li/MnO_2 cells are manufactured in a variety of sizes and configurations for low, medium, and high power consumer; military; and OEM electronic devices. Use of Li/MnO_2 cells in implantable devices was investigated for implantable cardiac pacemakers [32, 33].

Li/MnO_2 cells have a high energy density (230–270 Wh kg^{-1} and 535–520 Wh dm^{-3}) [19] that gives them the ability to provide long life for low current applications. They can also provide high power outputs required for device functions requiring high current levels. This makes Li/MnO_2 an excellent chemistry for implantable cardiac defibrillators. Indeed, Li/MnO_2 cells are now used to power implantable cardiac defibrillators from two manufacturers [34, 35].

The overall cathode reaction may be simply written as:

$$MnO_2 + xLi^+ + xe^- \rightarrow Li_xMnO_2$$

The MnO_2 positive material is similar to the type of MnO_2 used in alkaline Zn/MnO_2 cells. However, it must be heat treated at high temperatures, somewhere around 300–400°C to make it suitable for use in lithium cells. Heating removes water and modifies the structure of MnO_2. If this process is not done properly, the Li/MnO_2 cells will generate gas and internal pressure will build up, perhaps to the point where the cells will start to leak.

One or more conductive carbon powders, like acetylene black and graphite, and a binder such as PTFE are added to the MnO_2 to yield the cathode mix.

The discharge voltage of Li/MnO_2 cells exhibits two steps – rather flat during the first part of discharge, but curves downward in the last part of discharge (Fig. 11.11). Further, there is no remarkable decrease in pulse voltage as the internal resistance of the cell is relatively low and steady throughout discharge [36, 37].

Lithium-Carbon Monofluoride-Silver Vanadium Oxide

One of the challenges of designing batteries for some medical devices is the wide-ranging power requirements to support different device functions. The energy density should be sufficient to provide a long life battery in a small size, yet deliver high power, particularly in ICDs and CRT-Ds.

Pairing two different positive electrode materials – one possessing a high energy density, but a low rate capability and another capable of high rate output, but a lower energy density – is a concept that has been investigated for other applications. It is now starting to be applied to medical device batteries with diverse power requirements.

A straightforward way of accomplishing this (at least from a battery design perspective) is to put together two cells with different rate capabilities, such as Li/I_2 for low rate operations and Li/MnO_2 for high rate functions [38].

Another method is to combine the two positive electrode materials in the same cell. For example, some implantable cardiac defibrillators and pacemakers have started to use lithium cells containing both CF_x (higher energy density but a lower rate capability) and $Ag_2V_4O_{11}$ (high rate capability) positive electrode materials. The CF_x discharges at a voltage that is greater than $Ag_2V_4O_{11}$ throughout most of the cell life, when the device requires only low power from the cell, which is most of the time for implantable defibrillators and pacemakers. When higher power is occasionally required from the cell, such as during RF telemetry or charging high voltage capacitors, the CF_x cannot support the higher load but the $Ag_2V_4O_{11}$ electrode material can.

Dual CF_x-$Ag_2V_4O_{11}$ positive electrodes can be made by mixing the individual CF_x and $Ag_2V_4O_{11}$ materials in the same cathode mix [39] or assembling discrete CF_x and $Ag_2V_4O_{11}$ electrode layers to form the positive electrode [40].

Neuromodulation

Neurostimulators are much like cardiac pacemakers, except they apply small electrical signals to nerve tissue. Devices are available for spinal cord stimulation to manage chronic pain, vagal nerve stimulation to control epilepsy and deep depression, and deep brain stimulation to help relieve symptoms of Parkinson's disease and other neurological disorders.

According to the US Center for Disease Control and Prevention, in 2009 more than 25% of adults, 18 years and older, have experienced low back pain for 24 h or longer [41]. The treatment for chronic pain that does not respond well to drug therapy using implantable electrical neurostimulator devices is growing.

While primary batteries have been used for these devices in the past, the trend is toward using rechargeable batteries to extend the longevity of the device. For implantable devices, the longer the battery lasts, fewer surgeries are required to replace the device, which in turn means lower risks of surgical complications, like infections.

Neuromodulation Batteries

Lithium Ion

Secondary, or rechargeable, cells have been used in medical devices for some time. One of the first batteries to be used for implantable cardiac pacemakers was a custom rechargeable nickel-cadmium battery [6]. Systems to recharge the battery transcutaneously using inductive methods were developed starting in the late 1950s [42].

Some implantable medical devices use secondary cells as a way to minimize the number of device replacement surgeries compared to primary cells. An example is neurostimulation for pain management for which primary cells have almost been supplanted entirely by rechargeable cells. The result is longer device longevity in a small size.

Since they were first commercialized in 1991, lithium ion (Li ion) cells now have become the system of choice for high-end portable consumer applications (e.g., laptop and tablet computers, personal multimedia players, and cell phones). They are also seeing increased use in power tools, electric vehicles, and space satellites. Likewise, Li ion cells play an important role in some implantable medical devices.

The chemistry systems in the Li ion cells that are manufactured specifically for implantable medical devices are similar to those developed for consumer applications.

Reliability and safety are utmost concerns when designing Li ion cells for implantable medical devices. They are hermetically sealed and may contain redundant safety features in the batteries (as well as part of the charging circuitry in the device). Also, cell sizes and shapes are often specially designed for a particular device.

The negative electrode material for most Li ion cells is some form of carbon, such as graphite or coke. Cells that use lithium titanate ($Li_4Ti_5O_{12}$) negative electrode material are also available. These cells have cell voltages over 1 V lower than more traditional Li ion cells that use carbon negative electrode materials, though. The advantages are rapid charging, and they can provide thousands of charge–discharge cycles.

Unlike lithium metal, which oxidizes to soluble Li^+ during discharge of lithium primary cells, lithium ions are instead reversibly inserted, or intercalated, between graphite layers during the cell charging process and removed during discharge:

$$C_6 + xLi^+ + xe^- \underset{\text{discharge}}{\overset{\text{charge}}{\rightleftarrows}} C_6Li_x$$

The value for x is generally between 0 and 1. Carbons intercalated with Li^+ can have electrochemical potentials within tens of millivolts of the lithium metal potential when fully charged.

The structure of the carbon material remains intact after the complete charge and discharge cycle and so is able to repeat the charge–discharge process many times.

Positive electrode materials available for Li ion cells today include metal oxide or phosphate compounds, such as

- Lithium cobalt oxide ($LiCoO_2$)
- Lithium mixed metal oxide, which is a combination of cobalt (Co) and nickel (Ni) and perhaps other metal ions. Examples of mixed metal oxides are $LiNi_{0.8}Co_{0.2}O_2$, $LiNi_{0.8}Co_{0.15}Al_{0.5}O_2$, and $LiNi_{0.33}Co_{0.33}Mn_{0.33}O_2$
- Lithium manganese oxide ($LiMn_2O_4$), sometimes referred to as spinel after its structure type
- Lithium iron phosphate ($LiFePO_4$)

The first commercialized Li ion cells used $LiCoO_2$ as the positive electrode material. It consists of Li^+ inserted between the cobalt and oxygen layers. The charge and discharge reactions are similar to carbon in that Li^+ is removed and inserted leaving the basic $LiCoO_2$ unchanged:

$$\mathrm{LiCoO_2} \underset{\text{discharge}}{\overset{\text{charge}}{\rightleftarrows}} \mathrm{Li_{1-x}CO_2} + x\mathrm{Li^+} + xe^-$$

where x lies somewhere between 0 and 1.

The overall cell reactions are:

$$\mathrm{C_6} + \mathrm{LiCoO_2} \underset{\text{discharge}}{\overset{\text{charge}}{\rightleftarrows}} \mathrm{Li_xC_6} + \mathrm{Li_{1-x}CoO_2}$$

Some positive electrode materials have three-dimensional rather than layered structures. $LiMn_2O_4$ and $LiFePO_4$ are in this category. Lithium ions are inserted in tunnels rather than between layers.

Positive electrode materials are generally mixed with a conductive carbon powder and a polymer binder. Commonly used binders are polyvinylidene fluoride (PVDF) and a PVDF copolymer with hexafluorpropylene (PVDF-HFP). The positive electrode mix is coated as a slurry onto a thin aluminum foil current collector while the carbon negative electrode material, often using the same binders, is coated onto a thin copper foil current collector.

Cylindrical or prismatic spiral wound cells are assembled by winding the positive and negative electrodes together with a porous membrane separator between them. The electrolyte solution is typically a mixture of organic solvents containing a lithium salt, like lithium hexafluorophosphate ($LiPF_6$).

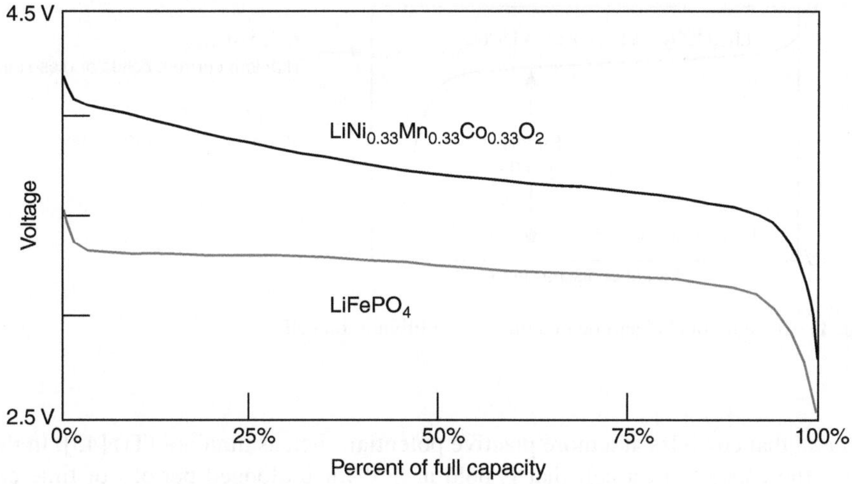

Fig. 11.12 Discharge voltage for two types of Li ion cells

A related Li ion cell type is the Li ion polymer cell that uses a polymer gel electrolyte. Li ion polymer cells are often designed as flat prismatic cells and can be made quite thin.

The discharge voltage can be sloped or flat depending on the type of cell chemistry used (for example, see Fig. 11.12).

Maximum voltages at the end of charge are usually in the range of 3.6–4.2 V depending on the chemistry. Energy densities can be as high as 240 Wh kg^{-1} and 640 Wh dm^{-3}.

Developers of secondary cells for implantable medical devices must consider the possibility of what happens when the battery is not charged for whatever reason in a timely fashion. If the cell becomes over discharged and its voltage is allowed to go below a certain threshold, the cell may not fully recover once it is charged and performance may be reduced.

One reason cell performance may be compromised in the event of the issues with typical Li ion cells is the use of a copper (Cu) current collector for the negative electrode. As a cell is discharged, the positive electrode potential decreases and the negative electrode potential increases. When the cell voltage drops to about 0 V, the potential of the negative electrode is the same as the positive electrode when measured against a reference electrode like Li. If discharge continues, the negative electrode potential could reach the potential at which the Cu current collector oxidizes or corrodes and copper ions dissolve (Fig. 11.13) [43]. This can lead to loss of cell capacity and degrade cell function. For this reason, the cell anode to cathode capacity balance is a critical design consideration. Cell performance is limited by the cathode capacity.

Selection of a positive electrode material that discharges at a lower voltage can minimize this issue, but an alternative is to replace the Cu current collector with

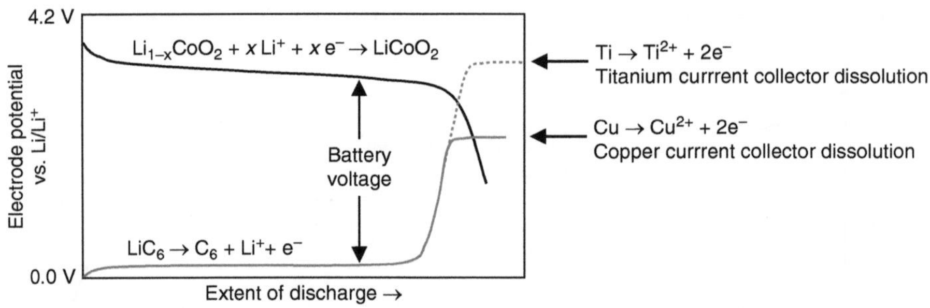

Fig. 11.13 Individual electrode potentials for a lithium ion cell

a metal that corrodes at a more positive potential, such as titanium (Ti) [43]. In this case, the capacity of a cell that is held at 0 V for prolonged periods of time can retain more of its capacity [44].

Hearing Loss

The US National Institutes of Health (NIH) reports that 36 million adults in the US report some level of hearing loss [45]. The World Health Organization puts the number at 278 million worldwide in 2005 [46]. Up to 0.3% of children born in the USA are deaf or hard of hearing [45].

Hearing aids are small devices that amplify the sounds picked up by a tiny microphone to enable the hearing-impaired to function. There are several types of hearing aid designs: behind the ear, in the ear, and completely in the canal. NIH finds that 1 out of 5 people in the USA who could benefit from a hearing aid actually wears one [45]. Worldwide, it is less than 1 out of 40 people [46].

The first portable radios operated by batteries were developed in the 1930s. The invention of the first portable hearing aids followed soon after in 1937 [47]. It could be carried by the hearing aid user, but since it used vacuum tubes it was the size of a lunch box. Wearable hearing aids became possible with the development of smaller electronic components that replaced vacuum tubes and the commercialization of small zinc-mercury oxide cells after World War II.

Most hearing aids today run on zinc-air cells because they have a high energy density and so last longer. Depending on the hearing needs of the user and the type of hearing aid, batteries may last from a few days to over 1 month. Recently, rechargeable nickel-metal hydride and zinc-silver oxide cells hearing aid batteries have been introduced.

Hearing Aid Batteries

Zinc-Air

The design of a zinc-air cell is different than most other battery types. Nearly all cells store all of the active materials required for the cell to function – negative and positive electrode materials, electrolyte – within a battery case or housing.

However, the positive electrode material for zinc-air cells is atmospheric oxygen, O_2. A thin electrode, or cathode, similar to a fuel cell electrode that provides a site for O_2 to be reduced during cell discharge is all that is necessary. As a result, there is more space within the cell that can be loaded with more of the zinc (Zn) negative electrode material.

The combination of a practically unlimited source of O_2 from the air with more available space inside the cell for additional Zn results in a very high energy density cell. Zinc-air button cells have the highest energy density of all commonly available button cells – up to 370 Wh kg^{-1} and 1,300 Wh dm^{-3} [14]. Of course, using atmospheric O_2 means zinc-air cells cannot be used for implantable devices.

Air was known to improve cell performance early in the nineteenth century, even though the electrochemical mechanism was unknown. Cells that used atmospheric O_2 as the positive electrode material were built in the late 1800s. Practical advances in 1930s through the 1950s made zinc-air cells commercially viable. Prismatic zinc-air cells were developed to operate hearing aids in the early 1950s [48, 49]. The early hearing aid cells were close to 22 cm^3 in volume with energy densities of about 200 Wh kg^{-1} [50].

Improvements in hearing aid electronics resulted in smaller devices that required less power. Lower power demands and advances in battery technology lead to smaller batteries, as well. Typical primary hearing aid cells today are small button cells that range from 5.8 mm in diameter and 2.15 mm in height to 11.5 mm in diameter and 5.4 mm in height (Table 11.1).

The largest hearing aid cell today is less than about 0.6 cm^3 – almost forty times less volume than the first zinc-air hearing aid batteries.

The zinc is in powder form that is mixed with a compound like polyacrylic acid or sodium carboxymethyl cellulose that is chemically stable and forms a gel in the alkaline electrolyte solution. The gelled anode is more stable mechanically.

The discharge reaction of Zn in aqueous potassium hydroxide (KOH) electrolyte solutions is somewhat complex, but may be simply written as:

$$Zn + 4OH^- \rightarrow Zn(OH)_4{}^{2-} + 2e^-$$

where the zincate product, $Zn(OH)_4{}^{2-}$, is dissolved in the electrolyte. Under certain conditions, especially as the battery is depleted, zinc ions can precipitate as zinc oxide (ZnO):

$$Zn + 2OH^- \rightarrow ZnO + H_2O + 2e^-$$

Table 11.1 Common primary hearing aid button cell sizes. IEC is the International Electrotechnical Commission and ANSI is the American National Standards Institute

Size designation			Dimensions		Available cells types
IEC	ANSI	Common	Diameter (mm)	Height (mm)	(IEC, ANSI)
63	7012ZD	5	5.8	2.15	Zinc-air (PR63, 7012ZD)
70	7005ZD	10	5.8	3.6	Zinc-air (PR63, 7012ZD)
	1191SO				Zinc-silver oxide (SR70, 1191SO)
41	7002ZD	312	7.9	3.6	Zinc-air (PR41, 7002ZD)
	1135SO				Zinc-silver oxide (SR41, 1135SO)
48	7000ZD	13	7.9	5.4	Zinc-air (PR48, 7000ZD)
	1137SO				Zinc-silver oxide (SR48, 1137SO)
44	7003ZD	675	11.6	5.4	Zinc-air (PR44, 7003ZD)
	1131SO				Zinc-silver oxide (SR44, 1131SO)

Typical alkaline, or basic, electrolyte solutions are comprised of 20–50% KOH (typically 28%) by weight in water. A small amount of ZnO, between a few percent up to saturation levels, is added to the KOH electrolyte to help reduce Zn self-discharge. ZnO dissolves to form potassium zincate in the electrolyte.

Zinc is used as the negative electrode material in aqueous primary cells because it has a high energy density. However, Zn is prone to corrosion, particularly when certain impurities are present, that can result in decreased battery life and compromised performance. A common contaminant is iron. When it comes into contact with the Zn and becomes galvanically coupled with it, hydrogen gas (H_2) will form at the iron surface from the reduction of water in the electrolyte. Zinc is oxidized and dissolves into the electrolyte as zincate or precipitates as ZnO.

If Zn corrosion proceeds at a sufficiently high rate for long periods of time, the gas pressure within the cell can build until the seals start to leak gas and electrolyte. This was a common problem in early cardiac pacemakers using Zn/HgO batteries. The gassing cannot be entirely avoided, but the corrosion reactions can be slowed and whatever gas is formed can be controlled to maximize the shelf life and minimize the loss of cell performance.

Alloying Zn with mercury metal to form an amalgam increases the electrochemical overpotential for hydrogen gas formation on Zn. Zinc amalgams in the range of 2–15% Hg by weight of Zn were common.

Since mercury (Hg) and its compounds are toxic, beginning in the early 1990s the European Union, followed by the United States, required battery manufacturers to reduce [51] and then eliminate [52] Hg added to batteries. This requirement addressed genuine concerns that cells discarded in landfills could release Hg into the environment.

Use of battery materials with higher purity levels (including Zn, electrolyte, and cathode), different alloying elements, seals that can withstand higher internal pressures, and cleaner manufacturing processes enabled the elimination of added Hg in batteries.

The positive electrode reaction involves the reduction of O_2 and proceeds in two basic steps:

$$O_2 + H_2O + 2e^- \rightarrow HO_2^- + OH^-$$

and then

$$2HO_2^- \rightarrow 2OH^- + O_2.$$

where HO_2^- is the hydrogen peroxide ion. The second step acts to curb the oxygen discharge reaction, so a catalyst, usually manganese dioxide, may be used to accelerate peroxide decomposition.

The overall cell reaction is:

$$2Zn + O_2 + 4KOH + 2H_2O \rightarrow 2K_2Zn(OH)_4$$

Or

$$2Zn + O_2 \rightarrow 2ZnO$$

The reduction of O_2 is thermodynamically favorable, but the reaction kinetics are relatively slow. Using an electrode material with a high surface area, such as activated carbon powder, can overcome the slow oxygen reduction kinetics by presenting a large number of sites at which O_2 can be reduced. A small amount of a conductive carbon material may also be added to increase the electrical conductivity of the cathode.

While it is important to maximize the electrode area accessible to O_2, the cathode must also allow for contact with the electrolyte. A three-phase boundary comprised of O_2, electrolyte and carbon with catalyst is central to the proper functioning of an air electrode. Maintaining a balance between access to atmospheric O_2 and electrolyte solution is achieved by adequately dispersing just the right amount of a hydrophobic material, polytetrafluoroethlyene (PTFE or Teflon®), in the carbon powder mix. Too little PTFE or it is not dispersed well enough and the electrolyte solution could saturate the cathode. This reduces the area at which O_2 can react with a resulting loss of performance. Too much PTFE and the electrolyte will not sufficiently wet the cathode. Again, this limits battery performance.

A zinc-air cathode must provide a surface at which O_2 can be reduced during battery discharge while not being consumed itself during cell discharge. The zinc-air cathode of today is comprised of multiple layers, each serving an important function. A carbon mix (activated carbon, conductive carbon, and PTFE) is applied to a nickel screen current collector that carries electrons to the cathode during cell discharge. There are two sides to an air cathode – the air side and the electrolyte side. The air side is laminated with a porous PTFE membrane. A polymeric separator sheet is pressed on the electrolyte.

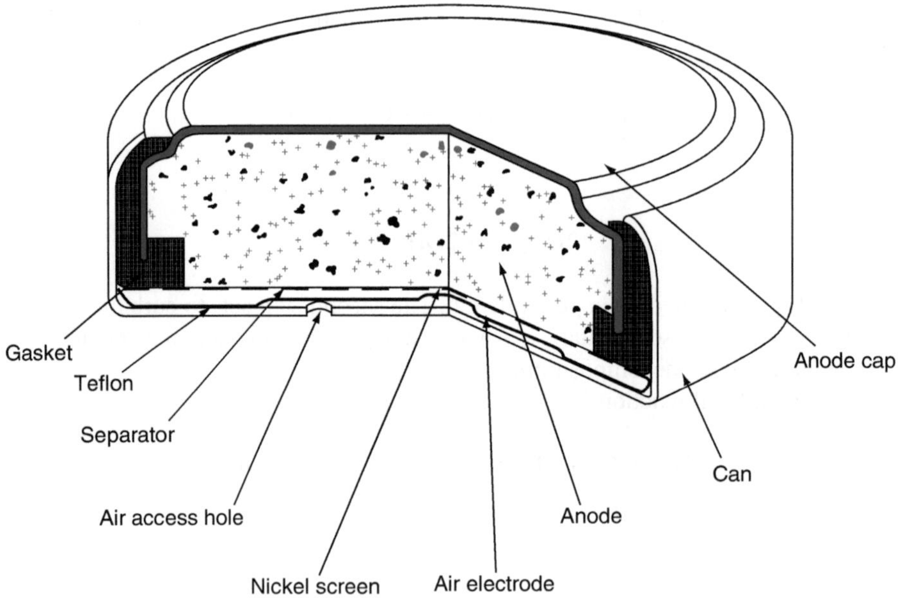

Fig. 11.14 Cross section illustrating the main components of a zinc-air cathode. © Eveready Battery Company, Inc (Reprinted with permission)

Important considerations for zinc-air cells are air and water management. Air must be allowed to enter the zinc-air cell, which is done by providing holes in the battery can near the cathode (see Fig. 11.14). There must be an adequate number of holes of a sufficient size to allow enough air to enter the cell such that the discharge reaction is not impeded by lack of O_2. An air diffusion layer may be included just inside the cell to disperse the air more uniformly.

The advantage of zinc-air cells also presents a big challenge for battery designers. Remaining open to the atmosphere renders zinc-air cells exposed to detrimental environmental conditions, especially humidity. Water in humid air can be absorbed by the basic electrolyte solution diluting it and subsequently flooding the cathode. Arid air may evaporate water from the electrolyte and dry the cathode. Both conditions lead to reduced cell performance and battery life. Carbon dioxide in the air can enter the cell, react with the basic electrolyte solution and precipitate carbonates, also decreasing performance.

A simple way to mitigate the effects of environmental exposure is to seal the holes until the cell is needed to operate. The holes are covered, usually with an adhesive tab, after manufacture to curtail exposure to the atmosphere during shipping and storage. The seal is removed prior to use, air enters the cell and it is ready for use. Hearing aid users can also replace the adhesive tab between uses, overnight while sleeping for example, to extend cell life.

The initial open circuit potential of the zinc-air cell is about 1.45 V. The voltage of a zinc-air cell is mostly constant throughout discharge (Fig. 11.15).

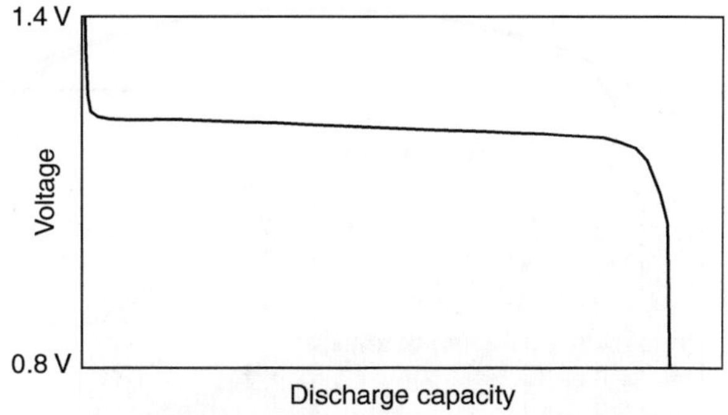

Fig. 11.15 Discharge voltage behavior for a zinc-air cell

Zinc-Silver Oxide

Zinc-silver oxide (Zn/Ag_2O) cells are used in medical applications like hearing aids, but also military, aerospace, watches, cameras, and calculators. Early implanted electrical bone growth stimulators [53] for fracture healing and spinal fusion used Zn/Ag_2O batteries. The Zn/Ag_2O cell is one of the alternatives to Zn/HgO cells (see above) and is used in applications where high energy and power density is required.

Originally dating back to 1883, [54] alkaline Zn/Ag_2O cells were eventually developed into commercially practical cells in the early 1960s.

The negative electrode material is zinc powder in a gelled KOH-ZnO-H_2O alkaline electrolyte solution. NaOH can be used instead of KOH for lower power applications.

The positive electrode reaction is:

$$Ag_2O + H_2O + 2e^- \rightarrow 2Ag + 2OH^-$$

The cathode pellet contains Ag_2O powder and 1–5% of a conductive carbon powder like graphite, to reduce internal resistance and provide good contact to all of the active silver oxide particles, mixed with a PTFE binder to maintain the mechanical integrity of the pellet.

A disadvantage of Ag_2O is its solubility in alkaline electrolyte. Silver ions dissolve into basic electrolyte solutions. The solubility of silver ions (Ag^+) from Ag_2O is on the order of 10^{-4} mol dm^{-3} in concentrated KOH [55]. Dissolved Ag^+ diffuses to the Zn and deposits there as Ag. If this continues, the Ag deposit will grow as dendrites through the separator and eventually create an internal short circuit by directly bridging the positive and negative electrode materials [56]. A big advance toward making this cell chemistry practical came when Henri André developed a cellophane separator that minimized diffusion of Ag^+ through the separator which mitigated this issue [54].

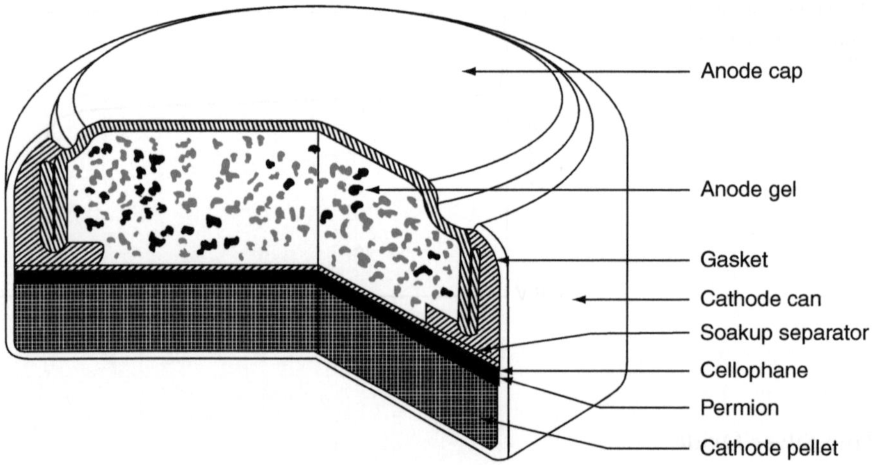

Fig. 11.16 Cross section illustrating the main components of a Zn/Ag_2O button cell. © Eveready Battery Company, Inc (Reprinted with permission)

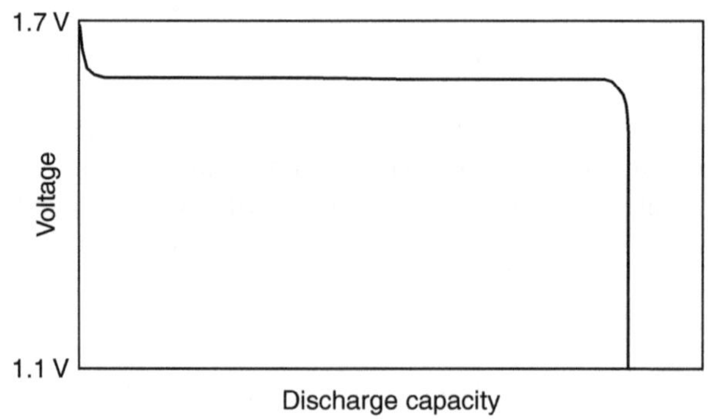

Fig. 11.17 Discharge voltage behavior for a Zn/Ag_2O cell

The overall cell reaction is:

$$Zn + Ag_2O + 2KOH + H_2O \rightarrow K_2Zn(OH)_4 + 2Ag$$

or

$$Zn + Ag_2O \rightarrow ZnO + 2Ag$$

The cross section of a Zn/Ag_2O button cell is pictured in Fig. 11.16.

The open circuit potential of Zn/Ag_2O cells is about 1.60 V and voltage remains relatively constant throughout discharge as seen in Fig. 11.17. Energy density for Zn/Ag_2O button cells is about 135 Wh kg^{-1} or 530 Wh dm^{-3} [14].

Future Directions

A wide variety of wearable, implantable, and even ingestible medical devices are under development or in the early stages of clinical use. Here are just a few.

There is an ongoing need to treat various heart conditions, hearing loss, and chronic pain. Improvements to electronic circuit components and designs, along with the batteries that make them function, will continue. New therapies and device features will be enabled by new battery technologies.

Additionally, there are many emerging indications for wearable or implantable medical devices, particularly new neuromodulation applications such as deep brain stimulation for various movement and neurological disorders [57] and occipital nerve stimulation to treat migraine and cluster headaches [58]. Implantable visual prostheses to restore sight are also under development.

There are ingestible devices for measurement of body core temperature and other sensors to help diagnose gastrointestinal tract disorders. Implantable sensors [59, 60], such as those for blood pH, oxygen, and glucose, along with remote telemetry, are expected to advance as well.

Bibliography

Primary Literature

1. Althaus J (1860) A treatise on medical electricity, theoretical and practical. Lindsay and Blakiston, Philadelphia, pp v–vi
2. Tibbits H (1873) A handbook of medical electricity. P. Blakiston, Son and Co., Philadelphia, p v
3. Steavenson WE, Jones HL (1892) Medical electricity. A practical handbook for students and practioners. P. Blakiston, Son and Co., Philadelphia, pp v–vi
4. Mohrman DE, Heller LJ (1996) Cardiovascular physiology, 4th edn. McGraw-Hill, New York
5. Lillehei CW, Gott VL, Hodges PC, Long DM, Bakken EE (1960) Transistor pacemaker for treatment of complete atrioventricular dissociation. JAMA 172:2006–2010
6. Greatbatch W, Holmes CF (1991) History of implantable devices. IEEE Eng Med Biol:38–41,49
7. Chardack W, Gage A, Greatbatch W (1960) A transistorized, self-contained, implantable pacemaker for the long-term correction of complete heart block. Surgery 48:543
8. Furman S (2003) The early history of cardiac pacing. Pacing Clin Electrophysiol 26:2023–2032
9. Liang CC, Holmes CF (1980) Lithium pacemaker batteries – an overview. In: Owens BB, Margalit N (eds) Proceedings of the symposia on power sources for biomedical implantable applications and ambient temperature batteries, vol 80–84, pp 27–33
10. Bilitch M, Parsonnet V, Furman S (1980) Clinical assessment of pacemaker power sources. In: Owens BB, Margalit N (eds) Proceedings of the symposia on power sources for biomedical implantable applications and ambient temperature batteries, vol 80–84, pp 18–26
11. Greatbatch W (1984) Pacemaker power sources. IEEE Eng Med Biol Mag:15–19

12. Greatbatch W, Lee JH, Mathias W, Eldridge M, Moser JR, Schneider AA (1971) The solid-state lithium battery: a new improved chemical power source for implantable cardiac pacemakers. IEEE Trans BioMed Eng BME 18:317–324
13. Kelly RG, Moran PJ (1987) The rate limiting mechanism of Li/I_2 (PV2P) batteries. J Electrochem Soc 134:25–30
14. Linden D (2002) Ch 7 Primary batteries – introduction. In: Linden D, Reddy TB (eds) Handbook of batteries, 3rd edn. McGraw-Hill, New York
15. Shellock FG, Hatfield M, Simon BJ, Block S, Wamboldt J, Starewicz PM, Punchard WFB (2000) Implantable spinal fusion stimulator: assessment of MR safety and artifacts. J Mag Res Imaging 12:214–223
16. Maurer J, Savvas E (2010) The Esteem system: a totally implantable hearing device. In: Böheim K (ed) Active middle ear implants. S Karger AG, Basel, p 59
17. Root MJ, Dumas R, Yazami R, Hamwi A (2001) The effect of carbon starting material on carbon fluoride synthesized at room temperature – characterization and electrochemistry. J Electrochem Soc 148:A339–A345
18. Spellman PJ, Dittberner KL, Pilarzyk JG, Root MJ (2000) Application of Li/CF_x system in portable electronics. In: Osaka T, Datta M (eds) Energy storage systems for electronics. Gordon and Breach Science, Amsterdam, pp 131–152
19. Linden D, Reddy TB (2002) Ch 14 Lithium batteries. In: Linden D, Reddy TB (eds) Handbook of batteries, 3rd edn. McGraw-Hill, New York
20. Parker B (1978) Obituary: a vindication of the zinc-mercury pacemaker battery. Pacing Clin Electrophysiol 1:148–149
21. Clarke CL (1884) US patent 298,175
22. Brønsted JN (1917) US patent 1,219,074
23. Friedman M, McCauley CE (1947) The Ruben cell – a new alkaline primary dry cell battery. Trans Electrochem Soc 92:195–215
24. Ruben S (1978) The evolution of electric batteries in response to industrial needs. Dorrance, Philadelphia, Ch VI
25. Mirowski M, Reid PR, Mower MM, Watkins L, Gott VL, Schauble JF, Langer A, Heilman MS, Kolenik SA, Fischell RE, Weisfeldt ML (1980) Termination of malignant ventricular arrhytmias with an implanted automatic defibrillator in human beings. N Engl J Med 303:322–324
26. Liang CC, Bolster ME, Murphy RM (1982) US patent 4,310,609
27. Leising RA, Takeuchi ES (1993) Solid-state cathode materials for lithium batteries:effect of synthesis temperature on the physical and electrochemical properties of silver vanadium oxide. Chem Mater 5:738–742
28. Crespi A (1993) US patent 5,221,453
29. Leising RA, Thiebolt WC, Takeuchi ES (1994) Solid state characterization of reduced silver vanadium oxide from the Li/SVO discharge reaction. Inorg Chem 33:5733–5740
30. Syracuse K, Waite N, Gan H, Takeuchi ES (2006) US patent 6,982,543
31. Ikeda H, Saito T, Tamura H (1975) Manganese dioxide as cathodes for lithium batteries. In: Proceedings of the manganese dioxide symposium, vol 1. The Electrochemical Society, Cleveland, pp 384–401
32. Gerbier G, Lehman G (1980) Mangalith: a new lithium pacemaker battery. In: Owens BB, Margalit N (eds) Proceedings of the symposia on power sources for biomedical implantable applications and ambient temperature batteries, vol 80–84. The Electrochemical Society, Pennington, pp 136–143
33. Merritt DR, Schmidt CL (1993) Impedance modeling of the lithium/manganese dioxide battery. In: Surampudi S, Koch VR (eds) Proceedings of the symposium on lithium batteries, vol 93–24. The Electrochemical Society, Pennington, pp 138–145
34. Drews J, Fehrmann G, Staub R, Wolf R (2001) Primary batteries for implantable pacemakers and defibrillators. J Power Sources 97–98:747–749

35. Root MJ (2008) Implantable cardiac rhythm device batteries. J Cardiovasc Trans Res 1:254–257
36. O'Phelan MJ, Victor TG, Haasl BJ, Swanson LD, Kavanagh RJ, Barr AG, Dillon RM (2009) US patent 7,479,349
37. Root MJ (2010) Lithium-manganese dioxide cells for implantable defibrillators – discharge voltage models. J Power Sources 195:5089–5093
38. Drews J, Wolf R, Fehrmann G, Straub R (1999) Development of a hybrid battery system for an implantable biomedical device, especially a defibrillator/cardioverter (ICD). J Power Sources 80:107–111
39. Weiss DJ, Cretzmeyer JW, Crespi AM, Howard WG, Skarstad PM (1993) US patent 5,180,642
40. Spillman DM, Takeuchi ES (1999) US patent 5,935,724
41. National Center for Health Statistics (2010) Health, United States, 2009. US Department of Health and Human Services Center for Disease Control and Prevention, Publication 2010–1232, Hyattsville, p 261
42. Fischell RE, Schulman JH (1976) A rechargeable power system for cardiac pacemakers. Proc 11th IECEC, pp 163–168
43. Tsukamoto H, Kishiyama C, Nagata M, Nakahara H, Piao T (2003) US patent 6,596,439
44. Kishiyama C, Nagata M, Piao T, Dodd J, Lam P, Tsukamoto H (2003) 204th Electrochemical Society Meeting, abstract 425
45. http://wwwnidcdnihgov/health/statistics/quickhtm. Accessed 27 Dec 2010
46. http://wwwwhoint/mediacentre/factsheets/fs300/en/indexhtml. Accessed 3 Jan 2011
47. Wengel AM (1940) US patent 2,192,669
48. Marsal PA, Fox RP (1952) US patent 2,597,116
49. Schumacher EA, Bennett RJ (1952) US patent 2,597,119
50. Brodd RJ, Kozawa A, Kordesch KV (1978) Primary batteries 1951–1976. J Electrochem Soc 125:271C–282C
51. Council Directive 91/157/EEC of 18 March 1991 on Batteries and Accumulators Containing Certain Dangerous Substances
52. "NEMA announces battery industry commitment to eliminating mercury in button cells". https://wwwnemaorg/media/pr/20060302acfm. Accessed 28 Dec 2010
53. Paterson DC, Lewis GN, Cass CA (1980) Treatment of delayed union and non-union with an implanted direct current stimulator. Clin Orthop Relat Res 148:117–128
54. Howard PL, Fleischer A (1971) Ch 1. Milestones in the electrochemistry of zinc-silver oxide batteries. In: Fleischer A, Lander JJ (eds) Zinc-silver oxide batteries. Wiley, New York
55. Amlie RF, Reutschi P (1961) J Electrochem Soc 108:813–819
56. Himy A (1986) Silver-zinc battery phenomena and design principles. Vantage, New York, p 7
57. Pereira EAC, Aziz TZ (2006) Surgical insights into Parkinson's disease. J R Soc Med 99:238–244
58. Schwedt TJ, Dodick DW, Trentman TL, Zimmerman RS (2006) Occipital nerve stimulation for chronic cluster headache and hemicrania continua: pain relief and persistence of autonomic features. Cephalagia 26:1025–1027
59. Van Laerhoven K, Lo BPL, Ng JWP, Thiemjarus S, King R, Kwan S, Gellersen H-W, Sloman M, Wells OW, Needham P, Peters N, Darzi A, Toumazou C, Yang G-Z (2004) Medical healthcare monitoring with wearable and implantable sensors. Proc 3rd international workshop on ubiquitous computing for healthcare applications. http://www.pervasivehealthcare.com/ubicomp2004/papers/final_papers/laerhoven.pdf
60. Fletter PC, Majerus S, Cong P, Damaser MS, Ko WH, Young DJ, Garverick SL (2009) Wireless micromanometer system for chronic bladder pressure monitoring. 6th international conference on networked sensing systems, IEEE, pp 1–4

Books and Reviews

Gabano J-P (ed) (1983) Lithium batteries. Academic, London

Holmes CF (1994) Implantable lithium power sources. In: Pistoia G (ed) Lithium batteries. Elsevier, Amsterdam, pp 377–416

Owens BB (ed) (1986) Batteries for implantable biomedical devices. Plenum, New York

Reddy TB (ed) (2010) Linden's handbook of batteries, 4th edn. McGraw-Hill, New York

Root M (2010) The TAB battery book. McGraw-Hill, New York

Schlesinger H (2010) The battery. HarperCollins, New York

Takeuchi ES, Leising RA, Spillman DM, Rubino R, Gan H, Takeuchi KJ, Marschilok AC (2004) Lithium batteries for medical applications. In: Nazri G-A, Pistoia G (eds) Lithium batteries science and technology. Kluwer, Boston, pp 686–700

Untereker DF, Crespi AM, Rorvick A, Schmidt CL, Skarstad PM (2007) Power systems for implantable pacemakers, cardioverters, and defibrillators. In: Ellenbogen KA, Kay GN, Lau C-P, Wilkoff BL (eds) Clinical cardiac pacing and defibrillation, 3rd edn. Saunders, Philadelphia, pp 235–259

Vincent C, Scrosati B (1997) Modern batteries, 2nd edn. Arnold, London

Chapter 12
Nanocarbons for Supercapacitors

François Béguin and Encarnación Raymundo-Piñero

Glossary

Carbon structure and texture	The structure is the arrangement of the carbon atoms in the space since the texture is the arrangement of the graphene layers in the space for giving porosity or empty space accessible for molecules or ions.
Electric double layer capacitor	Electrochemical capacitor in which the charge storage is achieved electrostatically because of the separation of charges in a double layer across the electrode/electrolyte interface.
Nanocarbon	Carbon material consisting in more or less disordered graphene layers which can be synthesized with different structures, porous texture, and surface functionality.
Pseudo-capacitor	Electrochemical capacitor in which the charge storage is achieved by an electron transfer that produces chemical or oxidation state changes in the electrode materials. As a difference from a battery, the electrode potential varies proportionally to the charge exchanged.
Supercapacitor or electrochemical capacitor	Electrochemical energy storage device in which the voltage declines linearly with the extent of charge. A supercapacitor consists of two electrodes separated by a porous membrane immersed in an electrolyte.

This chapter was originally published as part of the Encyclopedia of Sustainability Science and Technology edited by Robert A. Meyers. DOI:10.1007/978-1-4419-0851-3

F. Béguin (✉) • E. Raymundo-Piñero
Research Centre on Divided Matter, CRMD-CNRS & Orléans University,
1B rue de la Férollerie, 45071 Orléans, France
e-mail: beguin@cnrs-orleans.fr

R.J. Brodd (ed.), *Batteries for Sustainability: Selected Entries from the Encyclopedia of Sustainability Science and Technology*, DOI 10.1007/978-1-4614-5791-6_12,
© Springer Science+Business Media New York 2013

Surface functionality — Surface groups are ubiquitously present at the edges of graphene sheets in carbon materials, especially the high surface area ones. Oxygen-containing surface groups are the most occurring ones but carbon can contain other heteroatoms, such as nitrogen or sulfur.

Definition of the Subject

Supercapacitors (or electrochemical capacitors) are electrochemical energy storage devices having higher energy density than dielectric capacitors and higher power density than batteries. Actually, they are capable of delivering large amounts of energy in a very short time. These devices rely mainly on the characteristics of the electrical double layer that forms on all polarized conductors when immersed in an electrolyte. The double layer forms in less than 10^{-6} s and responds to changes in a similar time frame. This is about 1,000 times faster than an electrochemical reaction at a battery electrode that has a time constant in the range of 10^{-3} s. These devices find application where high-power delivery is required.

Introduction

Supercapacitors (or electrochemical capacitors or ultracapacitors) represent a relatively new energy storage technology which has known an important development during the recent years. Supercapacitors can be applied for stationary and mobile systems where high-power pulses are requested: cars acceleration, tramways, cranes, lifts, safety systems, etc. Moreover, owing to their low time constant, they can quickly harvest energy, for example during deceleration or braking of vehicles. Although supercapacitors are able to provide high power with a long cycle life compared to accumulators, they suffer of a relatively low energy density. Therefore, the main ongoing research direction concerns the optimization of the existing electrode materials and the development of new materials, both for protic electrolytes and aprotic organic electrolytes.

Industrial supercapacitors are essentially based on nanoporous carbon electrodes. The reasons of the choice lie in the high availability, low cost, chemical inertness, and good electrical conductivity of activated carbons, as well as a high versatility of texture and surface functionality. For these reasons, this chapter will present the capacitance properties of carbon-based electrodes showing optimization strategies playing on the structure/nanotexture of carbon and the nature of the electrolyte.

Brief Overview of Activated Carbons Properties

In industry, activated carbons are essentially produced by carbonization (pyrolysis at temperatures up to 900°C under neutral atmosphere) of various precursors (lignocellulosic, polymers, anthracites, …), followed by physical and chemical activation [1]. Physical activation is generally realized around 900°C through partial gasification of carbon, using CO_2 or steam, according to Eqs. 12.1 and 12.2:

$$C + CO_2 \rightarrow 2\,CO \tag{12.1}$$

$$C + H_2O \rightarrow CO + H_2 \tag{12.2}$$

For chemical activation, the main reagents used are KOH, $ZnCl_2$, and H_3PO_4 [1]. KOH activation is a complex process, involving several redox reactions with carbon, followed by potassium intercalation/insertion and expansion of the structure [2, 3]. In the particular case of H_3PO_4, carbonization and activation generally proceed simultaneously at temperatures lower than 600°C [4].

On the nanotextural point of view, activated carbons (AC) are characterized by high values of specific surface area, up to 3,000 $m^2\,g^{-1}$. Depending on the precursor and carbonization/activation conditions, they can be essentially microporous (pores < 2 nm) or microporous and mesoporous (2 nm $<$ pores $<$ 50 nm). Many structural/nanotextural models have been proposed, but none of them gives a perfect description of the properties. The HRTEM 002 lattice-fringe image of an AC in Fig. 12.1a shows short fringes, few nanometers long, representing graphene-type layers, highly misoriented, some of them forming stacks of few layers. The disorientation of the graphene-type units causes the porosity of the material. According to simulations (Fig. 12.1b, [5, 6]), the graphene layers are not perfectly planar, and curvatures might be produced by defects such as pentagons and heptagons in their structure.

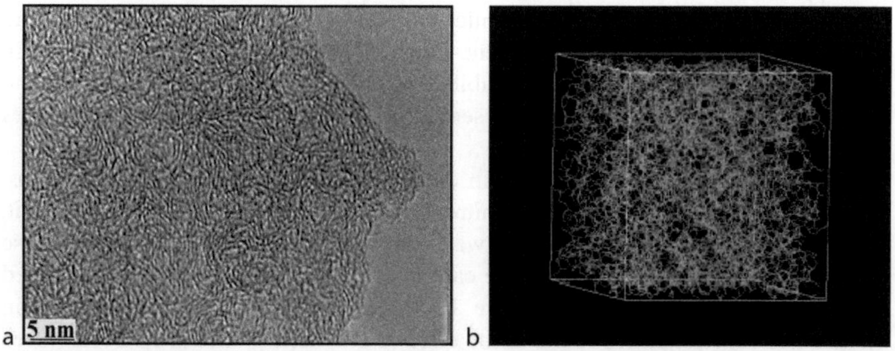

Fig. 12.1 (**a**) HRTEM 002 lattice fringe image of an activated carbon; (**b**) structure/nanotexture of an activated carbon obtained by coupling HRTEM and numerical simulation [6]

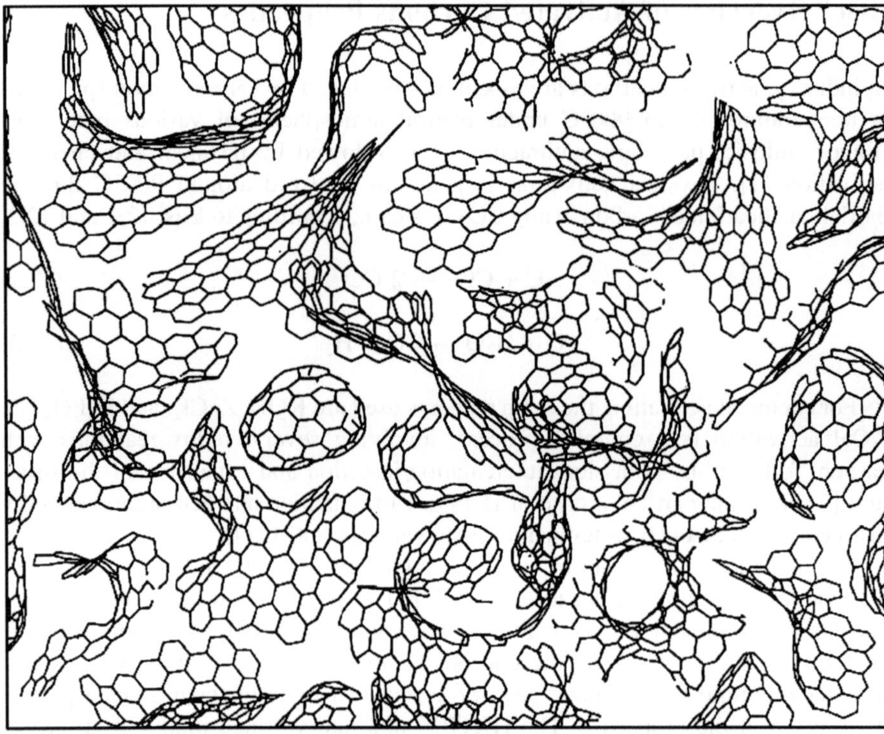

Fig. 12.2 Schematic nanotextural and structural model of an AC showing fullerene-like elements [7–9]

Based on various observations, Harris et al. proposed a model for the structure/nanotexture of non-graphitizable carbons that consists of discrete fragments of curved carbon sheets, in which pentagons and heptagons are dispersed randomly throughout networks of hexagons, as illustrated in Fig. 12.2 [7–9]. The size of the micropores in this model would be of the order of 0.5–1.0 nm, which is similar to the pore size observed in typical microporous activated carbons. Taking into account that the independent fullerene-like graphene units would be soluble in some solvents and that the thermal stability of activated carbons is high, one has to imagine that, in addition to this representation, the units are connected by bridges which keep the backbone.

Apart of carbon which is the main element constituting nanoporous carbons, heteroatoms (essentially oxygen and nitrogen) in the amount of few atomic percent, generally less than 5 at.%, are always present in activated carbons. These heteroatoms are either remains of the carbon precursor and/or they are introduced during the activation process. Figure 12.3 shows a representation of the main oxygenated functionalities found at the edge of graphene planes [10]. Beside, free edge sites are present together with unpaired electrons stabilized by resonance. In the case of C/C supercapacitors in organic electrolyte, the later are assumed to be

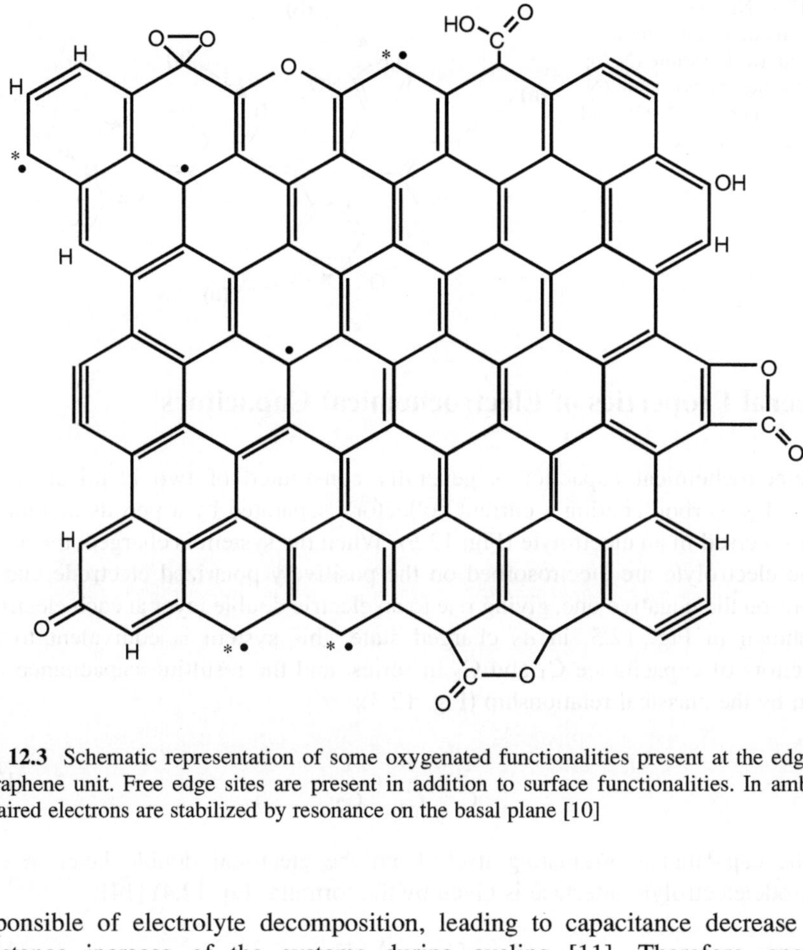

Fig. 12.3 Schematic representation of some oxygenated functionalities present at the edges of a graphene unit. Free edge sites are present in addition to surface functionalities. In ambient, unpaired electrons are stabilized by resonance on the basal plane [10]

responsible of electrolyte decomposition, leading to capacitance decrease and resistance increase of the systems during cycling [11]. Therefore, specific treatments of carbons are mandatory to partly deplete these sites and get rid of the drawbacks linked to their existence.

In the case of nitrogen, the variety of functionalities (Fig. 12.4) results both from the position occupied in the ring system and from the extent of association with oxygen, which hardly can be avoided during synthesis. Nitrogen is either substituted to carbon ("lattice nitrogen") or included in the form of functional groups ("chemical nitrogen") at the periphery of polyaromatic structural units [12, 13]. The surroundings of the nitrogen atom in a graphene layer obviously affect its charge, electron donor/acceptor properties, and the contribution to the delocalized π electron system.

For supercapacitor carbon electrodes, it will be further shown that (1) the developed surface area is responsible of an important electrical double-layer capacitance; (2) both the oxygenated and nitrogenated functionalities may be involved in redox reactions with the electrolyte, which enhance capacitance through a pseudo-capacitive contribution.

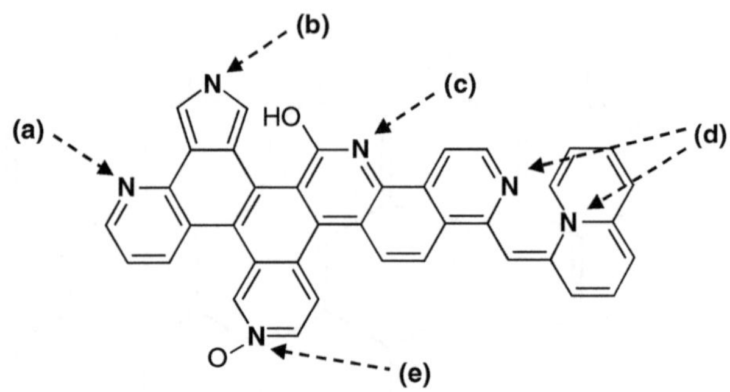

Fig. 12.4 Nitrogen functionalities in a carbon material. (**a**) Pyridinic (N-6); (**b**) pyrrolic; (**c**) pyridonic (N-5); (**d**) quaternary (N-Q); and (**e**) oxidized nitrogen (N-X)

General Properties of Electrochemical Capacitors

An electrochemical capacitor is generally constituted of two identical carbon electrodes (carbon coating a current collector), separated by a porous membrane, and immersed in an electrolyte (Fig. 12.5). When the system is charged, the anions of the electrolyte are electrosorbed on the positively polarized electrode and the cations on the negative one, giving rise to an electric double layer at each electrode. As shown in Fig. 12.5, in its charged state, this system is equivalent to two capacitors of capacitance C_1 and C_2 in series, and the resulting capacitance C is given by the classical relationship (Fig. 12.3):

$$\frac{1}{C} = \frac{1}{C_1} + \frac{1}{C_2} \tag{12.3}$$

The capacitance originating itself from the electrical double layer at each electrode/electrolyte interface is given by the formula (Eq. 12.4) [14]:

$$C_e = \epsilon A/d \tag{12.4}$$

where ε is the dielectric constant or permittivity, A the surface area of the electrode/electrolyte interface, and d the distance between the two charged layers of opposite sign. The value of d is controlled by the size of ions and is of the order of 1 nm. Consequently, the capacitance originating of the electrical double layer is very high, e.g., around 0.1 F m^{-2}. In electrical double-layer capacitors (EDLC), the electrodes are from high surface area–activated carbons; for this reason, the surface area of the interface is very high and the capacitance C_e reaches values as high as 100 F per gram of carbon.

As said in the introduction, the main research direction on supercapacitors is dedicated to enhancing the energy density of these systems in order to broaden their possible applications. The energy density of supercapacitors is given by formula (Eq. 12.5):

$$E = \frac{1}{2} CU^2 \tag{12.5}$$

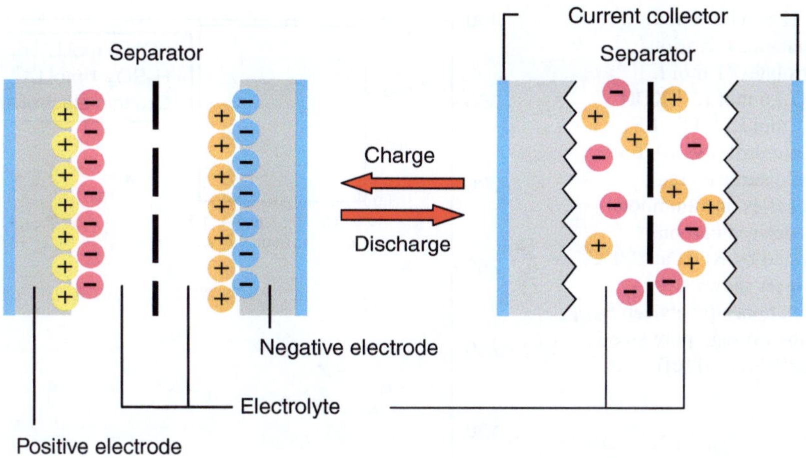

Fig. 12.5 Discharged (*right*) and charged (*left*) states of an electrochemical capacitor

where U represents the operating voltage. In supercapacitors, the later is strongly depending on the stability window of the electrolyte. In water medium, although the thermodynamic value is 1.23 V, practically it ranges from 0.7 to 1 V [15]. In organic electrolyte, the operating voltage can reach 2.7–2.8 V [15], and due to the square dependence of energy with voltage, this medium is definitively preferred for the industrial systems. Hence, the energy density of electrochemical capacitors is mainly controlled by the capacitance of the electrode materials. In aprotic electrolyte (e.g., organic electrolyte), the capacitance originates exclusively from the electrical double-layer formation, whereas an additional pseudo-capacitive contribution, due to fast pseudo-faradic reactions between the electrode surface and the electrolyte, may appear in protic electrolytic media. These two contributions will be considered separately.

Carbons for Electrical Double-Layer Capacitors

Considering Eq. 12.4 suggests that developing the surface area of activated carbons should lead to an increase of capacitance. For this reason, in the literature, many authors have suggested a proportional dependence between the gravimetric capacitance and the BET specific surface area measured by nitrogen adsorption at 77 K. An example is shown in Fig. 12.6 for a series of nanoporous carbons obtained by KOH activation of bituminous coal pyrolyzed at various temperatures [16]. Although these carbons were prepared from the same precursor using the same activation process and should be considered of identical nanotexture, the proportionality is observed only for the low values of BET surface area. Above 2,000 $m^2\ g^{-1}$, the observed saturation phenomenon is often interpreted by the fact

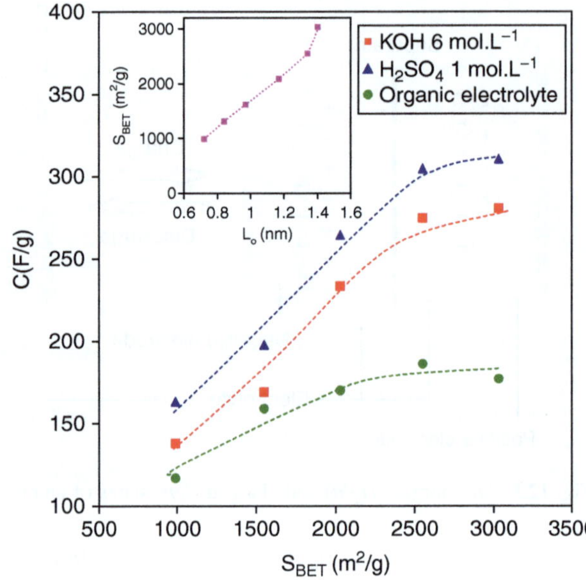

Fig. 12.6 Gravimetric capacitance in various electrolytes (1 mol L^{-1} H_2SO_4, 6 mol L^{-1} KOH and 1 mol L^{-1} $TEABF_4$ in acetonitrile) vs BET specific surface area for a series of bituminous coal–derived carbons activated by KOH at 800°C. The inset shows the relationship between S_{BET} and the average pore size L_0 (Adapted from [16])

that the BET model overestimates the values of surface area [17, 18]. Therefore, some plots have been proposed versus the DFT (density functional theory) surface area, but saturation is still observed for the high surface area carbons, i.e., for carbons with increasing activation degree. In order to explain the saturation, Barbieri et al. suggested that due to the decrease of average pore wall thickness in highly activated carbons, the electric field (and the corresponding charge density) no longer decays to zero within the pore walls [19].

Another explanation can be proposed from inset of Fig. 12.6, showing that the average pore size increases together with the specific surface area, i.e., when the activation degree of carbon increases. In high surface area carbons, with larger pores, one might suggest that the interaction of ions with pore walls is weaker, leading to smaller values of capacitance. Therefore, the normalized capacitance, e.g., the capacitance divided by the DFT surface area, has been plotted versus the average size L_0, both for the series of carbons from bituminous coal (Fig. 12.7a) [16] and for carbide-derived carbons (Fig. 12.7b) [20]. In the two cases, an increase of normalized capacitance is observed when the average pore size decreases, the highest values being observed for L_0 smaller than 1 nm. Taking into account the dimensions of solvated and non-solvated ions for tetraethylammonium tetrafluoroborate in acetonitrile, e.g., 1.30 and 0.67 nm for TEA^+, 1.16 and 0.48 nm for BF_4^-, respectively, suggests that the ions are at least partly desolvated in the pores [21]. Such property was already suggested for templated carbons which showed proportionality between capacitance and the ultramicropore (<0.7–0.8 nm) volume [22].

At this step, it is now clear that adapting the pore size of carbons to the size of ions is more important than developing the pore volume. Indeed, high surface area carbons display generally broader pores, and according to the data presented above

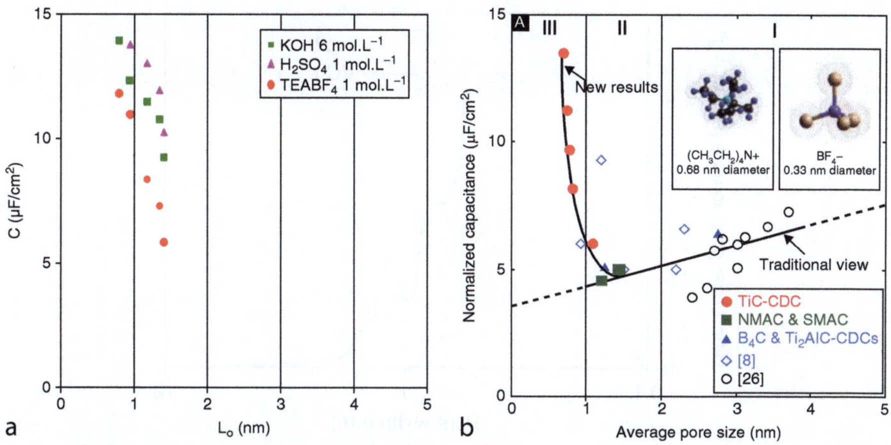

Fig. 12.7 Normalized capacitance versus average pore size (**a**) for bituminous coal–derived activated carbons in 1 mol L^{-1} H_2SO_4, 6 mol L^{-1} KOH, and 1 mol L^{-1} $TEABF_4$ in acetonitrile (Adapted from [16]) and (**b**) for carbide-derived carbons (CDC) in 1 mol L^{-1} $TEABF_4$ in acetonitrile [20]

the normalized capacitance is not optimized. Hence, the trend should be to develop carbons with pores ranging between 0.7 and 1 nm, which are able to accommodate desolvated ions, while allowing them to optimally interact with the pore walls. In addition, with such carbons, the useless volume would be reduced, and the volumetric capacitance optimized, fulfilling the industrial requirements for getting compact systems.

However, it has been shown that using carbons with pores in the optimal range, from 0.7 and 1 nm, may lead to porosity saturation before reaching the maximum voltage for the considered electrolyte [23]. Indeed, the comparison of two carbons, named PC (S_{DFT} = 1,434 m^2 g^{-1}) and VC (S_{DFT} = 2,160 m^2 g^{-1}), with different pore size distribution (Fig. 12.8), in 1.5 mol.L^{-1} tetraethylammonium tetrafluoroborate in acetonitrile, shows a perfectly rectangular voltammogram in the whole voltage range for VC while the capacitive current dramatically decreases above around 1.5 V for PC (Fig. 12.9). According to Fig. 12.8, all pores of PC are narrower than 1 nm and consequently able to optimally interact with ions. However, if one considers the pores larger than the diameter of desolvated TEA^+ cations (0.68 nm), the corresponding DFT values are 198 and 964 m^2 g^{-1}, respectively, for PC and VC. From these values, the maximum theoretical charge able to be accommodated in the pores has been calculated for both carbons and compared to the charge determined by integration of the respective voltammograms [23]. For PC the theoretical and experimental values are almost identical, confirming that the shape of the voltammetry curve is due to porosity saturation. By contrast, in the case of VC, the theoretical value is larger than the experimental one, demonstrating that for this carbon the porosity is not saturated, at least for the maximum voltage value reached in this experiment.

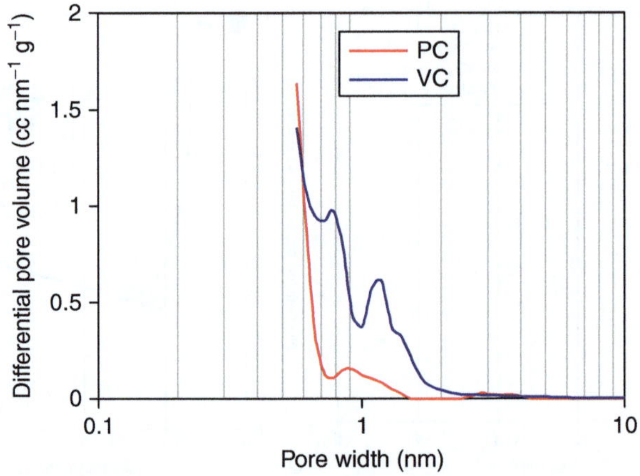

Fig. 12.8 Pore size distribution of carbons PC and VC [23]

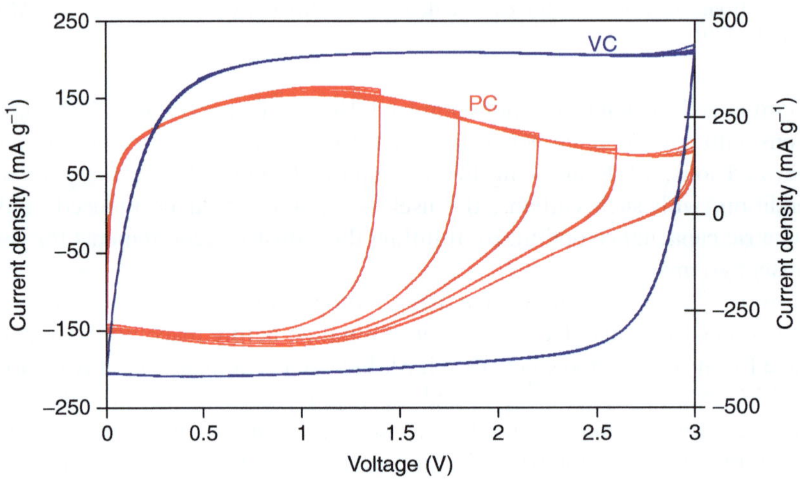

Fig. 12.9 Cyclic voltammograms for the EDLCs based on PC carbon (*left-hand side* Y-axis for current) and VC carbon (*right-hand side* Y-axis for current) [23]

In summary, the porous texture of carbons strongly influences their electrical double-layer properties. With $TEABF_4$ in acetonitrile, the normalized capacitance is optimal in subnanometric pores, which also indicates that ions are partly desolvated in charged electrodes. For obtaining high values of volumetric capacitance, nanoporous carbons should present high values of specific surface area (~2,000 $m^2\ g^{-1}$) with a narrow pore size distribution between 0.7 and 1 nm. Therefore, there is a need of new production techniques enabling to better reach this objective. Beside, taking into account that electrochemical intercalation is possible during charging [24], the structural parameters of carbons should also be more carefully considered in the future.

Carbons for Pseudo-capacitors

In general, two modes of energy storage are combined in electrochemical capacitors: (1) the electrostatic attraction between the surface charges and the ions of opposite charge (electrical double layer); (2) a pseudo-capacitive contribution which is related with quick redox reactions between the electrolyte and the electrode [14, 15]. Whereas the redox process occurs at almost constant potential in an accumulator, the electrode potential varies proportionally to the charge-exchanged dq in a pseudo-capacitor, what can be summarized by formula (Eq. 12.6):

$$dq = C * dU \tag{12.6}$$

where dU represents the potential variation.

The electrical response of such a system is comparable to that of a capacitor. Being of faradic origin and non-electrostatic, this capacitance is distinguished from the double-layer one and is called pseudo-capacitance. In summary, the electrical double-layer formation is a universal property of a polarized material surface, and pseudo-capacitance is an additional property which depends both on the type of electrode material and electrolyte. Compared to the double-layer normalized capacitance ($\sim$10 μF cm^{-2}), it has generally a high value (100–400 μF cm^{-2}), because it involves the bulk of the electrode and not only the surface. From a practical point of view, pseudo-capacitance contributes to enhancing the capacitance of materials and their energy density.

With nanoporous carbon electrodes in organic electrolyte, the double-layer capacitance is the major contribution. By contrast, in protic media (aqueous electrolytes or protic ionic liquids), an important pseudo-capacitive contribution is observed for nanotextured carbons enriched with heteroatoms as nitrogen and oxygen. Reversible hydrogen electrosorption is another kind of pseudo-faradic effect which might be also observed with nanoporous carbons in aqueous medium. In this case, nascent hydrogen formed by water reduction under negative polarization of carbon is adsorbed in the material; polarization reversal provokes its desorption by oxidation. The different types of pseudo-capacitive contributions for nanoporous carbons are reviewed in the following part of this chapter.

Pseudo-capacitance of Heteroatom-Enriched Carbons

The most reported dopants which confer pseudo-capacitive properties to carbons are oxygen (Fig. 12.3) and nitrogen (Fig. 12.4). Some other studies described below have recently considered boron or phosphorous as pseudo-capacitive dopants. The nitrogenated and oxygenated functionalities can undergo pseudo-faradic reactions, which can be pH-dependent or not, as presented in Fig. 12.10. The extent of the

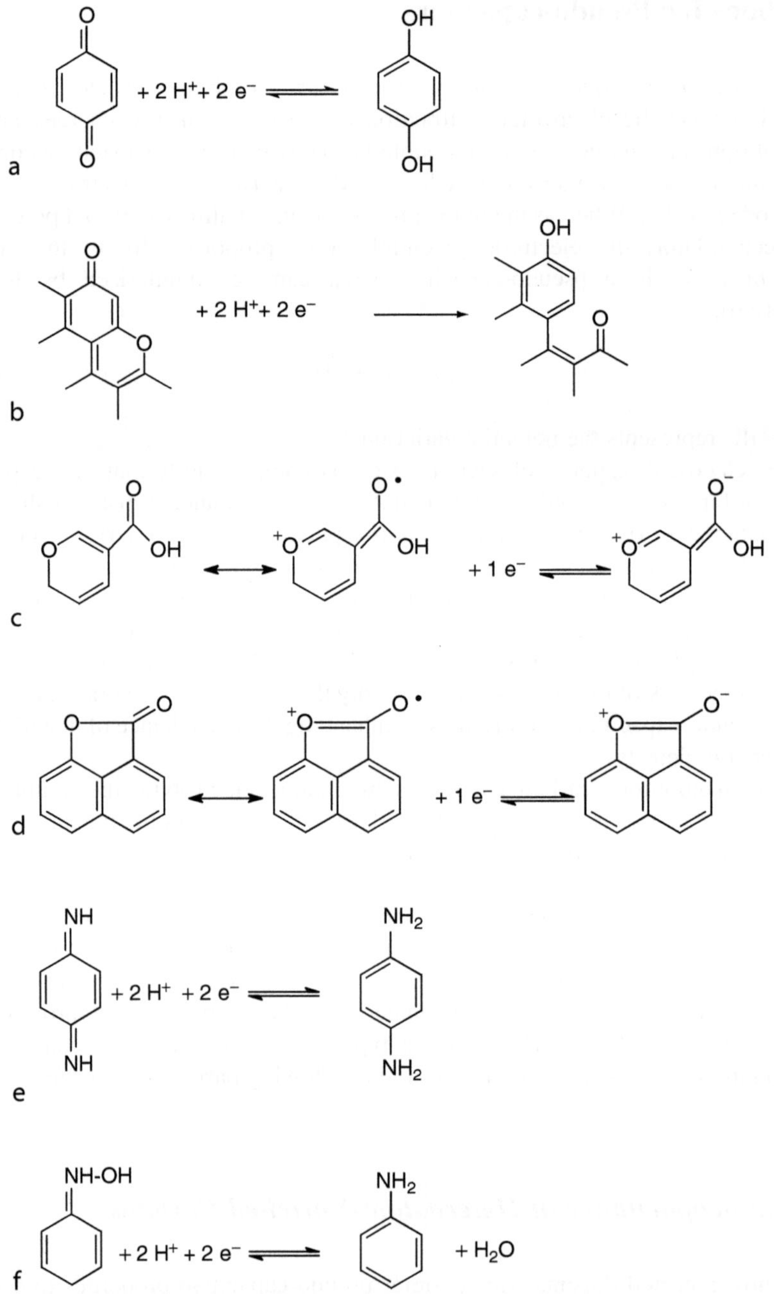

Fig. 12.10 Possible redox reactions involving oxygenated or nitrogenated surface functionalities

thereof extracted pseudocapacitance depends on the number of surface groups, their chemical nature, and accessibility.

Pseudocapacitance due to oxygenated surface functionalities is the most studied. In particular, the quinone/hydroquinone redox pair involves the exchange of one electron per oxygenated group (see Fig. 12.10a), against 0.18 electron stored per carbon atom in the electrical double-layer [14]. Other functionalities, such as pyrone-like structures (combination of non-neighboring carbonyl and ether oxygen atoms at the edges of the graphene layers – Fig. 12.10b), have been shown to be electrochemically active in the same potential range as the quinone/hydroquinone pair [26]. The activity of other functionalities also present in the complex chemistry of a carbon surface is not clear, even if some of them as the carboxylic or lactones could show resonance with other functional groups able to undergo electron transfer reactions (see Fig. 12.10c, d).

According to the equations in Fig. 12.10a and b, the electrochemical activity of oxygenated functionalities such as quinones or pyrones depends on the electrolyte pH [27]. The quinone/hydroquinone pair is undoubtedly active at low pH values [27, 28], and it has been claimed that the pyrone-like one is more active at high than at low pH [27]. Andreas et al. also suggested other unidentified pseudocapacitive reactions occurring on the carbon surface at basic pH. However, any oxygenated surface functionality being electrochemically active at neutral pH has been described. Due to this fact, and taking into account that smaller capacitance values are obtained at neutral pH than at acidic or basic ones [27], the main research efforts on the pseudocapacitive effect of the oxygenated surface functionalities have been focused on using KOH or H_2SO_4 as electrolytes.

Pseudo-capacitance in Basic and Acidic Media

Pseudo-capacitance due to Oxygenated Functionalities

The capacitance enhancement of O-enriched carbons in KOH electrolyte has been correlated with the oxygenated functionalities desorbing as CO during a thermal treatment in inert atmosphere, e.g., phenol, ether, carbonyl or quinone groups [29, 30]. A similar correlation between capacitance and the CO-type oxygenated groups has been shown for a series of activated carbons tested in H_2SO_4 [31]. By contrast, the functionalities desorbing as CO_2, e.g., carboxyl, anhydride, or lactone [30], are supposed to hinder the electrolyte migration in the micropores, thus leading to a delay in the electrochemical response.

Although much work is still necessary for determining the exact electrochemical behavior of the individual oxygenated surface functionalities, their contribution in enhancing capacitance of nanoporous carbons in acidic or basic media is a demonstrated fact. However, one of the drawbacks of most nanoporous carbons is their low density which leads to a low volumetric capacitance and consequently to insufficiently compact supercapacitors. Such disadvantage can be circumvented by producing nanotextured carbons through the pyrolysis of seaweed-based biopolymers [32] or seaweeds themselves [33] at mild temperature, e.g., 600°C. The carbons are characterized by a high electrical conductivity and a moderately developed porosity (S_{BET} = 200–700 $m^2\ g^{-1}$) and the presence of 10–15 wt% of oxygen. Due to the pseudo-faradic contribution of oxygenated functionalities, the capacitance reaches values as high as 260 F g^{-1} in 1 mol L^{-1} H_2SO_4 with these carbons, despite the poorly developed specific surface area. Figure 12.11a shows three-electrode cyclic voltammograms of the seaweed-based carbon (LN600) and a commercial-activated carbon AC1 (Norit Super 50; S_{BET} = 1,400 $m^2\ g^{-1}$; Oxygen = 4.6 at.%). AC1 shows an almost rectangular voltammogram, demonstrating that charging the electrical double layer is the dominant contribution, whereas LN600 presents very pronounced and reversible humps, at −0.4 and 0.1 V vs Hg/Hg_2SO_4, related with redox reactions probably involving the quinone and the pyrone-like groups. This surface functionality is suggested to enhance capacitance and to shift the irreversible and destructive oxidation of the material toward more positive potentials. On the negative side of the voltammogram, the potential of dihydrogen evolution is shifted to very negative values (ca. −1.2 V vs Hg/Hg_2SO_4) compared to the thermodynamic potential for water reduction, e.g., −0.62 V vs Hg/Hg_2SO_4 at pH = 0. Due to this overpotential, hydrogen is sorbed/inserted in the pores during the cathodic scan (hump at around −0.9 V vs Hg/Hg_2SO_4) and further oxidized (hump at around −0.2 V vs Hg/Hg_2SO_4) during the anodic one, giving rise to an additional pseudo-capacitive contribution [34]. This phenomenon will be discussed in detail in section "Pseudo-capacitance Related with Reversible Hydrogen Electrosorption in Nanoporous Carbons". As a result of broadening the stability window, a symmetric capacitor built with LN600 electrodes in 1 mol L^{-1} H_2SO_4 could be charged/discharged at voltage values as high as 1.4 V, whereas only 0.8–1.0 V can be applied with activated carbons [35].

The weakly porous seaweed carbons present a high density, which makes them more interesting than activated carbons in terms of volumetric energy density. Figure 12.11b shows that the volumetric energy density reached with the carbon LN600 is about one order of magnitude higher than for capacitors based on the commercial-activated carbons AC1 and AC2 (PX21, Kansai, Japan; S_{BET} =2,500 $m^2\ g^{-1}$; Oxygen = 5.0 at.%).

When seaweeds are carbonized in the presence of multiwalled carbon nanotubes (CNTs), the resulting composites are mesoporous, and the presence of CNTs dramatically enhances their conductivity [36]. Due to this open mesoporosity and the high conductivity of CNTs, the accessibility of ions to the active mass together with the charge propagation is improved. As it can be observed in Fig. 12.12, for the composites containing 10 wt% of CNTs, specific capacitance values as high as 180 F g^{-1} can still be obtained at a current density of 50 A g^{-1}, whereas the addition of carbon black does not give any benefit in comparison to the LN600 material alone. As a consequence, the highest extractable energy at high power density is obtained with CNT-based composite electrodes. In addition, the cycle life of the systems is improved owing to the resiliency introduced by the CNTs in the composite electrodes.

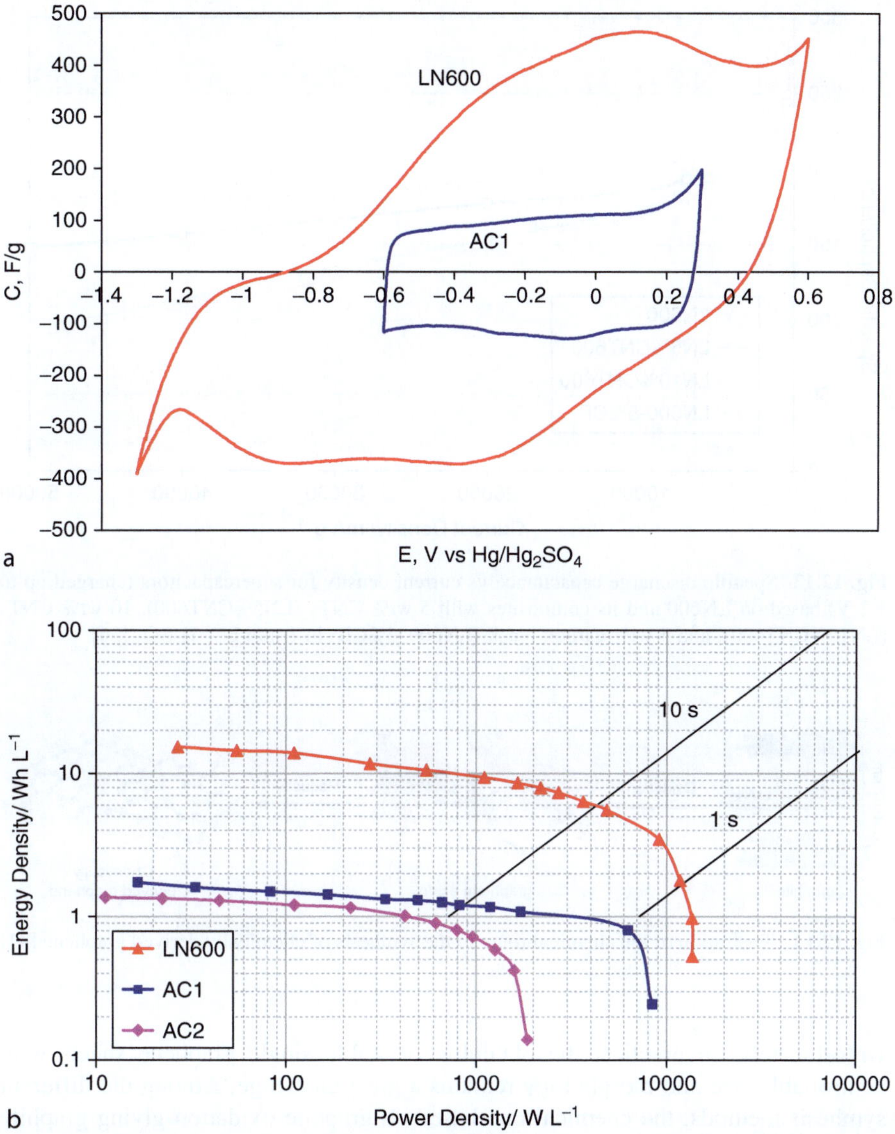

Fig. 12.11 (**a**) Three-electrode cyclic voltammograms of LN600- and AC1-based electrodes in 1 mol L^{-1} H_2SO_4; (**b**) Volumetric Ragone plots which compare the performance of supercapacitors with LN600-, AC1-, and AC2-based electrodes in 1 mol L^{-1} H_2SO_4 (Adapted from [33])

Graphene nanosheets are another kind of moderately developed surface area and highly oxygenated carbon material, which has attracted much attention during the last years [37–41]. In literature, perfectly dispersed individual graphene nanosheets are expected to display a high specific surface area, being promising for EDLC.

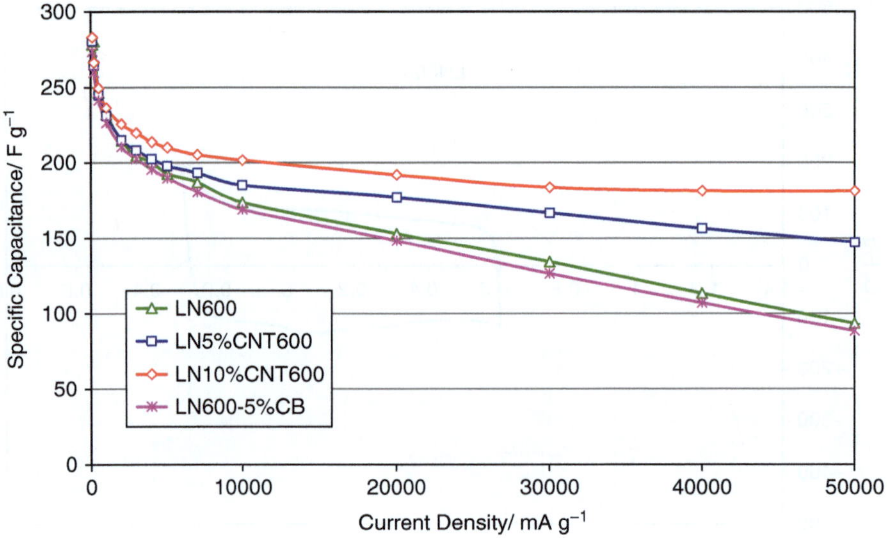

Fig. 12.12 Specific discharge capacitance vs current density for supercapacitors (charged up to 1.2 V) based on LN600 and its composites with 5 wt% CNTs (LN5%CNT600), 10 wt% CNTs (LN10%CNT600), and 5 wt% carbon black (LN600–5%CB) in 1 mol L^{-1} H_2SO_4 [36]

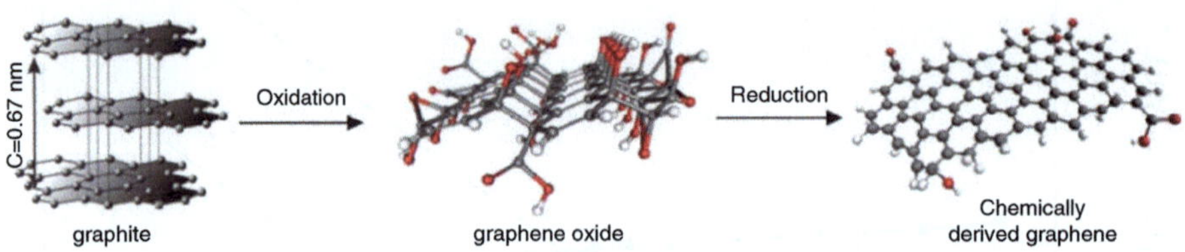

Fig. 12.13 Illustration of the chemical route to the synthesis of chemically derived graphene [41]

Unfortunately, in practice, the obtaining of stable single graphene sheets with controllable size and morphology remains a great challenge. Among the different synthesis methods, the chemical one based on graphite oxidation giving graphite oxide, which is subsequently reduced by hydrazine (Fig. 12.13), is the most proposed because of a possible upscaling, even if harmful chemicals are used. Whatever the preparation method, the resulting material consists in an agglomeration of highly functionalized graphene sheets. As a consequence, the specific surface area is relatively low (300–900 m^2 g^{-1}) compared to the theoretical one (2,620 m^2 g^{-1}), and the materials contain 10–20 at.% of oxygen and nitrogen. Obviously, the capacitance values, from 100 to 200 F g^{-1}, reported for these materials in KOH or H_2SO_4 electrolytes are not related to the nanotexture but mainly to the pseudo-faradic reactions involving the surface functionality at the edges of the graphene sheets. Hence, the literature claims that graphene exhibits good EDL properties are not correct. Moreover, despite the exponential growth of the number of publications on synthesis and applications of graphene for supercapacitors, a cost-effective and environment-friendly mass-production method still remains a challenge [41], and the electrochemical performance should be improved to compete with other existing carbon materials.

Pseudo-capacitance Due to Nitrogenated Functionalities

The specific introduction of nitrogen as a dopant of carbon is also extensively reported in the literature. Nitrogen-rich nanocarbons can be obtained either by post-treatment of carbons with N-containing reagents as the NH_3/air mixture [42] and urea [43] or by pyrolysis of N-containing polymers. Such pyrolysis can be performed on the polymer itself, and followed by an activation step for generating porosity [44–46], or on mixtures of polymer and nanotexturing agent as CNTs [47, 48], zeolites [49], or mica [50].

A correlation between capacitance in aqueous KOH or H_2SO_4 electrolytes and the total amount of surface nitrogen has been shown by Frackowiak et al. [45] for a series of N-doped carbons of similar nanoporous texture. The carbons were obtained by pyrolysis of nitrogen-containing precursors as polyvinylpyridine or polyacrylonitrile and their blends with coal tar pitch, followed by steam activation. Among the various nitrogenated functionalities, it seems that the ones located at the graphene edges, as pyridine and pyrrol (see Fig. 12.4), are the most active [46, 51]. The presence of nitrogen can also influence differently the capacitance of the positive and negative electrodes of a capacitor [43] due to the potential dependence of the pseudo-faradic reactions.

High-performance nitrogen-rich carbons have been developed by applying the "templating technique." Briefly a nitrogen-containing precursor is introduced in a nanoporous scaffold and subsequently pyrolyzed; afterward, the nitrogen-enriched carbon replica is obtained by etching the host with hydrofluoric acid. As an example, a weakly porous carbon material (S_{BET} = 400 m^2 g^{-1}) containing 24 at.% of nitrogen was synthesized through melamine intercalation in a lamellar aluminosilicate [50]. The specific capacitance, obtained in three-electrode cell, reaches 200 F g^{-1} in 1 mol L^{-1} H_2SO_4. Much higher gravimetric capacitance of 340 F g^{-1} was demonstrated in 1 mol L^{-1} H_2SO_4 by the carbon obtained through pyrolysis of acrylonitrile in NaY zeolite. This carbon displays a higher specific surface area of 1,680 m^2 g^{-1} and a moderate nitrogen amount of 6 at.%. The exceptionally high capacitance comes from the synergy of different contributions: (1) the ordered superstructure inherited from the scaffold which favors the diffusion of ions to the active surface; (2) micropores with a size perfectly adapted to the electrolyte which provide an EDL contribution; and (3) an important pseudo-capacitive effect related with the surface functionality [49]. Moreover, a symmetric capacitor from these carbons could operate up to 1.2 V during 10,000 charge/discharge cycles, which is much higher than the usual maximum voltage for systems based on activated carbon electrodes in aqueous media [35]. Hence, nitrogen enrichment seems to provide a profitable potential shift of the electrodes as it has been observed for oxygen [33].

Following the industrial requirements, the tendency now is to develop hetero-atom-enriched carbons enabling to operate at high current density. For example, carbon/CNTs nanocomposites have been prepared by one-step carbonization of polyacrylonitrile/carbon nanotube blends [47] and melamine-formaldehyde/carbon nanotube blends [48]. Although presenting a moderate specific surface area (S_{BET} $\sim$ 400 m^2 g^{-1}), the later carbons display gravimetric capacitance values of 170 and 130 F g^{-1} at current densities of 50 and 5 A g^{-1}, respectively [48]. Such exceptional performance is due, as for the previously presented LN600/CNTs composites, to the synergetic effects of nanotubes and surface functionality. The open mesoporosity of nanotubes and their good electrical conductivity enable good propagation of charges.

Recently, capacitance values around 100–200 F g^{-1} have been claimed for nonporous (specific surface area ranging from 6 to 86 m^2 g^{-1}) N-enriched carbons prepared by ammonia treatment of an already nitrogenated carbon derived from melamine [52]. However, the data were obtained at very low current density and the study does not provide any information on the performance of these materials in usual operating conditions of supercapacitors.

Even if the previously presented results seem to demonstrate the efficiency of nitrogen as carbon dopant for introducing a pseudocapacitive character, one must take into account that oxygen is always present on the surface of carbons materials. Studies performed on model compounds as crystalline oligomers of 1,5-diaminoanthraquinone (DAAQ) with a 3D-extended π-conjugated structure [53, 54] have shown that the redox peaks corresponding to quinone/hydroquinone and emeraldine/quinone diimide pairs arise in the same potential range. Therefore, the capacitance values normally assessed only to the nitrogenated functionalities may include a contribution due to the presence of oxygenated surface functionalities.

Conclusion

O- and N-enriched nanocarbons demonstrate high capacitance values in KOH or H_2SO_4 due to the contribution of pseudo-faradic reactions. Another positive effect of doping is the broadening of the potential stability window. Composites of these carbons with carbon nanotubes withstand remarkable capacitance values at high current load. Hence, the N- and O-doped carbons open the opportunity of developing high-performance supercapacitors in aqueous electrolytes.

Recently, boron [55] and phosphorous [56] have been tested as dopants for improving the electrochemical performance of carbons in acidic electrolytes. However, since the investigated boron-doped carbons contained also nitrogen and oxygen, the peculiar role of boron is not established. For the particular case of

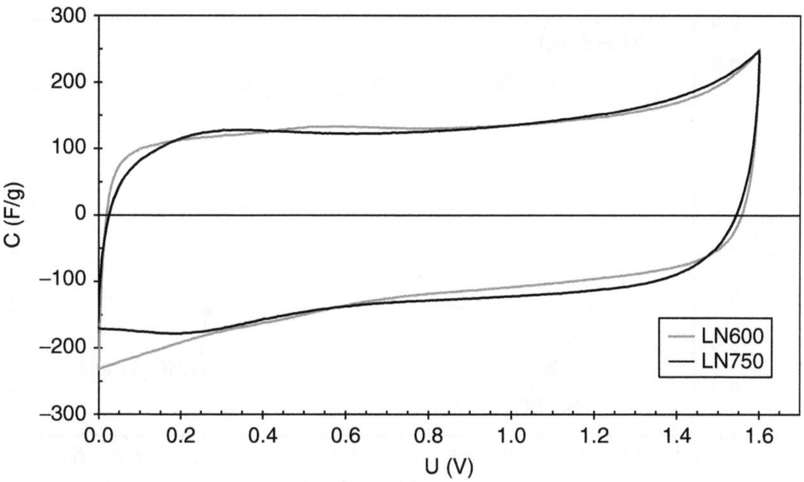

Fig. 12.14 Cyclic voltammograms of LN600- and LN750-based symmetric capacitors in 0.5 mol L^{-1} Na_2SO_4 [60]

phosphorous, it seems that small amounts affect positively the overall capacitance and improve the capacitance retention ratio at high current load in H_2SO_4.

Pseudo-capacitance in Neutral Medium

Neutral electrolytes have been scarcely applied for carbon-based supercapacitors. Indeed, the capacitance values of activated carbons in neutral electrolytes are lower than in other aqueous electrolytes, suggesting that the pseudo-faradic redox reactions of O and N electroactive surface groups are depressed in comparison with acidic or basic electrolytes [25, 27, 57–59]. Therefore, it was concluded that oxygenated or nitrogenated functionalities are electrochemically inactive in neutral electrolytes and that only the formation of the electrical double layer contributes to the capacitance in this media.

Nevertheless, a recent study has shown that a contribution of pseudo-faradic reactions is possible in a neutral electrolyte, e.g., Na_2SO_4, if the carbon material possesses an adequate surface functionality [60]. Figure 12.14 shows the cyclic voltammograms of supercapacitors built with nanotextured carbons obtained after pyrolysis of seaweeds at 600°C (LN600, S_{BET} = 746 m^2 g^{-1}) and at 750°C (LN750, S_{BET} = 1,082 m^2 g^{-1}). Despite the lower specific surface area of LN600 in comparison to LN750, the capacitive behavior is comparable for the two carbons. Taking into account that LN600 and LN750 contain 11.5 at.% and 7.3 at.% of oxygen, respectively, one can conclude that pseudo-faradic reactions contribute more markedly to the capacitance of LN600 in Na_2SO_4. In particular, the fact that quinone-like

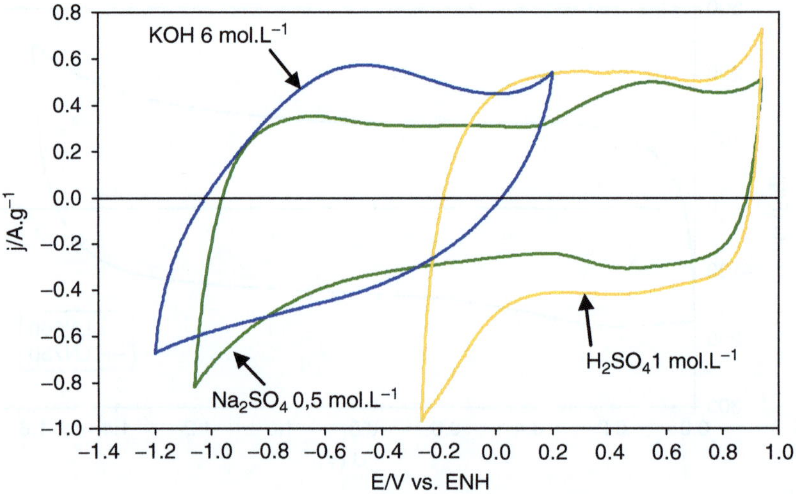

Fig. 12.15 Three-electrode cyclic voltammograms showing the potential stability window of AC in 6 mol L^{-1} KOH, 1 mol L^{-1} H_2SO_4, and 0.5 mol L^{-1} Na_2SO_4 [61]

groups are twice more present in LN600 than in LN750 suggests that the latter are the most active ones at pH = 6.4.

Notwithstanding these facts, the most striking result obtained in neutral electrolytes is the ability of symmetric carbon/carbon supercapacitors to operate up to voltages as high as 1.6 V during more than 10,000 cycles [60, 61]. According to Eq. 12.5, the energy density is higher in neutral than in other aqueous electrolytes.

The former result can be understood by comparing the three-electrode cyclic voltammograms of a commercial-activated carbon (AC, S_{BET} = 2,250 m^2 g^{-1}, O = 2.5 at.%; provided by MeadWestvaco, USA) in acidic (H_2SO_4), basic (KOH), and neutral (Na_2SO_4) electrolytes (Fig. 12.15) [61]. In Na_2SO_4, the stability potential window is about twice larger than in H_2SO_4 and KOH. Actually, in the neutral electrolyte, it reaches 2.0 V for AC and values of even 2.4 V have been found for seaweed-based carbons [60]. The reason of such a wide stability potential window is the high overpotential for dihydrogen evolution, e.g., 0.6 V, and the better oxidation resistance of carbons in the neutral medium. A pH dependence of the overpotential for H_2 evolution has been also shown with glassy carbon [62] or Pt [63] electrodes. In a neutral pH solution, the overpotential reaches 1.2 V with a glassy carbon electrode, whereas only values of 0.6 V or 0.8 V are attained in acidic or basic pH, respectively. With a Pt electrode, a small overpotential of 0.3 V is found in neutral electrolyte, whereas any overpotential can be detected in acidic or basic solutions. For the particular case of Pt [63], this property is attributed to the fact that the H_3O^+ concentration in neutral electrolyte is too low to induce gas evolution reactions. For nanoporous carbons, it is also related with their storage ability of nascent hydrogen below the thermodynamic potential for water reduction. This pseudo-capacitive contribution related to hydrogen sorption will be discussed in detail in the section

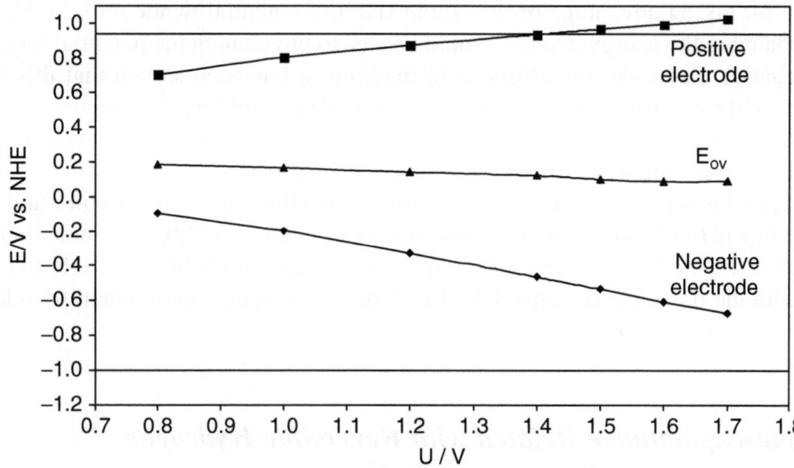

Fig. 12.16 Potential reached by the positive and negative electrodes during the operation of a 2-electrode cell (equipped with a reference electrode) at different values of maximum voltage. The E_{OV} values correspond to the electrodes potential when the cell voltage U is set to 0 V. Electrolyte: 0.5 mol L^{-1} Na_2SO_4. The *horizontal lines* represent the extreme potential values determined for AC in 3-electrode cell [61]

"Pseudo-capacitance Related with Reversible Hydrogen Electrosorption in Nanoporous Carbons".

Although 1.6 V is the highest voltage ever claimed for a symmetric carbon/carbon capacitor in an aqueous electrolyte, this value is smaller than the stability potential window of 2.0 V determined for AC in three-electrode cell. The reasons of the potential window underuse in Na_2SO_4 have been demonstrated after experiments using a special two-electrode cell equipped with a reference electrode. Such configuration allows the positive and negative electrode potentials to be recorded separately during cycling the supercapacitor between zero and a given maximum voltage. Figure 12.16 shows that, if the supercapacitor is charged between 0 V and the maximum stability voltage of 1.6 V, the operating potential window for the negative and the positive electrode is [0.09; −0.61] and [0.09; 0.99] V vs NHE, respectively. According to Fig. 12.15, in Na_2SO_4 electrolyte, the lowest potential for a negative electrode before dihydrogen production is around −1 V vs NHE and the highest one for avoiding an irreversible oxidation of a positive electrode is around 0.95 V vs NHE (see the horizontal lines included in Fig. 12.16). Hence, Fig. 12.16 clearly shows that the maximum voltage of the supercapacitor is limited by the positive electrode. In fact, the maximum potential reached by the positive electrode at U = 1.6 V is slightly beyond the oxidation limit, whereas the minimum potential reached by the negative electrode is still far away, ca. 0.4 V, from the limit imposed by dihydrogen evolution. During cycling, the maximum potential of the positive electrode slightly shifts to lower values, allowing a good cyclability of the capacitor.

From the foregoing, even if the AC/AC symmetric capacitor in Na_2SO_4 largely outperforms all the symmetric systems in other aqueous electrolytes, the strategy to follow for taking advantage of the whole stability potential window of 2.0 V, and for enhancing the energy density, would consist to broadening the potential range of the negative electrode. In sulfuric acid medium, it has been shown that this kind of shift of the electrodes potential range is possible by building asymmetric systems combining two different carbon materials with a surface functionality adapted for optimally operating as positive or negative electrode [64].

In conclusion, the nature of the surface functionalities of carbons and the electrolyte pH influence both the capacitance values and stability potential window. Actually, the lately developed supercapacitors in neutral medium are able to compete with the organic electrolyte-based systems while being environment friendly.

Pseudo-capacitance Related with Reversible Hydrogen Electrosorption in Nanoporous Carbons

As it has been introduced in the previous section, hydrogen electrosorption in carbon materials under negative polarization has a direct impact on the energy density of a supercapacitor operating in an aqueous electrolyte. Due to the overpotential of dihydrogen evolution, the electrochemical stability window can be extended to lower potential values. Moreover, the electro-desorption of hydrogen by anodic oxidation gives rise to a pseudo-faradic contribution in addition to the EDL capacitance of the material.

Under negative polarization of a nanotextured carbon electrode in aqueous electrolyte, hydrated cations are first adsorbed forming an electrical double layer. When the electrode potential is lower than the equilibrium value for water reduction, nascent hydrogen is formed (Eq. 12.7) and adsorbed onto the carbon surface [34] (Eq. 12.8):

$$H_3O^+ + e^- \leftrightarrow \langle H \rangle + H_2O \tag{12.7}$$

$$\langle C \rangle + \langle H \rangle \leftrightarrow \langle CH_x \rangle_{ads} \tag{12.8}$$

where $\langle H \rangle$ represents the nascent hydrogen, $\langle C \rangle$ the carbon host, and $\langle CH_x \rangle$ hydrogen adsorbed in the later.

It has been demonstrated that the nanoporous texture of the activated carbon plays an important role [22, 34, 65–67]. In particular for carbons prepared by a templating technique, the amount of sorbed hydrogen is proportional to the ultramicropore (0.6–0.7 nm) volume [22]. In this case, straight mesopores are very useful channels for the transport of ions. Galvanostatic charge/discharge experiments in alkaline medium on a series of activated carbons with similar porosity but with different surface chemistry show that the amount of hydrogen reversibly electrosorbed decreases with the increase of the oxygenated surface functionalities [68]. The electrosorption of hydrogen has been extensively studied

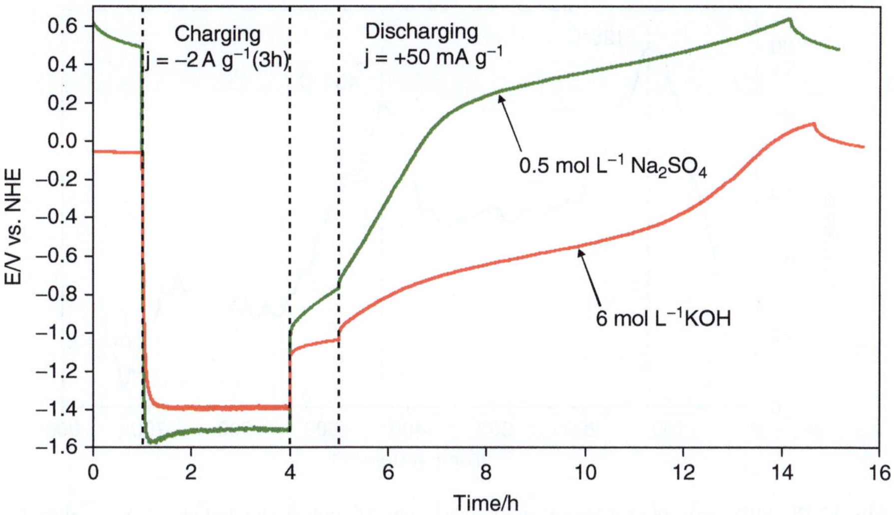

Fig. 12.17 Galvanostatic charge/discharge of AC in 6 mol L^{-1} KOH (*red curve*) and 0.5 mol L^{-1} Na_2SO_4 (*green curve*). After 1 h relaxation, a negative current of −2 A g^{-1} is applied during 3 h. Then, after 1 h relaxation, a positive current of +50 mA g^{-1} is applied in order to oxidize the sorbed hydrogen [70]

in alkaline or acidic medium [34]. In such media, the important polarization between reduction and oxidation potentials suggests that the interaction of sorbed hydrogen with the carbon substrate is stronger than physisorption. In the thermo programmed desorption (TPD) analyses realized on carbon samples submitted to galvanostatic charge in KOH medium, a hydrogen peak appears at around 200°C [69]. From this peak, the activation energy was estimated to 110 kJ mol^{-1} for hydrogen desorption, confirming a weakly chemisorbed state of hydrogen.

A recent study demonstrates the effectiveness of a neutral medium for electrochemical hydrogen bonding to the graphene units in nanocarbons [70]. From the galvanostatic charge/discharge curves (Fig. 12.17), the mechanism for hydrogen sorption seems different in neutral medium and in KOH. In KOH, a single plateau at ca 0.6 V vs NHE characterizes the anodic oxidation of hydrogen [34]. By contrast, in neutral medium (0.5 mol L^{-1} Na_2SO_4), the curve exhibits two different slopes, a first part where the stored hydrogen is oxidized at potentials lower than 0.1 V vs NHE, and a plateau at higher potentials corresponding to more strongly bonded hydrogen. It has also to be noted that the reduction overpotential is more important in neutral than in alkaline medium. The TPD analysis realized after the galvanostatic charge of AC in the neutral medium (Fig. 12.18) confirms two types of hydrogen bonding with peaks at 125°C and 430°C. The desorption activation energies calculated from these data are 110 and 205 kJ mol^{-1}, respectively. The highest value demonstrates that the main part of nascent hydrogen is strongly bonded to the graphene units in 0.5 mol L^{-1} Na_2SO_4.

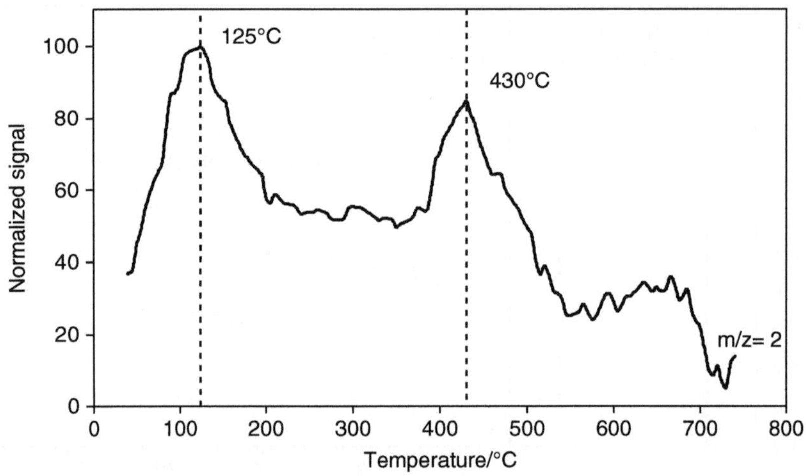

Fig. 12.18 TPD analysis of dihydrogen evolved from AC power charged at -2 A g^{-1} during 3 h at 50°C in 0.5 mol L^{-1} Na_2SO_4 [70]

The effect of temperature on the electrochemical performance in 0.5 mol L^{-1} Na_2SO_4 has been evaluated (Fig. 12.19). The oxidation time of stored hydrogen increases with temperature (1.5 wt% at 20°C, 2.1 wt% at 40°C and 2.4 wt% at 50°C), confirming a weakly chemisorbed state of hydrogen [69]. Meanwhile, the oxidation potential decreases when temperature increases, due to a better conductivity of the electrolytic medium. Hence, the energy density of a cell using activated carbon as negative electrode in neutral aqueous solution will be enhanced by operating at higher values of temperature.

To summarize, in comparison to acidic and basic electrolytes, a neutral medium has two positive impacts for supercapacitor applications: (1) the overpotential for dihydrogen evolution is higher, allowing the stability window to be enlarged; (2) the amount of hydrogen reversibly sorbed is enhanced enabling a higher pseudo-capacitive contribution related with its oxidation.

Protic Ionic Liquids as Electrolytes for Carbon-Based Pseudo-capacitors

Protic ionic liquids (PILs) are relatively cheap and environment friendly electrolytes [71]. Due to their proton-exchanging capabilities, a pseudo-faradaic enhancement of capacitance can be expected, as in aqueous electrolytes, while operating in a larger potential window. For these reasons, they have been proposed as electrolytes in carbon-based supercapacitors [72]. Cyclic voltammograms in Fig. 12.20 show the electrochemical performance of two pyrrolidinium-based PILs (Pyrrolidinium nitrate [Pyrr][NO3] (IL1) and pyrrolidinium formate [Pyrr][HCOO] (IL2)) [73] in presence of two different carbon electrodes:

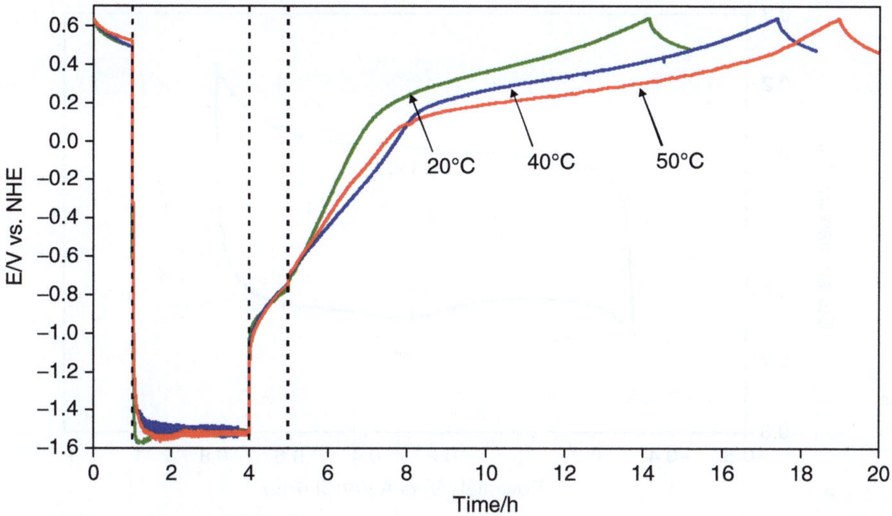

Fig. 12.19 Galvanostatic charge/discharge of AC in 0.5 mol L^{-1} Na_2SO_4 at 20°C, 40°C, and 50°C [70]

(1) a commercial-activated carbon (S50 from Norit, S_{BET} = 1,400 m^2 g^{-1}, O = 4.6 at.%); (2) the same oxidized in order to increase the amount of surface functionalities without modifying the porous texture ($S50_{ox}$, O = 9.7 at.%). For IL1 (Fig. 12.20a), redox peaks are seen at around 0.2 and 0.45 V vs Ag/AgCl(PIL). For activated carbons, peaks in such potential range are traditionally assessed to the electrochemical reactions of oxygenated surface functionalities such as quinone/hydroquinone pairs [14] or pyrone-like structures able to accept two protons and two electrons [26]. For both ILs, the most pronounced redox peaks are observed for $S50_{ox}$ (Fig. 12.20a, b), that is evidently caused by a stronger pseudo-capacitive effect linked to the higher amount of quinone-type groups and/or pyrone in $S50_{ox}$.

Two-electrode cells based on IL1 and IL2 with the two carbons S50 and $S50_{ox}$ are able to operate up to 1.2 V, which is a substantially higher value than the maximum of 0.7 – 1 V generally observed in aqueous electrolytes [35]. Distinct improvements in capacitance owing to the use of $S50_{ox}$ are noticeable for both ionic liquids; the capacitance increases from 99 to 126 F g^{-1} in IL1 and from 96 to 130 F g^{-1} in IL2, when using S50 with $S50_{ox}$, respectively. These values are comparable to those obtained in 1 mol.L^{-1} H_2SO_4 aqueous solution, e.g., 101 F g^{-1} for S50 and 137 F g^{-1} for $S50_{ox}$, demonstrating a remarkable pseudo-faradic efficiency of these ionic liquids.

Hence, protic ionic liquids represent a new kind of green electrolyte for developing high-performance capacitors involving both electrical double-layer charging and a pseudo-faradic mechanism. The electrochemical stability and conductivity of protic ILs, the relation between the carbon pore size and the electrolyte ion size, and the design of electrochemical cells are the most important matters for thorough investigations.

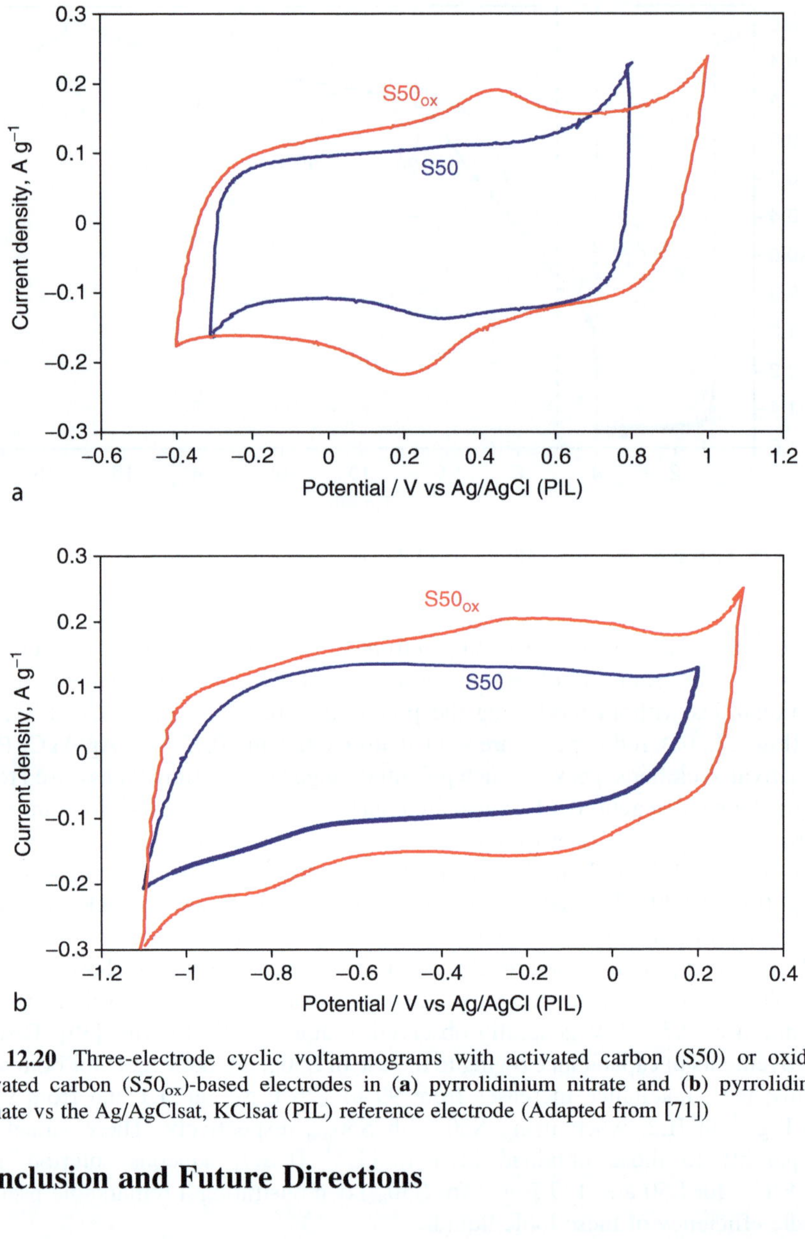

Fig. 12.20 Three-electrode cyclic voltammograms with activated carbon (S50) or oxidized activated carbon ($S50_{ox}$)-based electrodes in (**a**) pyrrolidinium nitrate and (**b**) pyrrolidinium formate vs the Ag/AgClsat, KClsat (PIL) reference electrode (Adapted from [71])

Conclusion and Future Directions

Nanocarbons are the most widely used material for supercapacitor electrodes because, among other characteristics, they are cheap, chemically inert, electrically conductive, and their nanoporous texture and surface functionality are highly versatile.

In the case of capacitors in organic electrolyte, it has been demonstrated that the electrical double-layer (EDL) properties are strongly influenced by the pore size.

With the tetraethylammonium tetrafluoroborate salt, the largest values of normalized capacitance are observed for pores smaller than 1 nm, with a maximum at around 0.7–0.8 nm, demonstrating that ions are essentially desolvated in the pores. In situ or ex situ techniques, such as small angle neutron scattering or nuclear magnetic resonance, should be applied for better understanding how ions occupy porosity during the charge/discharge of electrodes. Such investigations would help to better defining the optimal porous texture. Beside, taking into account that saturation of porosity can occur with carbons developing essentially subnanometric pores, new processes have to be investigated for developing the surface area of carbons while keeping only such pores. Finally, in light of some literature claims that the EDL properties are not only determined by the porous texture, another research direction should consider the structural parameters of carbons.

In protic media (aqueous electrolytes or protic ionic liquids) a pseudo-faradic mechanism can arise together with the formation of the double layer. In this case, the capacitance and the stability potential window are influenced both by the electrolyte pH and the nature of carbon functionalities. In the future, fundamental studies should be realized on carbons with controlled functionalities, using for example grafting techniques, in order to determine which functionalities are the most effective for pseudo-faradic processes. The demonstration that neutral aqueous electrolytes can offer capacitors with operating voltage values as high as 2 V is an important breakthrough of the recent years. The reasons for the important overpotential of dihydrogen evolution in these media must be elucidated, that could allow developing even higher voltage values. Different neutral electrolytes must also be more systematically screened versus carbons with various functionalities in order to enhance capacitance, electrical conductivity, and the operating temperature range. In this strategy, protic ionic liquids might be a new class of electrolytes opening new insights for optimizing electrochemical capacitors.

In summary, the high versatility of nanocarbons in terms of texture and surface functionality allows to perfectly adapt them to the different mechanisms depending on the electrolyte. Such versatility still offers high potentialities for developing high-performance supercapacitors.

Bibliography

1. Marsh H, Rodríguez-Reinoso F (2006) Activated carbons. Elsevier, London
2. Lillo-Ródenas MA, Juan-Juan J, Cazorla-Amorós D, Linares-Solano A (2004) Carbon 42:1371–1375
3. Raymundo-Piñero E, Azaïs P, Cacciaguerra T, Cazorla-Amorós D, Linares-Solano A, Béguin F (2005) Carbon 43:786–795
4. Jagtoyen M, Derbyshire F (1998) Carbon 36:1085–1097
5. Bandosz TJ, Biggs MJ, Gubbins KE, Hattori Y, Pikunic J, Thomson K (2003) Molecular models of porous carbons. In: Radovic LR (ed) Chemistry and physics of carbon. Marcel Dekker, New York, pp 137–199

6. Pikunic J, Gubbins KE, Pellenq RJM, Cohaut N, Rannou I, Guet JM, Clinard C, Rouzaud JN (2002) Appl Surf Sci 196:98–104
7. Harris PJF, Tsang SC (1997) Philos Mag A76:667–677
8. Harris PJF (1997) Inter Mater Rev 42:206–218
9. Harris PJF (2005) Critical Rev Solid State Mater Sci 30:235–253
10. Radovic LR, Bockrath B (2005) J Am Chem Soc 127:5917–5927
11. Azaïs P, Duclaux L, Florian P, Massiot D, Lillo-Ródenas MA, Linares-Solano A, Peres JP, Jehoulet C, Béguin F (2007) J Power Sourc 171:1046–1053
12. Pels JR, Kapteijn F, Moulijn JA, Zhu Q, Thomas KM (1995) Carbon 33:1641–1653
13. Kapteijn F, Moulijn JA, Matzner S, Boehm HP (1999) Carbon 37:1143–1150
14. Conway BE (1999) Electrochemical supercapacitors – scientific fundamentals and technological applications. Kluwer/Plenum, New York
15. Kötz R, Carlen M (2000) Electrochima Acta 45:2483–2498
16. Raymundo-Piñero E, Kierzek K, Machnikowski J, Béguin F (2006) Carbon 44:2498–2507
17. Gregg SJ, Sing KSW (1982) Adsorption, surface area and porosity. Academic, London, pp 103–104
18. Kaneko K, Ishii C (1992) Colloid Surf 67:203–212
19. Barbieri O, Hahn M, Herzog A, Kötz R (2005) Carbon 43:1303
20. Chmiola J, Yushin G, Gogotsi Y, Portet C, Simon P, Taberna PL (2006) Science 313:1760–1763
21. Chmiola J, Largeot C, Taberna PL, Simon P, Gogotsi Y (2008) Angew Chem 120:3440–3443
22. Vix-Guterl C, Frackowiak E, Jurewicz K, Friebe M, Parmentier J, Béguin F (2005) Carbon 43:1293–1302
23. Mysyk R, Raymundo-Piñero E, Béguin F (2009) Electrochem Commun 11:554–556
24. Ruch PW, Hahn M, Cericola D, Menzel A, Kötz R, Wokaun A (2010) Carbon 48:1880–1888
25. Frackowiak E, Béguin F (2001) Carbon 39:937
26. Montes-Moran MA, Suarez D, Menendez JA, Fuente E (2004) Carbon 42:1219–1225
27. Andreas HA, Conway BE (2006) Electrochima Acta 51:6510–6520
28. Okajima K, Ohta K, Sudoh M (2005) Electrochima Acta 50:2227–2231
29. Cheng PZ, Teng H (2003) Carbon 41:2057–2063
30. Figueiredo JL, Pereira MFR, Freitas MMA, Orfao JJM (1999) Carbon 37:1379–1389
31. Bleda-Martínez MJ, Maciá-Agulló JA, Lozano-Castelló D, Morallón E, Cazorla-Amorós D, Linares-Solano A (2005) Carbon 43:2677–2684
32. Raymundo-Piñero E, Leroux F, Béguin F (2006) Adv Mater 18:1877–1882
33. Raymundo-Piñero E, Cadek M, Béguin F (2009) Adv Funct Mater 19:1–8
34. Jurewicz K, Frackowiak E, Béguin F (2004) Appl Phys A 78:981–985
35. Ruiz V, Santamaría R, Granda M, Blanco C (2009) Electrochima Acta 54:4481–4486
36. Raymundo-Piñero E, Cadek M, Wachtler M, Béguin F (2011) ChemSusChem. doi:10.1002/cssc.201000376
37. Stoller MD, Park S, Zhu Y, An J, Ruoff RS (2008) Nano Lett 8:3498–3502
38. Vivekchand SRC, Rout CS, Subrahmanyam KS, Govindaraj A, Rao CNR (2008) J Chem Sci 20:9–13
39. Wang Y, Shi Z, Huang Y, Ma Y, Wang C, Chen M, Chen Y (2009) J Phys Chem C 113:13103–13107
40. Du X, Guo P, Song H, Chen X (2010) Electrochima Acta 55:4812–4819
41. Zhang LL, Zhou R, Zhao XS (2010) J Mater Chem 20:5983–5992
42. Jurewicz K, Babel K, Ziolkowski A, Wachowska H (2003) Electrochima Acta 48:1491–1498
43. Jurewicz K, Pietrzak R, Nowicki P, Wachowska H (2008) Electrochima Acta 53:5469–5475
44. Lota G, Grzyb B, Machnikowska H, Machnikowski J, Frackowiak E (2005) Chem Phys Lett 404:53–58
45. Frackowiak E, Lota G, Machnikowski J, Vix-Guterl C, Béguin F (2006) Electrochima Acta 51:2209–2214
46. Ra EJ, Raymundo-Piñero E, Lee YH, Béguin F (2009) Carbon 47:2984–2992

47. Béguin F, Szostak K, Lota G, Frackowiak E (2005) Adv Mat 17:2380–2384
48. Lota G, Lota K, Frackowiak E (2007) Electrochem Commun 9:1828–1832
49. Ania CO, Khomenko V, Raymundo-Piñero E, Parra JB, Béguin F (2007) Adv Funct Mater 17:1828–1836
50. Hulicova D, Kodama M, Hatori H (2006) Chem Mater 18:2318–2326
51. Hulicova-Jurcakova D, Seredych M, Lu GQ, Bandosz TJ (2009) Adv Funct Mater 19:438–447
52. Hulicova-Jurcakova D, Kodama M, Shiraishi S, Hatori H, Zhu ZH, Lu GQ (2009) Adv Funct Mater 19:1800–1809
53. Naoi K, Suematsu S, Hanada M, Takenouchi HJ (2002) J Electrochem Soc 149:472–476
54. Gao M, Yang F, Wang X, Zhang G, Liu L (2007) J Phys Chem C 111:17268–17274
55. Konno H, Ito T, Ushiro M, Fushimi K, Azumi K (2010) J Power Sourc 195:1739–1746
56. Hulicova-Jurcakova D, Puziy AM, Poddubnaya OI, Suarez-García F, Tascon JMD, Lu GQ (2009) J Am Chem Soc 131:5026–5027
57. Hu CC, Wang CC, Wu FC, Tseng RL (2007) Electrochima Acta 52:2498–2505
58. Subramanian V, Luo C, Stephan AM, Nahm KS, Thomas S, Wei B (2007) J Phys Chem C 111:7527–7531
59. Qu QT, Wang B, Yang LC, Shi Y, Tian S, Wu YP (2008) Electrochem Commun 10:1652–1655
60. Bichat MP, Raymundo-Piñero E, Béguin F (2010) Carbon 48:4351–4361
61. Demarconnay L, Raymundo-Piñero E, Béguin F (2010) Electrochem Commun 12:1275–1278
62. Wang J, Kirgoz UA, Mo JW, Lu J, Kawde AN, Muck A (2001) Electrochem Comm 3:203–208
63. Hong MS, Lee SH, Kim SW (2002) Electrochem Solid State Lett 5:A227–A230
64. Khomenko V, Raymundo-Piñero E, Béguin F (2010) J Power Sourc 195:4234–4241
65. Jurewicz K, Frackowiak E, Béguin F (2001) Electrochem Solid State Lett 4:A27–A29
66. Jurewicz K, Frackowiak E, Béguin F (2002) Fuel Process Technol 77–78:415–421
67. Fang B, Zhou H, Honma I (2006) J Phys Chem B 110:4875–4880
68. Bleda-Martínez MJ, Pérez JM, Linares-Solano A, Morallón E, Cazorla-Amorós D (2008) Carbon 46:1053–1059
69. Béguin F, Friebe M, Jurewicz K, Vix-Guterl C, Dentzer J, Frackowiak E (2006) Carbon 44:2392–2398
70. Demarconnay L, Raymundo-Piñero E, Béguin F (2010) CD Proceedings of the International Carbon Conference, Clemson, USA, 2010
71. Greaves TL, Drummond CJ (2008) Chem Rev 108:206–223
72. Mysyk R, Raymundo-Piñero E, Anouti M, Lemordant D, Béguin F (2010) Electrochem Comm 12:414–417
73. Anouti M, Caillon-Caravanier M, Dridi Y, Galiano H, Lemordant D (2008) J Phys Chem B 112:13335–13343

Chapter 13
Nickel-Based Battery Systems

Ralph J. Brodd

Glossary

β-β Transformation	The reaction of β-$Ni(OH)_2$ on charge to form β-NiOOH.
α-γ Transformation	The reaction of α-$Ni(OH)_2$ on charge to form γ-NiOOH.
Ah Ampere hour	3,600 coulombs, the quantity of current flow in 1 ampere in 1 h.
Charge reserve	Additional capacity of the negative or positive electrode to prevent gas evolution when the cell is overcharged or overdischarged.
Electrolyte	Electrical conducting liquid flow where charge is carried by positive and negative ions.
Flame-arresting vent	A cell vent designed to stop burning discharge from a vent.
Ni-MH	Nickel metal hydride cell containing a nickel hydroxide positive and a hydrogen-absorbing metal alloy negative electrode.
Plate	Electrode construction containing the active material and a current collector.
Pocket plate electrodes	Perforated nickel-plated steel pockets that contain the active materials in a cell with free access of electrolyte.
Self-discharge	Loss of capacity due to the chemical instability or a reaction of an active material in an electrode with the electrolyte.

This chapter was originally published as part of the Encyclopedia of Sustainability Science and Technology edited by Robert A. Meyers. DOI:10.1007/978-1-4419-0851-3

R.J. Brodd (✉)
Broddarp of Nevada, 2161 Fountain Springs Drive, Henderson, NV 89014, USA
e-mail: rbrodd@gmail.com

R.J. Brodd (ed.), *Batteries for Sustainability: Selected Entries from the Encyclopedia of Sustainability Science and Technology*, DOI 10.1007/978-1-4614-5791-6_13,
© Springer Science+Business Media New York 2013

Tubular plate electrodes	Perforated nickel-plated steel tubes used to contain the active material in a cell.
Misch metal	Containing a mixture of hydrogen-absorbing transition metals, Mm.
Nickel cadmium fiber plate	Batteries construction of the positive nickel battery plate that contains nickel metal fibers throughout the plate for more efficient current collection.
Sinter plate cells	Sealed battery cells using a nickel current collector structure usually produced by heating to a temperature where powdered nickel metal particles bond together to form a porous structure sinter current collector.
Trickle charging	The low-level current flow to maintain full charge in a battery.
Terminal	The external connection to the positive and negative electrodes containing the active material inside the battery case.
WH	Watt hour, energy content of a battery expressed as the product of ampere-hours times cell voltage.
Zircar	Trademark of nonwoven ceramic separator used in some Ni-MH cells.

Definition of the Subject

Nickel batteries are rechargeable batteries that are used in a variety of applications including portable electronic devices, electric and hybrid vehicles, aeronautics and aerospace and stationary energy storage among others. They operate over a wide temperature range, have a flat discharge curve and are available in sizes ranging from small coin cells to motive power batteries. Nickel batteries are physically and electrically rugged and abuse tolerant including over charge and overdischarge.

Introduction

The nickel electrode serves as cathode for several important commercial rechargeable battery systems. The characteristics of these systems are listed in Table 13.1. The first commercial nickel battery was the nickel-iron system which provided lighting in railroad cars due to its strong resistance to physical and electrical abuse. The electrode structure has a strong influence on the operating life of a battery system. The nickel systems are robust, both physically and chemically.

Table 13.1 Characteristics of Nickel batteries (Nominal values)

System	Cell voltage	Wh/kg	Wh/l	Cycle life, 100% DOD	Calendar life, years	Commercial introduction
Ni-Fe	1.37	30	55	2,000–4,000	25	1,898
Ni-Cd, Vented pocket plate	1.29	20	40	500–2,000	25	1,900
Ni-Cd, Vented sintered plate	1.29	30	58	500–2,000	10	1,940
Ni-Cd, Sealed sinter plate	1.29	35	100	300–700	5	1,950
Ni-Cd, Fiber	1.29	10	15	500–10,000	20	1,950
Ni-MH, Sealed sinter plate	1.2	110	430	300–1,000	5	1,980
$Ni\text{-}H_2$	1.4	60	65	1,500–10,000	25+	1,980
Ni-Zn	1.71	60	120	200–500	3–5	1,931, 2,009

Nickel battery systems compete directly with the lead acid battery in many commercial energy storage applications and with Li-Ion in portable electronic applications.

Advantages	Disadvantages
Long cycle life	Heavier than Li-Ion
High reliability	Higher cost than lead acid
Excellent rate capability	
Very robust cell constructions	

Characteristics of the Nickel Active Materials

The family of nickel batteries is based on the utility, strength, and reversibility of the nickel electrode reactions in alkaline media. The nickel active materials for use in batteries are produced, mainly, by chemical precipitation of $Ni(OH)_2$ with the addition of KOH to aqueous nickel sulfate solutions made by dissolving nickel metal in sulfuric acid. The nickel active material is usually produced by a precipitation process wherein a nickel sulfate solution is heated and pumped into hot NaOH. The process produces $\beta\text{-}Ni(OH)_2$. Fine graphite may be added, depending on the application. A small percentage of cobalt, ranging from 3% to 7%, depending on application, is co-precipitated with the nickel sulfate solution to improve charge acceptance, capacity, and operating life. The precipitate is filtered, dried, and ground. Care must be exercised to keep the drying temperature below 200°C as higher temperatures cause formation of electrochemically inactive bunsenite structure. Electrodes for cells used in satellites and other space applications are made by an in situ electrochemical precipitation technique described later.

Nickel Electrode Reaction Characteristics

The battery electrolyte is aqueous KOH solution in a concentration range of 25–30% KOH by weight. The mechanism that best describes the overall nickel electrode reactions in alkaline electrolyte is given in Eqs. 13.1 and 13.2 below.

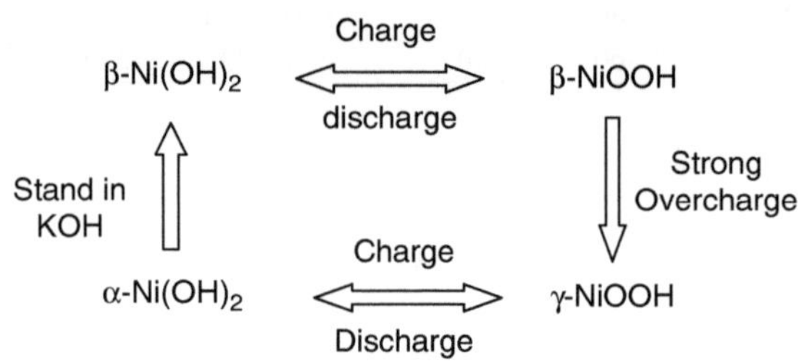

(13.1)

(13.2)

Equation 13.1 describes the classic β-β transformation reaction of the nickel electrode in alkaline media wherein the valence of the nickel shuttles between divalent Ni^{+2} and trivalent Ni^{+3}. The charged and discharged materials have similar structures and long cycle life. The battery reactions follow this regimen with little or no overcharge. On continued charging after reaching the Ni^{+3} valence state, the high oxygen overvoltage on the electrode surface permits formation of a Ni^{+4} valent nickel in the electrode structure. This transformation stops with oxygen evolution on the electrode surface and a nickel valence of 3.67, resulting in a higher capacity cell. On discharge the γ-NiOOH returns to α-$Ni(OH)_2$. Charge-discharge will continue on this higher capacity regime. However, if the cell is left on open circuit, the $Ni(OH)_2$ slowly transforms back to β-$Ni(OH)_2$. Transition back to the β-β cycle occurs only on open circuit stand as the α structure transforms slowly back to the β structure. The normal materials in electrode construction are produced by chemical or electrochemical precipitation, depending on the application for the cell. The uncharged β-$Ni(OH)_2$ is usually used in electrode fabrication and cells are initially assembled in the discharged state.

The four nickel materials in Eqs. 13.1 and 13.2 are distinct phases and have been characterized bothelectrochemically and by the XRD patterns exhibited in the chemically precipitated phases. A point defect, nonstoichiometric, structural model correctly describes the structure and interactions of nickel electrode active materials (22). In this model, the intercalated water is incorporated as interlamellar protons, with the oxygen effectively adding to the NiO_2 layers. The model introduces nickel vacancies that explain the empirical Ni/O ratios (less than 1/2). The Raman spectra gives no indication of water in the lattice as the excess protons and alkali cations occupy nickel vacancies or interlamellar sites and three protons can occupy a nickel site for charge compensation. This model correctly explained the empirical observation that Ni has a 3.67 oxidation state observed after overcharge. Raman spectra of these higher oxidation state materials are consistent with the simple rhombohedral space group and the simple γ-NiOOH-type crystal lattice (the ideal, stoichiometric crystal). The spectra are inconsistent with the larger space groups and those which contain molecular water. Thus, the structure is written ideally as γ-NiOOH.

There have been four generations of nickel electrode construction. The first generation included the pocket and tubular electrode structures that were developed independently by Junger in Sweden and Edison in the USA in 1900 for the original system introduction. The Sintered Plate is a second-generation technology and was developed in the 1940s by Varta. The Nickel Cadmium Fiber electrode battery, developed by DAUG, constitutes the third-generation (1985) technology. The nickel cathode electrodes used in nickel-hydrogen batteries for space applications constitute the fourth generation and are produced by an electrochemical deposition of the nickel hydroxide materials directly into the voids in the sintered nickel electrode structure. This is a more expensive process and is used mainly for space and special applications. Because of their ability to withstand physical and electrical abuse, nickel batteries are virtually indestructible. All four electrode constructions are articles of commerce in 2011. Their characteristics will be discussed in more detail below.

Nickel-Iron Batteries (Ni-Fe)

Pocket and Tubular Plate Cells

The iron-nickel battery system is one of the oldest rechargeable systems. It was invented about the turn of the century, independently in Sweden by Junger in 1899, and in the USA by Edison in 1900, each acting without specific knowledge of the other's activity. Common usage today is to call the system the Edison battery. It still finds application today to power materials handling vehicles, underground mining vehicles, miners' lamps, railway cars and signal systems, and emergency lighting. The Edison battery is virtually indestructible and can withstand both physical and electrical abuse situations, discharged stand, and short-circuiting. Edison used tubes for the positive and pockets for the negative electrode structures for his 1908 battery designs. Disadvantages include low power density, low temperature performance, poor charge retention, and hydrogen evolution on stand.

Both the positive and negative tubular and pocket current collectors are made of perforated nickel-plated steel. They are very robust and are virtually indestructible. The low energy density, poor charge retention, and poor low temperature performance, along with high cost of manufacture, have led to a decline in use of the nickel-iron battery system. The negative electrode, or anode, is iron and the positive electrode, or cathode, is nickel oxide with 6–8 molar potassium hydroxide (KOH) as the electrolyte. The cell reaction occurs in a two-step process. The system has a nominal voltage of 1.37 V.

The two-step cell reaction is:

$$Fe + 2\,NiOOH + 2H_2O = 2Ni(OH)_2 + Fe(OH)_2 \tag{13.3}$$

$$3\,Fe(OH)_2 + 2\,NiOOH = 2Ni(OH)_2 + Fe_3O_4 + 2H_2O \tag{13.4}$$

$$\text{Overall}: 3Fe + 8NiOOH + 4H_2O = 8Ni(OH)_2 + Fe_3O_4 \tag{13.5}$$

On discharge, iron initially forms a divalent hydroxide $Fe(OH)_2$ in the first plateau. Further discharge forms $Fe(OH)_3$, which then further reacts with Fe^{+2} to form Fe_3O_4.

Advantages	Disadvantages
Physically robust	High self-discharge
Long stand and cycle life	Hydrogen evolution on charge and discharge
Withstand electrical abuse	Low voltage and energy density
Very reliable	Poor low temperature performance

The construction of the tubular and pocket plate nickel-iron battery cell is essentially identical to that of the nickel cadmium battery and has not changed over the past 50 years. For good performance, special attention must be paid to use high purity materials and the particle size characteristics of the active materials. The long cycle life is a result of the low solubility of the reaction intermediates. On discharge, supersaturation of the discharge product results in the formation of minute crystallites of Fe_3O_4 near the reaction site. On recharge, these crystals form into a high surface area iron with excellent high rate performance on discharge.

Preparation of Negative Iron Electrode Materials

The active material is produced by dissolution of pure iron in sulfuric acid. The $FeSO_4$ is recrystallized, dried, and roasted at high temperature to produce Fe_2O_3. The materials are washed free of the sulfate, dried, and partially reduced in H_2 to produce a mixture of Fe_3O_4 and iron. The resulting mixture is blended with small amounts of FeS, sulfur, and HgO for use in negative plate assembly.

The anode current collector is an annealed, perforated, nickel-plated steel pocket plate assembly. The tubes for the positive electrode are produced from perforated nickel-plated strips that are wound to produce a tube. One end is crimped and the active material poured into the open end in layers and crimped again to close the tube. A machine automatically introduces the active material and tamps it into the pockets. After filling, the tubes are pressed into openings in the nickel steel electrode frame. The pocket plate follows a similar process. Rectangular pockets of perforated nickel-plated iron strips are filled with active material, crimped, closed, and fixed/bolted in a nickel-plated steel frame. The assembled cells are placed in polyethylene containers and filled with KOH electrolyte. Spacing of the negative and positive electrodes is maintained by the internal assembly structure. There are no separators as are common in the lead acid battery structure.

Preparation of Positive Nickel Electrode Materials

The preparation of the $Ni(OH)_2$ active material starts with dissolving a high purity nickel metal powder, or chips, in sulfuric acid. The hydrogen produced in this step is used in making the negative iron active material. The acidity is adjusted to pH 3 or 4 to remove iron and other insoluble materials. Further conditioning, to remove all of the ferrous iron and copper, may be needed. Cobalt sulfate is added at about the 1.5% level to enhance the charge acceptance of the nickel electrode performance. The resulting solution is heated and sprayed into hot 50% NaOH to precipitate the $Ni(OH)_2$ active material. The resultant is dried, crushed, and screened to pass 20- but not 200-mesh screen. The construction of the negative plate follows a similar process to that of the positive plate described above. The size of the tubes and configuration of the plates determine the capacity of the finished battery. The battery capacity is designed to be limited by the capacity of the positive plate for longer cycle life.

The nickel-iron battery cell fabrication process is essentially unchanged in over 50 years. Special attention must be paid to use high purity materials and particle size characteristics of the active materials. The iron negative active material is made from pure iron that is dissolved in sulfuric acid. The resulting $Fe(SO_4)_2$ is recrystallized and dried. This is washed free of sulfuric acid and roasted at 915°C to form a mixture of Fe_2O_3 and Fe metal and is, then, blended with small amounts of FeS, sulfur, and HgO for use in the negative plate assembly.

The negative current collector is a perforated nickel-plated steel pocket plate assembly. The tubes are produced from perforated nickel-plated strips that are wound to produce a tube. The tubular construction has one end crimped and the active material is poured into the open end in layers with nickel flakes. Alternatively, a machine automatically introduces the active material and tamps it into the pockets. After filling, the tubes/pockets are crimped and pressed into openings in the nickel-plated steel electrode frame. A similar process is used for the positive electrode.

The positive electrode consists of $Ni(OH)_2$ in alternate layers of nickel flake. High purity nickel metal powder or chip is dissolved in sulfuric acid. Hydrogen is used in making the negative iron active material. The acidity is adjusted to pH 3 or 4 to remove iron and other insoluble materials. Further conditioning to remove all of the ferrous iron and copper may be needed. Cobalt sulfate is added at about the 1.5% level to improve the nickel performance. The resulting sulfate solution is heated and sprayed into hot 50% NaOH to precipitate the $Ni(OH)_2$ active material. The resulting slurry is filtered, washed, dried, and screened to control the particle size to pass 20 but not 200 mesh.

The preferred electrolyte is 25–30% KOH solution with up to 50 g/l of LiOH added. The addition of LiOH improves cell capacity, prolongs cycle life, and facilitates the kinetics of the nickel electrode. It expands the discharge plateau and delays oxygen evolution. It also decreases the carbonate content of the electrolyte as Li_2CO_3 is not very soluble. The negative and positive plates are usually separated by hard rubber pins. The edges of the plates are insulated from the sides and bottom by hard rubber or polymer frames. The spring-loaded, hinged, filler cap protects against carbonation of the alkaline electrolyte by atmospheric CO_2.

The nickel flake is produced by the dissolution of high purity nickel powder in sulfuric acid and the H_2 used in the production of the negative iron active material. The acidity is adjusted to remove soluble iron and copper. A small amount of cobalt sulfate may be added to improve performance. The resulting solution is sprayed into a hot concentrated NaOH solution. The construction of the positive plate follows that for the negative plate.

The size of the tubes and configuration of the plates determine the capacity of the finished battery. The positive and negative electrode frames are bolted together with a spacer to separate the negative from the positive frames for strength and long life. The battery is designed to be limited by the capacity of the positive electrode for longer cycle life. Constant potential charging is not recommended as the system has a tendency to go into thermal runaway. A modified constant voltage with a current limit is acceptable.

Performance

The Ni-Fe batteries range in size from 5 Ah to over 1,200 Ah. The cell open circuit voltage of the Ni-Fe cell is 1.4 which drops quickly to 1.2 V during discharge. Tubular or pocket plate constructions are used. Active materials of high purity are contained inside the perforated nickel-plated steel tubes or rectangular pockets. The active materials are iron for the negative electrode and nickel oxide for the cathode and a KOH electrolyte.

The battery is assembled in the discharged state and a formation charge is used to activate it. The KOH electrolyte is invariant over the charge-discharge regime. The open circuit voltage is 1.4 V that drops quickly to 1.2 V on discharge. The cells vent hydrogen on charge and require a well-ventilated space to prevent accumulation of hydrogen that could ignite with a spark. The low solubility of iron on discharge results in the formation of minute crystallites of Fe_3O_4 near the reaction site. On charge, these crystallites result in the formation of a high surface area iron electrode deposition that increases the ability to deliver pulse and high current discharge.

Constant potential charging is not recommended as the system has a tendency to go into thermal runaway. A modified constant voltage with a current limit is preferred to prevent thermal runaway situations. The long cycle life of the Ni-Fe battery is a result of the low solubility of the iron electrode reaction intermediates. Series-connected cells require precautions in dealing with high voltage cell assembly connections. Batteries that will be inactive for a month or more should be discharged, short-circuited, and left in that condition. They can be stored in the discharged state with filling caps closed for long periods of time without damage.

Nickel Cadmium Batteries

Nickel Cadmium batteries are available in four different constructions: vented pocket plate, vented tubular plate, sealed sinter plate, and fiber plate constructions. The vented pocket and tubular electrode constructions followed directly from the nickel-iron batteries of Junger and Edison with a substitution of cadmium for iron in the electrode. Like the nickel-iron system, the nickel cadmium pocket and tubular systems are very sturdy, have long life and operate over a wide temperature range with good high rate performance, and require little maintenance. Because they can withstand overcharge, overdischarge, mechanical and electrical abuse, and require little maintenance, they find use in critical industrial applications including switchgear operations, telecommunications, aircraft power, uninterruptible power supplies (UPS), and emergency lighting even though their cost is higher than the lead acid battery system. The cell voltage is 1.29 V and the cell electrolyte is KOH with added LiOH to improve cycle life and high temperature operation.

$$\text{Negative}: \mathrm{Cd} + 2\mathrm{OH}^- - 2\mathrm{e}^- = \mathrm{Cd(OH)}_2 \tag{13.6}$$

$$\text{Positive}: 2\mathrm{NiOOH} + 2\mathrm{H_2O} + 2\mathrm{e}^- = 2\,\mathrm{Ni(OH)}_2 + 2\mathrm{OH}^- \tag{13.7}$$

$$\text{Overall: } \mathrm{Cd} + 2\mathrm{NiOOH} + 2\mathrm{H_2O} = 2\mathrm{Ni(OH)}_2 + \mathrm{Cd(OH)}_2 \quad E = 1.29\mathrm{V} \tag{13.8}$$

Advantages	Disadvantages
Very rugged	Low energy density
Long cycle life	Higher cost than lead acid
Reliable	Contains cadmium
Flat discharge profile	Oxidation of carbon conductor forms CO_2
	Vented cells need regular water service

Nickel cadmium batteries have long life as measured by the number of charge-discharge cycles a battery can sustain and still deliver useful capacity. Under normal conditions a typical Ni-Cd cell can deliver 2,000 cycles or more over a period of 5–20 years depending on application (drain rate, temperature, and ambient conditions of temperature, vibration, duration of high current demand and recharging conditions) as well as the internal design of the cell. Typically, the Ni-Cd cell has a strong internal mechanical design, chemically stable internal ambient and ability to withstand electrical abusive conditions. The sintered electrode cells do have a characteristic termed "memory effect." This arises if the cell has an exact repetitive charge-discharge regime. After a period of time, if the cell is asked to discharge longer than the specified regime, it loses voltage as if it "remembers" the discharge regime. The cell is not damaged. Discharge can continue at a lower voltage and the "effect" eliminated by reconditioning with a series of full capacity charge-discharges.

Nickel Cadmium Pocket and Tubular Plate Cells

Materials fabrication of the Ni-Cd pocket plate and tubular cells follow essentially an identical process to that described for the Ni-Fe cell construction but with cadmium substituted for the iron in the negative electrode construction. The preparation of the cadmium active material uses a co-precipitation process, usually starting with a $CdSO_4$ or $CdCl_2$ solution where KOH is added to precipitate the Cd $(OH)_2$ used in cell fabrication. Here, graphite is used to provide internal conductivity to contact the positive nickel active mass. On overcharge, the oxygen evolution attacks the graphite conductor to produce CO_2 which, in turn, reacts with the KOH electrolyte to produce K_2CO_3, lowering the concentration of the electrolyte. Each overcharge causes loss of graphite and decreases the available contact to the nickel active mass.

Because they can withstand overcharge, overdischarge, mechanical and electrical abuse, and require little maintenance, the Ni-Cd batteries find use in critical industrial applications including switchgear operations, telecommunications, uninterruptible power supplies (UPS), and emergency lighting even though their cost is higher than the lead acid battery system. The Ni-Cd vented pocket plate battery was used in the world's largest UPS power supply in Anchorage, Alaska.

Nickel Cadmium Fiber Plate Batteries

Fiber Plate Electrode Technology was developed by DAUG in Germany in 1983. The electrodes have an open three-dimensional nickel fiber structure that provides good conductivity throughout the electrode structure to ensure excellent electrical performance. The fibrous nickel matrix current collector matrix forms a lightweight flexible structure with about 90% of the electrode volume available for holding the active material. It is free from graphite and iron, found in other nickel-based cathode battery systems; so reduces water consumption by electrolysis and eliminates carbonation of the electrolyte on overcharge to ensure very long cycle life.

The elastic nature of a NCFP electrode and its fibrous current collector construction enable it to accommodate the volume changes and maintain good contact with the active materials during the charge and discharge cycles, as well as absorb shock and vibration. This feature results in longer cycle life compared to other Ni-Cd battery technologies. The NCFP technology is considerably lighter in weight and has higher energy density than the regular Ni-Cd constructions. Fiber plate batteries can be designed with very thin plates for ultra high rate discharge performance or with thick plates for long duration lower current discharge applications. Other features include no need for change of electrolyte throughout life due to graphite-free electrodes, low water consumption, long shelf life, and flame-arresting vent protection.

Applications include switchgear protection, emergency lighting, motive power, train lighting, instrumentation and process control, UPS, electric vehicles, diesel engine cranking, aircraft and helicopter ground starting.

Nickel Cadmium Sealed Sinter Plate Cells

The present sealed Ni-Cd sinter plate battery technology evolved from the sealed battery technology developed by Varta in the 1940's based on an IG Farben development in the 1930's. The sinter plate construction, developed in the 1940 time frame, has low internal resistance, superior high rate and low temperature performance compared to the pocket plate cell design. The batteries are sealed and can be used in any orientation. This battery powered the Luftwaffe in the 1940s, the early flights in the space program and in the development of portable electronic devices, such as cellular phones and portable notebook computers.

The small sealed sintered plate Ni-Cd cylindrical and prismatic cell constructions have the favorable characteristics of the pocket/tubular plate constructions with increased energy density. The electrodes are thin and the cells have low internal resistance with excellent high rate and low temperature performance. The stem is rugged and can withstand physical and electrical abuse. It requires minimal maintenance.

Advantages	Disadvantages
Sealed, no maintenance	Higher cost than lead acid
Long cycle and shelf life	Lower capacity than Li-Ion and Ni-MH
Excellent high rate performance	Environmental issues with cadmium
Flat discharge profile	
−40°C to +50°C operating range	
Operate in any orientation	

$$Cd + 2OH^- - 2e^- = Cd(OH)2 \quad (13.9)$$

$$2NiOOH - 2H2O = 2Ni(OH)2 + 2OH \quad (13.10)$$

$$Cd + 2NiOOH + 2H_2O = Cd(OH)_2 + 2Ni(OH)_2 \quad E = 1.29V \quad (13.11)$$

The electrodes for cell fabrication are produced by vacuum impregnation of the nickel sinter plate. The sinter plates are produced by coating slurry of carbonyl nickel powder in CMC onto a nickel-plated steel or nickel screen current collector and burn off the CMC at high temperature. This process produces a nickel current collector with about 80% porosity. The coils or plates (depending on cell construction) of the sinter plate are vacuum impregnated in an aqueous nickel nitrate solution, dipped into a NaOH bath to convert the nickel nitrate into $Ni(OH)_2$, and then dried. The process is repeated until the desired loading of $Ni(OH)_2$ is obtained. The negative electrode is produced by a similar process except cadmium nitrate solution is used to impregnate the sinter electrode. The impregnated strips of cobalt and nickel electrode stock are cut to size, stacked as plates or wound into a cylinder, and assembled in the appropriate cell container.

The sealed cylindrical and prismatic cells can operate in any orientation. They were instrumental in the electronic revolution that produced the portable computers, cellular phones, digital cameras, and the like. This requires specific design features that prevent pressure buildup from oxygen gas generation on overcharge, with no free electrolyte and no maintenance or servicing. The internal design of each electrode of the cell is critical to long life at full performance. The cell balance, ratio of the capacity of Cd to the capacity of NiOOH electrodes, is carefully controlled to prevent any hydrogen evolution from the cadmium electrode. Sealed cells have a different balance (ratio of the negative and positive electrode capacities) from that for vented cell constructions. In sealed Ni-Cd cells, it is essential to prevent hydrogen evolution on charge as it causes the cell to dry out and develop high internal pressure. This is accomplished by balancing the capacity of each electrode, positive and negative, during cell assembly to ensure that oxygen is the only gas evolved, as depicted in Fig. 13.1. As the cells reach full charge and enter overcharge, oxygen is evolved from the nickel positive. The oxygen diffuses through/around the porous separator to the cadmium electrode where it reacts directly with the active cadmium metal, just deposited on the negative electrode, to produce $Cd(OH)_2$ with heat evolution. To that end, the negative cadmium plate is designed with an excess of uncharged active mass.

Nickel Metal Hydride

Nickel metal hydride (Ni-MH) batteries have application in hybrid electric vehicles, portable electronic devices such as cameras, shavers, toothbrushes, etc. They offer a higher energy option than Ni-Cd batteries with the hydrogen-absorbing metal hydride alloy (MH) replacing the cadmium electrode in Ni-Cd cell construction. Like Ni-Cd, the Ni-MH also shows the "memory effect" on repetitive cycle regimes. The nickel metal hydride battery was introduced commercially in 1989. The technology is based on the development of rare earth alloys with nickel that have the ability to reversibly absorb and desorb hydrogen. The

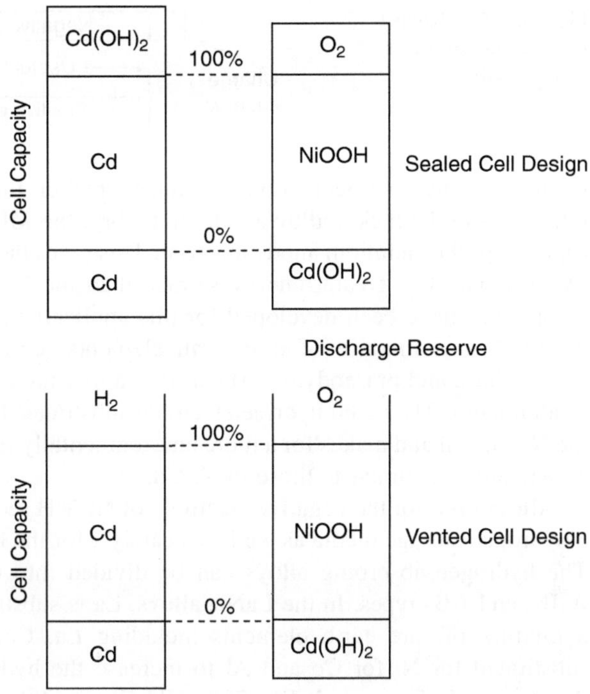

Fig. 13.1 Cell balance for vented and sealed nickel cadmium cells. It is important to avoid hydrogen gas evolution in the sealed cell

nickel metal hydride (MH) electrode replaces the cadmium electrode in the Ni-Cd cell construction. It stores, and can deliver, more energy than an equivalent-sized Ni-Cd cell.

The Ni-MH battery has found wide application as a replacement for rechargeable Ni-Cd cells and primary alkaline manganese cells, especially in portable electronic devices. It offers a rechargeable replacement for the primary alkaline cell and gives superior performance. Although it initially has a lower voltage, it has a constant voltage discharge as compared to the sloping voltage of the alkaline cell that is very desirable for cameras, portable electronics, etc. More recently, it has found application in hybrid electric automobiles.

$$\text{Negative:}\ \ MH + H_2O + 2e^- = M + 2OH^- \tag{13.12}$$

$$\text{Positive:}\ \ 2NiOOH + 2e^- = 2Ni(OH)_2 + 2OH^- \tag{13.13}$$

$$\text{Overall:}\ \ MH + 2NiOOH = M + 2Ni(OH)_2\ \ E = 1.35V \tag{13.14}$$

The "metal" M in the equations above refers to the negative electrode of a Ni-MH cell and is an intermetallic alloy or compound capable of storing and releasing hydrogen. These compounds can store many times their volume of

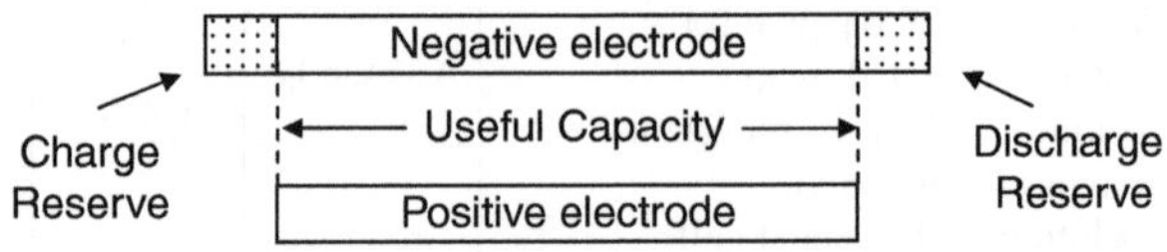

Fig. 13.2 Cell balance in sealed nickel metal hydride cells

hydrogen. The hydrogen atom is much smaller than nickel and can occupy interstices in the nickel alloy anode. In theory, the MH alloy has almost twice the capacity of the cadmium anode material. This gives the Ni-MH battery the capability of about 20% greater energy storage than the Ni-Cd battery. Many different compounds have been developed for this application. The most common is AB_5, where A is a mixture of rare earth elements (chromium, lanthanum, cerium, neodymium, and praseodymium) and B is a mixture of nickel, cobalt, manganese, or aluminum. The metal hydride electrode substitutes for the cadmium electrode in the Ni-Cd cell and makes for a more environmentally friendly system. Many design parameters are similar to those in Ni-Cd.

Alloys used for the negative electrode of Ni-MH secondary batteries function as a hydrogen storage media as well as a catalyst for the hydrogen electrode reaction. The hydrogen-absorbing alloys can be divided into three main types: the AB_5, A_2B_7, and AB_2 types. In the $LaNi_5$ alloys, La is substituted for misch metal, Mm, a mixture of rare earth elements including La, Ce, Pr, and Nd and partially substituted for Ni for Co and Al to increase the hydrogen storage and lengthen the charge-discharge cycle life. The cell balance of the capacity of the positive and negative electrodes for best performance are given in Fig. 13.2.

Ni-MH cells can be practically charged at 1 C rate using a constant current voltage charge regime and discharged at 4 C. Its 1 C charge-discharge cycle life expectancy is more than 500 cycles enabling commercial applications of Ni-MH to a wide variety of cordless equipment. The Ni-MH cells do not have the memory effect characteristic, where cells discharged on the exact same charge-discharge regime "memorize" the cycle and if discharged, further, act as if fully discharged.

As with Ni-Cd, when overcharged at low rates, the oxygen produced at the positive electrode on overcharge diffuses through the porous separator and recombines at the surface of the negative to suppress hydrogen evolution and converts the charging energy into heat. This process prevents internal pressure buildup to allow Ni-MH cells to remain sealed if overcharged during normal operation. Ni-MH cells use a KOH electrolyte and the same hydrophilic polyolefin nonwoven separator materials used in the Ni-Cd cell.

The internal construction of the Ni-MH cell mimics that of the Ni-Cd cells. The Ni-MH cells are available in a wide variety of sizes ranging from the AA-size for digital cameras, up to 50 Ah or larger used in hybrid car applications. Useful discharge capacity is constant up to the 1 h discharge rate and does not differ significantly from the rated capacity.

Nickel metal hydride batteries consist of a positive electrode containing a mixture of carbon/graphite conductive diluent and nickel hydroxide as its principal active material. The negative electrode consists mainly of hydrogen-absorbing conducting metal alloys, a porous polymer separator filled with KOH electrolyte, a metal case and a resealing vent. The anode-separator-cathode foils are wound in a cylinder or stacked in a prismatic format and inserted into a nickel-plated steel can. The prismatic cell construction may also be formed by stacking alternate positive and negative individual electrodes or by winding the electrode strips into a cylindrical form and then inserted into a prismatic can or wound and inserted into a cylindrical can construction.

A complete discharge of a Ni-MH cell causes permanent damage to the negative electrode. This situation can occur in a group of series-connected cells if one low-capacity cell in the series is completely discharged and driven into reverse due to its lower capacity than surrounding cells. When this happens, the good cells will start to drive the discharged cell in reverse and can cause permanent damage to that cell.

Ni-MH cells have a 5–10% self-discharge rate at the beginning of life but stabilize around 0.5–1% per day. Cooler storage temperatures have lower discharge rate and longer battery life. Continual trickle charging of small Ni-MH cells can cause permanent deterioration in cell performance.

Ni-MH batteries provide a relatively constant voltage during the discharge cycle, unlike a standard alkaline battery where the voltage falls steadily during discharge. This characteristic gives the Ni-MH longer useful service to power portable electronic devices than the standard alkaline zinc manganese cell and can be substituted for the AA-size alkaline cell in most devices.

Ni-MH batteries have application in hybrid electric vehicles and portable electronic devices, such as cameras, shavers, toothbrushes, etc. The Ni-MH battery is also used in hybrid cars to improve stop-start and acceleration performance. It absorbs braking energy to assist the breaking operations and during the acceleration, returns that energy to supplement the gas engine.

Improper disposal of Ni-MH batteries poses less environmental hazard than that of Ni-Cd because of the absence of toxic cadmium.

Nickel-Zinc Batteries

The performance of nickel-zinc (Ni-Zn) batteries lies between that of high-energy silver-zinc and nickel cadmium. The nickel-zinc cells have excellent high rate performance and about twice the energy density of Ni-Cd but the cycle life is severely limited due to "shape change" problems that limit recharge of the zinc electrode. While Junger and Edison were both awarded patents on the rechargeable Ni-Zn battery, the first commercial cell was developed by J. Drumm in Dublin, Ireland, in 1931 to power the lights on a passenger train between Bray and Dublin. It ran until 1949. Later, Yuasa introduced a 10 Ah sealed Ni-Zn battery for use in lawn mowers and light duty electric vehicles without success. Their cells limited cycle life of 200 cycles at the two-hour rate and were not sufficient for a commercial rechargeable battery system. PowerGenix successfully demonstrated that the Ni-Zn battery could substitute for the Ni-MH battery in a Toyota Prius car. A potentially new large application for Ni-Zn is stationary energy storage.

Although the zinc electrode is a difficult system and can deliver excellent high rate performance, it suffers from problems with the surface morphology, dendrite formation on charge, shape change, agglomeration/redistribution and passivation, etc., that limit cycle life. The problems directly result from the characteristics of the reactions of the zinc ions in solution and the reactivity of zinc metal with the electrolyte.

$$\text{Negative: } Zn + 2OH^- - 2e^- = ZnO + 2OH^-$$

$$\text{Positive: } NiOOH + 2H_2O = 2Ni(OH)_2 + 2OH^-$$

$$\text{Overall: } Zn + NiOOH = Ni(OH)_2 + ZnOE = 1.73V$$

In 2008, PowerGenex introduced a new sealed cell for the power tool market with a modified KOH electrolyte formulation. The higher voltage of the zinc electrode results in a unit cell voltage of 1.73 versus 1.2 for Ni-Cd, giving the Ni-Zn a distinct advantage in energy density. In addition, the facile kinetics of the zinc electrode gives the system superior performance for high rate applications. The reaction of the active materials with the electrolyte limits the cell life.

$$\text{Zinc electrode: } Zn + 2OH^- - 2e = Zn(OH)_2 Zn(OH)_2 + 2OH^- = ZnO_2{}^{-2} + 2H_2O$$
$$ZnO_2{}^{-2} + H_2O = ZnO + 2OH^-$$

$$\text{Nickel electrode: } 2NiOOH + 2H_2O + 2e^- = 2Ni(OH)_2 + 2OH^-$$

$$\text{Overall: } Zn + 2NiOOH + H_2O = 2Ni(OH)_2 + ZnO$$

Cell construction for the Ni-Zn cell is similar to that for the sealed spiral wrap cylindrical and prismatic Ni-Cd and Ni-MH cells for portable electronic applications but with some differences. The Ni-Zn cell incorporates a multilayer micro-porous separator with "wicking" components used to maintain uniform electrolyte distribution, as well as acting as a buffer/barrier to prevent shorting from zinc dendrite formation. The barrier also slows oxygen gas diffusion and lowers the recombination rate that can lead to pressure buildup inside the cell. The KOH electrolyte contains CaO additions to calcium zincate that can control the free zinc ions in solution. In the cell electrolyte, the CaO reacts with zinc ions to form calcium zincate. Other improvements center on the use of fiber (steel wool) current collectors for the zinc electrode. The Ni-Zn cells have excellent high rate and low temperature performance.

Nickel-Hydrogen Batteries

The nickel-hydrogen ($Ni\text{-}H_2$) battery was developed in the late 1960s as a new, long-lived, highly reliable power source to replace Ni-Cd for space applications. It won out over strong competition from regenerative fuel cells mainly due to issues related to life expectancy and system engineering problems in that resulted in increased size, weight, complexity, loss of catalytic activity, and cost to assure the required long life.

The $Ni\text{-}H_2$ system characteristics of its lightweight, extremely long cycle life and simple, yet, accurate state of charge determination based on a pressure measurement, makes it an ideal system for all space applications ranging from low orbits, mid-altitude orbits, to the higher geosynchronous orbits. It replaced the Ni-Cd system because of its lighter weight and longer life. It has powered the space shuttle, space stations, and interplanetary probes including extending planetary exploration stations on the moon and Mars. $Ni\text{-}H_2$ battery cells are by far the longest-lived and most reliable of all sealed rechargeable battery systems.

Advantages	Disadvantages
Long calendar life	Expensive
10+ year's cycle life	Low energy density
Accurate state-of-charge measure	High self-discharge
Abuse-tolerant	

$$\text{Negative}: \quad H_2 - 2e = 2H^+ \tag{13.15}$$

$$\text{Positive}: \quad 2NiOOH + 2e^- = 2Ni(OH)_2 + H_2O \tag{13.16}$$

$$\text{Cell}: \quad H_2 + 2NiOOH = 2Ni(OH)_2 + H_2OE = 1.2V \tag{13.17}$$

Internal cell design considerations include the depth of overdischarge, positive pre-charge, or negative pre-charge cell designs. The type of pre-charge depends on if there is excess capacity associated with the positive nickel electrodes (positive pre-charge) or the negative hydrogen material (negative pre-charge). In this configuration, there is a capacity loss mechanism caused by the reaction of hydrogen with the cobalt additive in the active material of the positive electrodes. Newer cell designs specify positive pre-charge designs where there is capacity remaining in the positive electrodes of the cell at the end of discharge.

The hydrogen electrodes are Teflon bonded, platinum black catalyzed, supported on a nickel substrate. A high surface area platinum black hydrogen catalyst loading, in the range of 7 mg/cm^2, ensures long life. The electrodes have a porous Teflon backing to stop water loss from the back side while allowing full access for gas transport. The system is designed so that on overcharge, O_2 is forced out of the backside of the electrode to prevent direct contact with the active surface of the hydrogen electrode. Usually, a fibrous Zircar nonwoven ceramic separator holds the KOH electrolyte.

The sintered nickel substrate for the cathode is similar to that used for Ni-Cd and Ni-MH cell constructions. The nickel active materials are loaded into the sinter plate using either an aqueous (Bellcore) or alcoholic-based (Air Force or Pickett) electrochemical impregnation process. A 5–10% cobalt additive is deposited with nickel hydroxide to improve charge acceptance. These electrodes have a significantly longer cycle life over the standard vacuum impregnated or pasted nickel electrodes used in commercial Ni-Cd cells.

In the "pineapple" cell design, two of these nickel sinter electrodes are stacked back-to-back with asbestos cloth separators on each face of the electrodes. The cell construction is named after an exploded view of the slices (electrode components) in a can of pineapple. The recirculating stack design permits a natural concentration gradient across the stack. A thin flame-sprayed zirconium oxide coating on the inner wall of the pressure vessel is contacted by the protruding edges of the separator to redistribute the electrolyte through the diffusive gradient set up by the recirculation pattern. The hydrogen electrodes are platinum-catalyst-based electrodes with a hydrophilic platinum catalyst on the side facing the separator, and the side facing the gas screen consisting of a hydrophobic porous Teflon layer. The electrolyte is 31% by weight aqueous potassium hydroxide. The cell components are stacked on a polysulfone core that fits into the center hole of the pineapple slice components.

All nickel-hydrogen cell designs have the electrolyte contained within the porous stack components. A hydrophilic wall wick provides a path for diffusion of the electrolyte by capillary forces from wet to dry regions of the stack. The finished Ni-H_2 cell is a sealed pressure vessel with positive and negative terminals.

The second-generation cell, 5.5 in. diameter cell design, has capacities up to 350 Ah. Eagle Picher developed a cell line based on the electrochemical impregnation method for nickel electrodes employing either slurry sinter or dry sinter. The nominal thickness of the nickel electrode was 35 mils, as opposed to the 30-mil thickness of the electrodes used in the COMSAT cells. The cells were built with a back-to-back stacking arrangement that included two layers of Zircar separator, using a 31% potassium hydroxide electrolyte. They were built with about a 15% nickel pre-charge. A nylon compression seal was used to seal the terminals into the pressure vessel. A flame-sprayed zirconium oxide wall wick was included in the design to allow electrolyte gradients to equilibrate by diffusion. The cells had no free electrolyte.

A dual-stack cell design that enclosed two stacks within the same pressure vessel was developed to reduce the number of pressure vessels required in some power systems. This configuration was termed a common pressure vessel. These powered the Iridium constellation of 88 satellites (98 satellites, including orbital spares). While these batteries have functioned reasonably well, there are some difficulties in maintaining all the cells at a similar state of charge. However, these single pressure vessel batteries continue to perform well in orbit.

Bibliography

1. Anderman M, Baker C, Cohen F (1997) Proceedings of the 32nd intersociety energy conversion conference, Honolulu, Hawaii, vol 1, p 97465 Honolulu, Hawaii
2. Anderson B, Ojefors L (1979) In: Thompson JF (ed) Power sources, vol 7. Academic, London, p 329
3. Baker C, Barekatien M (2000) Proceedings of the SAE power systems conference, San Diego, 2000
4. Beauchamp RL (1971) US Patent 3,573,101
5. Beauchamp RL (1972) US Patent 2,653,967
6. Beccu K (1972) US Patent 3,669,745
7. Buzzelli E (1978) Silver-iron battery performance characteristics. In: Proceedings of the 28th power sources symposium, Electrochemical Society, Pennington, p 160
8. (2004) Cadnica sealed type Nickel-Cadmium batteries engineering handbook, Sanyo Electric Company, Osaka
9. Casellato U, Comisso N, Mengoli G (2006) Effects of Li ions on reduction of Fe oxides in aqueous alkaline medium. Electrochemica Acta 51:5669–5681
10. Cook J (1999) Separator-hidden talent Electric and hybrid vehicle technology
11. Corrigan DA, Venkatesan S, Gifford P, Holland A, Fetcenko MA, Dhar SK, Ovshinsky SR (1997) Proceedings of the 14th international electric vehicle symposium, Orlando
12. Corrigan DA, Knight SK (1996) J Electrochem Soc 143(5):1613
13. Ettel V, Ambrose J, Cushnie K, Bell JAE, Paserin V, Kalal PJ (1997) US Patent 5,700,363
14. Falk SU, Salkind AJ (1969) Alkaline storage batteries. Wiley, New York
15. Feduska W, Rosy R (1980) An advanced technology Iron-Nickel battery for electric vehicle propulsion. In: Proceedings of the 15th IECEC, Seattle, p 1192
16. Fetcenko MA, Ovshinsky SR, Chao B, Reichman B (1996) US Patent 5,536,591
17. Fetchenko MA, Venkatesan S, Ovshinsky S (1991) Proceedings of the symposium on hydrogen storage materials, batteries and electrochemistry, Electrochemical Society, Pennington, p 141
18. Fleischer A (1948) J Electrochem Soc 94:289
19. Ford FE (1994) Handbook for handling and storage of Nickel-Cadmium batteries: lessons learned, NASA Ref. Publ. 1326
20. Gutjahr MA, Buchner H, Beccu KD, Saufferer H (1973) In: Collins DH (ed) Power sources, vol 4. Oriel, Newcastle Upon Tyne, p 79
21. Halpert G (1984) J Power Sources 12:117
22. Halpert G (1990) Proceedings of the symposium on nickel hydroxide electrodes, The Electrochemical Society, Hollywood, Oct 1989, J Electrochem Soc, Pennington, pp 3–17
23. Hill TE, Rosy R, Vaill RE (1978) Performance characteristics of iron nickel batteries. In: Proceedings of the 28th power sources symposium, Electrochemical Society, Pennington, p 149
24. Hudson R, Broglio E (1980) Development of nickel-iron battery system for electric vehicle propulsion. In: Proceedings of the 29th power sources conference, Electrochemical Society, Pennington
25. Ishiwa K, Ito T, Miyamoto K, Takano K, Suzuki S (1999) Evolution and extension of NiMH technology. In: 16th international seminar on primary and secondary batteries, Ft. Lauderdale
26. Kanagawa I (1998) 15th international seminar on primary and secondary batteries, Ft. Lauderdale
27. Kruger FJ (1998) 15th international seminar on primary and secondary batteries, Ft. Lauderdale
28. Kulin TM (1998) 33rd intersociety engineering conference on energy conversion, ICECE-98-145, Colorado Springs, 2–6 Aug 1998
29. Reddy TB (2011) Linden's handbook of batteries, 4th edn. McGraw Hill, New York
30. Lindstrom O (1975) In: Collins DH (ed) Power sources, vol 5. Academic, London, p 283

31. McBreen J (1990) The nickel oxide electrode. In: White RE, Bockris JO'M, Conway BE (eds) Modern aspect of electrochemistry, vol 21. Plenum, New York, p 29
32. McRae B, Nary D (1998) Proceedings of the 38th power sources conference, pp 123–126
33. Matsumoto I, Ogawa H, Iwaki T, Ikeyama M (1988) 16th international power sources symposium, 1988, Bournmouth, England
34. Mil-B-81757 (1984) Performance specifications, batteries and cells, storage, nickel cadmium, Aircraft General Specification, Crane Division, NSWC, 1 July 1984
35. Mishima R, Miyamura H, Sakai T, Kuriyama N, Ishikawa H, Uehara I (1993) J Alloys Compd 192:176–178
36. Notten PHL, Hokkeling P (1991) J Electrochem Soc 138(7):1877
37. Notten PHL, Daams JLC, Einerhand REF (1992) Ber Bunsenges Phys Chem 96:5
38. Nickel-Cadmium batteries (2004) charge system guide, Panasonic Industrial Company, Secaucus
39. Ohta K, Matsuda H, Ikoma M, Morishita N, Toyoguchi Y (1996) US Patent 5,571,636
40. Ojefors L, Carlson L (1977) An iron-air vehicle battery. J Power Sources 2:287
41. Oshitani M, Yufu H, Takashima K, Tsuji S, Matsumaru Y (1989) J Electrochem Soc 136:6
42. Oshitani M, Yufu H (1989) US Patent 4,844,999
43. Ovshinsky SR, Dhar SK, Fetcenko MA, Young K, Reichman B, Fierro C, Koch J, Martin F, Mays W, Sommers B, Ouchi T, Zallen A, Young R (2000) 17th international seminar and exhibit on primary and secondary batteries, Ft. Lauderdale, 6–9 Mar 2000
44. Ovshinsky SR (1998) Materials research society fall meeting, Boston
45. Ovshinsky SR, Fetcenko M, Ross J (1993) Science 260:176
46. Ovshinsky SR (1991) In: Adler D, Schwartz B, Silver M (eds) Disordered materials: science and technology. Institute for Amorphous Studies Series/Plenum Publishing Corporation, New York
47. Ovshinsky SR, Corrigan DA, Venkatesan S, Young R, Fierro C, Fetcenko M (1994) US Patent 5,348,822, 14 Apr 1994
48. Pell MB, Blossom RW (1970) US Patent 3,507,699
49. Picket DF, Maloy JT (1978) J Electrochem Soc 12:1026
50. Picket DF (1974) US Patent 3,827,911
51. Picket DF (1975) US Patent 3,873,368
52. Puglisi V (2000) 17th international seminar and exhibit on primary and secondary batteries, Ft. Lauderdale 6–9 Mar 2000
53. Reichman B, Mays W, Fetcenko MA, Ovshinsky SR (1999) Electrochemical society proceedings, vol 97–16, Oct 1999
54. Salkind AJ, Venuto CJ, Falk SU (1964) The reaction at the iron alkaline electrode. J Electrochem Soc 111:493
55. Sapru SR, Reichman B, Reger A, Ovshinsky SR (1986) US Patent 4,623,597
56. Singh D, Wu T, Wendling M, Bendale P, Ware J, Ritter D, Zhang L (1998) Mater Res Soc Proc 496:25–36
57. Souza A, Carlos IA, Lopes M, Finazzi GA, de Almeida MRH (2004) Self-discharge of Fe-Ni alkaline batteries. J Power Sources 132:288–290
58. Stempel RC, Ovshinsky WR, Gifford PR, Corrigan DA (1998) IEEC spectrum 35(11):29–34
59. Takagi S, Minohara T (2000) Society of automotive engineers, 2000-01-1060, Mar 2000
60. Tuomi D (1976) The forming process in nickel positive electrodes. J Electrochem Soc 123:1691
61. van Beek JR, Donkersloot HC, Willems JJG (1984) Proceedings of the 14th international power sources symposium (1984), Waikola, Hawaii
62. Watanabe K, Koseki M, Kumagai N (1996) J Power Sources 58:23–28
63. Weininger JL (1982) In: Gunther RG, Gross S (eds) The nickel electrode, vol 82–84. Electrochemical Society, Pennington, pp 1–19
64. Weizhong T, Guangfei S (1994) J Alloy Compd 203:195–198

65. Young K, Fetcenko MA, Reichman B, Mays W, Ovshinsky SR (2000) Proceedings of the 197th electrochemical society meeting, May 2000
66. Yu X, Licht S (2007) Advances in Fe(VI) charge storage part I, primary alkaline super-iron batteries. J Power Sources 171:966–980
67. Yu X, Licht S (2007) Advances in Fe(VI) charge storage part ii, reversible alkaline super-iron batteries and nonaqueous super-iron batteries. J Power Sources 171:1010–1022

Chapter 14
Olivine Phosphate Cathode Materials, Reactivity and Reaction Mechanisms

Atsuo Yamada

Glossary

Cathode	Oxidizing agent in Battery.
Diffusion	Spontaneous movement of some species (ions in the present case).
$LiFePO_4$	A promising low-cost and safe cathode materials for LIB.
Lithium-ion battery (LIB)	A rechargeable battery with highest energy density.
Miscibility gap	A compositional region where two species cannot mix.
Olivine	A name of mineral Mg_2SiO_4 with identical structure of $LiFePO_4$.
Phase diagram	Stable phase map as a function of composition, temperature.

Definition of the Subject and Its Importance

Crystalline size has tremendous effect to the thermodynamics and kinetics in intercalation compounds. This includes diffusion/transport length, effective surface area for exchange current, surface energy, and interphase energy. The focus now is on Li_xFePO_4, where the phase changes by simple two-phase separation but with controllable miscible character by raising temperature and/or reducing crystalline size, thereby provides simple model system to rationalize thermodynamics and

This chapter was originally published as part of the Encyclopedia of Sustainability Science and Technology edited by Robert A. Meyers. DOI:10.1007/978-1-4419-0851-3

A. Yamada (✉)
Department of Chemical System Engineering, The University of Tokyo, Tokyo, Japan
e-mail: yamada@chemsys.t.u-tokyo.ac.jp

R.J. Brodd (ed.), *Batteries for Sustainability: Selected Entries from the Encyclopedia of Sustainability Science and Technology*, DOI 10.1007/978-1-4614-5791-6_14,
© Springer Science+Business Media New York 2013

electrochemistry in electrode reaction. Systematic experimental results on this issue are reviewed and summarized. Negative aspects of nanoparticle including surface poisoning by air contact and its diagnosis will be also demonstrated. As an atomic-scale phenomenon, experimental verification of one-dimensional lithium diffusion in Li_xFePO_4 will be also demonstrated.

Introduction

The crystal chemistry of phospho-olivine Li_xFePO_4 ($0 < x < 1$) has received much attention ever since it was introduced as an alternative cathode material for new generation lithium-ion batteries. The material provides several advantages over the conventional cathodes; lower cost, much safer performance, lower toxicity, and extremely flat charge-discharge profile at reasonably high potential of ca. 3.4 V versus Li/Li^+ [1, 2]. Nevertheless, at an early stage of the development, its poor intrinsic electronic conductivity, on the order of 10^{-9} $S{\cdot}m^{-1}$, and small tap density had been addressed as major problems to be solved before it can be deployed on a commercial scale [3–5]. Now, at least the first problem has been overcome by using small particle of <100 nm and the efficient formation of composites with conductive phase, typically carbon. Surprisingly high specification for the power density of more than 5 kW/kg is reported on the commercial cells with $LiFePO_4$ cathode.

One of the essential reasons for the very low conductivity measured for the stoichiometric $LiFePO_4$ seems to be in the pre-factor rather than in the activation energy. In addition, a key feature of the Li_xFePO_4 cathode is its extremely flat charge-discharge profile, typical of two-phase reaction systems. Conventional belief based on the Gibbs phase rule was that the reaction proceeds in a topotactic manner, varying the $FePO_4/LiFePO_4$ ratio throughout the range of x in Li_xFePO_4 [1]. However, carrier (mobile electron or hole) densities in the monovalent stoichiometric end members should be negligible, and this evidently accounts for the very low electronic conductivity of $<10^{-9}$ S/cm in $LiFePO_4$, despite its relatively small binding energy for polaron localization of 390–500 meV.

It is true that within the polaronic conduction mechanism and the conventional $FePO_4/LiFePO_4$ two-phase model, the number of charge carriers in monovalent systems such as $LiFePO_4$ or $FePO_4$ is determined extrinsically by impurities. Defect chemistry studies of related $Li_{1-3x}Fe_xNiPO_4$ phases suggest that cation vacancies can be accommodated in the olivine structure [6]. However, without any external doping in $LiFePO_4$, a rapid electrochemical response as well as a good rate capability has been achieved in the lab with an optimized electrode composite of pure $LiFePO_4$. These motivated a closer look at the phase change of Li_xFePO_4 at room temperature to determine if the creation of charge career is spontaneous in non-doped Li_xFePO_4 during the electrochemical charge/discharge process (i.e., does lithium or lithium defect function as a carrier-doping impurity).

Once the mechanisms of charge carrier formation are understood, the next important information is the geometry of the charge movement. Successful visualization of lithium diffusion path in olivine Li_xFePO_4 will be demonstrated together with its methodology by combination of neutron diffraction and maximum entropy method. This is a first experimental visualization of ion motion in any battery electrode.

Binary Phase Diagram

As in many binary systems, thermodynamics suggests there are two solid-solution regions outside the miscibility gap at finite temperatures. Generally, these two solid-solution regions grow as a function of temperature finalized by the continuous solid solution over the whole composition between the two end members as there is an increasing contribution of entropy to the total free energy at elevated temperatures. This is typically seen in spinodal decomposition in binary liquid systems with an inverse-parabolic phase boundary in the temperature-composition diagram. For intercalation solid electrode materials, this behavior was intensively studied in the 1980s using a lattice gas model based on mean field theory, where the attractive interaction between lithium ions is essential to induce phase separation. Onset temperature of the miscibility gap varies from ca. 270 K for metallic Sheverel $Li_xMo_6Se_8$ $(0 < x < 1)$ [7, 8] to ca. 1,200 K for insulating NASICON-type $Na_{1+x}Ti_2(PO_4)_3$ $(0 < x < 2)$ [9]. Analogous to these compounds, the intermediate phase in the Li_xFePO_4 cathode at room temperature may be dominated by the mixture of the $Li_\alpha FePO_4$ and $Li_{1-\beta}FePO_4$ phases with the narrow monophase regions, $0 < x < \alpha$ and $1-\beta < x < 1$, close to the stoichiometric end members of $LiFePO_4$ and $FePO_4$ at each end of the region.

This was first proved by the refinement of x-ray diffraction profiles measured for Li_xFePO_4 $(0 \leq x \leq 1)$ phases at room temperature [10]. In the bi-phase reaction of Li_xFePO_4 at room temperature, the orthorhombic lattice constants in each phase slightly deviate from those of the stoichiometric end components of $LiFePO_4$ and $FePO_4$. The deviation can be explained by assuming a model in which a most of the intermediate phase in the Li_xFePO_4 cathode is dominated by the mixture of $Li_\alpha FePO_4$ and $Li_{1-\beta}FePO_4$ phases (Fig. 14.1).

Size Effect Versus Surface Effect

Meethong et al. have suggested that the miscibility gap shrinks systematically with decreasing particle size and increasing temperature [11]. Such a size-dependent miscibility gap is a general phenomena observed in other two-phase intercalation reactions such as Li_xTiO_2 or MgH_x systems [12, 13]. Thermodynamically, the dependence can be explained by the increasing contribution of the elastic energy

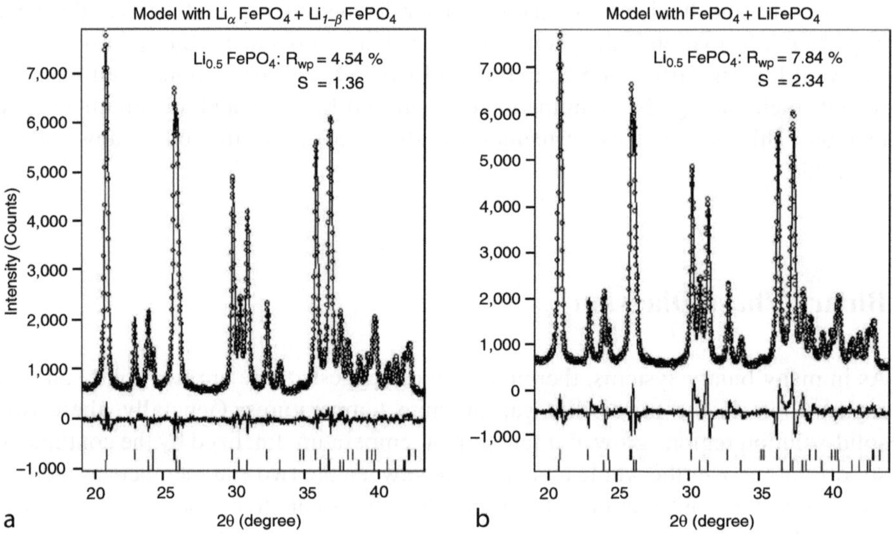

Fig. 14.1 Difference in the goodness of fit in the Reitveld refinement of $Li_{0.5}FePO_4$ under the model with (**a**) a mixture of the stoichiometric $FePO_4$ and $LiFePO_4$, and (**b**) a mixture of the nonstoichiometric $Li_\alpha FePO_4$ and $Li_{1-\beta}FePO_4$ (α = ca. 0.030, β = ca. 0.036) phases

induced along coherent two-phase interphase in the smaller particles by a factor of $r \sim A(x)/V(x)$, where r is the particle radius, $A(x)$ is the interface area, and $V(x)$ is the particle volume. The coexistence of two crystallographic phases within one particle leads to a domain wall related energy penalty, which is determined by the strain mainly due to a difference in lattice parameters. Such a strain-induced energy penalty can destabilize the two-phase coexistence in smaller particles. As a result, the energy gain due to phase separation, represented by the free energy of mixing ΔG_{mix}, will decrease for smaller particle sizes. This will gradually close the miscibility gap for decreasing particle size, which explains the enhanced solubility of Li in the smaller particles. More recently, Gibot et al. reported the inherent crystalline disorder coupled with nano-crystallization below 40 nm leads to the peculiar situation, where solid solution dominates most of the lithium compositional region in Li_xFePO_4 even at room temperature [14]. Such a thermodynamic concept have been verified in detail by Wagemaker et al. [15], including the effect of particle shape. Meethong et al. [16] related this intriguing findings to the kinetics of phase-boundary movement, where the reduction in lattice strain caused by the reduction in lattice mismatches in smaller particles was claimed to be suitable to form coherent interface between $Li_\alpha FePO_4$ and $Li_{1-\beta}FePO_4$ so as to enhance the cooperative phase-boundary movement upon electrochemical reaction with minimum formation of dislocation or clacks. This logic was regarded to give comprehensive understanding of the enhanced electrochemical activity of the nano-sized $LiFePO_4$ particle as compared to the other slower two-phase electrode system with larger mismatch.

However, attention should be paid that the surface portion of the particle abruptly increases when the size is smaller than 100 nm. A simple estimation based on the sphere particle gives the surface (one unit cell) portions to be ca. 4%, 7% and 17% for the particles with 100 nm, 50 nm, and 20 nm diameters, respectively. In such nanoscale regime, there is an increasing possibility to observe the simultaneous effects both from bulk crystal and from particle surface, which may lead to some confusion. For example, surface redox potential was calculated to be dependent on the crystallographic plane and was shown a wide dispersion over the range of 2.95–3.84 V [17]. Both of the surface impurities and the redox on the specific crystallographic facet plane would be important as the perturbation that may cause parasitic capacity above or below the bulk two-phase equilibrium potential in the voltage curve. The capacity outside the two-phase region for Li_xFePO_4 have been used as one of the indicator of the bulk solubility limit of lithium and lithium defect in $Li_\alpha FePO_4$ and $Li_{1-\beta}FePO_4$ and, then, may need careful reinvestigation.

More importantly, the side reaction upon ambient exposure is known to be pronounced for smaller particle, leading to a spontaneous lithium extraction and surface oxidation [18, 19]. Before going to any further discussions, such a side reaction should be removed; then, this phenomenon is explained in the next section.

Air Poisoning

XRD has been carried out to investigate the effect of different atmosphere exposure conditions. Table 14.1 summarizes the lattice dimensions. In all cases, the diffractograms display a single phase with an ordered olivine structure indexed to the orthorhombic Pnmb space group. The Rietveld refinement permits to approximate the olivine crystallites size to 75 nm, which is verified on the SEM image of the powder presented.

Systematic decrease of the cell volume upon atmosphere exposure was clearly observed together with the corresponding changes in lattice parameters. It begins to be visible even after a short time (1 h) of ambient air exposure at room temperature and become more evident after longer time (10 days) or higher temperature (120°C) exposure. Water and oxygen should be included in this side reaction as the effect is significantly suppressed by using dry O_2 20%/N_2 80% and become negligible by using dry CO_2. These volume changes can be linked to the solid-solution compositional domain in $Li_{1-x}FePO_4$ ($x < \beta$) [10, 11, 19, 20], in which olivine can accommodate the corresponding lithium defect without first-order phase transformation.

Figure 14.2 demonstrates the linear relative dependence of the *a* and *b* lattice constants following the Vegard's law in the hypothesis of a solid solution between $LiFePO_4$ prepared with no contact with air and $FePO_4$. These results show that some lithium is certainly extracted when $LiFePO_4$ is put in contact with atmosphere. The Vegard's law gives 2.3% of missing lithium after ambient air exposure

Table 14.1 Parameters and cell volumes of $LiFePO_4$ calculated from XRD refinement after different atmosphere exposure conditions

Stage condition		a(Å)	b(Å)	c(Å)	V(Å3)	x
Dry air standard	Room temperature	10.3308(2)	6.00813(11)	4.69444(10)	291.378(6)	0
Ambient air	Room temperature,1 h	10.3275(2)	6.00656(13)	4.69271(12)	291.1040(12)	0.007
	Room temperature,1 day	10.3254(2)	6000573(11)	4.69329(11)	291.037(10)	0.011
	Room temperature, 21 days	10.3227(2)	6.00428(15)	4.69346(14)	290.903(14)	0.017
	120°C, 1 day	10.3188(2)	6.00303(15)	4.69451(14)	290.79(4)	0.023
Dry air (N_2/O_2)	Room temperature,1 day	10.3294(4)	6.0072(2)	4.6932(2)	291.22(2)	0.004
	120°C, 1 day	10.3254(3)	6.0054(2)	4.69373(19)	291.048(19)	0.012
Dry CO_2	Room temperature,1 h	10.3302(4)	6.0078(2)	4.6936(2)	291.29(2)	0.001
	120°C, 1 day	10.3290(3)	6.00716(18)	4.696361(17)	291.231(16)	0.004

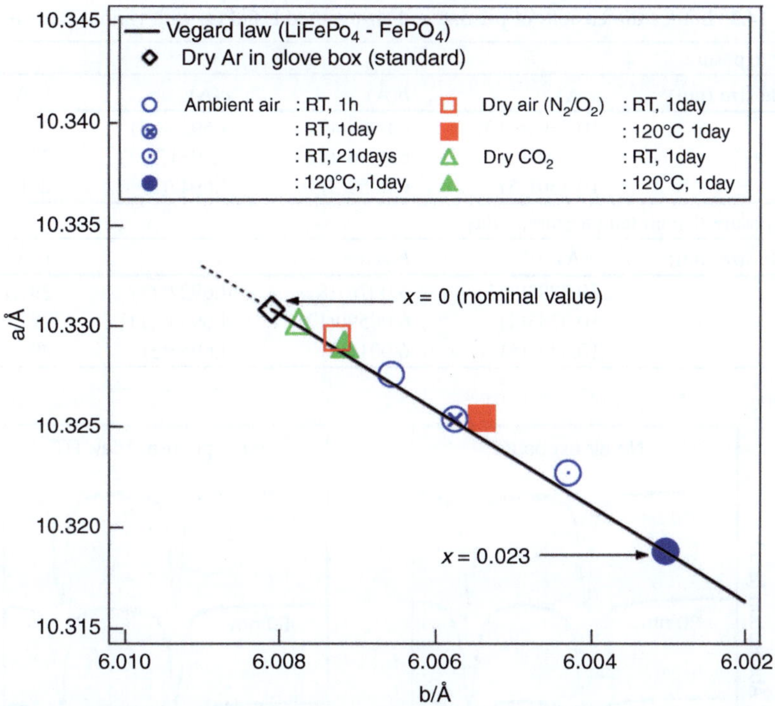

Fig. 14.2 Lattice constants change for $Li_{1-x}FePO_4$ after the different air contact experiments compared to Vegard's law (*arrow*). The values for x are calculated using Vegard's law and $FePO_4$ cell parameters ($a = 9.8142$ Å and $b = 5.7893$ Å)

for 24 h at 120°C. It has to be noted that the lattice constants obtained by avoiding air contact are very close to the ones reported for pure, much larger particles [21], which present less possibilities of reactivity with air due to smaller solid-solution domain and smaller exposed surface. It tends to prove that the sample has been well protected during the experiment and that possible air pollution during storage in glove box has been minimized.

For a subsequent study of size-dependent phase diagram, effort was taken to eliminate such fluctuant factor in the pristine samples because it would degrade the quality of any crystallographic and/or electrochemical data and, hence, mislead the subsequent quantitative discussions. The controlled pristine $LiFePO_4$ samples used in this study are free from such poisoning by air exposure, and the particle size is the only varied parameter. The samples used in this study are free from any disorder as the lattice dimensions are almost identical to the single crystal values ($a = 10.332(4)$ Å, $b = 6.010(5)$ Å, $c = 4.692(2)$ Å, $V = 291.4(3)$ Å^3) [21] and showed negligible dependence on the particles size, while air-exposed samples suffered from significant volume contraction particularly on smaller particles as in Table 14.2. With the same trend in variations of the lattice dimensions, Mössbauer

Table 14.2 Lattice dimensions of pristine and air-exposed $LiFePO_4$ with various particle sizes

Particle size (nm)	*a*(Å)	*b*(Å)	*c*(Å)	*V*(Å^3)
No air exposure				
200	10.33046(13)	6.00846(7)	4.69238(7)	291.257(7)
80	10.3303(2)	6.00764(12)	4.69313(11)	291.259(11)
40	10.3301(5)	6.0052(3)	4.6947(3)	291.23(2)
Air exposure: Room temperature, 1 day				
Particle size (nm)	*a*(Å)	*b*(Å)	*c*(Å)	*V*(Å^3)
200	10.32839(15)	6.00767(8)	4.69228(8)	291.157(8)
80	10.3253(2)	6.00590(12)	4.69335(11)	291.047(11)
40	10.3223(5)	6.0015(3)	4.6955(3)	290.88(2)

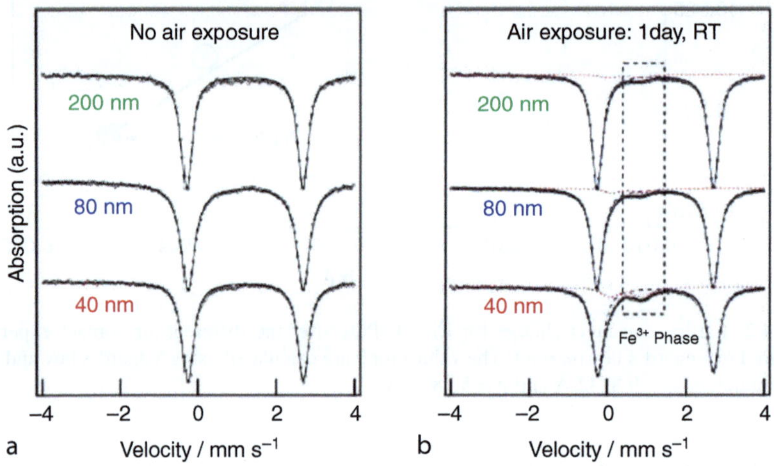

Fig. 14.3 Mössbauer spectra measured for pristine (**a**) and air-exposed (**b**) $LiFePO_4$ with various particle sizes of 40, 80, and 200 nm

spectra for the controlled pristine samples were analyzed by single Fe^{2+} doublet inherent to $LiFePO_4$, while parasitic Fe^{3+} sub-signal was always observed for the air-exposed samples increasing its intensity for smaller particles, as shown in Fig. 14.3.

Careful Verification of Phase Diagram

Electrochemical Approach

The first step for the analysis is the accurate determination of the open-circuit voltage (OCV) profile versus lithium composition *x* for three samples with different

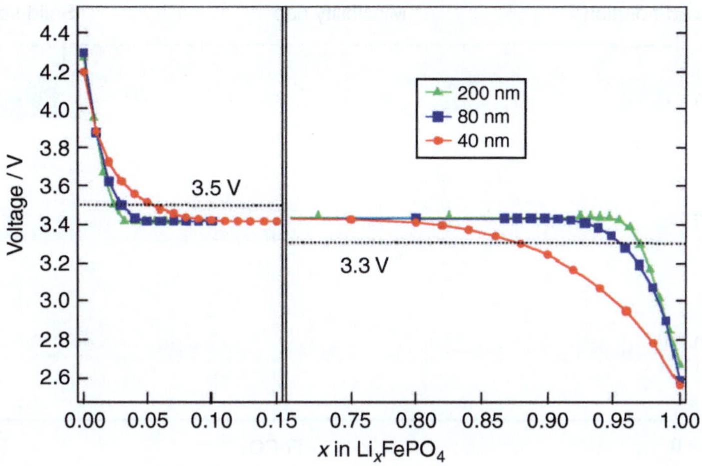

Fig. 14.4 Open-circuit voltage (OCV) curves measured for Li_xFePO_4 with various mean particle sizes of 200, 80, and 40 nm

particle sizes of 40 nm, 80 nm and 200 nm. Initially, the cell was charged galvanostatically at C/20 rate at 25°C followed by maintenance at 4.5 V for 24 h forming $FePO_4$. Capacity close to the theoretical capacity of 170 mAh/g was confirmed for all samples. Then, it was discharged under a rate of C/20 for 12 min, which corresponds to ca. 1% of the theoretical capacity. The open-circuit voltage at 25°C was measured after 24 h. Again, the cell was discharged under a rate of C/20 for 12 min. These procedures were repeated several times until the open-circuit voltage at 25°C become constant at ca. 3.42 V in the biphasic region. A similar experiment was performed starting from $LiFePO_4$ toward the anodic direction and confirmed the identical OCV values at same lithium compositions. Size-dependent OCV curves for Li_xFePO_4 are given in Fig. 14.4, where the capacity observed below or above the two-phase equilibrium potential becomes larger as particle size decreases.

Structural Approach

Two target lithium compositions of x in Li_xFePO_4 ($x = 0.6$, $x = 0.93$) were chosen and were adjusted by the chemical way, allowing more homogeneous Li^+ incorporation over electrochemical reactions. Based on the room temperature phase diagram reported previously [10, 19, 20], the composition near an end member, $Li_{0.93}FePO_4$, was selected as an experimental criterion phase since it would be around the phase transition zone between the two-phase region and the single-phase region in the $FePO_4 - LiFePO_4$ binary phase diagram, where the two-phase (larger particle) to single-phase (smaller particle) transformation is expected as a function of the particle size. The composition $x = 0.6$, $Li_{0.60}FePO_4$, was

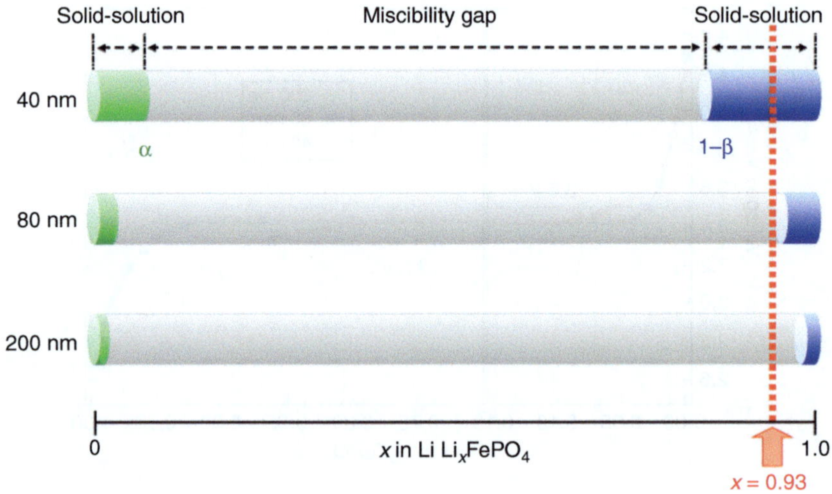

Fig. 14.5 Schematic derivation of the size-dependent $FePO_4$–$LiFePO_4$ binary phase diagram using the XRD data for $Li_{0.6}FePO_4$

selected as a reference phase because it should be composed of a mixture of two intermediate phases, analysis of which should lead to the definitive positions of the solid solubility limits, α and β, in $Li_{\alpha}FePO_4$ and $Li_{1-\beta}FePO_4$. In order to set the lithium composition in Li_xFePO_4 to $x = 0.6$ or $x = 0.93$, all of the lithium in pristine $LiFePO_4$ was first removed to form isostructural $FePO_4$ using a strong oxidizing agent NO_2BF_4 in acetonitrile. Then, $FePO_4$ was reacted with the specific amount of mild reducing agent LiI in acetonitrile to form Li_xFePO_4 with $x = 0.6$ or $x = 0.93$. The chemical treatment was performed for each pristine $LiFePO_4$ sample with three different particle sizes of 40, 80, and 200 nm for the subsequent high-resolution synchrotron x-ray diffraction.

In order to overview the change in the room-temperature phase diagram with decreasing particle size, the lattice parameters refined for $Li_{0.6}FePO_4$ (two-phase mixture of $Li_{\alpha}FePO_4$ and $Li_{1-\beta}FePO_4$) were analyzed. For all particle sizes, $Li_{0.6}FePO_4$ is evidently in two-phase state as typically seen in the two diffraction peaks clearly observed at around 13.5° and 14.5°, corresponding to the (200) reflection of $Li_{1-\beta}FePO_4$ and the (200) reflection of $Li_{\alpha}FePO_4$, respectively. All of the lattice parameters refined for the two separated phases in $Li_{0.6}FePO_4$ deviates from those of stoichiometric $FePO_4$ and $LiFePO_4$, and can be explained by assuming partial lithium occupancy α to $FePO_4$ and partial lithium deficiency β from $LiFePO_4$, as well as Vegard's low along $FePO_4$–$LiFePO_4$ solid solution.

Schematic derivation of the size-dependent $FePO_4$–$LiFePO_4$ binary phase diagram using the XRD data for $Li_{0.6}FePO_4$ is illustrated in Fig. 14.5, which highlights the fact that the solid solubility limits α, β, in $Li_{\alpha}FePO_4$, and $Li_{1-\beta}FePO_4$ systematically increases along with the reduction in particle size. Most probably, these systematic variations originate from the elastic thermodynamic effects on the miscibility gap as discussed in the various other binary systems. The solid–solid phase transitions are thermodynamically first order and involve a change in enthalpy (ΔH), which is highly related to the particle size as depicted by $\Delta H = \Delta H_{intinsic} + \Delta H_D$, where ΔH_D represents the particle size contribution of opposite sign [15]. If the particle size decreases, the Gibbs free energy (ΔG) of mixing would be decreased by the increasing contribution of the penalty in elastic energy at the two-phase boundary, resulting in extension of the single-phase regions near the end members on the composition phase diagram.

The extended solid solution has been claimed to be beneficial both for the charge movement and the subsequent cooperative phase-boundary movement because the diffusion/hopping pre-factor along the one-dimensional diffusion path is increased while the lattice mismatch along the two-phase boundary is reduced. However, experimental evidences given so far for the existence of the small solid-solution compositional domain are not direct ones and led some controversy whether they really originate from the bulk effect or from the artifact by the enhanced surface effect in nanoregime [17]. Then, isolation of the solid-solution phase would give more definitive and straightforward experimental evidence for the size-dependent miscibility gap model.

Toward this end, the focus is on the sample with $x = 0.93$, which is around the phase transition compositional zone between the two-phase region and the single-phase region in the $FePO_4$–$LiFePO_4$ binary phase diagram (see Fig. 14.5). At the composition of $x = 0.93$, the two-phase (larger particle) to single-phase (smaller particle) transformation is expected as a function of the particle size as easily speculated from the phase behavior along the vertical compositional line at $x = 0.93$ in Fig. 14.5. The changes in the lattice parameter a are summarized in Fig. 14.6 together with the low-angle peaks of $Li_{0.93}FePO_4$ indexed as $(200)_T$ (lithium-rich triphylite phase) and $(200)_H$ (lithium-poor heterosite phase) at different particle sizes. The intensity of $(200)_H$ is gradually decreased when the particle size is reduced and finally disappeared for the 40 nm particle. More importantly, the deviation of the lattice constant a at $x = 0.93$ from those of the phase-separated $Li_{0.6}FePO_4$ was observed only for 40 nm particle. Furthermore, it is noteworthy that the lattice constant a at $x = 0.93$ for 40 nm particle positions strictly on the line of Vegard's low assuming the complete solid-solution of $FePO_4$ and $LiFePO_4$. The same behavior was identified also for the b-, c-axes. Clearly, the immiscible–-miscible transition is induced by the particle size reduction at the identical lithium composition $x = 0.93$, and the solid-solution in Li_xFePO_4 is isolated at $x = 0.93$ in the 40 nm particle, details of which are described below.

As for the 200 nm particle at $x = 0.93$, two independent peaks were clearly observed as indexed $(200)_T$ and $(200)_H$, respectively, which indicates that the $Li_{0.93}FePO_4$ phase consists of two intermediate phases, $Li_\alpha FePO_4$ and $Li_{1-\beta}FePO_4$. This is a reasonable result because the lithium composition for the immiscible–-miscible transition was estimated to be $1 - \beta_{200} = 0.98$, which is much larger than the present lithium composition, $x = 0.93$. The lattice parameters are almost identical with those of $Li_{0.6}FePO_4$, which also reveal that the $Li_{0.93}FePO_4$ with a particle size of 200 nm is inside the miscibility gap at room temperature.

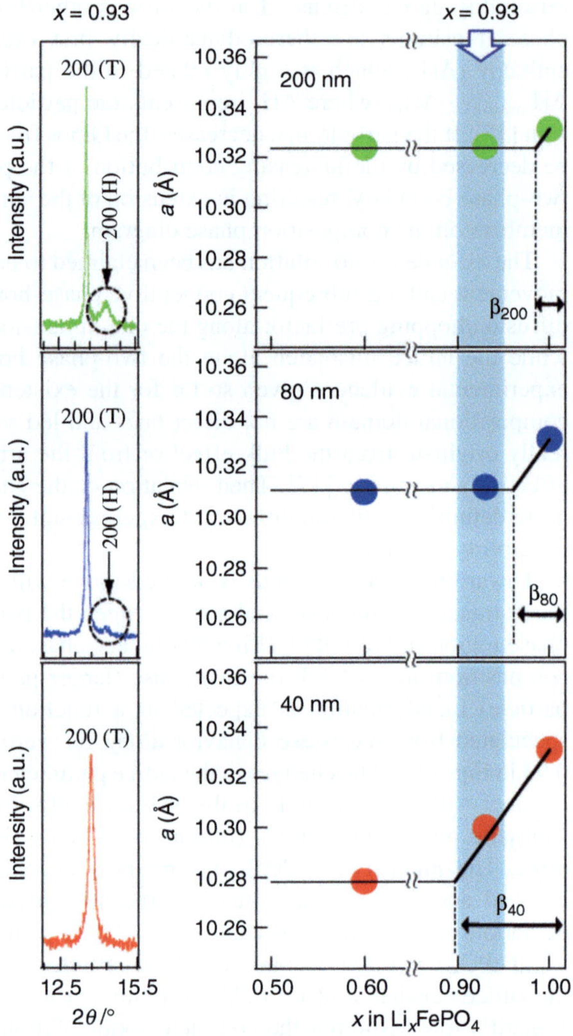

Fig. 14.6 *Left*: magnified view of the 200 peak measured for $Li_{0.93}FePO_4$ with different particle sizes of 200, 80, and 40 nm at room temperature. *Right*: derivation of changes in *a*-axis lattice parameter as a function of lithium composition *x* for different particle sizes

When the particle size is reduced to 80 nm, the phase at $x = 0.93$ is still in the two-phase separation, but the intensity of the $(200)_H$ peak has significantly decreased in the synchrotron XRD profile, which can be explained by the reduction of the $Li_{\alpha}FePO_4$ phase fraction in the two phase mixture of $Li_{\alpha}FePO_4$ and $Li_{1-\beta}FePO_4$, following the size-dependent miscibility gap model [11]. The estimated lithium composition for the immiscible-miscible transition, $1-\beta_{80}$, has now decreased to 0.96 and becomes closer to the present lithium composition, $x = 0.93$. The $Li_{0.93}FePO_4$ phase with an average particle size of 80 nm is in close to the transition zone from immiscible to miscible state and contains smaller amount of $Li_{\alpha}FePO_4$ at room temperature. The lattice constants are still identical with those of bi-phased $Li_{0.6}FePO_4$; the $Li_{0.93}FePO_4$ with a particle size of 80 nm is still inside

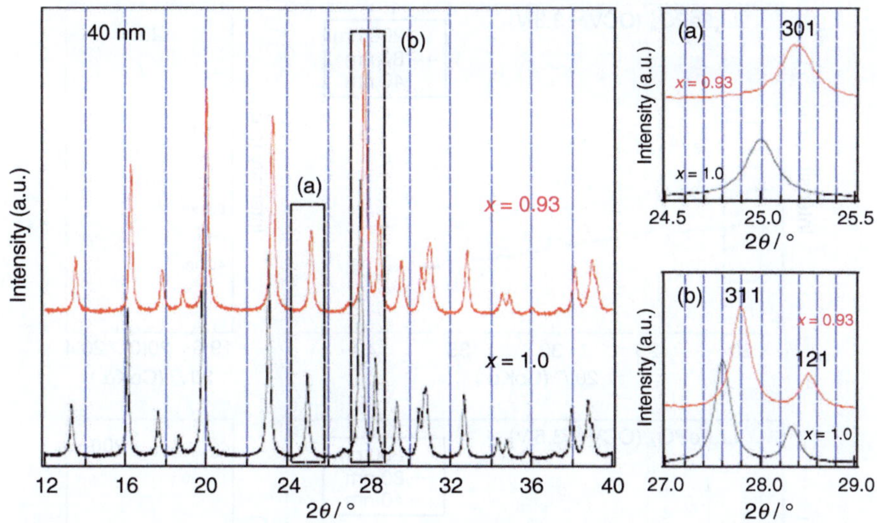

Fig. 14.7 Comparison of the synchrotron x-ray diffraction profiles measured for stoichiometric $LiFePO_4$ and the isolated solid solution of $Li_{0.93}FePO_4$ with an identical particle size of 40 nm with magnified presentations of 301, 311, and 121 as the typical peaks

the miscibility gap at room temperature but in closer access to the immiscible––miscible transition zone.

At the 40 nm particle size with $x = 0.93$, the $(200)_H$ peak has disappeared, and the discrepancies in the lattice parameters from the two-phase values are prominent. Instead, the lattice constants are exactly on the line of Vegard's low assuming the complete solid solution between $FePO_4$ and $LiFePO_4$. This behavior proves that the $Li_{0.93}FePO_4$ phase with 40 nm size is in the single-phase solid-solution range and allows to conclude that a solid solution of Li_xFePO_4 could be isolated as a single phase with the solid solubility of $\beta_{40} > 0.1$.

For a direct comparison between the stoichiometric $LiFePO_4$ and the isolated solid solution of $Li_{0.93}FePO_4$ with an identical particle size of 40 nm, synchrotron x-ray diffraction profiles of the two samples are shown in Fig. 14.7 with some magnified presentations of the typical peaks. Peak shifts are visually evident and large enough to confirm the wide lithium compositional range of at least 7% for solid solution. The shift observations without appearing of any second phase further support the room-temperature isolation of solid solution in Li_xFePO_4.

Consistency Between Electrochemical and Structural Data

For a further confirmation of the inverse relationship between the solid solubility and particle size, the composition x has been electrochemically controlled by galvanostatically charging the cells in a potential range of 2.0–4.5 V and then

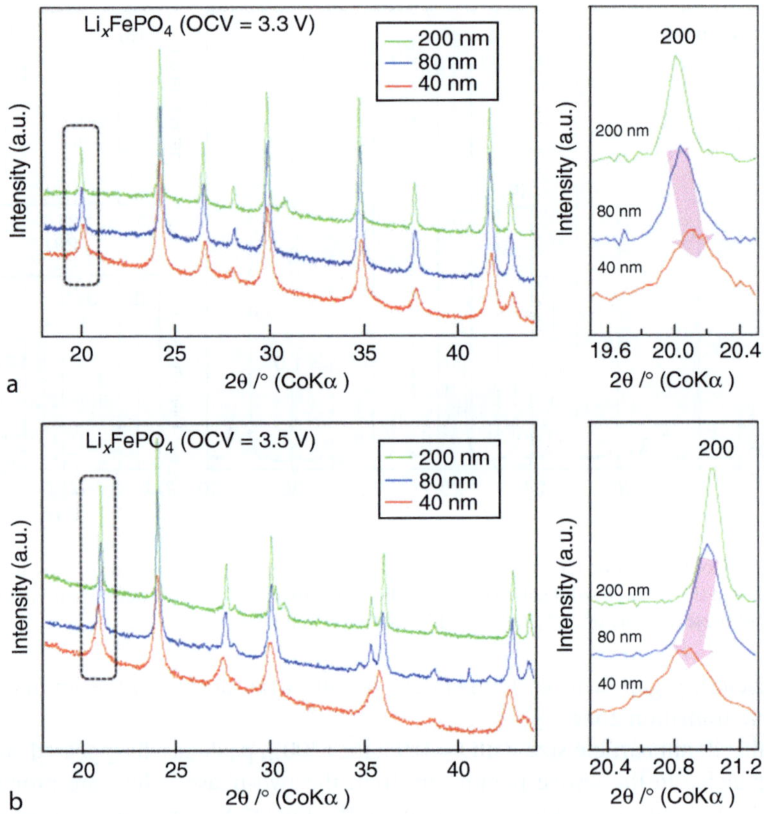

Fig. 14.8 X-ray diffraction profiles for Li_xFePO_4 electrode with various mean particle sizes of 200, 80, and 40 nm, equilibrated at 3.3 V (**a**) and 3.5 V (**b**), with magnified presentation of the 200 diffraction peaks

potentiostatically discharging them at a constant potential below the equilibrium potential at 3.30 V, which corresponds to $Li_{0.98}FePO_4$ (200 nm), $Li_{0.96}FePO_4$ (80 nm), and $Li_{0.88}FePO_4$ (40 nm) in OCV curves in Fig. 14.4. Similar experiment was performed with higher constant potential above equilibrium potential at 3.50 V, which corresponds to $Li_{0.02}FePO_4$ (200 nm), $Li_{0.03}FePO_4$ (80 nm), and $Li_{0.06}FePO_4$ (40 nm) in OCV curves in Fig. 14.4. The cell was dismantled in an Ar glove box, and then the cathode electrode was washed in dimethyl carbonate (DMC) for subsequent XRD measurements.

Size-dependent peak shift is clear both for samples in equilibrium at 3.30 V and at 3.50 V as shown in Fig. 14.8. The lattice parameters were refined for all samples by Rietveld analyses and are summarized in Table 14.3. The extent of lithium nonstoichiometry x in Li_xFePO_4 were calculated for each samples assuming the Vegard's low between $FePO_4$ and $LiFePO_4$, showing excellent agreement with the electrochemical estimation of x, i.e., x (Vegard's low) = x (OCV), for all of the

Table 14.3 Lattice dimensions of Li_xFePO_4 electrode with various mean particle sizes of 200, 80, and 40 nm, equilibrated at 3.3 V (a) and 3.5 V (b). Lithium nonstoichiometry from $x = 1$ and $x = 0$ estimated by two different method, Vegard's law and OCV curves, show excellent agreement. The extent of the lithium nonstoichiomery is enhanced as mean particle size decreases

Li-rich (Triphylite, OCV = 3.3 V)						
Particle size (nm)	*a*(Å)	*b*(Å)	*c*(Å)	V(Å^3)	*x* (Vegard's law)	*x* (OCV)
200	10.3224(4)	6.0039(2)	4.6909(2)	290.67(2)	0.975	0.98
80	10.3184(7)	6.0016(4)	4.6904(4)	490.46(3)	0.965	0.96
40	10.2790(13)	5.9525(6)	4.7054(7)	289.35(6)	0.891	0.88
Li-poor (Heterosite phase, OCV = 3.5 V)						
Particle size (nm)	*a*(Å)	*b*(Å)	*c*(Å)	V(Å^3)	*x* (Vegard's law)	*x* (OCV)
200	9.8159(9)	5.7891(4)	4.7814(4)	271.70(4)	–	0.022
80	9.8292(19)	5.7994(5)	4.7792(11)	272.43(5)	0.030	0.029
40	9.8486(13)	5.8138(6)	4.7831(7)	273.90(6)	0.074	0.056

six cases with three different particle sizes (40, 80, 200 nm) and two different OCV (3.30, 3.50 V). This provides firm experimental evidence that the capacity observed below or above the two-phase equilibrium potential is not caused by some impurities or the potential distribution dependent on the crystalline surface but largely due to the bulk lithium nonstoichiometry at least for particle sizes larger than 40 nm.

Further inspections on the solid-solution compositional domain were performed for the electrochemically adjusted Li_xFePO_4 with a particle size of 40 nm, which allow largest miscibility in the present study. The potential was set to 3.8, 3.6, 3.5, 3.3, 3.2, 3.1, 3.0, and 2.8 V versus Li^+/Li. They correspond to the lithium composition of 0.02, 0.04, 0.06, 0.88, 0.91, 0.93, 0.95, and 0.95, respectively, with an eye to the OCV data shown in Fig. 14.9 and are inside the solid-solution domains. All the samples were subjected to the X-ray diffraction experiments with the subsequent Rietveld analysis, and the plots of lattice parameters versus lithium compositions are given in Fig. 14.9. All the data are on the line of Vegard's Law of $FePO_4$–$LiFePO_4$ solid solution without any systematic deviation and further support the two main conclusions of this paper with no ambiguity: (1) Validity of the size-dependent miscibility gap model in Li_xFePO_4, and (2) Bulk miscibility dominates the electrochemical behavior of Li_xFePO_4 at least for particle sizes larger than 40 nm.

Thermodynamic Approach

The thermodynamic experimental evidences for the solid solution outside of the miscibility gap are addressed now [20]. Microcalorimetry can directly measure entropy changes versus x at fixed temperature T during the electrochemical reaction. Observed heat flow P is related to entropy change ΔS by $P = -IT\Delta S/nF$, where I is the input current and F is the Faraday constant. The classical entropy of mixing $S_{cf}(x)$ can be calculated for a fully disordered lattice of lithium atoms and vacant

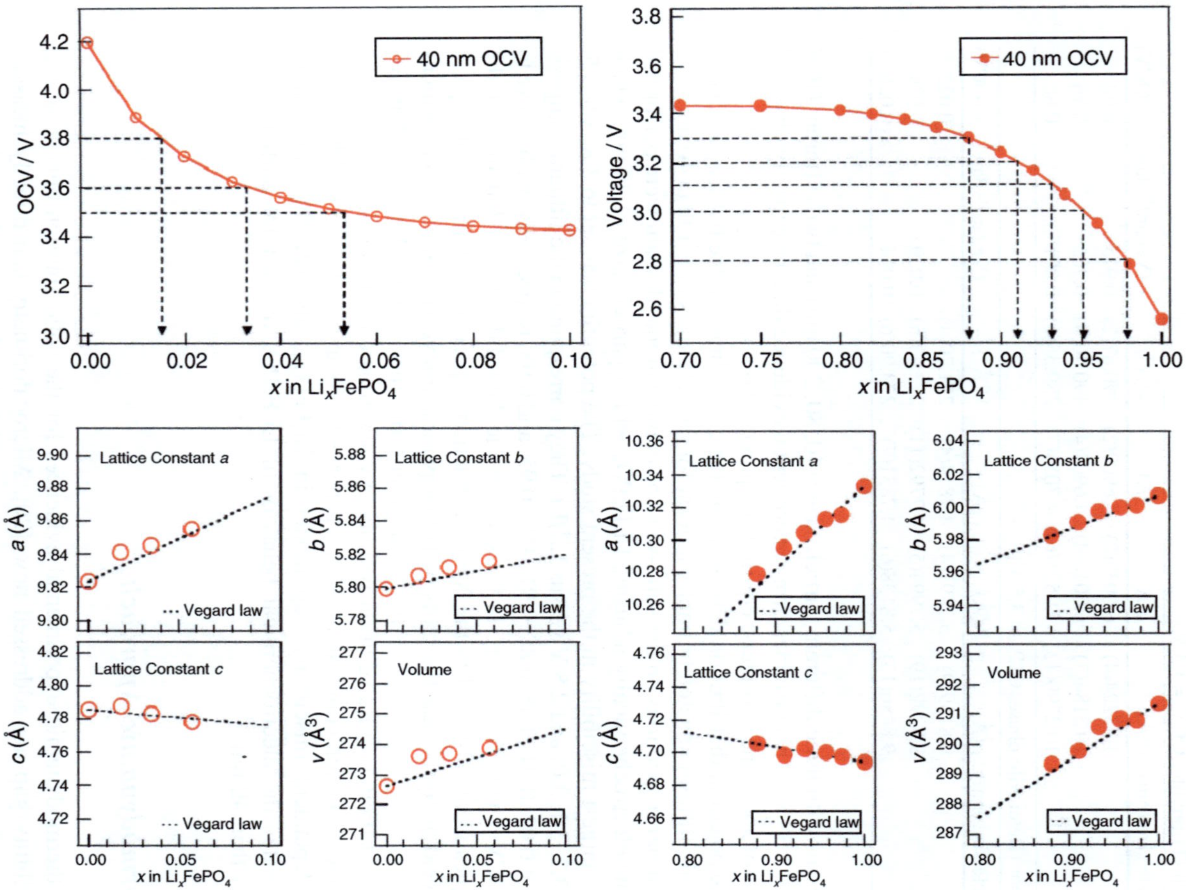

Fig. 14.9 Open-circuit voltage curve and change in lattice dimensions above (**a**) and below (**b**) the two-phase equilibrium potential. Lithium composition was adjusted potentiostatically (equilibrium condition: <0.1 μA) and estimated by the OCV curves as indicated by dashed lines. The linear lines along the plot of lattice constants show the Vegard's law between $FePO_4$ and $LiFePO_4$

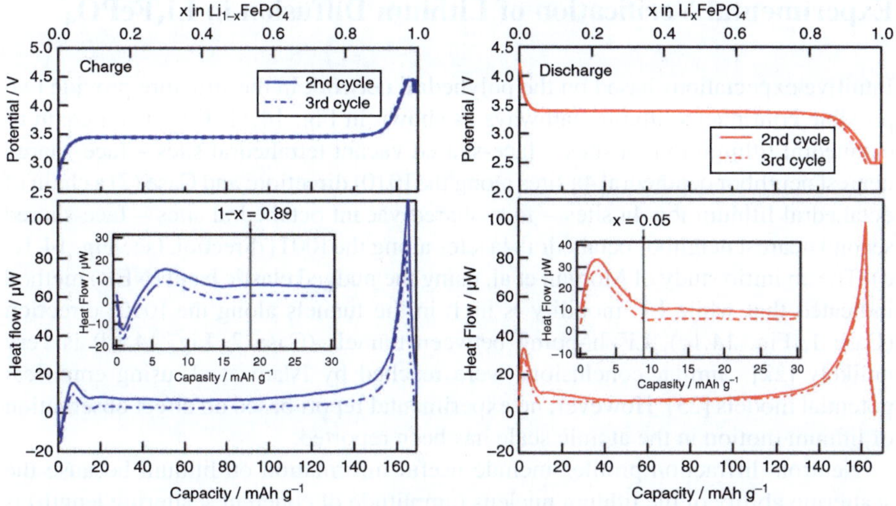

Fig. 14.10 Calorimetric measurements for a 2032 type coin cell for charging and discharging processes of Li_xFePO_4. A clear anomaly typical for disordered region exists at the beginning of both charge and discharge. Vertical lines are the monophase/bi-phase boundary estimated from the Li site occupancy number shown in Table 14.1. Except for these two monophase regions, heat flow exhibited almost constant exothermic behavior with no significant anomalies, explained by a simple two-phase reaction scheme with a continuous phase-boundary movement without any structural or electronic phase transition. Abrupt increase of the exothermic heat flow at the end of charge/discharge processes in Li_xFePO_4 is due to the increasing polarization and hinders the thermal anomaly for one of the two monophase regions

sites, and heat flow varies linearly with $\Delta S_{cf}(x) = k_B \ln[1/x]$, where lithium atom–vacancy disorder when $0 < x < 1$. Although this expression is singular at $x = 0$ and $x = 1$, an important overall trend demonstrates that the heat flow changes symmetrically from negative at initial composition $x = 0$ to positive at final composition $x = 1$. Typical anomalies can be clearly seen at the very beginning of charging $LiFePO_4$, as shown in Fig. 14.10. However, the endothermic region is narrow compared to the subsequent anomalous exothermic region. Similar anomalies were also detected at the very beginning of $FePO_4$ discharge, but again with a much narrower, negligible endothermic region. Such strong asymmetries for endothermic/exothermic behavior may be due to the dynamic character of calorimetric measurements. These two thermally anomalous compositional regions close to $LiFePO_4$ and $FePO_4$ in Li_xFePO_4 show good agreement with the values of α and $1 - \beta$ in $Li_\alpha FePO_4$ and $Li_{1-\beta}FePO_4$ obtained by structural data also with the region in which the deviation of open-circuit voltage from the constant value of V_{oc} = 3.42 V is evident (see Fig. 14.4). Furthermore, these initial anomalies in entropy upon charge/discharge were not observed when measurement was initiated at any point inside the miscibility gap. Therefore, it is likely they represent the change in configurational entropy in solid solution outside the miscibility gap in Li_xFePO_4.

Experimental Verification of Lithium Diffusion in Li_xFePO_4

Intuitive expectations based on the polyhedral network in the structure provide two possible continuous lithium pathways as shown in Fig. 14.11: Case (1) a chain of octahedral lithium ion 4a sites – face-shared vacant tetrahedral sites – face-shared nearest neighbor octahedral 4a sites along the [010] direction; and Case (2) a chain of octahedral lithium ion 4a sites – face-shared vacant octahedral sites – face-shared second nearest neighbor octahedral 4a sites along the [001] direction (see Fig. 14.1c, d). The ab initio study of Morgan et al. using the nudged elastic band (NEB) method indicated that while Li^+ mobility is high in the tunnels along the [010] direction (Case 1, Fig. 14.1c), Li^+ hopping between tunnels (Case 2, Fig. 14.1d) is very unlikely [22]. Similar conclusions were reached by Islam et al. using empirical potential models [23]. However, no experimental report based on direct observation of lithium motion in the atomic scale has been reported.

Neutron diffraction profiles include useful information on lithium because the scattering ability of the lithium nucleus (amplitude of coherent scattering length) is relatively large and independent of scattering vector $Q = 4\pi \sin\theta/\lambda$. This nature is

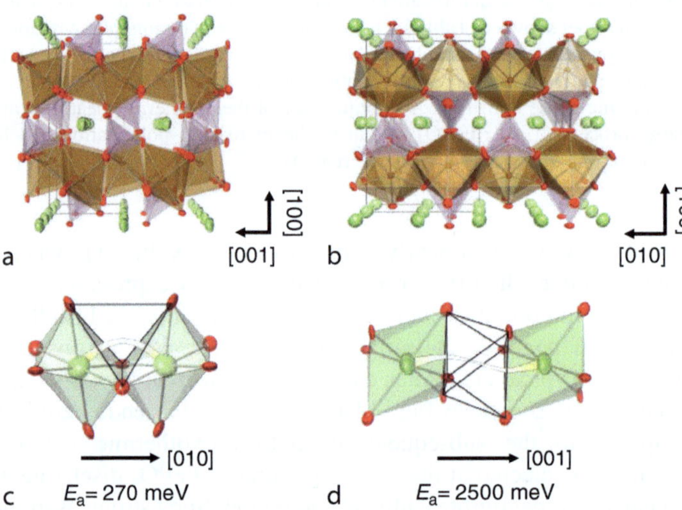

Fig. 14.11 The crystal structure of $LiFePO_4$ projected along the (**a**) [010] and (**b**) [001] directions; (**c**) and (**d**) Possible Li diffusion pathways are parallel to these directions. The structure can be described as a distorted hexagonal close-packed (hcp) oxygen sub-array, in which Li, Fe, and P atoms occupy interstitial sites to form (1) corner-sharing FeO_6 octahedra that are nearly coplanar to form a distorted two-dimensional square lattice perpendicular to the *a* axis, (2) edge-sharing LiO_6 octahedra aligned in parallel chains along the *b* axis, and (3) tetrahedral PO_4 groups connecting neighboring planes or arrays. The green, brown, purple, and red ellipsoids indicate Li, Fe, P, and O atoms, respectively. Intuition leads to two possible lithium migration paths: (**c**) along the [010] direction through face-shared vacant tetrahedral sites; and (**d**) along the [001] direction through face-shared octahedral sites. One-dimensional diffusion along the [010] direction (**c**) was predicted by the computational method [22, 23]

amenable for detailed analysis of the thermal motion of the lithium nucleus in contrast to the negligible X-ray scattering ability of lithium or lithium ions with only two or three electrons. To enhance this advantage $^7LiFePO_4$ was prepared using ^{7}Li-enriched Li_2CO_3 as the raw material. This greatly improved data quality since the natural existence of 7.5% ^{6}Li causes: (1) ca. 2×10^5 larger absorption, (2) ca. 15% decrease of coherent scattering length, and, hence, (3) larger incoherent scattering.

Figure 14.12a demonstrates the Rietveld refinement pattern for time-of-flight (TOF) neutron diffraction data measured at room temperature for $^7LiFePO_4$. Fitting was satisfactory ($R_{wp} = 2.66\%$, $R_f = 0.46\%$, $S = 1.34$) with accurately refined atomic positions as well as anisotropic atomic displacement parameters for all atoms under the classical harmonic oscillation model.

Important information is included in the anisotropic atomic displacement parameters for lithium, which determine the overall anisotropy of the thermal vibration by the shape of ellipsoid. Green ellipsoids shown in Figs. 14.11a, c and 13 represent the refined lithium vibration. The preferable direction of thermal displacement is toward the face-shared vacant tetrahedra. The expected curved one-dimensional continuous chain of lithium atoms is drawn in Fig. 14.13 and is consistent with the computational prediction by Morgan et al. [22] and Islam et al. [23]. Such anisotropic thermal vibrations of lithium were further supported by the Fourier synthesis of the model-independent nuclear distribution of lithium (see Fig. 14.14).

The subsequent experimental direction was significant enhancement of lithium motion by introducing a large number of lithium defects at elevated temperatures to show how the motions of Li atoms evolve from vibrations to diffusion. This was possible with respect to the phase diagram of Li_xFePO_4 reported in the literature, which is shown in the inset of Fig. 14.12. Delacourt et al. [24] and Dodd et al. [25] confirmed the small miscibility at low temperatures [10, 20], but also reported an unusual eutectoid point at ca. 500 K where the solid-solution phase emerges at approximately $x = 0.6$. At temperatures higher than 570 K, solid solution dominates all compositions [24, 25]. Rapid hopping and delocalization of lithium ions coupled with small polarons were confirmed by motional narrowing of Mössbauer spectra in the solid-solution phases formed at elevated temperatures [26, 27].

Based on the above binary phase diagram and corresponding lithium dynamics, the composition and temperature of the choice for further neutron diffraction study was $x = 0.6$ and $T = 620$ K, as shown in Fig. 14.12b. A solid-solution phase of $^7Li_{0.6}FePO_4$ was formed simply by mixing the end members, $^7LiFePO_4$ and $FePO_4$, in a 6:4 ratio and heating to 620 K in vacuum. Significant difference of the neutron diffraction pattern with that measured at room temperature (0.6 $LiFePO_4$ + 0.4 $FePO_4$) is shown in supplemental Fig. 14.15. Temperature-dependent X-ray diffraction profiles were measured in a helium-sealed cell prior to the neutron diffraction experiment and confirm the formation of a single phase of compositionally homogeneous $Li_{0.6}FePO_4$ solid solution only in the very narrow temperature range of 600–630 K, as shown in supplemental Fig. 14.16.

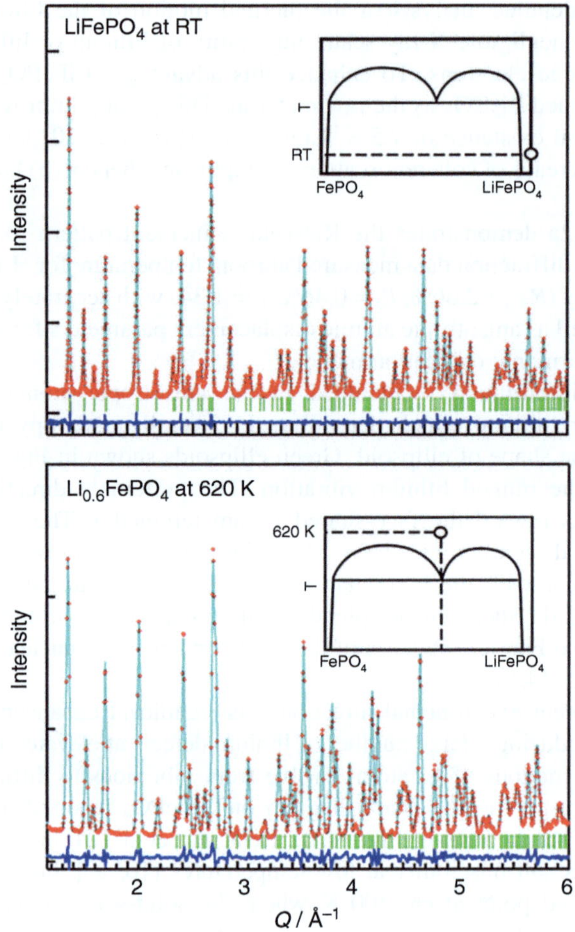

Fig. 14.12 Rietveld refinement patterns for: (**a**) the time-of-flight neutron diffraction profile measured for $LiFePO_4$ at room temperature; and (**b**) angle dispersive neutron diffraction profile measured for $Li_{0.6}FePO_4$ at 620 K. Two different neutron diffractometers were used for different target information for each measurement as explained in the experimental section. The data points are plotted using the common scale $Q = 4\pi\sin\theta/\lambda$ and the common Q range for VEGA and HERMES for comparison. Specific points of measured composition and temperature are given in the inset phase diagram reported by Delacourt [24] and Dodd [25]. Observed intensity $Y_{obs.}$ and calculated intensity $Y_{calc.}$ are represented by red plus signs and the green solid line, respectively. The blue curve at the bottom represents the residual difference, $Y_{obs}–Y_{calc}$. No impurity phase was identified, and the crystal structure was successfully refined with the space group *Pnma*

The first analysis performed for the $Li_{0.6}FePO_4$ solid-solution phase at 620 K was the Rietveld refinement for the neutron diffraction profile and the resultant pattern is summarized in Fig. 14.12b. To evaluate the dynamic disorder of lithium, maximum entropy method (MEM) was applied to estimate the neutron scattering length density distribution, which corresponds to the nuclear density distribution.

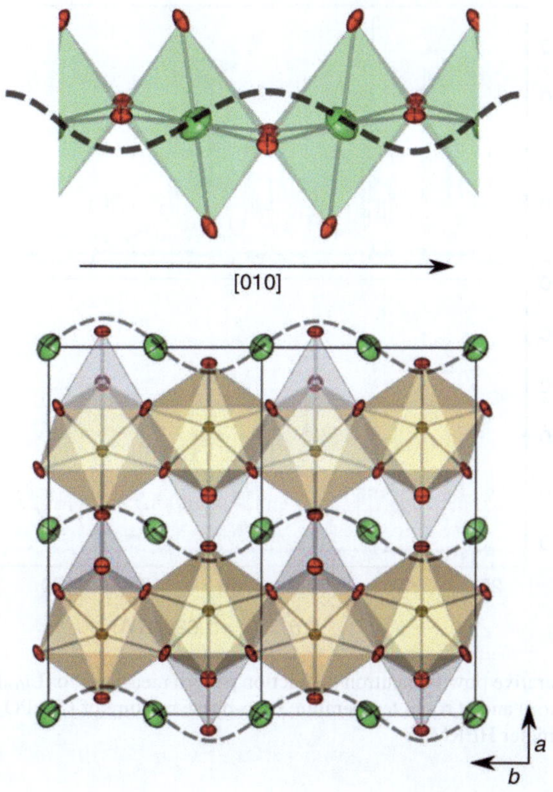

Fig. 14.13 Anisotropic harmonic lithium vibration shown as the green thermal ellipsoids with 95% probability refined by Rietveld analysis for room-temperature neutron diffraction data measured for $LiFePO_4$. Expected curved one-dimensional continuous chains of lithium motion were drawn by the dashed lines to show how the motions of Li atoms evolve from vibrations to diffusion

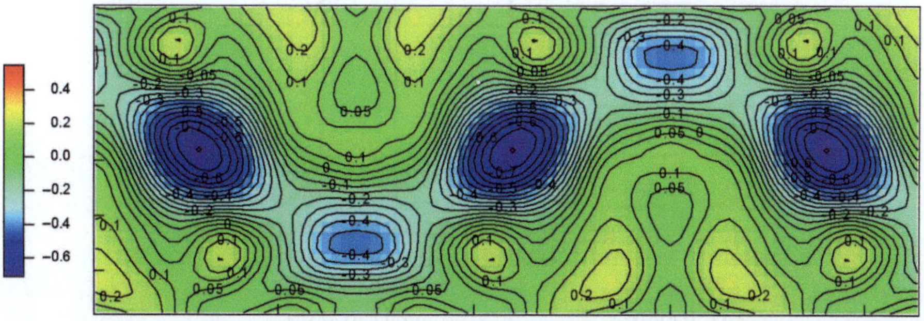

Fig. 14.14 The 010 plane slice of difference Fourier scattering length density plot of $^7LiFePO_4$ with contours in 0.05 fm Å^{-3} steps. The map was calculated by $F_o(Li) = F_o(^7LiFePO_4) - F_{calc.}(Li_0FePO_4)$, where F_o and $F_{calc.}$ are the observed and calculated structure factors, respectively, and Li_0FePO_4 expresses the $FePO_4$ framework having identical structural parameters with $LiFePO_4$. The nuclear density distribution of lithium itself is anisotropic with the same direction as the refined thermal vibration

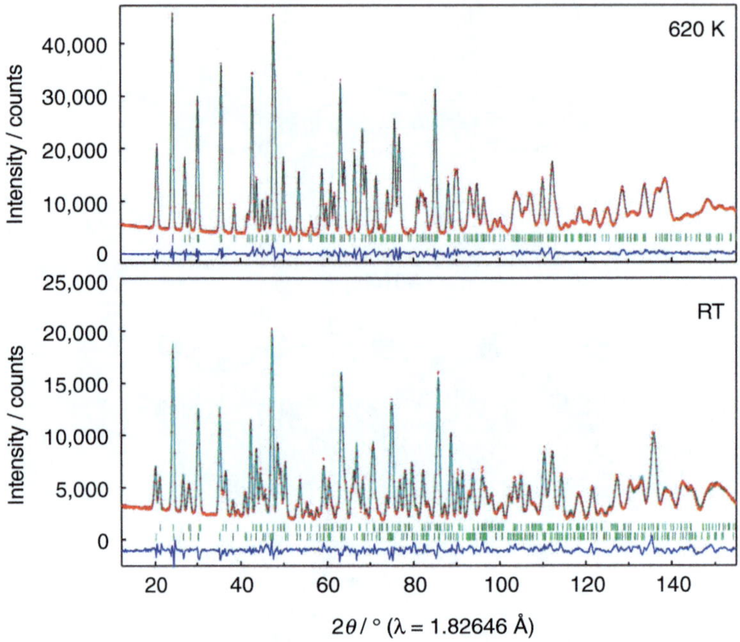

Fig. 14.15 Comparative powder neutron diffraction pattern measured for $Li_{0.6}FePO_4$ measured at 620 K (solid-solution) and at room temperature (two-phase mixture of $LiFePO_4$ and $FePO_4$) using the same diffractometer HERMES

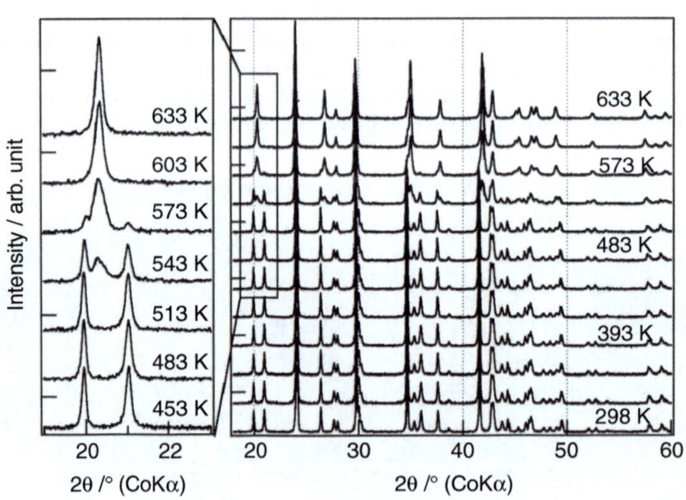

Fig. 14.16 X-ray diffraction patterns of a mixture of 0.6 $LiFePO_4$ and 0.4 $FePO_4$ recorded at 30 K steps from 298 to 633 K with magnification of 200 reflections. Bruker AXS D8 ADVANCE powder diffractometer was used with Co-Kα radiation and linear position-sensitive detector Vantec-1. Measurement ranges were from 15° to 120°. The measurements were conducted under a high-purity He atmosphere in an Anton Paar HTK 450 temperature-controlled chamber

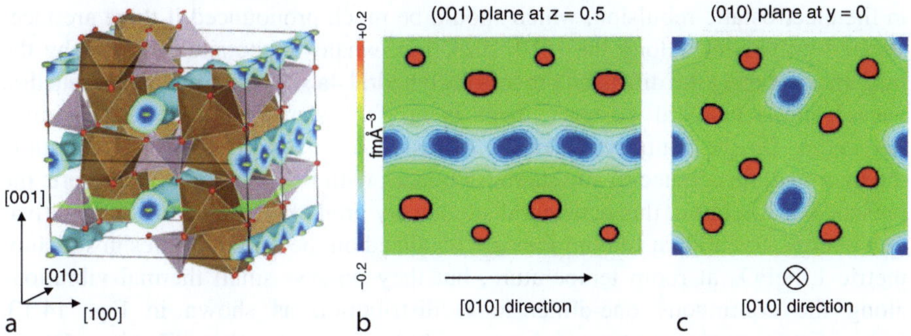

Fig. 14.17 Nuclear distribution of lithium calculated by maximum entropy method (MEM) using neutron powder diffraction data measured for $Li_{0.6}FePO_4$ at 620 K. The classic static atom models with harmonic vibration were no longer appropriate to describe the dynamic disorder of lithium ion $Li_{0.6}FePO_4$ at 620 K; the MEM nuclear density distribution provided much information on the time and spatially averaged complicated dynamic disorder upon lithium diffusion. (**a**) Three-dimensional Li nuclear density data shown as blue contours (equi-value 0.15 fm $Å^{-3}$ of the negative portion of the coherent nuclear scattering density distribution). The brown octahedra represent FeO_6 and the purple tetrahedra represent PO_4 units. (**b**) Two-dimensional contour map sliced on the (001) plane at $z = 0.5$; lithium delocalizes along the curved one-dimensional chain along the [010] direction, while Fe, P, and O remain near their original positions. (**c**) Two-dimensional contour map sliced on the (010) plane at $y = 0$; all atoms remain near their original positions

MEM is a model-free method used to calculate precise nuclear densities in solids, including some disorder and/or an-harmonic vibrations using experimentally obtained structure factors as an initial input. The MEM is primarily an information-theory-based technique that was first developed by Gull and Daniel [28] in the field of radioastronomy to enhance the information from noisy data. Afterward, Collins et al. [29] applied its methodology to crystallography for electron density enhancement from X-ray diffraction. In the theory of this methodology, information entropy, which deals with the most probable distribution of numerical quantities over the ensemble of pixels, is considered. Successful MEM enhancement makes it possible to evaluate not only the missing and heavily overlapped reflections but also any type of complicated electron or nuclear distribution, which is hard to describe with classical structure model. By applying this method, possible bias imposed by the empirical static structural model is reduced, allowing any type of complicated nuclear distribution as long as it satisfies the symmetry requirements. The validity of such methodology has been well established for various ionic conductors [30–32].

A three-dimensional contour surface (0.15 fm $Å^{-3}$) of nuclear distribution of lithium atoms is shown in Fig. 14.17. The probability density of lithium nuclei strictly distributes into the continuous curved one-dimensional chain along the [010] direction, which is consistent with the computational predictions by Morgan et al. [22] and Islam et al. [23]. Other atoms, Fe, P, and O remained to be localized at the initial positions even after MEM analysis. Given the two possible diffusion paths in Fig. 14.11, the microscopic reason of the diffusion anisotropy can be the difference in the electrostatic repulsion, which should be much pronounced if there are face-sheared polyhedra. Along the [010] direction, whatever the site occupied by the lithium ion during the diffusion process (octahedral 4a site–intermediate tetrahedral vacant site–octahedral 4a site), there is no face sharing with other occupied polyhedra. On the contrary, when the diffusion occurs along the [001] direction, the intermediate octahedral site shares two faces with PO_4 tetrahedra; therefore, the presence of lithium in this octahedral site is very unlikely, leading to a high activation energy. Recall that lithium ions are localized on the initial 4a sites in stoichiometric $LiFePO_4$ at room temperature, but they possess small thermal vibrations along the continuous one-dimensional distribution as shown in Fig. 14.13. In $Li_{0.6}FePO_4$ at temperatures as high as 620 K, a large number of lithium defects as lithium acceptor are thermodynamically stabilized, and enough kinetic motional energy is given to each lithium ion to overcome the hopping barrier of the excitonic Li^+–e^- pair.

Summary and Future Directions

These systematic findings on nano-size effects for the phase diagram of important electrode material of Li_xFePO_4 redefine guidelines for understanding the polaronic transport, cooperative phase-boundary movement, and its thermodynamics in biphasic intercalation processes. Polaronic insulators of the transition-metal polyanion family show a wide variety of framework structure and accommodate various transition metals and guest atoms. Recent advances in carbon coating technology have been applied to many of these compounds to shed light on their potential application as intercalation battery cathodes. There is, therefore, still plenty of scope for further investigations on this issue. The long-awaited experimental evidence of strong dimensional restriction of lithium motion in Li_xFePO_4 was provided. Beyond Li_xFePO_4, this is, to our knowledge, the first visual demonstration of the ion diffusion path in a battery electrode. Delocalization of mobile ions at elevated temperatures is a universal phenomena occurring in any intercalation electrode material, and can be utilized to shed light on the ion dynamics. Particularly, such temperature-driven motional enhancement should be more significant in the electron/hole localized system, which recently has been recognized to form an important emerging materials group that may yield new electrode materials. Therefore, high-temperature neutron diffraction coupled with the maximum entropy method is a powerful tool for future battery electrode research.

Bibliography

1. Padhi AK, Nanjundaswamy KS, Goodenough JB (1997) Phospho-olivines as positive-electrode materials for rechargeable lithium batteries. J Electrochem Soc 144:1188

2. Yamada A, Chung SC, Hinokuma K (2001) Optimized $LiFePO_4$ for lithium battery cathodes. J Electrochem Soc 148:A224
3. Huang H, Yin S-C, Nazar LF (2001) Approaching theoretical capacity of $LiFePO_4$ at room temperature at high rates. Electrochem Solid State Lett 4:A170
4. Chung SY, Bloking JT, Chiang YM (2002) Electronically conductive phospho-olivines as lithium storage electrodes. Nat Mater 1:123–128
5. Chen Z, Dahn JR (2002) J Electrochem Soc 149:A1184
6. Goñi A, Lezama L, Arriortua MI, Barberis GE, Rojo T (2000) J Mater Chem 10:423
7. Dahn JR, McKinnon WR (1985) Phys Rev B32:3003
8. Dahn JR, McKinnon WR, Murray JJ, Haering RR, McMillan RS, Rivers-Bowerman AH (1985) Phys Rev B32:3316
9. Delmas C, Nadiri A, Soubeyroux JL (1988) Solid State Ionics 28–30:419
10. Yamada A, Koizumi H, Sonoyama N, Kanno R (2005) Phase change in Li_xFePO_4. Electrochem Solid State Lett 8:A409
11. Meethong N, Huang H-YS, Carter WC, Chiang Y-M (2007) Size-dependent lithium miscibility gap in nanoscale $Li_{1-x}FePO_4$. Electrochem Solid State Lett 10:A134
12. Wagemaker M, Borghols WJH, Mulder FM (2007) J Am Chem Soc 129:4323
13. Schimmel HG, Huot J, Chapon LC, Tichelaar FD, Mulder FM (2005) J Am Chem Soc 127:14348
14. Gibot P, Casa-Cabanas M, Laffont L, Levasseur S, Carlach P, Hamelet S, Tarascon JM, Masquelier C (2008) Nat Mater 7:741
15. Wagemaker M, Mulder FM, Van der Ven A (2009) Adv Mater 25:2703
16. Meethong N, Huang H-YS, Speakman SA, Carter WC, Chiang Y-M (2007) Strain accommodation during phase transformations in olivine-based cathodes as a materials selection criterion for high-power rechargeable batteries. Adv Func Mater 17:1115
17. Wang L, Zhou F, Meng YS, Ceder G (2007) Phys Rev B 76:165435
18. Martin JF, Yamada A, Kobayashi G, Nishimura S, Kanno R, Guyomard D, Dupre N (2008) Electrochem Solid State Lett 11:A12
19. Kobayashi G, Nishimura S, Park M-S, Kanno R, Yashima M, Ida T, Yamada A (2009) Adv Func Mater 19:395
20. Yamada A, Koizumi H, Nishimura S, Sonoyama N, Kanno R, Yonemura M, Nakamura T, Kobayashi Y (2006) Room-temperature miscibility gap in Li_xFePO_4. Nat Mater 5:357
21. Streltsov VA, Belokoneva EL, Tsirelson VG, Hansen NK (1993) Acta Crystallogr B49:147
22. Morgan D, Van der Ven A, Ceder G (2004) Li conductivity in Li_xMPO_4 (M = Mn, Fe, Co, Ni) olivine materials. Electrochem Solid State Lett 7:A30
23. Islam M, Driscoll D, Fisher C, Slater P (2005) Atomic-scale investigation of defects, dopants, and lithium transport in the $LiFePO_4$ olivine-type battery material. Chem Mater 17:5085
24. Delacourt C, Poizot P, Tarascon JM, Masquelier C (2005) The existence of a temperature-driven solid solution in Li_xFePO_4 for $0<x<1$. Nat Mater 4:254
25. Dodd J, Yazami R, Fultz B (2006) Phase diagram of Li_xFePO_4. Electrochem Solid State Lett 9:A151
26. Ellis B, Perry LK, Ryan DH, Nazar LF (2006) Small polaron hopping in Li_xFePO_4 solid solutions: coupled lithium-ion and electron mobility. J Am Chem Soc 128:11416
27. Dodd J, Halevy I, Fultz B (2007) Valence fluctuations of ^{57}Fe in disordered $Li_{0.6}FePO_4$. J Phys Chem C 111:1563
28. Gull SF, Daniel GJ (1978) Image reconstruction from incomplete and noisy data. Nature 272:686
29. Collins DM (1982) Electron density images from imperfect data by iterative entropy maximization. Nature 298:49
30. Shikanai F (2006) Neutron powder diffraction study on the high-temperature phase of $K_3H(SeO_4)_2$. Physica B 385–386:156
31. Yashima M, Itoh M, Inaguma Y, Morii Y (2005) Crystal structure and diffusion path in the fast lithium-ion conductor $La_{0.62}Li_{0.16}TiO_3$. J Am Chem Soc 127:3491

32. Yashima M (2003) Conduction path and disorder in the fast oxide-ion conductor $(La_{0.8}Sr_{0.2})(Ga_{0.8}Mg_{0.15}Co_{0.05})O_{2.8}$. Chem Phys Lett 380:391
33. Nishimura S, Kobayashi G, Ohoyama K, Kanno R, Yashima M, Yamada A (2008) Experimental visualization of lithium diffusion in Li_xFePO_4. Nat Mater 7(9):707–711

Chapter 15
Silicon-Based Anodes for Li-Ion Batteries

Ji-Guang Zhang, Wei Wang, Jie Xiao, Wu Xu, Gordon L. Graff, Gary Yang, Daiwon Choi, Deyu Wang, Xiaolin Li, and Jun Liu

Glossary

Li-ion battery	A family of rechargeable batteries in which lithium ions move from the negative electrode to the positive electrode during discharge, and back to the anode when charging.
Electric vehicle	Vehicle propelled by an electric motor (or motors) powered by rechargeable battery packs.
PHEV	Plug-in hybrid electrical vehicle. This is a hybrid vehicle with rechargeable batteries that can be restored to full charge by connecting a plug to an external electric power source.
Nanowire	A nanowire is a nanostructure, with the diameter of the order of a nanometer and aspect ratio greater than 10:1.
CVD	Chemical vapor deposition.
HEMM	High-energy mechanical milling.
Coulombic efficiency	The efficiency with which charge (electrons) is transferred in a system facilitating an electrochemical reaction.
Specific capacity	Capacity per unit weight of a battery (Ah/kg or mAh/g).
Specific energy	Energy per unit weight of a battery (Wh/kg).
Energy density	Energy per unit volume of a battery (Wh/l).

This chapter was originally published as part of the Encyclopedia of Sustainability Science and Technology edited by Robert A. Meyers. DOI:10.1007/978-1-4419-0851-3

J.-G. Zhang • W. Wang • J. Xiao • W. Xu • G.L. Graff (✉) • G. Yang • D. Choi • D. Wang • X. Li • J. Liu
Pacific Northwest National Laboratory,
Richland, WA, USA
e-mail: gl.graff@pnl.gov

R.J. Brodd (ed.), *Batteries for Sustainability: Selected Entries from the Encyclopedia of Sustainability Science and Technology*, DOI 10.1007/978-1-4614-5791-6_15,
© Springer Science+Business Media New York 2013

Definition of the Subject

Silicon is environmentally benign and ubiquitous. Because of its high specific capacity, it is considered one of the most promising candidates to replace the conventional graphite negative electrode used in today's Li-ion batteries. The theoretical specific capacity of silicon is 4,212 mAh/g ($Li_{22}Si_5$) [1], which is 10 times greater than the specific capacity of graphite (LiC_6, 372 mAh/g). However, the high specific capacity of silicon is associated with large volume changes (more than 300%) when alloyed with lithium. These extreme volume changes can cause severe cracking and disintegration of the electrode and lead to significant capacity loss. Significant scientific research has been conducted to circumvent the deterioration of silicon-based anode materials during cycling. Various strategies, such as reduction of particle size, generation of active/inactive composites, fabrication of silicon-based thin films, use of alternative binders, and the synthesis of one-dimensional silicon nanostructures, have been implemented by a number of research groups. Fundamental mechanistic research also has been performed to better understand the electrochemical lithiation and delithiation processes during cycling in terms of crystal structure, phase transitions, morphological changes, and reaction kinetics. Although efforts to date have not attained a commercially viable silicon anode, further development is expected to produce anodes with three to five times the capacity of graphite. In this entry, an overview of research on silicon-based anodes used for lithium-ion battery applications is presented. The overview covers electrochemical alloying of the silicon with lithium, mechanisms responsible for capacity fade, and methodologies adapted to overcome capacity degradation observed during cycling. The recent development of silicon nanowires and nanoparticles with significantly improved electrochemical performance also is discussed relative to the mechanistic understanding. Finally, future directions on the development of silicon-based anodes are considered.

Introduction

Energy-storage technologies, particularly lithium-ion batteries, have been a focal point for development of advanced, fuel-efficient vehicles, especially plug-in hybrid electric vehicles (PHEVs). Although significant progress has been made during the last 20 years in various battery systems, existing systems do not satisfy all of the energy-storage needs for PHEV applications. More improvements are required with respect to energy density, power density, cycle life, safety, and cost. Commercial lithium-ion batteries primarily use graphite-based anodes, which have a theoretical specific capacity of 372 mAh/g (LiC_6). Continuous efforts have been made to find alternative anode materials, such as elemental metals, to replace graphite-based anodes since the early 1960s [2], when Dey first reported that lithium could be electrochemically alloyed with a number of metals at room temperature. Many elements such as aluminum (Al), silicon (Si), tin (Sn), antimony

Table 15.1 Specific capacities and molar volume changes exhibited by different candidate elements during cycling [5]

Elements	C	Al	Si	Sn	Bi
Lithiated phase	LiC_6	Li_9Al_3	$Li_{21}Si_5$	$Li_{17}Sn_4$	Li_3Bi
Theoretical specific capacity (Ah/kg)	372	2235	4012	959	385
Theoretical volumetric capacity (Ah/l)	833	6035	9340	7000	3773
Molar volume change (%)	12	238	297	257	115

(Sb), bismuth (Bi), magnesium (Mg), and zinc (Zn) are known to react with lithium to form alloys (see Eq. 15.1) by undergoing partially reversible electrochemical reactions that result in high specific and volumetric capacities.

$$Li_xM \leftrightarrow xLi^+ + xe^- + M \tag{15.1}$$

Among possible alternative alloying elements, silicon is the most attractive and widely investigated [3, 4] because of its high gravimetric and volumetric capacities and abundance in the natural environment. Silicon in the fully lithiated form of $Li_{4.4}Si$ provides a theoretical specific capacity of 4,212 mAh/g which is 10 times more than the capacity of graphite. The specific capacities and volume changes of the different electrochemically active elements are shown in Table 15.1.

Although the silicon-based anode has great potential, the alloying and subsequent de-alloying reactions during lithium insertion and lithium de-insertion result in severe crystallographic volume changes (~300% for silicon as shown in Table 15.1) because of various phase transitions. The mechanical strain generated during these phase transformations leads to cracking and disintegration of the electrode that, in turn, leads to failure of the anode after only a few cycles from loss of electrical contact between particles and the current collector. More details on the electrochemical alloying/de-alloying processes between lithium and silicon and the related failure mechanisms are discussed below. To prevent confusion, the following conventions are used in this chapter: discharge of silicon means lithium is inserted into silicon (lithiation or alloying of lithium with silicon); charge of silicon means lithium is de-inserted from silicon (delithiation or de-alloying of silicon).

Electrochemical Alloys of Lithium with Silicon

Studies of electrochemical alloys of lithium and silicon began with elemental lithium electrodes in high-temperature lithium/metal-sulfide secondary batteries [1, 6–8]. Lithium-silicon alloys with different stoichiometry were reported during the electrochemical lithiation of silicon at elevated temperatures (~400–500°C). The formation of $Li_{12}Si_7$, $Li_{14}Si_6$, $Li_{13}Si_4$, and $Li_{22}Si_5$ alloys was identified at distinct voltage plateaus according to each two-phase region, which follows the equilibrium lithium-silicon phase diagram as shown in Fig. 15.1. However,

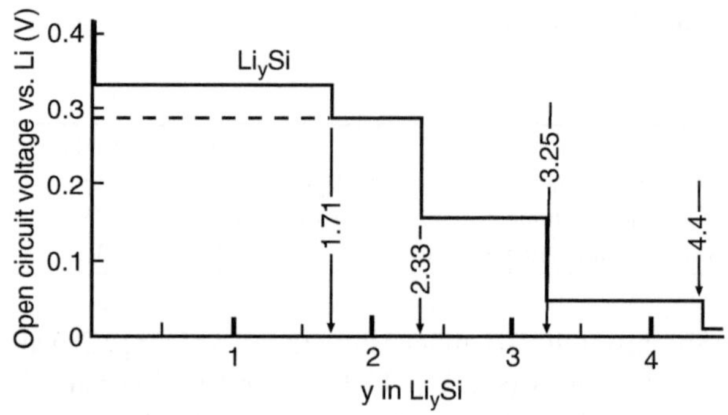

Fig. 15.1 Silicon lithiation showing two-phase plateaus at 415°C. (Reprinted with permission from Boukamp et al. [1]. Copyright 1981 The Electrochemical Society)

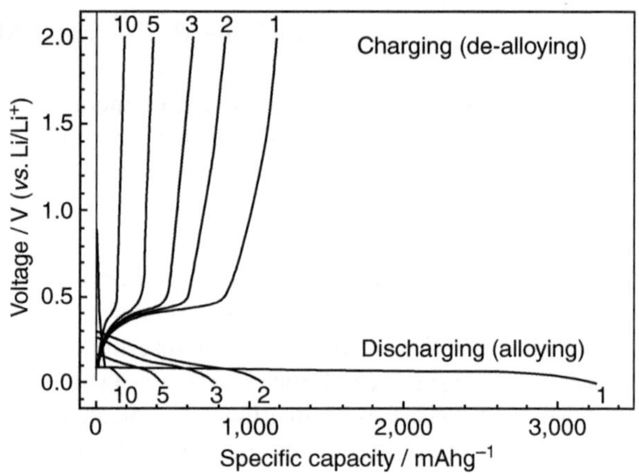

Fig. 15.2 Galvanostatic charge/discharge profiles for micro-silicon (10 μm) cycling at a current density of 100 mA/g in the voltage window of 0.0 –2.0 V (versus Li/Li^+) (Reprinted with permission from Ryu et al. [13]. Copyright 2004 The Electrochemical Society)

the electrochemical alloying process of lithium with silicon at room temperature was not readily understood because only one voltage plateau was observed for the silicon anode cycled between 0.0 V and ~1.2 V. Additionally, only ~3,500 mAh/g of capacity was obtained during the first discharge of a silicon anode at room temperature. Very few reports describing fundamental research on room-temperature electrochemical lithiation and delithiation of the silicon anode were published prior to the year 2000. Between 2000 and 2003, several groups observed that crystalline silicon undergoes electrochemical solid-state amorphization upon reacting with lithium at room temperature to form a metastable, lithium-silicon-based, amorphous phase where crystallization of the equilibrium intermetallic phases is kinetically suppressed [9–12]. The thermodynamically favored, high-temperature intermetallics do not easily crystallize at room temperature because of the sluggish kinetics.

Figure 15.2 shows the typical cycling performance of a pure crystalline silicon anode [13]. This sample has a discharge capacity of ~3,260 mAh/g, and a charge capacity of 1,170 mAh/g, which gives an irreversible loss of ~64%. Such a large irreversible capacity loss in the first cycle is commonly observed for silicon-based anodes when discharged to near-zero potential. The coexistence of an amorphous lithium-silicon phase with the unreacted crystalline silicon leads to a single voltage plateau that is long and flat. This finding is consistent with results reported by several researchers who observed similar single, long, flat plateaus at ~0.1 V during electrochemical lithiation of crystalline silicon [3, 7, 9]. Though contrary to the equilibrium lithium-silicon phase diagram that shows multiple lithium-silicon phases [10, 14, 15], both ex situ and in situ X-ray diffraction (XRD) studies revealed a decrease and disappearance of the crystalline silicon peaks during the first lithium insertion, while no peaks corresponding to the equilibrium lithium-silicon phases were detected [10, 14, 15]. High-resolution transmission electron microscopy (HR-TEM) results confirmed that the lithiated silicon is formed primarily as an amorphous phase [10]. A change in the slope of the cell potential profile has been observed toward the end of the first discharge if the silicon anode is lithiated at an extremely low rate (~C/100), indicating formation of a new lithium-silicon phase. Obrovac and Christensen first reported the formation of a new crystalline phase when the highly lithiated amorphous silicon was electrochemically driven to ~50 mV versus Li/Li^+. Using XRD analysis, this new phase was identified as $Li_{15}Si_4$ [15]. Dahn et al. also confirmed the formation of $Li_{15}Si_4$ in both crystalline and amorphous silicon thin-film anodes when discharged below ~50 mV using in situ XRD analysis [14, 16]. Therefore, rather than forming the equilibrium $Li_{22}Si_5$ phase, $Li_{15}Si_4$ is the terminal phase that is achieved when the silicon anode is discharged to near-zero potential at ambient temperature. This phase renders a theoretical capacity of 3,579 mAh/g, a value that is in good agreement with most published reports on the silicon anode. It is worth noting that the formation of the $Li_{15}Si_4$ phase is observed only when the silicon anode is discharged at an extremely low rate (C/100). When the electrode is cycled at a moderate current rate (~C/10), the change of slope toward the end of the discharge is not observed. This characteristic provides further evidence of kinetically suppressed crystallization of lithium-silicon phases at low temperatures [17, 18].

Upon charge, the $Li_{15}Si_4$ phase delithiates to form an amorphous Li_ySi phase. The coexistence of these phases leads to a narrow potential window between ~0.3 V and 0.4 V, as shown in Fig. 15.2, which corresponds to the potential range of the de-alloying reaction. The in situ XRD study of the charge process and the XRD results from the chemical delithiation of the $Li_{12}Si_7$ phase have confirmed that the lithiated silicon becomes amorphous after charging [14, 15]. During the first discharge/charge process of the silicon-based anode, it is important to note that the formation of the $Li_{15}Si_4$ phase occurs only when the potential is driven below ~50 mV. Control of the cycling voltage above this value will bypass the formation of the $Li_{15}Si_4$ phase, leading to a different potential profile. For example, Obrovac [15] showed that a cell discharged to 0 V showed a change of slope at the end of the first discharge and a two-phase region during charging, both of which can be attributed to the formation of $Li_{15}Si_4$ phase. These two features are not present in the voltage profile of the cell cycled above 50 mV.

Amorphous silicon demonstrates different features than those of crystalline silicon when subjected to lithiation at room temperature. Studies on an amorphous silicon thin-film anode prepared by magnetron sputtering and an amorphous silicon anode obtained through chemical delithiation of $Li_{12}Si_7$ phase show two distinctive, slightly sloping plateaus during the first discharge. These plateaus correspond to the two peaks observed in the dQ/dV curves [12, 15, 16]. The first peak is observed at ~0.2 V, which is about 0.1 V higher than the starting potential of lithiation observed in crystalline silicon. The second sloping plateau appears at ~0.1 V. The formation of the $Li_{15}Si_4$ phase also was identified in the amorphous silicon anode at a voltage ~30 mV. However, Dahn et al. reported that the crystalline phase only forms when the amorphous silicon thin films have a thickness greater than 2 μm. During the first lithiation, the amorphous silicon thin film that was cycled at a very low rate of C/48 exhibited a flat plateau at ~0.4 V, which may correspond to the conversion of the crystalline $Li_{15}Si_4$ phase to the amorphous Li_xSi. In comparison, the amorphous silicon thin-film anode cycled at a higher rate generally produced two broad humps at 0.3 V and 0.5 V, respectively. The phase formation of amorphous silicon during the first discharge/charge is not entirely understood. It generally is believed that the different peaks measured on the dQ/dV curves are related to the transition between different amorphous Li_xSi phases [17]. Because of the solid-state amorphorization discussed above, the voltage profile of the crystalline silicon anode and amorphous silicon anode exhibit similar features after the first charge/discharge cycle. The potential window in which silicon anodes operate is low enough to be useful in high-energy-density Li-ion batteries; however, Fig. 15.2 also reveals a large irreversible loss and rapid capacity fade during the lithium alloying/de-alloying process with silicon. These two problems have been reported by many researchers and are discussed in more detail below.

Failure Mechanisms of Bulk Silicon Anodes

The main challenge with the bulk silicon anode is the poor cycling stability as shown in Fig. 15.2. The reversible capacity decreases to less than 500 mAh/g by the fifth cycle. Several observations can be made concerning the rapid capacity fade from the charge/discharge profile. First, the irreversible loss is observed in every cycle, in contrast to conventional graphite anodes, where the irreversible losses cease after the initial one to two cycles. Second, although capacity fade occurs with every charge (i.e., de-alloying) cycle, upon re-discharging (alloying), the capacity recovers to approximately the same value measured in the previous charge cycle. This observation suggests that the degradation of the silicon anode is much more severe in the de-alloying cycle than in the alloying cycle.

There is strong evidence that failure of the silicon anode is caused by massive volume variations during lithiation and delithiation of the lithium-silicon alloys because the intermetallics formed have greater molar volume than the parent pure silicon phase. For example, $Li_{22}Si_5$ has a ~300% volume expansion over pure silicon, which causes substantial internal stress during lithiation or alloying. The fully lithiated phase of $Li_{15}Si_4$ exhibits a capacity of 3,579 mAh/g and has a density of 1.179 g/cm^3 calculated from XRD data, which represents

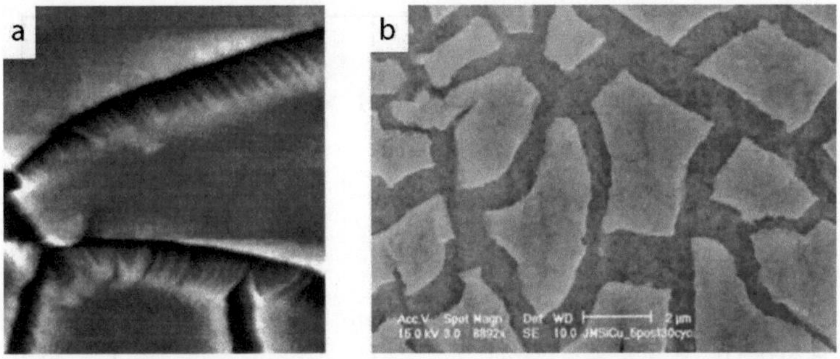

Fig. 15.3 (**a**) In situ AFM image showing the cracks of a silicon thin-film anode during cycling. (Reprinted with permission from Beaulieu et al. [20]. Copyright 2001 The Electrochemical Society) (**b**) SEM morphology of a 250-nm silicon thin film after 30 charges at a C/2.5 rate. (Reprinted with permission from Maranchi et al. [12]. Copyright 2003 The Electrochemical Society)

a volume expansion of 280% [19]. The amorphous Li_ySi alloy formed during discharging progressively undergoes volume expansion until formation of a final $Li_{15}Si_4$ intermetallic, which has nearly three times the volume per atom than the parent silicon. During the subsequent delithiation (de-alloying) process, the volume then contracts as lithium is extracted from the lithium-silicon intermetallic and amorphous phases [17, 20]. Crack formation and disintegration caused by the repetitive volume expansion and contraction are evident from the in situ atomic force microscopy (AFM) and scanning electron microscopy (SEM) images shown in Fig. 15.3 [12, 20]. The repeated volume expansion/contraction results in mechanical cracking and disintegration of the electrode, loss of electronic contact between active materials, and between active materials and current collector, thus rapid capacity fading.

A galvanostatic intermittent titration (GITT) study of the silicon anode shed more light on the poor cycling stability of the bulk silicon anode by measuring the variation in internal resistance during charge/discharge cycling [13]. In that study, a current of 100 mA/g was applied for 10 min, and then turned off for 20 min. A closed-circuit voltage (CCV) was measured while the current was on. A quasi-open-circuit voltage (QOCV) was obtained while the current was off. The internal resistance was then calculated from the difference between QOCV and CCV for each voltage transient. Figure 15.4 shows that, upon discharge, the internal resistance decreases, which is attributed to better electronic contact between the silicon particles resulting from the volume expansion during formation of Li_xSi alloys. One also could argue that the Li_xSi alloys have a slightly higher electronic conductivity than the pure silicon [19]. A drop in the internal resistance is observed below 0.3 V, which suggests that the alloying process primarily occurs below that voltage. During charge, the internal resistance increases because of increased contact resistance resulting from contraction of silicon particles/grains. Loss of electrical contact between particles leaves lithium ions trapped inside the anode (incomplete de-alloying) as the charge cycle is prematurely terminated when the electrode

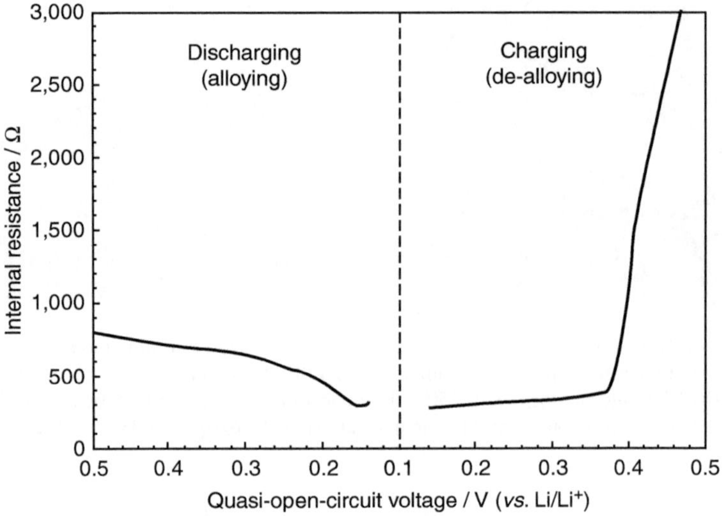

Fig. 15.4 Variation of internal resistance in a silicon anode (Reprinted with permission from Ryu et al. [13]. Copyright 2004 The Electrochemical Society)

potential exceeds the cycling potential window. Upon charge, an abrupt increase in internal resistance is seen above 0.4 V, which suggests that the de-alloying reaction occurs primarily above that voltage. These results confirm that failure of silicon anodes is caused by breakdown of the electronically conductive network resulting from the large volume changes and cracking of the native anode structures.

Strategies to Improve the Cycle Life of Silicon-Based Anodes

In recent years, tremendous research efforts have been made to improve the electrochemical performance of silicon-based materials. One strategy is to create a fine composite microstructure comprising an active lithium alloy phase uniformly dispersed in an inert host matrix, such as silicon/carbon, Si/TiB_2, or Si/TiN composite materials [22, 23]. Composites of silicon and carbonaceous materials have been reported to have enhanced cycle life. These can be made by chemical vapor deposition (CVD), pyrolysis of silicon-containing organic compounds, or by mechanochemical methods. The carbon increases the electrical conduction and also may accommodate the volume change of silicon and buffer the mechanical stress, thereby improving the cycle life. Another strategy is to develop various nanoscale silicon or silicon/carbon architectures, including one-dimensional nanowires, two-dimensional thin films, and three-dimensional composite architectures. This section will provide a brief review of various approaches used to improve the capacity and cycling stability of silicon-based anodes [24–26].

Active/Inactive Nanocomposites

Although nanostructured materials are known to exhibit superplastic behavior, it is difficult to envisage that even nanostructured silicon would survive mechanical degradation (cracking) after long-term cycling because of the large volume expansion/contraction. A concept that has been developed to capitalize on the attributes of nanostructured materials is the generation of electrochemically active/inactive composites [4, 27–29] in which the material reactive to lithium (termed as "active") is dispersed within materials that are nonreactive or less reactive to lithium (termed "inactive"). To generate a successful nanocomposite, both the active and inactive components need to be carefully selected to satisfy the requirement of low irreversible loss (20–30%), high capacity (>500 mAh/g), and good cyclability (~0.01% loss/cycle). Large volume changes are expected during the lithiation of silicon; therefore, the selection of matrix materials is of paramount importance to maintain the structural integrity.

The inactive components acting as the matrix usually consist of soft and ductile materials that play a very important role as "buffers" to alleviate the mechanical stresses arising from the large volumetric change of the active components. Increasing the inactive components in the electrode also was reported to suppress the aggregation of the active components upon cycling [30]. In this regard, inactive matrix components act as a skeleton or a network to form a microstructure that not only accommodates the volume changes, but also provides reaction sites for the active elements. Matrix materials also must exhibit good electronic and lithium-ion conductivity. Furthermore, the inactive matrix elements should be lightweight so that high gravimetric capacity can be achieved. Finally, the matrix materials should be inert to the active components because any reaction between matrix materials and active components will greatly reduce the useable capacity. The ratio between the active and inactive components is important to achieve the desirable cyclability and capacity. Previous reports show TiO_2 and ZrO_2 nanoparticles exhibit superplastic deformation [31], suggesting that other nanometer-size particles/phases may behave similarly and function as matrix components that accommodate structural changes upon cycling.

A schematic model of the "active/inactive" nanocomposite is shown in Fig. 15.5 [32]. Fuji Photo Film Celltec Company reported a tin-based amorphous composite oxide (TCO) [33] in 1997, which is the first demonstration of the "active/inactive" nanocomposite concept. In their results, the tin compound is the only active component reacting with lithium. The volume expansion of tin during cycling is reduced by the presence of other electrochemically inert oxide components including B_2O_3, P_2O_5, and Al_2O_3, which leads to good reversibility at a specific capacity of 800 mAh/g. The limitation of TCO is the large irreversible loss of >50% resulting from the electrochemical reaction depicted below.

$$SnO_2 + 8.4Li \rightarrow Li_{4.4}Sn + 2Li_2O \quad (15.2)$$

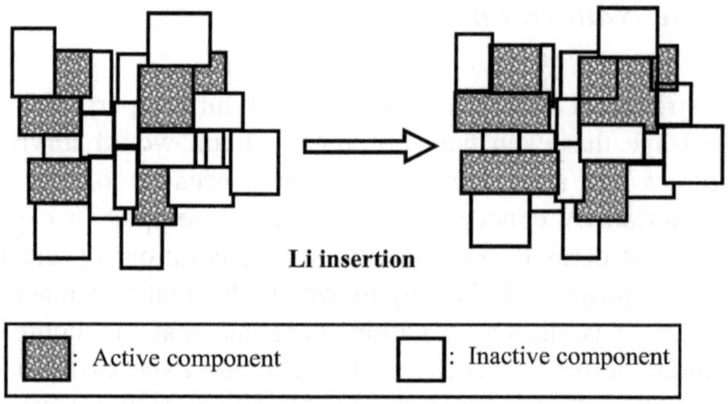

Fig. 15.5 Schematic model of the active/inactive nanocomposite and the morphological change occurring during lithium insertion Wang et al. [32]

The Li_2O formed in situ helps to bond the $Li_{4.4}Sn$ species but does cause a large irreversible loss of Li. Efforts have subsequently been made to synthesize composites containing the pure active phase (such as tin) in the electrochemically inactive phase, thus eliminating or minimizing the irreversible loss. Over the years, many different combinations of active and inactive systems have been investigated using two general approaches: (1) in situ nanocomposites and (2) ex situ nanocomposites. In the in situ approach, alloys or compounds such as Mg_2Si and SiAg are used as starting materials, and upon lithium insertion, the active phases (lithium-silicon alloys) are formed in situ with the less active phases or compounds [34, 35]. XRD and atomic emission spectroscopy (AES) studies showed that the lithium-silicon and lithium-magnesium alloys were formed during the first insertion [35]. The in situ approach relies on phase instability of the parent phases Mg_2Si and SiAg upon lithium insertion. Because magnesium and silver are high-density elements with large atomic masses, the net specific capacity that can be attained is only moderate. The availability of suitable compounds and the ability to synthesize such nanocomposites in the case of silicon also are very limited.

Like SnO_2, SiO was also investigated as a lithium-ion battery anode through the reduction of SiO in the first-discharge process to form, in situ, a silicon active phase and the compound Li_2O as the matrix, possibly according to the following reaction.

$$5SiO + 6Li \rightarrow 4Si + Li_2O + Li_4SiO_4 \qquad (15.3)$$

Electrodes produced using this approach typically deliver a capacity of ~600 mAh/g with stable cycling, and could become an effective and low-cost route for preparing silicon-based anodes.

The ex situ approach relies on generation of active and inactive phases that are known to be thermodynamically stable. Thus, the lithium-ion reactions are confined to the active elements while the inactive phase(s) act as buffering agents to prevent electrode cracking/crumbling. The ex situ approach typically uses mechanical milling to form nanocomposites through mechanical alloying. The ex situ approach is attractive because of the ease of sample preparation and the lack of chemical reactions between active and matrix components, even upon cycling. A number of systems studied as inactive materials have demonstrated limited improvements in the performance of the silicon anode [5]. Hard materials such as SiC have been used as the inactive material to curtail volume expansion during cycling. Kim et al. successfully achieved stable reversible capacities up to ~600 mAh/g by ex situ generation of amorphous silicon particles embedded within nanocrystalline matrices of SiC, TiN, TiB_2, and C [36–38].

Both inactive and active forms of carbon have been investigated as possible host matrices with finely divided silicon [36–38]. As with the inactive phases, active matrices must accommodate the volume changes in silicon and provide good mechanical strength and facile transport for both electrons and lithium ions. Of the various materials investigated in this approach, graphite is a good candidate as an active matrix material because of its good electronic conductivity and lubrication characteristics. In recent years, active/inactive and active/active silicon/carbon composites have attracted considerable interest and the electrochemical performance reported appears quite promising. The graphite matrix acts as a buffer to accommodate the large volume expansions of silicon while the silicon contributes to the overall capacity because of its large gravimetric and volumetric capacity. Various synthesis methods have been reported to prepare silicon/carbon composites and nanocomposites, ranging from high-energy mechanical milling (HEMM), pyrolysis, and physical mixing. The raw materials used include a variety of phases, such as crystalline and amorphous silicon, graphite, disordered carbon, mesoporous carbon, carbon nanotubes, and fullerenes. The following sections provide more detail on the various approaches used to produce silicon/carbon electrodes.

Silicon/Carbon Composites Prepared through Ball Milling

Mechanical milling can readily generate alloy systems with desirable compositions, structures, and particle sizes; therefore, HEMM has become a common approach for the synthesis of silicon/carbon composites. Results from different starting carbon materials such as graphite and disordered carbon [39] show a high first discharge and charge capacity (~800–1,400 mAh/g). However, the cycling stability is not as good as silicon/carbon composites derived from thermal pyrolysis of silicon-containing polymers. Gross et al. [40] reported a first-lithiation capacity of ~800 mAh/g with a fading rate of ~1.25% for a composite made by 15-min mechanical milling of graphite and a pre-milled silicon powder. The poor capacity retention is likely caused by the large silicon particles (>1 μm) in the composite. These particles disintegrate because of volume changes during lithium alloying and de-alloying processes, as discussed above. This hypothesis suggests that extended milling may improve the cycling stability by decreasing the silicon particle size and homogeneously distributing silicon particles in the graphite matrix. However, extended milling leads to formation of electrochemically inactive SiC because of mechanochemical reactions between silicon and carbon [37]. In addition to the formation of SiC, amorphization of graphite during prolonged milling also prevents stable high-capacity cycling of the silicon/carbon composite. Several polymer additives have been identified that serve as effective diffusion barriers to circumvent the formation of SiC and preserve the graphite structure during prolonged milling. Kim et al. [41] reported nanocomposites of silicon/carbon with a capacity as high as ~850 mAh/g and reasonable capacity retention (~1.1% loss/cycle). These nanocomposites were synthesized from silicon and polystyrene (PA resin) using HEMM. Datta et al. [22] recently reported methods to stabilize a composite using silicon, graphite, and polyacrylonitrile-based disordered carbon through HEMM, followed by subsequent heat treatment.

Si/C Composites Prepared Through High-Temperature Pyrolysis and CVDProcesses

In addition to the selection of appropriate starting materials to form active/inactive or active/active composites, the synthesis method must be carefully chosen to generate materials with desired properties. Recently, several novel methods including CVD, physical vapor deposition (PVD), and templated growth have been developed to generate nanoscale and nanostructured silicon-based materials. Control of structure, particle size, and composition of silicon/carbon composites using CVD or pyrolysis reactions is difficult. In addition, maintaining a homogeneous distribution of silicon throughout the sample, which is a critical attribute for good cycling performance, is particularly challenging. In general, the chosen method should synthesize composites containing discrete phases without any chemical interaction to prevent loss of the active component and minimizing subsequent loss of electrochemical capacity.

Decomposition of organic precursors is one of the most common approaches investigated. Dahn et al. [42] systematically studied the pyrolysis of different silicon-containing polymers during the mid-1990s. More recent results include thermal pyrolysis of polyvinyl chloride dispersed with nanosized silicon and fine graphite particles to achieve a reversible capacity of ~700 mAh/g as reported by Yang et al. [43]. The same group also reported similar results from pyrolysis of pitch embedded with graphite and silicon powders [43]. However, silicon/carbon composites synthesized by pyrolysis of silicon-containing polymers suffer from a large irreversible loss (~50–60%) in the first cycle mainly because of the presence of disordered carbon and sulfur/oxygen/hydrogen impurities generated during high-temperature decomposition.

Another high-temperature approach for generating silicon/carbon composites is to deposit silicon particles on carbon or vice versa using CVD. Xie et al. [44] reported the deposition of silicon on mesocarbon microbeads (MCMB) by CVD of silane at 450°C and 500°C, but only a very small amount of silicon actually deposited on the MCMB. The material also demonstrated a very high (55%) irreversible loss. Vacuum deposition of nanometer-sized silicon particles on graphite surfaces has also been reported [45, 46]. The as-prepared anode showed an irreversible capacity loss of 26% with a capacity fade rate of less than 1% over 100 cycles. This cycling performance probably is one of the best reported for a silicon/carbon composite, and is attributed to the nanoscale silicon particle size (~50 nm) and strong adhesion of the silicon particles with the graphite matrix. In an attempt to reduce irreversible loss, carbon also has been deposited on pure silicon powders [47, 48]. However, the electrochemical performance of these materials was inferior to the silicon/carbon composite generated by depositing silicon on the carbon surface.

Silicon/Carbon Nanotube Composites

Recently, carbon nanotubes (CNTs) also were investigated as candidate matrix materials. CNTs possess certain unparalleled characteristics that could be valuable in producing stable silicon/carbon composite anodes. The diameter of single-wall nanotubes (SWNTs) is in the range of 1–10 nm, while that of multiwall nanotubes (MWNTs) is in the range of ~100 nm. CNTs can be several microns in length, resulting in aspect ratios greater than ~1,000. Their specific surface area (SSA) is usually greater than 500 m^2/g. Strong C=C bonding to three adjacent neighboring atoms results in excellent mechanical properties in CNTs. The Young's modulus of SWNTs is approximately 1 TPa, and the maximum tensile strength is ~30 GPa [49]. Treacy et al. [50] reported an elastic modulus as high as 1.25 TPa. These values compare well with the modulus of MWNTs (1.28 TPa) measured by Wong et al. [51]. High tensile strengths up to ~63 GPa have been reported for MWNTs [52], which is an order of magnitude higher than that of carbon fibers and stainless steel (~5 GPa and ~3 GPa, respectively). The unique bonding structure also enables exceptional flexibility. The theoretical maximum elongation of an SWNT is almost 20% [53]. CNTs also are highly resistant to damage from external forces. Application of force to a nanotube typically results in bending, but not permanent damage. When the force is removed, the tube returns to its original state [54]. With respect to electrical conductivity, both MWNTs and SWNTs are excellent electrical conductors. The resistivity of SWNT ropes is approximately 10^{-4} Ω-cm at 300 K or roughly on the same order of magnitude as graphite [55]. MWNTs can carry extremely high current densities that exceed 107 Acm^{-2}, while a current density of 109 Acm^{-2} has been reported in SWNTs [56, 57]. These values clearly exceed the electrical property requirement of lithium-ion anodes. All of these attributes suggest that CNTs would be good candidates for matrix materials. The high-aspect-ratio, one-dimensional structures can form entangled three-dimensional networks that maintain electrical contact with silicon particles during cycling. The unusual flexibility and elongation could further facilitate stability, particularly after the active materials undergo cracking and disintegration. Finally, the high electrical conductivity and current density, combined with the low material density, could produce unprecedented matrix properties.

Although promising, in practice, the high irreversible loss and large voltage hysteresis during electrochemical cycling greatly limit the use of CNTs as active anode materials in lithium-ion batteries. Shu et al. [58] reported the synthesis of CNT-based silicon/carbon composites by decomposition of acetylene at 800°C using a nickel-phosphorus catalyst deposited on the silicon particles to seed nanotube growth. This silicon/carbon composite anode demonstrated a first cycle discharge capacity of 1,120 mAh/g and an irreversible loss of 20%. Wang et al. [59] achieved a capacity of ~1,000 mAh/g by homogenously dispersing a pre-milled silicon/graphite powder into a single-wall carbon nanotube matrix using ultrasonication. Si et al. [60] reported that the direct mixing of carbon nanofibers with carbon-coated silicon nanoparticles helped to maintain a stable cycling for the silicon-based anode. When carbon nanofibers are used as the current collector, superior overall capacity, cyclability, and rate capability have been demonstrated [61]. Lee et al. [62] also reported the use of wet ball milling to thoroughly mix silicon starting materials with different carbon sources, including single-walled and multiwalled CNTs. Although a high first-discharge capacity of ~2,000 mAh/g was achieved, rapid capacity fade was observed after 15 cycles [62]. Generally, anodes generated from simple mixtures of silicon particles and CNTs show poor capacity retention, but significant improvements were observed when silicon was directly deposited, using silane-based CVD, onto the CNTs or carbon nanofibers. The performance improvements were attributed to the formation of strong bonding layers between the silicon cluster and the CNT surface, which helped facilitate transport of lithium ions and electrons, in addition to the relaxation of stresses generated from the repetitive volume changes [18, 63].

Nanoscale Architectures

Several research groups have reported that reduction of the particle size leads to better accommodation of the strain generated during lithium insertion/de-insertion, therefore resulting in improved cycling performance. Although this work initiated with tin-based anode materials, the same strategy soon was applied to silicon-based systems by Yang et al. [64–66]. Huggins et al. [67] also suggested that decrepitation leads to a critical particle size more tolerant to mechanical stress. Researchers have reported superior electrochemical performance of nanoscale, over bulk, silicon, and attributed the performance to better strain accommodation in the smaller particles [68].

Generation of nanosized particles (~1–100 nm) results in a relatively low number of atoms per grain, less volume change upon cycling, and therefore, reduced mechanical stresses within the particles. In addition, the large number of grain boundaries existing in nanosized materials may help stabilize the particles and also act as channels for lithium insertion and de-insertion [69]. Other advantages include increased contact area between particles, within the electrode, and between the electrode and electrolyte, all leading to higher charge/discharge rates. Nanosized systems also provide a shorter diffusion path for the transport of both electrons and lithium ions, which improves performance for materials exhibiting poor electronic or lithium-ion conductivity. Another possible benefit of nanostructured materials is the potential to undergo superplastic deformation during cycling to better accommodate strain generated by large volumetric changes. Equation 15.4 describes the relationship between applied stress (σ), temperature (T), grain size (d), and the resulting strain rate ($\dot{\varepsilon}$) [31].

$$\dot{\varepsilon} = A\sigma^{n}d^{-p}D_{0}\exp(-Q/RT) \tag{15.4}$$

where A is a constant, n is the stress exponent ($n \approx 2$), p is the grain size exponent ($p \approx 2$–3), Q is the activation energy, and R is the gas constant. An examination of Eq. 15.4 reveals that the strain rate for nanocrystalline materials will be several orders of magnitude higher than that of conventional-sized (i.e., micron-sized) materials at the same temperature. Therefore, nanostructured materials could possibly achieve appreciable superplastic strain at room temperature and are more suitable for use as anode materials. Nanocrystalline materials can be either single-phase or multiphase materials, which are in the size range of about 1 nm to 100 nm. Compared to materials containing larger grain sizes, the nanocrystalline materials generally exhibit improved mechanical hardness, fracture toughness, and ductility because of the large fraction of the atoms located at surfaces or grain boundaries [70]. Therefore, using nanocrystalline materials as anodes will enhance not only the diffusion rates because of higher grain-boundary diffusion, but also the overall mechanical strength of the anode. Recently, this nanoscale synthesis approach has expanded into the growth of one-dimensional silicon nanowires and silicon nanoparticles with demonstrated superior electrochemical cycling performance, as will be discussed in a later section.

Two-Dimensional Silicon Thin-Film Anodes

Inspired by the improved performance of nanoscale-sized silicon particles over bulk silicon particles, many groups have tested silicon thin films deposited on metallic substrates (e.g., copper or nickel), in which the thin-film silicon contains nanometer-sized silicon features. In general, thin-film silicon anodes can be classified into nanocrystalline thin-film anodes and amorphous thin-film anodes. Graetz et al. [71] prepared nanocrystalline silicon thin-film anodes using PVD. These anodes exhibited specific capacities of approximately 1,100 $mAhg^{-1}$ with a 50% capacity retention after 50 cycles. The improved electrochemical performance of the thin-film anode was linked to good adhesion between deposited silicon particles and the substrate current collector.

Because of isotropic expansion of the particles/grains, amorphous thin-film silicon anodes typically perform better than crystalline thin films. Maranchi et al. [12, 72] demonstrated that 250-nm-thick amorphous silicon thin films deposited by radio-frequency magnetron sputtering on copper substrates achieved near theoretical capacity for a limited number of cycles. The authors stated that growth of

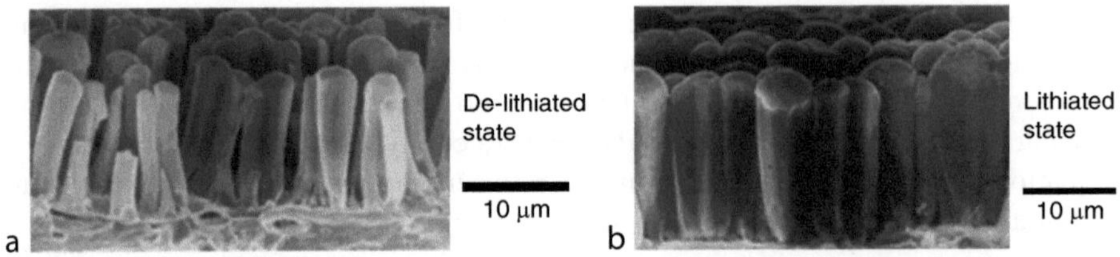

Fig. 15.6 Cross-sectional SEM images of an a-silicon thin-film electrode in the (**a**) de-lithiated and (**b**) lithiated states (Reprinted with permission from Yonezu et al. [77]. Copyright 2004 The Electrochemical Society)

a lithium-copper-silicon phase at the interface between silicon and copper, and the incremental plastic strain in the copper substrate upon cycling, resulted in a reduction of adhesion between the silicon film and the current collector, and progressive failure of the silicon film. Good electrochemical cycling performance also has been reported for multicomponent thin films, such as submicron-sized iron/silicon multilayer films and alternating cobalt/silicon layers [73, 74].

Surface roughness of the substrate can significantly influence the performance of silicon thin films [75, 76]. Yonezu et al. [77] reported that an amorphous silicon (a-silicon) film sputtered on a moderately roughened copper foil achieved virtually 100% reversibility with a corresponding capacity greater than 3,000 mAh/g. The as-deposited a-silicon film (5-μm thick) grew to a thickness of 17 μm during the first full charge. After the first cycle, the silicon thin film divided into micro-columns.

Figure 15.6a, b show that the micro-columns of lithium-silicon predominantly expand/contract along the column diameter. This preferential expansion may directly relate to minimization of surface tension and surface energy in the given structures. Although these sputtered a-silicon thin films show excellent capacity and cycle performance, materials prepared by sputtering are not economical for large-scale, high-volume applications. Another serious limitation is that the practical electrode thickness attainable using vacuum deposition is not useful for anodes in large-scale applications such as PHEVs. If silicon nanorods can be prepared, using scalable synthesis routes, and that retain the high aspect ratio (similar to those shown in Fig. 15.6), it is reasonable to expect that the lithium-silicon nanorod would expand/contract preferentially along its diameter and maintain mechanical integrity during lithium insertion/de-insertion. Anodes with capacities >600 mAh/g and improved mechanical/structural stability over silicon thin films or bulk powders would be possible. Such developments could enable the use of silicon nanorod anodes in large-scale applications such as PHEVs and other commercial applications.

Another approach involves dispersing silicon nanoparticles with graphene sheets to form another class of two-dimensional, silicon-based nanostructures in which the silicon particles are in intimate contact with the graphene. In this method, regions of graphene restack to form a graphite network that anchors the more flexible graphene and provides a highly electronically conductive matrix. The highly

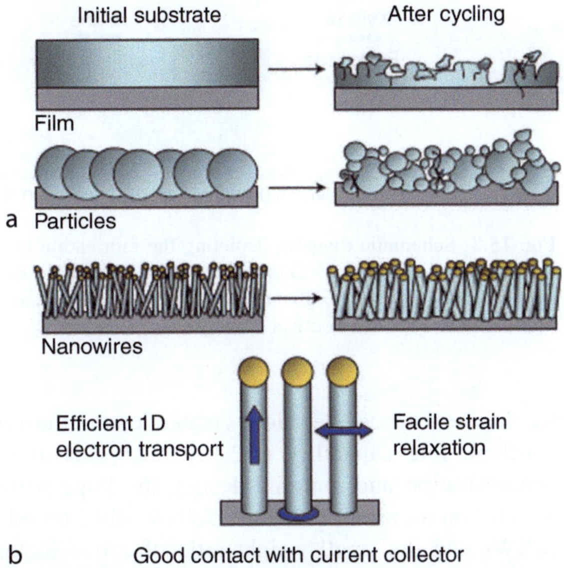

Fig. 15.7 Schematic of morphological changes of silicon during cycling. (**a**) The volume of silicon anodes changes by about 400% during cycling. (**b**) SiNWs do not disintegrate or break into smaller particles after cycling. This SiNW anode design has each nanowire connecting with the current collector (Reprinted with permission from Chan et al. [80]. Copyright 2008)

dispersed graphene/graphite matrix then buffers the volume expansion/contraction of silicon particles during repeated alloying and de-alloying with lithium [78, 79]. Specific capacities exceeding 1,500 mAh/g after 200 cycles with a capacity fading rate of less than 0.5% per cycle have been achieved using this approach.

One-Dimensional Silicon Nanowires

The previous section clearly established that the morphology, size, and structure of the silicon phase can have a dramatic effect on the ultimate performance and stability of a silicon-based anode. Figure 15.7 graphically depicts observed morphological changes that occur in thin-film silicon during electrochemical cycling when different starting morphologies are used. The rod-like microstructures formed during initial charge processes of silicon films (Fig. 15.6) suggest that one-dimensional geometries (such as silicon nanowires [SiNWs]) may provide a stable structure that could be synthesized from a more viable manufacturing route. In 2008, Chan et al. [80] prepared SiNWs (diameter < 100 nm) using a CVD method based on the vapor-liquid-solid mechanism using silane gas as the silicon source, gold as the catalyst with stainless steel substrates. The one-dimensional SiNWs grown directly on the stainless steel current collector had a measured capacity of 4,277 mAh/g (based on the weight of silicon) during the first discharging process; this capacity is essentially the theoretical value for silicon within the experimental error. In subsequent cycles, the SiNWs maintained a charge/discharge capacity of 75% of the initial capacity, with little fading during cycling. The shortened lithium transport distances in the silicon nanostructure and

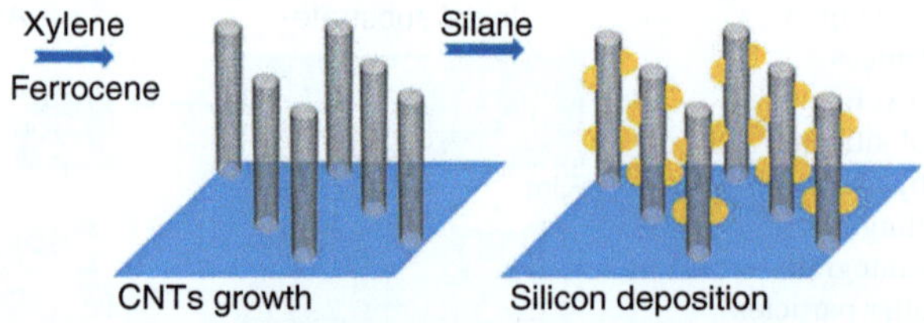

Fig. 15.8 Schematic diagram depicting the fabrication of silicon/carbon hybrid nanostructures using a liquid-injection CVD process to grow the initial vertically aligned CNTs, followed by the subsequent deposition of silicon (Reprinted with permission from Wang and Kumta [18]. Copyright 2010 American Chemical Society)

the low-resistance electrical connection to the current collector also led to an excellent rate capability (>2,100 mAh/g at 1C). While sufficient for concept demonstration and material design, the large-scale application of SiNWs grown directly on a current collector (foil) is still limited by the high production cost of SiNWs and the small weight ratio of active material to inactive material in the complete anode (which includes the substrate/current collector).

Several other approaches have been used to grow silicon-based nanowires. For example, Yu et al. [81] prepared SiNWs on silicon wafers using the solid-liquid-solid (SLS) mechanism. Kolb et al. [82] prepared silicon-based nanowires by evaporating silicon monoxide (SiO) in an inert gas atmosphere using a gold-coated silicon wafer as a substrate. Chang et al. [83] reported growth of silicon-based nanowires by heat treatment of an iron-catalyst–coated silicon nanopowder at 980°C. However, no electrochemical performances of these silicon-based nanowires were reported. Recently, Zhang et al. [84] prepared free-standing silicon-based nanowires from commercial silicon powders in a three-dimensional manner rather than on the surface of the substrate (i.e., two-dimensional growth). A vapor-induced SLS mechanism was proposed to explain the observed growth of the resulting multicomponent nanowires from solid-powder sources. The composition, morphology, crystal structure, and electrochemical performance of the nanowires also were investigated.

In addition to growth of SiNWs, one-dimensional composite structures prepared by vacuum deposition of silicon clusters on CNTs also have been investigated. Hybrid silicon/carbon nanotube one-dimensional nanostructures were synthesized using a two-step CVD process (see Fig. 15.8) [18]. The spaces between the CNTs, formed through controlled nucleation in the first deposition step, allow for subsequent penetration of silane gas and a homogenous deposition of silicon clusters on the surfaces of the CNTs during the second CVD deposition step. The hybrid silicon/CNTs exhibit a high reversible capacity of 2,000 mAh/g with a 0.15% capacity loss per cycle over 25 cycles. When compared with commercially available silicon particles, in situ prepared silicon by CVD always has a smaller irreversible capacity loss in the first cycle. This may be related to the reduced amount of SiO_x on the surface of the silicon particles which usually traps lithium in

Fig. 15.9 Schematic diagram of the lithium-ion pathway in silicon nanotubes (Reprinted with permission from Park et al. [21]. Copyright 2009 American Chemical Society)

the form of Li_2O and silicon in the first cycle. Coating silicon clusters on graphite nanofibers may also provide promising improvements in the cycling performance of silicon-based anodes.

Three-Dimensional Nanostructured Silicon Anodes

In addition to one-dimensional and two-dimensional silicon anodes, several forms of three-dimensional nanostructured silicon have been explored. For example, silicon nanotubes (Fig. 15.9) were investigated by Cho et al. [21] as an anode material for lithium-ion batteries. Both interior and exterior surfaces of the nanotubes are accessible to the electrolyte and lithium ions. Through carbon coating, a stable solid electrolyte interface (SEI) was generated on the inner and outer surfaces of the silicon nanotubes. These silicon/carbon assemblies showed a reversible capacity as high as 3,247 mAh/g (based on the weight of silicon) and good capacity retention.

Three-dimensional porous silicon was recently investigated as a high-performance anode material for lithium-ion batteries [85]. The three-dimensional porous silicon particles with in situ coated carbon on the surface contained many "octopus foot"-like voids as shown in Fig. 15.10. This highly porous, interconnected structure not only facilitates the transport of lithium ions but also significantly alleviates the detrimental effects of volume expansion and shrinkage. At a 3C rate (1C rate = 2,000 mA/g or 40 mA/cm^2), a discharge capacity of 2,158 mAh/g (based on the weight of silicon) with a 72% capacity retention was obtained after 100 cycles. Earlier reports on nanoporous silicon/carbon composites showed similar cycling stability [86]. A reversible capacity of 700 mAh/g based on the total weight of nanoporous silicon/graphite/carbon electrode was obtained with negligible capacity loss up to 120 cycles. Commercially available micrometer-sized silicon particles with nanoporous surface structures (Fig. 15.11) have also been investigated. Stable cycling behavior could be achieved if CVD carbon coatings were applied to individual silicon particles and an elastic carbon additive was used

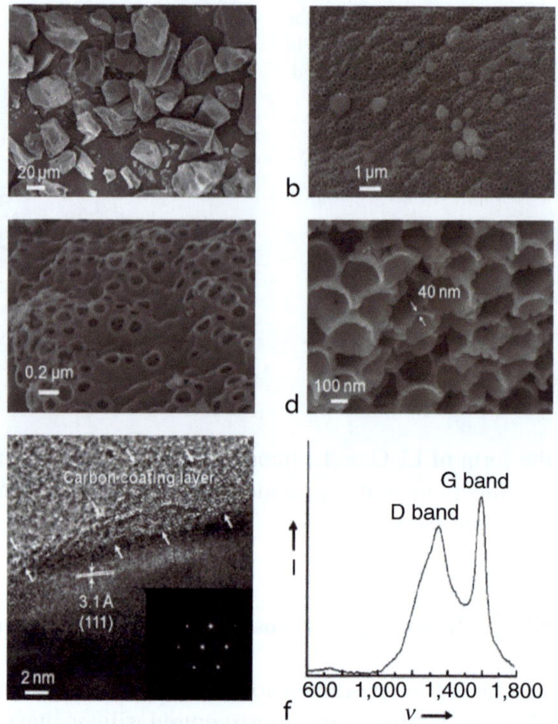

Fig. 15.10 (**a**), (**b**), (**c**), (**d**) SEM images of the three-dimensional porous c-silicon particles after etching ([**d**] is the cross-sectioned image of [**e**]). (**e**) TEM image of the cross-sectioned, three-dimensional porous c-silicon particle (the inset shows a selected area diffraction pattern. (**f**) Raman spectrum of three-dimensional porous c-silicon particles after etching. I = intensity (Reproduced with permission from Kim et al. [85]. Copyright Wiley-VCH Verlag GmbH & Co. KGaA)

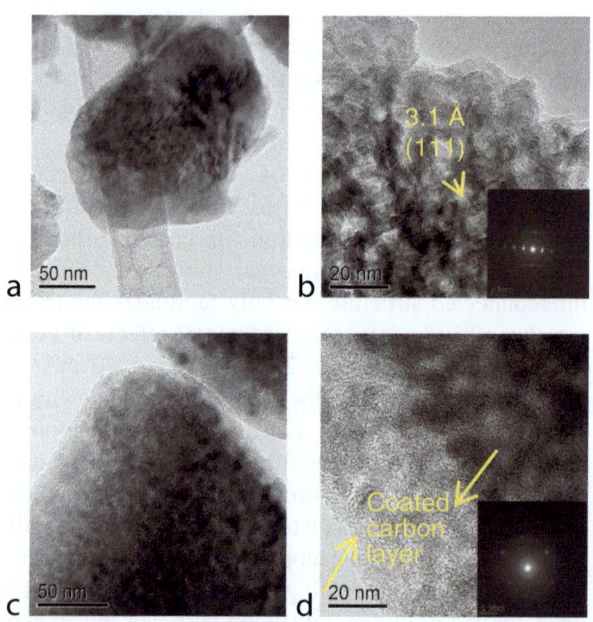

Fig. 15.11 (**a**) and (**b**) TEM images of porous silicon at different magnifications before carbon coating. (**c**) and (**d**) TEM images of silicon at different magnifications after carbon coating using CVD [1] (Reprinted with permission from Xiao et al. [87]. Copyright 2010 The Electrochemical Society)

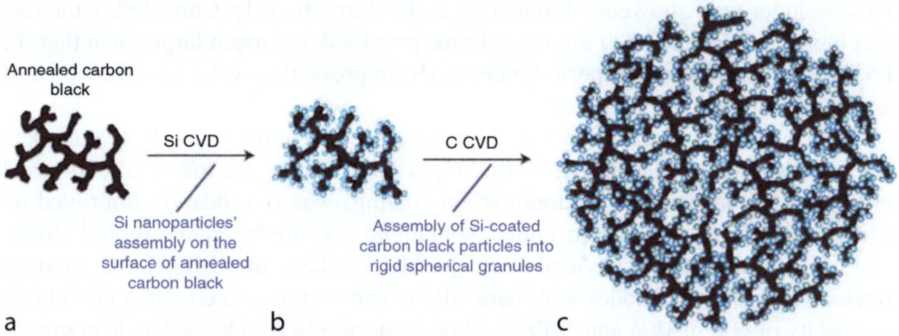

Fig 15.12 Schematic of silicon/carbon nanocomposite granule formation through hierarchical bottom-up assembly. (**a–c**). Annealed carbon-black dendritic particles (**a**) are coated by silicon nanoparticles (**b**) and then assembled into rigid spheres with open interconnected internal channels during C deposition (**c**) (Reprinted with permission from Magasinski et al. [89]. Copyright 2010)

to improve interparticle contact in the anode [87]. Nest-like silicon structures also are reported to have better electrochemical performance than that of coil-like silicon nanospheres, probably through a mechanism similar to that used to form three-dimensional nanostructured silicon [88].

To further improve the mechanical and electrical stability of silicon-based anodes, a hierarchical bottom-up approach (Fig. 15.12) was successfully utilized to develop a three-dimensional nanostructured silicon/carbon porous composite [89]. The existence of pores in the composite granules provides sufficient space to accommodate silicon expansion during lithium insertion. CVD deposition of silicon clusters (Fig. 15.12b) avoids formation of SiO_x, thus reducing the first cycle, irreversible capacity. A high specific capacity of ~1,950 mAh/g (C/20 rate) based on the total weight of the silicon/carbon composite was reported. In addition, the composite anodes had negligible capacity fade after 100 cycles at 1C rate and excellent rate capability (870 mAh/g at 8C rate).

Effect of Binder on the Performance of Silicon-Based Anodes

Sodium Carboxylmethylcellulose–Based Binder

Polyvinylidenedifluoride (PVDF) has been used widely as the binder for graphite-based anodes in lithium-ion batteries. However, silicon-based anodes with a PVDF-based binder exhibit poor cyclability. Therefore, significant efforts have been made during the last 10 years to develop new binders that can improve the performance of silicon-based anodes. Chen et al. [90] first reported that a poly(vinylidene fluoride-–tetrafluoroethylene–propylene)-based elastomeric binder system kept the capacity of the amorphous $Si_{0.6}Sn_{0.4}$ alloy at around 800 mAh/g for 40 cycles, while the

PVDF binder only showed ~490 mAh/g at the thirtieth cycle. Considering the fact that breakage elongation of the new elastomeric binder is much larger than that for PVDF [91, 92], the elastomeric binder could improve the cycling performance of a silicon-based alloy electrode.

In recent years, several new materials have been developed as effective binders for silicon-based anodes. Liu et al. [93] reported that the cycle life of silicon-based electrodes (either with or without carbon coating) was significantly improved by using a modified elastomeric binder containing styrene-butadiene-rubber (SBR) and sodium carboxylmethylcellulose (SCMC). When an SBR-SCMC mixture binder was used, the anodes with bare silicon and carbon-coated silicon exhibited a capacity of 600 mAh/g and 1,000 mAh/g, respectively at >50 cycles. In comparison, the capacity of a silicon electrode with a PVDF binder quickly decreased to several mAh/g after eight cycles. This is because the SBR-SCMC binder has a lower elastic modulus, a larger maximum elongation, better adhesion to the copper current collector, and much less solvent absorption in organic carbonate electrolytes. Several other research groups [46, 94–97] also have reported improved performance on silicon/carbon composite electrodes using an SBR-SCMC mixture binder over PVDF binder (sometimes combined with SBR [98]). Although a recent report shows that a thermally annealed PVDF can reconstruct the compact morphology of the electrode, the long-term cycling stability of a silicon electrode using an annealed PVDF binder is still inferior to that using CMC as the binder [99]. Li et al. [94] showed that the cycling performance of the crystalline silicon-powder electrode was improved further by using only SCMC as the binder over the SBR-SCMC mixture binder. Li et al. suggests that the stiff SCMC polymer might restrict the volume change of the silicon particles to only about 100% and also might modify the silicon particle surface. Thus, the binding mechanism of SCMC for a silicon electrode should be different from that of PVDF.

Hochgatterer et al. [100] reported that the improved long-term cycling behavior in an anode with an SCMC binder can be attributed to the formation of a covalent chemical bond between the SCMC binder and the silicon particles. SCMC exhibits a porous scaffold structure with cavities or pores that allow the silicon particles to expand without severe deformation of the electrode. Therefore, the cycle stability can be improved. Lestrie et al. [96] suggested that the efficiency of SCMC was likely attributed to the formation of an efficient conductive silicon/carbon-SCMC network because of its extended conformation in solution. Moreover, the silicon/carbon-SCMC electrode prepared at pH 3 showed prolonged cyclability improvement, which probably resulted from the physical cross-linking of the SCMC chains in the acid solution. Guo and Wang [101] recently reported a polymer scaffold structure based on an SCMC binder for a silicon electrode. The SCMC scaffold-based silicon electrode had much higher surface area and total pore volume but a reduced mean pore diameter than the SCMC-based silicon electrode prepared using a conventional method. Therefore, silicon nanoparticle electrodes with a SCMC scaffold binder structure could accommodate greater volume changes from the silicon, enhance lithium-ion transport in the electrodes, and improve the electrochemical reaction kinetics. In addition, high energy and power densities and superb cycle stability were achieved for porous SCMC, scaffold-based silicon electrodes. After 150 cycles at a rate of 250 mA/g, the anode capacity remained at 1,685 mAh/g based on the weight of silicon. The improved rate performance of silicon is attributed to the SCMC scaffold, which facilitates lithium-ion transport in the porous electrode and accelerates the charge transfer reaction kinetics [101].

Although the binding mechanism of SCMC in silicon electrodes is unclear, the significant performance improvements observed drives efforts to optimize SCMC/silicon-based electrodes. In fact, some researchers believe that the binder plays a more important role than the nanostructure of the silicon particles on the performance of silicon-based anodes. Beattie et al. [102] reported that silicon-based electrodes with a relatively low silicon content (20–34 wt.%) and a high SCMC content (33–56 wt.%) had large capacities (660 mAh/g) for hundreds of cycles. They believed that, if an appropriate binder was used with silicon electrodes, no special electrode processing or cycling procedures were required to achieve high capacities with good cyclability.

Other Binders

In addition to SCMC, several other polymeric binders have been developed for silicon-based anodes in recent years. Zheng et al. [103] reported that a nanoporous silicon/graphite/carbon composite electrode containing polyacrylic latex LA132 binder (10 wt.%) exhibited a reversible capacity of 650 mAh/g after 200 cycles with negligible capacity loss. Chen et al. [104] reported a significant cycling-performance enhancement for silicon/carbon composite electrodes using an acrylic adhesive (polyacrylic latex, LA132) and a modified acrylic adhesive with SCMC. The capacity retentions of the silicon/carbon composite electrodes were 79% and 90% after 50 cycles, respectively, while the silicon/carbon composite electrode with PVDF retained only 67% of its initial capacity. This was ascribed to improved adhesion between the coating and the copper current collector, as well as reduced solvent absorption of the electrolyte solvent than with PVDF. Polyamide imide (PAI) also can significantly improve the initial coulombic efficiency of a silicon-based anode because of PAI's excellent mechanical properties and ease of processing, which maintains the stability of the electrical-conducting network during charge/discharge processes [105].

Recently, Liu et al. [106] reported on conductive polymers used as both the binder and the conductive matrix. The electronic conductivity of the reported binder increases during the lithium-insertion process. Stable cycling above ~1,200 mAh/g (based on silicon) was reported for the silicon-based anode using these binders. The cycling stability of silicon improved significantly, especially within the reduced electrochemical window (0.17–0.9 V) [107]. However, the initial irreversible capacity loss in these anodes still requires further improvement (i.e., reduction) before these silicon-based anodes can be used in a full cell. Cámer et al. [108] reported that silicon-composite electrodes with very good electrochemical performance could be prepared by simply mixing appropriate amounts of nanosized silicon and cellulose fibers without including conductive carbon and other binders. A silicon/carbon cellulose composite electrode at a weight ratio of 34:66 exhibited very good cycling performance. A specific capacity of ~1,400 mAh/g after 50 cycles was obtained, compared to only 400 mAh/g for an electrode made from pure silicon. Cámer et al. believe that the improved performance resulted from mitigation of the large volume changes of silicon particles during lithium insertion and de-insertion and the retention of connectivity between particles by the binder-like fibers.

Effect of Electrolytes on the Performance of Silicon-Based Anodes

The effect of nonaqueous electrolytes and additives on the performance of silicon electrodes has not been systematically investigated. The electrolytes used in the investigation of silicon-based electrodes were mainly adopted from electrolytes for graphite-based lithium-ion batteries (i.e., $LiPF_6$ in carbonate mixtures of EC and DMC, ethyl methyl carbonate [EMC] or diethyl carbonate [DEC]). Alternative electrolytes used in the study of silicon-based anodes include (1) adding SEI formation additives into the regular electrolytes of $LiPF_6$ in carbonates, (2) using different lithium salts, (3) using different solvent systems, (4) using ionic-liquid-based electrolytes, (5) using polymer electrolytes, and (6) using solid-state electrolytes. The effects of these electrolytes on the performance of a silicon-based electrode are reviewed in this section.

SEI Formation Additives

Formation of the SEI layer plays an important role on the cycling stability of silicon-based anodes. Kulova and Skundin [109] reported a method used to preform the SEI layer on an electrode surface (prior to initial cathodic polarization) by direct contact of silicon and lithium metal in the electrolyte. The electrolyte was 1-M $LiClO_4$ in a mixture of propylene carbonate (PC) and dimethoxyethane. This method effectively reduced the irreversible capacity of the amorphous silicon electrode. Pretreatment of nanometer-sized silicon in ethanol also can form functionalized surfaces on the silicon particles that improve adhesion of silicon-based electrodes. A stable capacity of 2,500 mAh/g after 25 cycles has been reported for ethanol-treated silicon electrodes [110].

Several electrolyte additives have been used to form stable SEI layers and improve cyclability of silicon anodes. Doh et al. [111] used 5% 4-fluoroethylene carbonate (FEC) in the electrolyte of 1.0-M $LiPF_6$ in EC/DMC/EMC/PC at a volume ratio of 4:3:3:1 when investigating silicon/carbon composites formed by polyaniline carbonization. They found that the addition of FEC to the electrolyte increased the initial discharge capacity of the silicon/carbon composites when compared with the electrolyte without FEC. Choi et al. [112] reported that the discharge capacity retention and coulombic efficiency of a silicon thin-film electrode could be significantly improved by adding 3% of FEC into the electrolyte of 1.3-M $LiPF_6$ in EC/DEC (3:7 by volume). They used SEM and X-ray photoelectron spectroscopy (XPS) to investigate the surface morphology and chemical composition of the SEI layers formed on the surface of the silicon thin-film electrode after cycling. The SEI layer structure formed in the FEC-containing electrolyte was much smoother and more stable than those formed in electrolytes that did not contain the FEC additive. Chen et al. [113] investigated the effect of vinylene carbonate (VC) as an electrolyte additive on the electrochemical performance of a silicon thin-film anode. They found that the cycle performance and efficiency of the silicon thin-film anode were significantly enhanced with the addition of 1% VC into the electrolyte of 1-M $LiPF_6$ in EC/DMC. The capacity of the silicon thin film shows minimal fade after 500 cycles, which was attributed to the advanced properties of the SEI layer formed during the initial cycles. Actually, in both VC-containing and VC-free electrolytes, the major components of the SEI layer on the silicon thin-film anode were quite similar and primarily consist of lithium salts (e.g., $ROCO_2Li$, Li_2CO_3, LiF), polycarbonates, and silicon oxide formed through the reaction of lithiated silicon with permeated electrolyte. However, the morphology of the SEI layer was smoother and more uniform in VC-containing electrolytes than in VC-free electrolytes. The impedance of the SEI layer in VC-containing electrolytes did not change significantly upon cycling because of the presence of VC-reduced products and less LiF content in the SEI layer, which led to better properties of the SEI layer. However, the impedance of the SEI layer in a VC-free electrolyte increased constantly upon cycling because of the increasing thickness and high LiF content in the SEI layer, resulting in increased electrode polarization and degradation in the cycling performance of the silicon thin-film anode.

Han et al. [114] studied the effect of succinic anhydride (SA) as an electrolyte additive on the electrochemical performances of a silicon thin-film electrode. They found that addition of a small amount (3 wt.%) of SA into the electrolyte of 1-M $LiPF_6$ in EC/DEC could significantly enhance the capacity retention and coulombic efficiency of the silicon electrode. SA also prevented decomposition of the $LiPF_6$ salt, and the SEI layer contained higher levels of hydrocarbon and Li_2CO_3 on the silicon surface. Modification to the SEI layer by SA was probably the primary factor for the enhanced performance of the silicon thin-film electrode. Baggetto et al. [115] used a poly-silicon thin film as the active anode material and cycled the poly-silicon anodes in one solid and two liquid electrolytes to investigate the thermodynamic and kinetic properties of the anodes and the growth of SEI layers on top of poly-silicon anodes. They studied the electrochemical and material characteristics of a potential planar anode stack (active anode material/barrier layer/silicon substrate) for all-solid-state, three-dimensional, integrated batteries. The solid-state electrolyte was amorphous lithium phosphorus oxynitride (LiPON), and the two liquid electrolytes were 1-M $LiPF_6$ in EC/DEC and 1-M $LiClO_4$ in PC. The silicon electrodes cycled in the two liquid electrolytes showed stable storage capacities up to about 30 or 40 cycles, respectively, then the capacities decreased sharply. When the inorganic solid-state electrolyte LiPON was used to overcoat the silicon, no degradation in capacity was observed for 60 cycles. Cross-sectional SEM analysis revealed thick, porous SEI layer formation in the case of $LiPF_6$ and $LiClO_4$ based electrolytes. The silicon layer was almost invisible, and it seemed to be "dissolved" within the SEI layer, implying that the active silicon clusters slowly became electronically isolated resulting in the loss of reversible capacity. In contrast, the silicon covered by a LiPON layer had no visible SEI layer formation and improved cycle life over cells that were LiPON-free.

Arie et al. [116] investigated the electrochemical characteristics of phosphorus- and boron-doped silicon thin-film (n-type and p-type silicon) anodes integrated with a solid polymer electrolyte in lithium-polymer batteries. The doped silicon electrodes showed enhanced discharge capacity and coulombic efficiency over the un-doped silicon electrode, and the phosphorus-doped, n-type silicon electrode showed the most stable cyclic performance after 40 cycles with a reversible specific capacity of about 2,500 mAh/g. The improved electrochemical performance of the doped silicon electrode was mainly due to enhancement of its electrical and lithium-ion conductivities and stable SEI layer formation on the surface of the electrode. In the case of the un-doped silicon electrode, an unstable surface layer formed on the electrode surface, and the interfacial impedance was relatively high, resulting in high electrode polarization and poor cycling performance.

Alternative Solvents and Salts

Inose et al. [117] studied the influence of glyme-based electrolyte solutions on charge-discharge properties of two silicon-based anodes prepared from (1) a silicon/carbon mixture and (2) a mixture of a silicon-SiO_2-carbon composite and carbon (Si-C/C). After glymes of poly(ethylene glycol) dimethyl ethers [$(CH_3O(CH_2CH_2O)_nCH_3$, $n = 1, 2, 3$ and 4)] were added into an electrolyte solution of 1 M $LiPF_6$ in EC and EMC (3:7 by vol), the discharge capacity tended to increase relative to electrolytes without glyme additions. Carbon coating of the silicon particles in the Si-C/C electrode improved cycle life when compared to the silicon/carbon electrode. Physical breakdown of the electrode was suppressed by the thin carbon surface layer, and reactivity of electrolyte toward lithium metal also was reduced. They also found that the discharge capacity of the silicon/carbon electrode was dependent on the reduction potential of the glymes, while the discharge capacity of the Si-C/C electrode depended on electrolyte conductivity.

Choi et al. [118] compared the effect of two different lithium salts on the cycling performance of a 200-nm silicon thin-film electrode. The electrolytes tested were 1.3-M $LiPF_6$ in EC/DEC (3:7 by vol) and 0.7-M lithium bis(oxalato) borate (LiBOB) in the same solvent mixture. They found that the LiBOB-based electrolyte markedly improved the discharge capacity retention of the lithium-silicon half-cell, over the $LiPF_6$-based electrolyte. The surface layer on the silicon electrode in the LiBOB-based electrolyte was less porous and effectively limited the formation of electrochemically inactive silicon phases. The capacity fading of the lithium-silicon half-cell was strongly related to the physical morphology of the silicon surface and also to the formation of inactive silicon phases in the surface layer.

Ionic liquids have also been studied as possible electrolytes in silicon-based anode cells. Lux et al. [119] evaluated the use of N-butyl-N-methylpyrrolidinium bis (trifluoromethylsulfonyl)imide (PYR14TFSI) as a solvent for silicon electrodes at room temperature. The electrolyte contained 0.3-M lithium bis(trifluoromethylsulfonyl)imide ($LiN(SO_2CF_3)_2$, LiTFSI) in PYR14TFSI + 5 wt.% VC. The performance of silicon electrodes in PYR14TFSI-based electrolytes was comparable with that of the same electrodes in more conventional organic solvent-based electrolytes. The degree of affinity between ionic liquids and active materials was suggested as an important parameter for the selection and use of ionic liquids in lithium batteries.

Full Cells Containing Silicon-Based Anodes

To fully evaluate the efficacy of silicon-based anodes, testing in a full-cell configuration against a lithium intercalation cathode, such as $LiCoO_2$, is required because of possible detrimental interactions between the cathode and anode (through electrolyte as the media), as was discovered in the case of graphite/$LiMn_2O_4$. In addition, the management of volume change in the complete cells is not a trivial challenge for silicon-based cells where the anode exhibits large volume variations during cycling. Nonetheless, studies of electrochemical performance for silicon-anode-based full batteries are rare, possibly because of the poor cycling performance of these anodes, which remains a hurdle to practical application of the technology.

Cui et al. [120] reported a full-cell test using a $LiCoO_2$ cathode and carbon-silicon core-shell nanowire anode. This anode has about 10 times the specific capacity of the cathode and, therefore, a mass loading approximately 10 times lower than that of the cathode. The full cell operated at $\sim$3.3 V with a sloping voltage profile and a fast fade of $\sim$0.7% per cycle. The same group also reported full battery performance of the silicon nanowire anode and sulfur-based cathode pair [121]. Although the voltage profile of the full-cell test revealed large overpotentials because of poor electrical conductivity of the Li_2S cathode, the specific energy density of the battery was reported to be approximately four times higher than the current $LiCoO_2$/graphite-based, lithium-ion battery mostly because of the high specific capacity of the sulfur cathode. High-voltage $LiNi_{0.5}Mn_{1.5}O_4$ also has been paired with silicon nanoparticles to form a full-cell battery [122]. The cycling performance at increased operating voltage resulted in higher specific capacity than a conventional $LiCoO_2$/graphite-based commercial battery. However, the cycling stability of this battery is still problematic. There also are reports of thin-film batteries using amorphous silicon anodes paired with $LiMn_2O_4$ [123] or $LiCoO_2$ cathodes [124–126] with better cyclability than the similar bulk batteries, but large-scale manufacturing of thin-film batteries is hindered by the high manufacturing cost of these materials.

Future Directions

Although the silicon-based anode has great potential to significantly improve the capacity of lithium-ion batteries, there are still many obstacles that prevent its practical application. Long-term cycling stability remains the foremost challenge facing the battery research community. The cyclability of full cells using silicon-based anodes is complicated by multiple factors, such as diffusion-induced stress and fracture [127], loss of electrical contact among silicon particles and between silicon and current collector, and the breakdown of SEI layers during volume expansion/contraction processes. A detailed study of the design and engineering of a full cell with a silicon-based anode still needs to be conducted after a stable silicon-based anode structure is developed. Critical research remaining in this area includes, but is not limited to, the following:

- Further understanding of the effect of SEI formation on the cyclability of silicon-based anodes. Electrolytes and additives that can produce a stable SEI layer need to be developed. In situ SEM of working electrochemical test cells could be a very useful tool to directly observe the formation and evolution of SEI layers.
- The effect of particle size and the mechanical stability of silicon particles need to be understood. A theoretical model needs to be developed to find the critical dimension at which silicon particles are stable under long-term expansion/shrinking cycles.
- A good binder and conductive matrix (such as carbon) need to be developed. They should provide flexible but stable electrical contacts among silicon particles and between particles and the current collector under repeated volume changes during charge/discharge processes. The specific capacity of the complete anode (including the silicon, binder, and conductive matrix) needs further optimization. An acceptable balance between the specific capacity and the long-term stability of the anode also is needed. Considering the fact that the specific capacity of state-of-the-art cathodes is between ~140 mAh/g and 200 mAh/g, Kasavajjula et al. [24] suggest that improvement in the total capacity of an 18,650 cell is negligible after the anode specific capacity reaches more than 1,200 mAh/g if other components of the batteries remain unchanged. Based on this analysis, a specific capacity of ~1,200 mAh/g for the complete anode with a loading of 5–10 mg/cm^2 would be a reasonable technical target. Improvement in cycling stability remains the primary challenge for silicon-based anodes.
- The performance of full cells needs to be studied and optimized. Parameters to be optimized include cell geometry, anode and cathode balancing, accommodation of volume expansion, and electrolyte/additive selection.

Last, it should be recognized that the overall effect of the high-capacity anode on the performance of lithium-ion batteries is limited primarily because the anode only occupies a relatively small portion of the total weight (15–20%) and volume of the lithium-ion battery [128]. An anode with infinite specific capacity (or zero weight) could only lead to less than a 20% reduction in the total weight of the system, if other components of the battery remain the same. Many factors affect the total specific capacity and weight/volume of a full battery, such as the specific capacities of the anode and cathode, electrolyte, separator, current collectors, and case. Therefore, significant improvements in both the anode and the cathode, as well as other battery components, must be achieved to realize the goal of producing widely applied, high-energy batteries.

Bibliography

1. Boukamp BA, Lesh GC, Huggins RA (1981) All-solid lithium electrodes with mixed-conductor matrix. J Electrochem Soc 128:725–729
2. Dey AN (1971) Electrochemical, alloying of lithium in organic electrolytes. J Electrochem Soc 118:1547–1549
3. Tirado JL (2003) Inorganic, materials for the negative electrode of lithium-ion batteries: state-of-the-art and future prospects. Mater Sci Eng R Rep 40:103–136
4. Winter M, Besenhard JO (1999) Electrochemical lithiation of tin and tin-based intermetallics and composites. Electrochim Acta 45:31–50
5. Kim I-S (2003) Synthesis, structure and properties of electrochemically active nanocomposites. Ph.D thesis, Carnegie Mellon University
6. Sharma RA, Seefurth RN (1976) Thermodynamic properties of the lithium-silicon system. J Electrochem Soc 123:1763–1768
7. van der Marel C, Vinke GJB, van der Lugt W (1985) The phase diagram of the system lithium-silicon. Solid State Commun 54:917–919
8. Lai S-C (1976) Solid lithium-silicon electrode. J Electrochem Soc 123:1196–1197
9. Li H et al (2000) The crystal structural evolution of nano-Si anode caused by lithium insertion and extraction at room temperature. Solid State Ionics 135:181–191
10. Limthongkul P, Jang Y-I, Dudney NJ, Chiang Y-M (2003) Electrochemically-driven solid-state amorphization in lithium-silicon alloys and implications for lithium storage. Acta Mater 51:1103–1113
11. Limthongkul P, Jang Y-I, Dudney NJ, Chiang Y-M (2003) Electrochemically-driven solid-state amorphization in lithium-metal anodes. J Power Sources 119–121:604–609
12. Maranchi JP, Hepp AF, Kumta PN (2003) High capacity reversible silicon thin film anodes lithium ion batteries. Electrochem Solid-State Lett 6:A198–A201
13. Ryu JH, Kim JW, Sung Y-E, Oh SM (2004) Failure modes of silicon powder negative electrode in lithium secondary batteries. Electrochem Solid-State Lett 7:A306–A309
14. Li J, Dahn JR (2007) An in situ x-ray diffraction study of the reaction of Li with crystalline Si. J Electrochem Soc 154:A156–A161
15. Obrovac MN, Christensen L (2004) Structural changes in silicon anodes during lithium insertion/extraction. Electrochem Solid-State Lett 7:A93–A96
16. Hatchard TD, Dahn JR (2004) In situ XRD and electrochemical study of the reaction of lithium with amorphous silicon. J Electrochem Soc 151:A838–A842
17. Datta MK, Kumta PN (2009) In situ electrochemical synthesis of lithiated silicon-carbon based composites anode materials for lithium ion batteries. J Power Sources 194:1043–1052

18. Wang W, Kumta PN (2010) Nanostructured hybrid silicon/carbon nanotube heterostructures: reversible high-capacity lithium-ion anodes. ACS Nano 4:2233–2241
19. Obrovac MN, Krause LJ (2007) Reversible cycling of crystalline silicon powder. J Electrochem Soc 154:A103–A108
20. Beaulieu LY, Eberman KW, Turner RL, Krause LJ, Dahn JR (2001) Colossal reversible volume changes in lithium alloys. Electrochem Solid-State Lett 4:A137–A140
21. Park MH et al (2009) Silicon nanotube battery anodes. Nano Lett 9:3844–3847
22. Datta MK, Kumta PN (2006) Silicon and carbon based composite anodes for lithium ion batteries. J Power Sources 158:557–563
23. Sandia National Laboratories. Sandia National Laboratories News Releases. Sandia National Laboratories, Livermore, CA, 6 March 2003
24. Kasavajjula U, Wang C, Appleby AJ (2007) Nano- and bulk-silicon-based insertion anodes for lithium-ion secondary cells. J Power Sources 163:1003–1039
25. Larcher D et al (2007) Recent findings and prospects in the field of pure metals as negative electrodes for Li-ion batteries. J Mater Chem 17:3759–3772
26. Timmons A et al (2007) Studies of Si[sub 1 − x]C[sub x] electrode materials prepared by high-energy mechanical milling and combinatorial sputter deposition. J Electrochem Soc 154:A865–A874
27. Huggins RA (1999) Lithium, alloy negative electrodes. J Power Sources 81–82:13–19
28. Mao O et al (1999) Active/inactive nanocomposites as anodes for Li-ion batteries. Electrochem Solid-State Lett 2:3–5
29. Weydanz WJ, Wohlfahrt-Mehrens M, Huggins RA (1999) A room temperature study of the binary lithium-silicon and the ternary lithium-chromium-silicon system for use in rechargeable lithium batteries. J Power Sources 81–82:237–242
30. Courtney IA, McKinnon WR, Dahn JR (1999) On the aggregation of tin in SnO composite glasses caused by the reversible reaction with lithium. J Electrochem Soc 146:59–68
31. Mayo MJ (1997) High, and low temperature superplasticity in nanocrystalline materials. Nanostruct Mater 9:717–726
32. Wang W (2009) Silicon Based Nanocomposites as Lithium-ion Battery Anodes. PhD dissertation, Carnegie Mellon University
33. Idota Y, Kubota T, Matsufuji A, Maekawa Y, Miyasaka T (1997) Tin-based amorphous oxide: a high-capacity lithium-ion-storage material. Science 276:1395–1397
34. Hwang S-M et al (2001) Lithium insertion in SiAg powders produced by mechanical alloying. Electrochem Solid-State Lett 4:A97–A100
35. Kim H, Choi J, Sohn H-J, Kang T (1999) The insertion mechanism of lithium into Mg[sub 2] Si anode material for Li-ion batteries. J Electrochem Soc 146:4401–4405
36. Kim I-S, Kumta PN, Blomgren GE (2000) Si/TiN nanocomposites novel anode materials for Li-ion batteries. Electrochem Solid-State Lett 3:493–496
37. Kim I-S, Blomgren GE, Kumta PN (2004) Si-SiC nanocomposite anodes synthesized using high-energy mechanical milling. J Power Sources 130:275–280
38. Kim I-S, Blomgren GE, Kumta PN (2003) Nanostructured Si/TiB_2 composite anodes for Li-ion batteries. Electrochem Solid-State Lett 6:A157–A161
39. Wang CS, Wu GT, Zhang XB, Qi ZF, Li WZ (1998) Lithium insertion in carbon-silicon composite materials produced by mechanical milling. J Electrochem Soc 145:2751–2758
40. Gross KJ, Wang JCF, Roberts GA (2004) Synthesis of carbon/silicon composites. US Patent 2004/137,327 (2004)
41. Kim I-S, Kumta PN (2004) High capacity Si/C nanocomposite anodes for Li-ion batteries. J Power Sources 136:145–149
42. Wilson AM, Reimers JN, Fuller EW, Dahn JR (1994) Lithium insertion in pyrolyzed siloxane polymers. Solid State Ionics 74:249–254
43. Yang J et al (2003) Si/C composites for high capacity lithium storage materials. Electrochem Solid-State Lett 6:A154–A156

44. Xie J, Cao GS, Zhao XB (2004) Electrochemical performances of Si-coated MCMB as anode material in lithium-ion cells. Mater Chem Phys 88:295–299
45. Holzapfel M et al (2005) Chemical vapor deposited silicon/graphite compound material as negative electrode for lithium-ion batteries. Electrochem Solid-State Lett 8:A516–A520
46. Holzapfel M, Buqa H, Scheifele W, Novak P, Petrat F-M (2005) A new type of nano-sized silicon/carbon composite electrode for reversible lithium insertion. Chem Commun 1566–1568
47. Dimov N, Fukuda K, Umeno T, Kugino S, Yoshio M (2003) Characterization of carbon-coated silicon: structural evolution and possible limitations. J Power Sources 114:88–95
48. Liu W-R et al (2005) Electrochemical characterizations on Si and C-coated Si particle electrodes for lithium-ion batteries. J Electrochem Soc 152:A1719–A1725
49. Yu M-F, Files BS, Arepalli S, Ruoff RS (2000) Tensile loading of ropes of single wall carbon nanotubes and their mechanical properties. Phys Rev Lett 84:5552
50. Krishnan A, Dujardin E, Ebbesen TW, Yianilos PN, Treacy MMJ (1998) Young's modulus of single-walled nanotubes. Phys Rev B 58:14013
51. Wong EW, Sheehan PE, Lieber CM (1997) Nanobeam mechanics: elasticity, strength, and toughness of nanorods and nanotubes. Science 277:1971–1975
52. Roche S (2000) Carbon nanotubes: exceptional mechanical and electronic properties. Ann Chim Sci Matériaux 25:529–532
53. Zhao Q, Nardelli MB, Bernholc J (2002) Ultimate strength of carbon nanotubes: a theoretical study. Phys Rev B 65:144105
54. Demczyk BG et al (2002) Direct mechanical measurement of the tensile strength and elastic modulus of multiwalled carbon nanotubes. Mater Sci Eng A 334:173–178
55. Thess A et al (1996) Crystalline ropes of metallic carbon nanotubes. Science 273:483–487
56. Yao Z, Kane CL, Dekker C (2000) High-field electrical transport in single-wall carbon nanotubes. Phys Rev Lett 84:2941
57. Frank S et al (1998) Carbon nanotube quantum resistors. Science 280:1744–1746
58. Shu J, Li H, Yang R, Shi Y, Huang X (2006) Cage-like carbon nanotubes/Si composite as anode material for lithium ion batteries. Electrochem Commun 8:51–54
59. Wang W, Kumta PN (2007) Reversible high capacity nanocomposite anodes of Si/C/SWNTs for rechargeable Li-ion batteries. J Power Sources 172:650–658
60. Si Q et al (2010) A high performance silicon/carbon composite anode with carbon nanofiber for lithium-ion batteries. J Power Sources 195:1720–1725
61. Guo JC, Sun A, Wang CS (2010) A porous silicon-carbon anode with high overall capacity on carbon fiber current collector. Electrochem Commun 12:981–984
62. Lee J et al (2009) Effect of randomly networked carbon nanotubes in silicon-based anodes for lithium-ion batteries. J Electrochem Soc 156:A905–A910
63. Jang S-M, Miyawaki J, Tsuji M, Mochida I, Yoon S-H (2009) The preparation of a novel Si-CNF composite as an effective anodic material for lithium-ion batteries. Carbon 47:3383–3391
64. Yang J, Winter M, Besenhard JO (1996) Small particle size multiphase Li-alloy anodes for lithium-ionbatteries. Solid State Ionics 90:281–287
65. Yang J, Takeda Y, Imanishi N, Ichikawa T, Yamamoto O (2000) SnSbx-based composite electrodes for lithium ion cells. Solid State Ionics 135:175–180
66. Yang J, Takeda Y, Imanishi N, Yamamoto O (1999) Ultrafine Sn and $SnSb_{0.14}$ Powders for lithium storage matrices in lithium-ion batteries. J Electrochem Society 146:4009–4013
67. Huggins R, Nix W (2000) Decrepitation model for capacity loss during cycling of alloys in rechargeable electrochemical systems. Solid State Ionics 6:57–63
68. Li H, Huang X, Chen L, Wu Z, Liang Y (1999) A high capacity nano-Si composite anode material for lithium rechargeable batteries. Electrochem Solid-State Lett 2:547–549
69. Beaulieu LY, Dahn JR (2000) The reaction of lithium with Sn-Mn-C intermetallics prepared by mechanical alloying. J Electrochem Soc 147:3237–3241
70. Gleiter H (1989) Nanocrystalline materials. Prog Mater Sci 33:223–315

71. Graetz J, Ahn CC, Yazami R, Fultz B (2003) Highly reversible lithium storage in nanostructured silicon. Electrochem Solid-State Lett 6:A194–A197
72. Maranchi JP, Hepp AF, Evans AG, Nuhfer NT, Kumta PN (2006) Interfacial properties of the a-Si/Cu:active–inactive thin-film anode system for lithium-ion batteries. J Electrochem Soc 153:A1246–A1253
73. Kim J-B, Lee H-Y, Lee K-S, Lim S-H, Lee S-M (2003) Fe/Si multi-layer thin film anodes for lithium rechargeable thin film batteries. Electrochem Commun 5:544–548
74. Kim Y-L et al (2003) Electrochemical characteristics of Co-Si alloy and multilayer films as anodes for lithium ion microbatteries. Electrochim Acta 48:2593–2597
75. Lee KL, Jung JY, Lee SW, Moon HS, Park JW (2004) Electrochemical characteristics of a-Si thin film anode for Li-ion rechargeable batteries. J Power Sources 129:270–274
76. Kim YL, Sun YK, Lee SM (2008) Enhanced electrochemical performance of silicon-based anode material by using current collector with modified surface morphology. Electrochim Acta 53:4500–4504
77. Yonezu I, Tarui H, Yoshimura S, Fujitani S, Nohma T (2004) Abstracts of the 12th International Meeting on Lithium Batteries, vol 58. Electrochemical Society, Nara, Japan, 2004
78. Lee JK, Smith KB, Hayner CM, Kung HH (2010) Silicon nanoparticles-graphene paper composites for Li ion battery anodes. Chem Commun 46:2025–2027
79. Chou SL et al (2010) Enhanced reversible lithium storage in a nanosize silicon/graphene composite. Electrochem Commun 2:303–306
80. Chan CK et al (2008) High-performance lithium battery anodes using silicon nanowires. Nat Nanotechnol 3:31–35
81. Yu DP et al (2001) Controlled growth of oriented amorphous silicon nanowires via a solid-liquid-solid (SLS) mechanism. Physica E 9:305–309
82. Kolb FM et al (2004) Analysis of silicon nanowires grown by combining SiO evaporation with the VLS mechanism. J Electrochem Soc 151:G472–G475
83. Chang JB et al (2006) Ultrafast growth of single-crystalline Si nanowires. Mater Lett 60:2125–2128
84. Zhang JG et al (2010) Vapor-induced solid–liquid–solid process for silicon-based nanowire growth. J Power Sources 195:1691–1697
85. Kim H, Han B, Choo J, Cho J (2008) Three-dimensional porous silicon particles for use in high-performance lithium secondary batteries. Angew Chem Int Ed 47:10151–10154
86. Zheng Y, Yang J, Wang JL, NuLi YN (2007) Nano-porous Si/C composites for anode material of lithium-ion batteries. Electrochim Acta 52:5863–5867
87. Xiao J et al (2010) Stabilization of silicon anode for Li-ion batteries. J Electrochem Soc 157: A1047–A1051
88. Ma H et al (2007) Nest-like silicon nanospheres for high-capacity lithium storage. Adv Mater 19:4067–4070
89. Magasinski A et al (2010) High-performance lithium-ion anodes using a hierarchical bottom-up approach. Nat Mat 9:353–358
90. Chen ZH, Christensen L, Dahn JR (2003) Large-volume-change electrodes for Li-ion batteries of amorphous alloy particles held by elastomeric tethers. Electrochem Commun 5:919–923
91. Chen ZH, Christensen L, Dahn JR (2003) Comparison of PVDF and PVDF-TFE-P as binders for electrode materials showing large volume changes in lithium-ion batteries. J Electrochem Soc 150:A1073–A1078
92. Chen ZH, Christensen L, Dahn JR (2004) Mechanical and electrical properties of poly (vinylidene fluoride-tetrafluoroethylene-propylene)/super-S carbon black swelled in liquid solvent as an electrode binder for lithium-ion batteries. J Appl Polym Sci 91:2958–2965
93. Liu WR, Yang MH, Wu HC, Chiao SM, Wu NL (2005) Enhanced cycle life of Si anode for Li-ion batteries by using modified elastomeric binder. Electrochem Solid-State Lett 8: A100–A103

94. Li J, Lewis RB, Dahn JR (2007) Sodium carboxymethyl cellulose – a potential binder for Si negative electrodes for Li-ion batteries. Electrochem Solid-State Lett 10:A17–A20
95. Dimov N, Xia Y, Yoshio M (2007) Practical silicon-based composite anodes for lithium-ion batteries: Fundamental and technological features. J Power Sources 171:886–893
96. Lestrie B, Bahri S, Sandu I, Roue L, Guyomard D (2007) On the binding mechanism of CMC in Si negative electrodes for Li-ion batteries. Electrochem Commun 9:2801–2806
97. Key B et al (2009) Real-time NMR investigations of structural changes in silicon electrodes for lithium-ion batteries. J Am Chem Soc 131:9239–9249
98. Buqa H, Holzapfel M, Krumeich F, Veit C, Novak P (2006) Study of styrene butadiene rubber and sodium methyl cellulose as binder for negative electrodes in lithium-ion batteries. J Power Sources 161:617–622
99. Xu YH, Yin GP, Ma YL, Zuo PJ, Cheng XQ (2010) Simple annealing process for performance improvement of silicon anode based on polyvinylidene fluoride binder. J Power Sources 195:2069–2073
100. Hochgatterer NS et al (2008) Silicon/graphite composite electrodes for high-capacity anodes: Influence of binder chemistry on cycling stability. Electrochem Solid-State Lett 11:A76–A80
101. Guo JC, Wang CS (2010) A polymer scaffold binder structure for high capacity silicon anode of lithium-ion battery. Chem Commun 46:1428–1430
102. Beattie SD, Larcher D, Morcrette M, Simon B, Tarascon JM (2008) Si electrodes for Li-ion batteries – a new way to look at an old problem. J Electrochem Soc 155:A158–A163
103. Zheng Y, Yang J, Tao L, Nuli YN, Wang JL (2007) Study of nano-porous Si/Graphite/C composite anode materials for Li-ion batteries. Chin J Inorg Chem 23:1882–1886
104. Chen LB, Xie XH, Xie JY, Wang K, Yang J (2006) Binder effect on cycling performance of silicon/carbon composite anodes for lithium ion batteries. J Appl Electrochem 36:1099–1104
105. Choi NS, Yew KH, Choi WU, Kim SS (2008) Enhanced electrochemical properties of a Si-based anode using an electrochemically active polyamide imide binder. J Power Sources 177:590–594
106. Liu G (2010) DOE hydrogen program and vehicle technologies program annual merit review and peer evaluation meeting. Department of Energy, Office of Energy Efficiency & Renewable Energy, Washington DC, 2010
107. Zhang JG, Liu J (2010) DOE hydrogen program and vehicle technologies program annual merit review and peer evaluation meeting. Department of Energy, Office of Energy Efficiency & Renewable Energy, Washington DC, 2010
108. Carmer JLG, Morales J, Sanchez L (2008) Nano-Si/cellulose composites as anode materials for lithium-ion batteries. Electrochem Solid-State Lett 11:A101–A104
109. Kulova TL, Skundin AM (2010) Elimination of irreversible capacity of amorphous silicon: direct contact of the silicon and lithium metal. Rus J Electrochem 46:470–475
110. Urbonaite S, Baglien I, Ensling D, Edstrom K (2010) Effect of ethanol-assisted electrode fabrication on the performance of silicon anodes. J Power Sources 195:5370–5373
111. Doh CH et al (2006) Synthesis of silicon-carbon by polyaniline coating and electrochemical properties of the Si-C vertical bar Li cell. Bull Korean Chem Soc 27:1175–1180
112. Choi NS et al (2006) Effect of fluoroethylene carbonate additive on interfacial properties of silicon thin-film electrode. J Power Sources 161:1254–1259
113. Chen LB, Wang K, Xie XH, Xie JY (2007) Effect of vinylene carbonate (VC) as electrolyte additive on electrochemical performance of Si film anode for lithium ion batteries. J Power Sources 174:538–543
114. Han GB, Ryou MH, Cho KY, Lee YM, Park JK (2010) Effect of succinic anhydride as an electrolyte additive on electrochemical characteristics of silicon thin-film electrode. J Power Sources 195:3709–3714
115. Baggetto L et al (2009) On the electrochemistry of an anode stack for all-solid-state 3D-integrated batteries. J Power Sources 189:402–410
116. Arie AA, Chang W, Lee JK (2010) Electrochemical characteristics of semi conductive silicon anode for lithium polymer batteries. J Electroceramics 24:308–312

117. Inose T, Watanabe D, Morimoto H, Tobishima SI (2006) Influence of glyme-based nonaqueous electrolyte solutions on electrochemical properties of Si-based anodes for rechargeable lithium cells. J Power Sources 162:1297–1303
118. Choi NS, Yew KH, Kim H, Kim SS, Choi WU (2007) Surface layer formed on silicon thin-film electrode in lithium bis(oxalato) borate-based electrolyte. J Power Sources 172:404–409
119. Lux SF et al (2010) Li-ion anodes in air-stable and hydrophobic ionic liquid-based electrolyte for safer and greener batteries. Int J Energy Res 34:97–106
120. Cui LF, Yang Y, Hsu CM, Cui Y (2009) Carbon – silicon core – shell nanowires as high capacity electrode for lithium ion batteries. Nano Lett 9:3370–3374
121. Yang Y et al (2010) New nanostructured Li2S/silicon rechargeable battery with high specific energy. Nano Lett 10:1486–1491
122. Arrebola JC et al (2009) Combining 5 V LiNi0.5Mn1.5O4 spinel and Si nanoparticles for advanced Li-ion batteries. Electrochem Commun 11:1061–1064
123. Lee K-L, Jung J-Y, Lee S-W, Moon H-S, Park J-W (2004) Electrochemical characteristics and cycle performance of LiMn2O4/a-Si microbattery. J Power Sources 130:241–246
124. Yin J et al (2006) Micrometer-scale amorphous Si thin-film electrodes fabricated by electron-beam deposition for Li-ion batteries. J Electrochem Soc 153:A472–A477
125. Baranchugov V, Markevich E, Pollak E, Salitra G, Aurbach D (2007) Amorphous silicon thin films as a high capacity anodes for Li-ion batteries in ionic liquid electrolytes. Electrochem Commun 9:796–800
126. Yang H et al (2007) Amorphous Si film anode coupled with $LiCoO_2$ cathode in Li-ion cell. J Power Sources 174:533–537
127. Christensen J (2010) Modeling diffusion-induced stress in Li-Ion cells with porous electrodes. J Electrochem Soc 157:A366–A380
128. Gaines L, a.C., Roy in http://www.transportation.anl.gov/pdfs/TA/149.pdf

Index

A
Abaqus, 300
absorption spectroscopy, 251
abuse tolerance, lithium ion cell, 342
accelerating rate calorimetry (ARC), 300
accumulator, 54
acetonitrile (AN), 401
  electrolytes, 98
action by galvanism, 363
activated carbon (AC), 99
  nanoporous texture, 414
  properties, 395
adenosine triphosphate (ATP), 96
air poisoning, 449
alloy system, high-energy mechanical milling (HEMM), 481
aluminum, 15
amorphous silicon, 475
anion mobility, 217
anode, 240–241, 326–327
  anode–electrolyte interface, 201
  deposition of nickel, 296
  materials, 273, 311
  silicon-based, 345
  with Li^+-ion liquid electrolytes, 71
anolyte, 66
aqueous batteries, separator, 177
Arrhenius behavior, 261
asymmetric hybrid capacitor, 99
Avogadro's number, 255, 261

B
Batteries, 65, 361
  sustainability, 1
battery separators, 153, 158, 161 battery
  absorptive glass mat (AGM), 125
  active materials, 52
  anode, 52
  capacity, 53
  cathode, 5, 52
    materials, 9
  chemistries, 136
  current, 53
  designs, 118, 121, 125
  discharge
    rates, 128
    reactions, 129
  dual Li-ion intercalation system, 8
  Effect of Temperature, 131
  electrical-energy storage, 91
  electrolyte, 130
  energy density, 53
  flat and tubular electrodes, 122
  gel batteries, 126
  hermetic seals, 361
  industry, 2, 136
  layered transition metal oxides, 9
  life, 131
  maintenance, 131
  material safety data sheets (MSDS), 121
  materials, 384
  medical devices, 361
  monopolar and bipolar electrodes, 121
  nail penetration test, 171
  negative terminal, 54
  nickel systems, 183
  overcharge reactions, 130
  positive terminal, 54
  power capability, 53

R.J. Brodd (ed.), *Batteries for Sustainability: Selected Entries from the Encyclopedia of Sustainability Science and Technology*, DOI 10.1007/978-1-4614-5791-6,
© Springer Science+Business Media New York 2013

battery separators (*cont.*)
  rechargeable, 54
  redox-flow, 89
  separator
    battery, 157
    characterization, 153
    classification, 139
    electrical resistance, 154
    Gurley number, 154
    mathematical modeling, 185
    melt integrity, 164
    mix penetration strength, 161
    pore size, 157
    porosimetry, 158
    porosity, 156
    properties, 153
    puncture strength, 160
    requirements, 150
    shrinkage, 161
    shutdown, 162
    tensile strength, 161
    tortuosity, 186
    wettability, 165
    wetting speed, 165
  silver–zinc, 65
  solar energy storage, 120
  stacked cells, 138
  thermal runaway, 171
  zinc–air, 65
  zinc–manganese, 63
Beer-Lambert Law, 249
binary copper compounds, 268
binary ionic liquid electrolyte, 218
bipolar batteries, 133
Bohr magneton, 255
Boltzmann constant, 261
bradycardia, 364
Bragg's law, 244
Bruggeman exponent, 185
bulk silicon anode, failure mechanisms, 476

C
cadmium, 432
  cadmium–nickel cell, 67
capacitor, Energy Density, 98
capillary flow porometry, 158
carbohydrate, Protic Ionic Liquids, 416
carbonate-containing liquid electrolytes, 265
carbonate molecules, 228
carbon-based pseudo-capacitors, 416
carbonization, 395
carbon
  electrodes, 94
  for electrical double-layer capacitors, 399
  for pseudo-capacitors, 403
  heteroatom-enriched, 403, 410
  materials, 327
  monofluoride (CFx), 368
  nanofibers, 484
  nanotubes, 483
  nitrogen-rich, 409
  seaweed-based, 412
  thermo programmed desorption (TPD) analyses, 415
carbonyl oxygen atom, 232
carboxymethylcellulose (CMC), 275, 334
cardiac pacemakers, 365, 384
cardiac resynchronization therapy defibrillator device (CRT-D), 373
cathode, 7, 34, 198, 240–241, 326, 329, 371
  Li-ion liquid electrolytes, 72
  materials, 266, 349
  nickel metal particle, 297
cathode/electrolyte interface, 226
  Li-ion liquid electrolytes, 72
CdTe solar cell
  cell voltage
    overvoltage, 57
    polarization, 57
  effect on cell performance and safety, 166
  performance and safety, 166
  separators, 149
Celgard membranes, 156
cellulose/cellulosic fibers, 148
charcoal, 95
chemical/chemicals nitrogen, 397
cobalt, 11, 15, 77
  sulfate, 429
comb-branch
  copolymer, 216
  polyethers, 214
  polymer, 216
conversion electrode, 34
crystalline silicon anode, 474
Curie-Weiss behavior, 266, 273

D
deep-cycle battery, 119
defibrillators, 378
delithiation, 11, 27, 29
density functional theory (DFT), 199, 263, 400
diethyl carbonate (DEC), 326
differential scanning calorimetry (DSC), 300

diffraction method, 244
dual carbon cell, 100

E
elastomeric binder, 492
electric
  double-layer capacitor (EDLC), 94, 96
  vehicle (EV), 8, 21, 353
    battery management, 353
electrical/electrically
  capacitors (EDLC), 398
  double-layer (EDL), 55, 419
electricity
  as medicine, 362
  physiological effects, 362
electrocardiogram, 364
electrochemical
  capacitors (EC), 94, 398
    operation voltage, 97
  cell, 53–54
    electrolyte, 54
    separator, 54
  energy storage, 1
  supercapacitors, 93
electro-conductivity, 105
electrode–electrolyte contact, 55
electrode-electrolyte interface, 196
  molecular dynamics simulations, 223
electrode, 240, 326
  fabrication, winding process, 334
  reactions, 61
  surfaces, decomposition, 206
  tap density, 77
electrolyte
  additives, 332, 350
  EC-based, 202
  electrochemical stability, 232
  electrode surfaces, 264
  electrolyte systems, 125
  ionic liquid–based, 226
  materials, 326
  modeling of reductive decomposition, 201
  non-aqueous, 57, 69, 240
  oxidative stability, 198
  single-ion conductor, 218
  solvents, 199, 226
  stability, 196
  window, 56
electrolyte/electrode interactions, 232
electron energy loss spectroscopy (EELS), 26
electronic
  conductivity, 111
  insulator, 54
electrostatic repulsive energy, 75
energy
  density, 59
  storage, 2, 94
    systems (ESS), 323
ester-carbonate compound, 203, 205
ethyl methyl carbonate (EMC), 326
ethylene carbonate (EC), 199, 202, 205, 243, 264, 326
  electrolytes, 264
  two-electron reduction, 205
ethylene-propylene-diene (EPDM) copolymer, 176
extended X-ray absorption fine structure (EXAFS), 251

F
Faraday constant, 60
Fermi
  constant, 255
  contact shifts, 257
  energy, 73
float/floating voltage, 119
flooded electrolyte lead acid, 178
fluorine, 369
fluorophosphate, 31
Fourier transformation, 247
fuel batteries, 97
Full Cells, 497

G
galvanic cell, 53
"gel" battery, 125
gel-type polymer electrolyte, 173
Gibbs free energy, 57, 60
graphene nanosheet, 407
graphite
  intercalation compound (GIC), 7
    staging phenomenon, 343
  interface, 223
  lithium anode, 2
Graphite intercalation compounds (GIC), 343
Graphite Interface, 223
gravimetric energy, 53
Grotthus motion, 57
Gurley value, 160

H
Hahn-echo pulse, 262
Hamiltonian term, 258
hard carbon, 343
hearing aid battery, 383
hearing loss, 382
heart failure devices, 373
heteroatom, 396
hexafluorpropylene, 380
high-density polyethylene (HDPE), 145
highest occupied molecular orbital (HOMO) calculations, 200
hybrid
  capacitor systems, 111, 113
  electric vehicle (HEV), 8
  systems, 93
hybrid electric vehicles (HEVs), 119–120
hydrazine, 408
hydrogen electrode, 127, 440

I
implantable cardiac defibrillator (ICD), 267, 378
  batteries, 374
implantable cardiac pacemaker, 364, 372
  batteries, 366
information entropy, 467
ingestible medical devices, 389
intercalation electrode, 321
intercrystallite ionic transport, 27
International Lead Association (ILA), 121
ion-exchange membrane, 141
ion/ionic
  conductivity, 57
  correlation, molecular dynamics simulation, 208
  diffusion, 111
  dynamic correlation, 209
  liquid, 265
    electrolyte (ILE), 218, 226, 232
  transfer, 55
iron-III fluoride, 269
iron nitrate, 23

J
Jahn–Teller effect, 19
  distortion of the Mn3+ ion, 29
  Mn3+ ions, 13

K
Kirchhoff' circuit law, 353
knight shift, 258, 273

L
Larmor time scale, 258
lattice nitrogen, 397
lead
  Alloys, 123
  grids, 123
lead-acid battery, 2, 67
  commercial designs, 118
  electrodes, 125
  hybrid electric vehicle, 120
  markets, 119
  separators, 178
  for sustainable energy, 117
  uninterruptible power systems (UPS), 119
lead-battery electrode, 122
Leclanché cell, 63
LFP, 11
life-enhancing device, 360
life-sustaining device, 360
Li-ion battery see also lithium-ion battery, 7
  configuration, 12
  silicon-based anodes, 471
Li-ion capacitor (LIC), 113
liquid electrolyte, 332
liquid electrolyte systems, 136
liquid-metal-cooled fast reactors, lithiated graphite, 273
liquid electrolyte, 54, 136, 240, 332
LISICON, 83
lithiation, 16
lithium
  anisotropic thermal vibrations, 463
  anode, 17, 66, 73
  battery, 120, 140
    electrode/electrolyte interfaces, 196
    electrolyte stability, 195
    electrolyte, 196
    medical devices, 366
  bis(oxalato)borate (LiBOB), 20, 496
  cell technologies, 366
  chloride, 371
  cobalt oxide (LCO), 347
  composition, 452
  compounds, 242
  desolvation, 224–225
  electrochemical alloys, 473
  FePO4, 25
  fluoride (LiF), 369

free-energy profile, 224
in nanoparticles, 34
ion, 270, 379
iron phosphate (LiFePO4), 226
LiFePO4, 23, 25, 28, 34, 86
LiMnPO4/MnPO4 system, 29
LiNiO2, 11
LiNi0.5Mn0.5O2, 14
LiNi1/2Mn3/2O4, 22
Li2CoSiO4, 33
Li2FeSiO4, 32
Li2MnO3, 13
Li2MnSiO4, 33
Li2NiSiO4, 33
Li2CoO2, 11
manganese oxide (LiMn2O4), 12
spinels, 20
maximum entropy method (MEM), 464
metal oxides, 240
nanoparticulate LiMnPO4, 28
neutron diffraction, 462
nickel oxide, 347
non-stoichiometry, 25
nuclear distribution, 467
phosphates, 30
phosphorus oxynitride, 495
polymers, separator, 172
primary batteries, 65
salts, 70, 208, 242
shifts, 273
spinel framework, 80
tetrafluoroborate (LiBF4), 369
thionyl chloride, 370
titanate anode, 345
transference number, 262
transition metal-containing poly-anionic compounds, 30
transport, 220, 266
lithium-air battery, 89
lithium-carbon monofluoride, monofluoride-silver vanadium oxide, 377
lithium-iodine, 366
lithium-ion battery
abuse safety tests, 287, 313
abuse tolerance, 311
anode, 480
decomposition reactions, 303
materials, 344
cell
balancing, 353
components, 324
level safety, 341
cell-operating voltage, 324
commercial success, 322
cylindrical cell, 310
electrochemical reactions, 239, 241
energy storage capability, 322
field-failures, 287, 293, 311
graphite-free, 31
heat release, 289
history, 321
internal shorts, 314
circuits, 293
formation in a field-failure, 295
LiMn2O4, 17
manufacturing
process, 334
quality, 295
market survey reports, 323
mechanisms of internal shorting, 298
prismatic cell, 310
safe zone concept, 306
safety, 285–286, 288, 314, 339
safety testing procedures, 168
development, 145
self-heating of the cell, 304
separators, 142, 145, 333, 351
short initiation, 294
system level safety, 340
technology, 319
thermal runaway, (TR) 287, 341
from internal shorts, 299, 301
threshold energy, 305
transportation applications, 314
working mechanism, 321
lithium-ion capacitor (LIC), 100
lithium-ion cell
autoignition, 290
cell balance, 331
cyclability, 328
exothermal reactions, 290
external abuses, 287
triggers, 291
fabrication process, 335
field-failures, 316
flat plate pouch configuration, 337
formation and aging, 338
hot box test, 291
ideal electrolyte solute, 332
liquid electrolytes, 332
mechanism, 326
safety failures, 287
solid electrolyte interface (SEI), 328
thermal runaway, 291–292, 315
lithium-ion gel polymer batteries
separator, 173

lithium-manganese dioxide, 376
lithium-silicon alloy, 473, 476
lithium-silver vanadium oxide, 374
lithium-sulfur battery, 89
low-maintenance battery, 123

M
MacMullin number, 150, 155, 160
Madelung ionic electrostatic energy, 80
magic angle–spinning (MAS) spectra, 255
magnetic
  reflections, 245
  scattering, 245
manganese oxide spinel, 16, 19, 28
massive artificial graphite (MAG) battery application, 345
medical device battery design, 361
memory effect, 434
mercury, 384
metal
  hydride (MH) alloy, 434
  hydride–nickel cell, 68
  oxide cathodes, 289
microcalorimetry, 459
microfiber glass separator, 183
microglass separator, 183
microporous polyolefin separator, 143
microporous separator, 140
mixed oxide (MOX) cathodes, 348
mobile phone, 137
molecular dynamics (MD) simulation, 197, 206
molecular dynamics simulation, 207
molten salt, 218
  electrolyte, 221
motional narrowing, 261

N
Nafion, 149
nanocarbon, 419
  for supercapacitors, 393
nanocomposite concept, 479
nanocrystalline materials, 485
nanogate capacitor, 100
nanohybrid capacitor (NHC), 102, 107, 113
nanoporous carbon, 405
  electrodes, 403
nano-sizing cathode powder, 35
nanostorage capacitor, 100
nanostructured silicon anodes, 479, 489
  three-dimensional, 489
nanowire, silicon-based, 488
Nasicon, 30–31, 36
  framework, 83
natural graphite, 344
NCA, 11
near-edge X-ray absorption fine structure (NEXAFS), 251
negative electrode material, 327, 379, 428
  lithium ion cell, 343
Nernst-Einstein equation, 262
neuromodulation batteries, 379
neurostimulator batteries, 378
neutron diffraction (ND), 243–245, 463
nickel
  active materials, 425
  batteries, 424–425
  cadmium batteries, 431–433
    fiber plate batteries, 432
    pocket and tubular plate cells, 432
    sealed sinter plate cells, 433
  cadmium, 431–433
  cobalt manganese oxide, 349
  electrode, 424, 426–427
    construction, 427
    reaction, 426
  metal hydride (NiMH) electrode, 435
  metal hydride (NiMH) battery, 3, 139, 436
Nickel Electrode, 426
Nickel metal hydride (Ni-MH) batteries, 434
Nickel-Based Battery Systems, 423
nickel-hydrogen battery, 2, 427, 439
nickel-hydrogen, 427
nickel-iron battery (Ni-Fe), 427
  fabrication process, 429
  performance, 430
nickel-zinc battery, 437
nitrogen-rich nanocarbon, 409
non-wovens, 140
  materials, 148
nuclear magnetic resonance (NMR) spectroscopy, 254

O
octahedral lithium, 462
Ohmic polarization, 59
oligoether, 214
olivine phosphate
  cathode materials, 445
  data, 457
  diffractograms, 449
  electrochemical and structural, 457
  electrode material, 468

lattice parameters, 455
lithium compositions, 453
miscibility gap, 459
verification of lithium diffusion, 462
verification of phase diagram, 452
open-circuit voltage, 55
organic liquid electrolyte (OLE), 206, 226, 232
oxygen cycle, 124

P
pacemaker, 364
pair distribution function (PDF) analysis, 247
pentaglyme, 211
Perdew–Burke–Ernzerhof (PBE) functional, 206
personal digital assistant (PDA), 137
phospho-olivine, 23, 29
binary phase diagram, 447
crystal chemistry, 446
Gibbs phase rule, 446
miscibility gap, 447
plastic, lithium-ion battery, 174, 274
plug-in hybrid electric vehicle (PHEV), 8, 472
Poisson potential, 229
polyacrylonitril, 409
polyamide imide (PAI), 493
polyannaline, 86
polyethylene (PE), 145
oxide (PEO), 70, 176, 242
based polymers, 211
separators, 180
polyfurfuryl alcohol resin (PFA), 343
polymer/polymeric, 210, 213–214, 493
architectures, 214
binders, 493
electrolytes, 142, 210, 242
relaxation, 213
polymorph, 32
polyolefin, 151, 158, 177
resin, 145, 148
separator, 164
polypropylene, 145, 163
microporous films, 147
separators, 151
polypyrrole, 86
polyvinyl chloride (PVC), separators, 180
polyvinylidene difluoride (PVdF), 334
polyvinylidene fluoride (PVDF), 380, 491
polyvinylpyridine, 409
positive electrode material, 329, 347, 371, 378
positive nickel electrode material, 429
potential of mean force (POMF), 229
powder neutron diffraction, 246
powder X-ray diffraction (PXRD), 243
power system regulation, 2
primary battery, 54, 63
solution-precipitation reaction, 61
propargyl methanesulfonate (PMS), 200
propylene carbonate (PC), 97, 494
protected lithium electrode (PLE), 36
protic ionic liquids (PILs), 416–417
pseudo-capacitance, 403, 405, 409, 411, 414
in basic and acidic media, 405
due to nitrogenated functionalities, 409
in neutral medium, 411
related with reversible hydrogen electrosorption in nanoporous carbons, 414
pseudocapacitor, 94
pseudo-faradic effect, 403
pseudo-faradic reactions, 411

Q
QC calculation, 197
quantum chemistry, 196
quinone/hydroquinone pair, 405
quinone, 406

R
reactions, 126
ReaxFF, 197
simulations, 204
rechargeable battery, 135–136
separator membranes, 136
redox-flow battery, 67, 90
redox, 84, 241, 411
energy, 84
reactions
pseudo-faradic, 411
reduction-oxidation reaction, 241
Rietveld refinement, 449, 463
room temperature ionic liquid (RTIL) electrolyte, 218
rubber separator, 180

S
Scofield factor, 250
secondary battery
with aqueous electrolytes, 67
cathode, 79
cost, 61
cycle life, 54

secondary battery (*cont.*)
efficiency, 54
insertion compounds, 57
lithium–metal salt, 89
secondary lithium batteries, 142
Separion separator, 149
silicon anode, 474
silicon/carbon, 481–483
composites, 481–483
ball milling, 481
deposition of silicon on mesocarbon microbeads (MCMB), 482
high-temperature pyrolysis, 482
nanotube composites, 483
silicon, 274, 345
anode, 473–474
closed-circuit voltage (CCV), 477
cycle life, 478
cycling stability, 498
delithiation, 474
effect of electrolytes, 494
effect of succinic anhydride, 495
electrochemical lithiation, 474
galvanostatic intermittent titration (GITT), 477
quasi-open-circuit voltage (QOCV), 477
salts, 496
SEI formation additives, 494
sodium carboxylmethylcellulose–based binder, 491
solvents, 496
electrode, 492
monoxide (SiO), 488
nanoparticles, 472, 485
one-dimensional, 487
nanostructures, 472
nanowires, 472, 485
silver vanadium oxide, 267, 376
silver–zinc battery, 65
single-ion conductor, 217
SLI battery, 119
sodium, 87
sodium-sulfur battery, 87
solid electrodes, 60
solid electrolyte interface (SEI), 8, 338, 489
model compounds, 222
solid electrolyte interphase (SEI), 243
solid ion conductors, 142
solid polymer electrolyte (SPE), 172, 210
solid-electrolyte interphase (SEI), 57, 201
solvent-free molten salts electrolyte, 221
spinal cord stimulation, 378
spin-orbit coupling, 249
standard hydrogen electrode (SHE), 198
stoichiometric LiNiO2, 11
styrene butadiene rubber (SBR), 334
sulfite compounds, 350
sulfuric acid, 125
sulfur, 87, 371
supercapacitor, 94, 96, 113, 396–398
carbon electrodes, 397
nanoporous carbon electrodes, 394
in neutral medium, 414
in organic electrolyte, 396
supported liquid membrane, 142
sustainability/sustainable energy, 118
synthetic
graphite, 344
synthetic pulp separators (SPG), 182

T

tachycardia, 373
Teflon, 439
telecommunication back-up power systems, 119
thermal mechanical analysis (TMA) test, 162
thermodynamic(s) potential, 230
thermogravimetric (TG) measurement, 107
thin-film silicon anodes
amorphous, 485
nanocrystalline, 485
thionyl chloride, 370
thiospinel framework, 80
tin-based amorphous composite oxide (TCO), 479
transmission electron microscopy (TEM), 26
transportation applications, field-failures, 314

U

ultra battery, 132
design, 120
ultracapacitor, 94, 96
ultracentrifugal force (UC), nanohybridization method, 105
ultrahigh-molecular-weight polyethylene (UHMWPE), 145, 147

V

valve-regulated lead-acid (VRLA) battery, 124, 182
Van der Waals bonding, 72, 75
vanadium, 268
oxides, 78

Vegard's law, 449, 455, 458
vehicle starting, lighting and ignition (SLI), 119
vinylene carbonate (VC), 350
vinylethylene carbonate (VEC), 201
voltaic pile, 52
voltammogram, 401, 406, 416
volumetric
  energy density, 406
  power density, 53
VRLA battery, 130
wet process method, 351
wooden separators, 180
working ion, 54
  transfer, 57

X
X ray
  absorption spectroscopy (XAS), 250–251
    pre-edge, 251
    rising edge, 251
  diffraction (XRD), 244–245
    structure factor, 245
  photoelectron spectroscopy (XPS)
    ultrahigh vacuum (UHV)
    conditions, 248

Y
Young's Modulus, 152

Z
ZEBRA cell, 88
zinc-air cathode, 385
zinc-air cell, 383, 386
Zinc–manganese batteries, 63
zinc-mercuric oxide, 371
zinc, 438
  electrode, 438
  ion, 438
zinc-silver oxide, 387

MIX
Papier aus verantwortungsvollen Quellen
Paper from responsible sources
FSC® C105338
FSC
www.fsc.org

If you have any concerns about our products,
you can contact us on
ProductSafety@springernature.com

In case Publisher is established outside the EU,
the EU authorized representative is:
Springer Nature Customer Service Center GmbH
Europaplatz 3, 69115 Heidelberg, Germany

Printed by Libri Plureos GmbH
in Hamburg, Germany